Linear Algebra

M. Thamban Nair · Arindama Singh

Linear Algebra

M. Thamban Nair
Department of Mathematics
Indian Institute of Technology Madras
Chennai, Tamil Nadu, India

Arindama Singh
Department of Mathematics
Indian Institute of Technology Madras
Chennai, Tamil Nadu, India

ISBN 978-981-13-4533-3 ISBN 978-981-13-0926-7 (eBook)
https://doi.org/10.1007/978-981-13-0926-7

Softcover re-print of the Hardcover 1st edition 2018

Printed on acid-free paper

This Springer imprint is published by the registered company Springer Nature Singapore Pte Ltd.
The registered company address is: 152 Beach Road, #21-01/04 Gateway East, Singapore 189721, Singapore

Preface

Linear Algebra deals with the most fundamental ideas of mathematics in an abstract but easily understood form. The notions and techniques employed in Linear Algebra are widely spread across various topics and are found in almost every branch of mathematics, more prominently, in Differential Equations, Functional Analysis, and Optimization, which have wide applications in science and engineering. The ideas and techniques from Linear Algebra have a ubiquitous presence in Statistics, Commerce, and Management where problems of solving systems of linear equations come naturally. Thus, for anyone who carries out a theoretical or computational investigation of mathematical problems, it is more than a necessity to equip oneself with the concepts and results in Linear Algebra, and apply them with confidence.

Overview and Goals

This book provides background materials which encompass the fundamental notions, techniques, and results in Linear Algebra that form the basis for analysis and applied mathematics, and thereby its applications in other branches of study. It gives an introduction to the concepts that scientists and engineers of our day use to model, to argue about, and to predict the behaviour of systems that come up often from applications. It also lays the foundation for the language and framework for modern analysis. The topics chosen here have shown remarkable persistence over the years and are very much in current use.

The book realizes the following goals:

- To introduce to the students of mathematics, science, and engineering the elegant and useful abstractions that have been created over the years for solving problems in the presence of linearity
- To help the students develop the ability to form abstract notions of their own and to reason about them

- To strengthen the students' capability of carrying out formal and rigorous arguments about vector spaces and maps between them
- To make the essential elements of linear transformations and matrices accessible to not-so-matured students with a little background in a rigorous way
- To lead the students realize that mathematical rigour in arguing about linear objects can be very attractive
- To provide proper motivation for enlarging the vocabulary, and slowly take the students to a deeper study of the notions involved
- To let the students use matrices not as static array of numbers but as dynamic maps that act on vectors
- To acquaint the students with the language and powers of the operator theoretic methods used in modern mathematics at present

Organization

Chapter 1 lays the foundation of Linear Algebra by introducing the notions of vector spaces, subspaces, span, linear independence, basis and dimension, and the quotient space. It is done at a leisurely pace. Sometimes, the steps have been made elaborate intentionally so that students would not jump to conclusions and develop the habit of hand waving. Both finite and infinite dimensional vector spaces are discussed so that a proper foundation is laid for taking up Functional Analysis as a natural extension, if the student so desires to pursue in his later years of study.

Chapter 2 introduces linear transformations as structure-preserving maps between vector spaces. It naturally leads to treating the matrices as linear transformations and linear transformations as matrices. Particular attention is given to bijective linear transformations and how they act on bases. So, change of bases is considered by looking at the identity linear transformation. The space of linear transformations is introduced along with the composition of maps. It raises the issue of equivalence and similarity of matrices leading to the rank theorem.

Chapter 3 deals with elementary operations in detail. Starting from the computation of rank, elementary operations are used to evaluate determinants, compute the inverse of a matrix, and solve linear systems. The issue of solvability of linear systems is treated via linear transformations showing to the students how abstraction helps.

Chapter 4 brings in the so far neglected notion of direction, or angle between vectors by introducing inner products. The notion of orthogonality and the necessary elegance and ease it ensues are discussed at length. The geometric notion of an orthogonal projection is given a lead for solving minimization problems such as the best approximation of a vector from a subspace and least squares solutions of linear systems. Constructing a linear functional from the inner product via Fourier expansion leads to Riesz representation theorem. The existence of the adjoint of a linear transformation is also shown as an application of Riesz representation.

Chapter 5 asks a question of how and when a linear operator on a vector space may fix a line while acting on the vectors. This naturally leads to the concepts of eigenvalues and eigenvectors. The notion of fixing a line is further generalized to invariant subspaces and generalized eigenvectors. It gives rise to polynomials that annihilate a linear operator, and the ascent of an eigenvalue of a linear operator. Various estimates involving the ascent, the geometric, and algebraic multiplicities of an eigenvalue are derived to present a clear view.

Chapter 6 takes up the issue of representing a linear operator as a matrix by using the information on its eigenvalues. Starting with diagonalization, it goes for Schur triangularization, block-diagonalization, and Jordan canonical form characterizing similarity of matrices.

Chapter 7 tackles the spectral representation of linear operators on inner product spaces. It proves the spectral theorem for normal operators in a finite-dimensional setting, and once more, that of self-adjoint operators with somewhat a different flavour. It also discusses the singular value decomposition and polar decomposition of matrices that have much significance in application.

Special Features

There are places where the approach has become non-conventional. For example, the rank theorem is proved even before elementary operations are introduced; the relation between ascent, geometric multiplicity, and algebraic multiplicity are derived in the main text, and information on the dimensions of generalized eigenspaces is used to construct the Jordan form. Instead of proving results on matrices, first a result of the linear transformation is proved, and then it is interpreted for matrices as a particular case. Some of the other features are:

- Each definition is preceded by a motivating dialogue and succeeded by one or more examples
- The treatment is fairly elaborate and lively
- Exercises are collected at the end of the section so that a student is not distracted from the main topic. The sole aim of these exercises is to reinforce the notions discussed so far
- Each chapter ends with a section listing problems. Unlike the exercises at the end of each section, these problems are theoretical, and sometimes unusual and hard requiring the guidance of a teacher
- It puts emphasis on the underlying geometric idea leading to specific results noted down as theorems
- It lays stress on using the already discussed material by recalling and referring back to a similar situation or a known result
- It promotes interactive learning building the confidence of the student

- It uses operator theoretic method rather than the elementary row operations. The latter is primarily used as a computational tool reinforcing and realizing the conceptual understanding

Target Audience

This is a textbook primarily meant for a one- or two-semester course at the junior level. At IIT Madras, such a course is offered to master's students, at the fourth year after their schooling, and some portions of this are also offered to undergraduate engineering students at their third semester. Naturally, the problems at the end of each chapter are tried by such master's students and sometimes by unusually bright engineering students.

Notes to the Instructor

The book contains a bit more than that can be worked out (not just covered) in a semester. The primary reason is: these topics form a prerequisite for undertaking any meaningful research in analysis and applied mathematics. The secondary reason is the variety of syllabi followed at universities across the globe. Thus different courses on Linear Algebra can be offered by giving stress on suitable topics and mentioning others. The authors have taught different courses at different levels from it sticking to the core topics.

The core topics include vector spaces, up to dimension (Sects. 1.1–1.5), linear transformation, up to change of basis (Sects. 2.1–2.5), a quick review of determinant (Sect. 3.5), linear equations (Sect. 3.6), inner product space, up to orthogonal and orthonormal bases (Sects. 4.1–4.5), eigenvalues and eigenvectors, up to eigenspaces (Sects. 5.1–5.3), the characteristic polynomial in Sect. 5.5, and diagonalizability in Sect. 6.1. Depending on the stress in certain aspects, some of the proofs from these core topics can be omitted and other topics can be added.

Chennai, India
March 2018

M. Thamban Nair
Arindama Singh

Contents

1 Vector Spaces 1
1.1 Vector Space 1
1.2 Subspaces 10
1.3 Linear Span 14
1.4 Linear Independence 21
1.5 Basis and Dimension 27
1.6 Basis of Any Vector Space 32
1.7 Sums of Subspaces 36
1.8 Quotient Space 42
1.9 Problems 47

2 Linear Transformations 51
2.1 Linearity 51
2.2 Rank and Nullity 64
2.3 Isomorphisms 71
2.4 Matrix Representation 76
2.5 Change of Basis 85
2.6 Space of Linear Transformations 93
2.7 Problems 98

3 Elementary Operations 107
3.1 Elementary Row Operations 107
3.2 Row Echelon Form 111
3.3 Row Reduced Echelon Form 118
3.4 Reduction to Rank Echelon Form 128
3.5 Determinant 132
3.6 Linear Equations 141
3.7 Gaussian and Gauss–Jordan Elimination 147
3.8 Problems 156

4 Inner Product Spaces 163
4.1 Inner Products 163
4.2 Norm and Angle 168
4.3 Orthogonal and Orthonormal Sets 172
4.4 Gram–Schmidt Orthogonalization 176
4.5 Orthogonal and Orthonormal Bases 182
4.6 Orthogonal Complement 186
4.7 Best Approximation and Least Squares 190
4.8 Riesz Representation and Adjoint 196
4.9 Problems 205

5 Eigenvalues and Eigenvectors 209
5.1 Existence of Eigenvalues 209
5.2 Characteristic Polynomial 213
5.3 Eigenspace 218
5.4 Generalized Eigenvectors 224
5.5 Two Annihilating Polynomials 234
5.6 Problems 238

6 Block-Diagonal Representation 243
6.1 Diagonalizability 243
6.2 Triangularizability and Block-Diagonalization 248
6.3 Schur Triangularization 252
6.4 Jordan Block 257
6.5 Jordan Normal Form 264
6.6 Problems 271

7 Spectral Representation 277
7.1 Playing with the Adjoint 277
7.2 Projections 285
7.3 Normal Operators 295
7.4 Self-adjoint Operators 302
7.5 Singular Value Decomposition 308
7.6 Polar Decomposition 317
7.7 Problems 324

References 335

Index 337

About the Authors

M. Thamban Nair is a professor of mathematics at the Indian Institute of Technology Madras, Chennai, India. He completed his Ph.D. at the Indian Institute of Technology Bombay, Mumbai, India, in 1986. His research interests include functional analysis and operator theory, specifically spectral approximation, the approximate solution of integral and operator equations, regularization of inverse and ill-posed problems. He has published three books, including a textbook, *Functional Analysis: A First Course* (PHI Learning), and a text-cum-monograph, *Linear Operator Equations: Approximation and Regularization* (World Scientific), and over 90 papers in reputed journals and refereed conference proceedings. He has guided six Ph.D. students and is an editorial board member of the *Journal of Analysis and Number Theory*, and *Journal of Mathematical Analysis*. He is a life member of academic bodies such as Indian Mathematical Society and Ramanujan Mathematical Society.

Arindama Singh is a professor of mathematics at the Indian Institute of Technology Madras, Chennai, India. He completed his Ph.D. at the Indian Institute of Technology Kanpur, India, in 1990. His research interests include knowledge compilation, singular perturbation, mathematical learning theory, image processing, and numerical linear algebra. He has published five books, including *Elements of Computation Theory* (Springer), and over 47 papers in reputed journals and refereed conference proceedings. He has guided five Ph.D. students and is a life member of many academic bodies, including Indian Society for Industrial and Applied Mathematics, Indian Society of Technical Education, Ramanujan Mathematical Society, Indian Mathematical Society, and The Association of Mathematics Teachers of India.

Chapter 1
Vector Spaces

1.1 Vector Space

A vector in the plane is an object with certain length and certain direction. Conventionally it is represented by an arrow with an initial point and an endpoint; the endpoint being the arrow head. We work with plane vectors by adding them, subtracting one from the other, and by multiplying them with a number. We see that the plane vectors have a structure, which is revealed through the two operations, namely addition and multiplication by a number, also called scalar multiplication. These operations can be seen in an alternate way by identifying the vectors with points in the plane. The identification goes as follows.

Since only length and direction matter and not exactly the initial or the endpoints, we may think of each vector having its initial point at the origin. The endpoint can then be identified with the vector itself. With O as the origin with Cartesian coordinates $(0, 0)$ and P as the point with Cartesian coordinates (a, b), the vector $\overrightarrow{OP}$ is identified with the point (a, b) in the plane

$$\mathbb{R}^2 = \{(\alpha, \beta) : \alpha \in \mathbb{R},\ \beta \in \mathbb{R}\}.$$

Then the familiar parallelogram law for addition of vectors translates to component-wise addition. If u, v are vectors with initial point $(0, 0)$ and endpoints (a, b) and (c, d), respectively, then the vector $u + v$ has initial point $(0, 0)$ and endpoint $(a + c, b + d)$. Similarly, for a real number α, the vector αu has the initial point $(0, 0)$ and endpoint $(\alpha a, \alpha b)$.

M. T. Nair and A. Singh, *Linear Algebra*,
https://doi.org/10.1007/978-981-13-0926-7_1

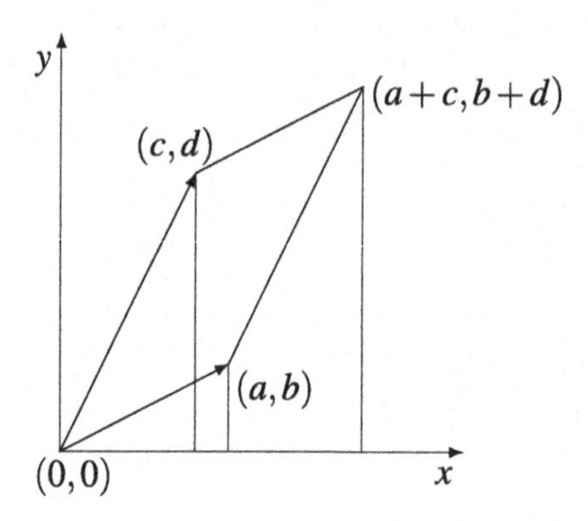

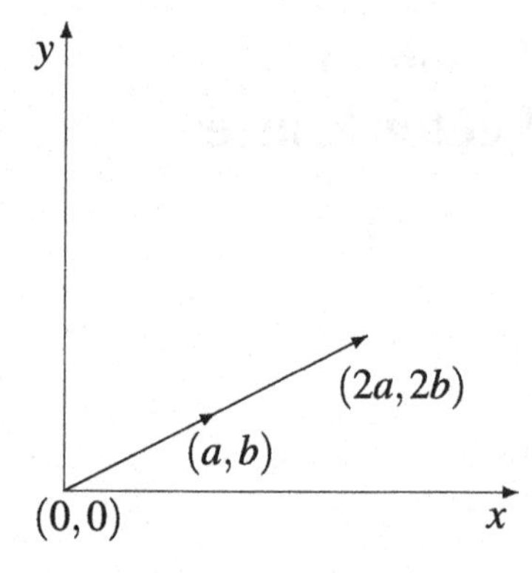

Thus, $(-1)\,u$, which equals $(-a, -b)$, represents the additive inverse $-u$ of the vector u; the direction of $-u$ is opposite to that of u. Now, the plane is simply viewed as a set of all plane vectors.

Similarly, in the three-dimensional space, you may identify a vector with a point by first translating the vector to have its initial point as the origin and its arrow head as the required point. The sum of two vectors in three dimensions gives rise to the component-wise sum of two points. A real number α times a vector gives a vector whose components are multiplied by α. That is, if $u = (a_1, b_1, c_1)$ and $v = (a_2, b_2, c_2)$, then

$$u + v = (a_1 + a_2, b_1 + b_2, c_1 + c_2), \quad \alpha u = (\alpha a_1, \alpha b_1, \alpha c_1).$$

Notice that the zero vector, written as 0, is identified with the point $(0, 0, 0)$, and the vector $-u = (-a_1, -b_1, -c_1)$ satisfies $u + (-u) = 0$.

The notion of a *vector space* is an abstraction of the familiar set of vectors in two or three dimensions. The idea is to keep the familiar properties of addition of vectors and multiplication of a vector by a scalar. The set of scalars can be any field. For obtaining interesting geometrical results, we may have to restrict the field of scalars. In this book, the field $\mathbb{F}$ denotes either the field $\mathbb{R}$ of real numbers or the field $\mathbb{C}$ of complex numbers.

Definition 1.1 A **vector space** over $\mathbb{F}$ is a nonempty set V along with two operations, namely

(a) **addition,** which associates each pair (x, y) of elements $x, y \in V$ with a unique element in V, denoted by $x + y$, and
(b) **scalar multiplication,** which associates each pair (α, x), for $\alpha \in \mathbb{F}$ and $x \in V$, with a unique element in V, denoted by αx,

satisfying the following conditions:

(1) For all $x, y \in V,\ x + y = y + x$.
(2) For all $x, y, z \in V,\ (x + y) + z = x + (y + z)$.
(3) There exists an element in V, called a **zero vector**, denoted by 0, such that for all $x \in V,\ x + 0 = x$.

(4) For each $x \in V$, there exists an element in V, denoted by $-x$, and called an **additive inverse of** x, such that $x + (-x) = 0$.
(5) For all $\alpha \in \mathbb{F}$ and for all $x, y \in V$, $\alpha(x + y) = \alpha x + \alpha y$.
(6) For all $\alpha, \beta \in \mathbb{F}$ and for all $x \in V$, $(\alpha + \beta)x = \alpha x + \beta x$.
(7) For all $\alpha, \beta \in \mathbb{F}$ and for all $x \in V$, $(\alpha\beta)x = \alpha(\beta x)$.
(8) For all $x \in V$, $1x = x$.

Elements of $\mathbb{F}$ are called **scalars**, and elements of a vector space V are called **vectors**. A vector space V over $\mathbb{R}$ is called a **real vector space**, and a vector space over $\mathbb{C}$ is called a **complex vector space**. As a convention, we shorten the expression "a vector space over $\mathbb{F}$" to "a vector space". We denote vectors by the letters u, v, w, x, y, z with or without subscripts, and scalars by the letters $a, b, c, d, \alpha, \beta, \gamma, \delta$ with or without subscripts.

You have ready-made examples of vector spaces. The plane

$$\mathbb{R}^2 = \{(a, b) : a, b \in \mathbb{R}\}$$

and the familiar three-dimensional space

$$\mathbb{R}^3 = \{(a, b, c) : a, b, c \in \mathbb{R}\}$$

are real vector spaces. Notice that $\mathbb{R}$ is a vector space over $\mathbb{R}$, and $\mathbb{C}$ is a vector space over $\mathbb{C}$ as well as over $\mathbb{R}$. Before presenting more examples of vector spaces, we observe some subtleties about the conditions (3) and (4) in Definition 1.1. It is unusual to write a particular symbol such as 0 for all zero vectors. It is also unusual to write $-x$ for all additive inverses of x. The philosophical hurdle will be over once we prove that a zero vector is unique and an additive inverse of a vector is also unique.

Theorem 1.2 *In any vector space the following statements are true:*

(1) *There exists exactly one zero vector.*
(2) *Each vector has exactly one additive inverse.*

Proof Let V be a vector space.
(1) Suppose 0 and $\tilde{0}$ in V are zero vectors. Then for all $x \in V$, $x + 0 = x$ and $x + \tilde{0} = x$. Using the condition (1) in Definition 1.1, we have

$$\tilde{0} = \tilde{0} + 0 = 0 + \tilde{0} = 0.$$

(2) Let $x \in V$. Let x' and $\tilde{x}$ be additive inverses of x. Let 0 be the zero vector. Then $x + x' = 0$ and $x + \tilde{x} = 0$. Therefore,

$$\tilde{x} = \tilde{x} + 0 = \tilde{x} + (x + x') = (\tilde{x} + x) + x' = (x + \tilde{x}) + x' = 0 + x' = x' + 0 = x'.$$ ∎

Theorem 1.2 justifies the use of the symbols 0 for *the zero vector* and $-x$ for *the additive inverse* of the vector x. Of course, we could have used any other symbol, say, θ for the zero vector and x' for the additive inverse of x; but the symbols 0 and $-x$ follow the custom. Note that $-0 = 0$. We also write $y - x$ instead of $y + (-x)$ for all vectors x and y.

Notice the double meanings used in Definition 1.1. The addition of scalars as well as of vectors is denoted by the same symbol $+$, and the multiplication of scalars as well as of a vector with a scalar is written by just concatenating the elements. Similarly, 0 denotes the zero vector as well as the scalar zero. Even the notation for the additive inverse of vector x is $-x$; just the way we write $-\alpha$ for the additive inverse of a scalar α. You should get acquainted with the double meanings.

It is easy to check that in every vector space V over $\mathbb{F}$,

$$0 + x = x, \quad x + (y - x) = y \quad \text{for all } x, y \in V.$$

Every vector space contains at least one element, the zero vector. On the other hand, the singleton $\{0\}$ is a vector space; it is called the **zero space** or the **trivial** vector space. In general, we will be concerned with *nonzero* vector spaces, which contain nonzero elements. A nonzero vector space is also called a *nontrivial* vector space.

In a vector space, addition of two elements is allowed. This is generalized by induction to a sum of any finite number of vectors. But an infinite sum of vectors is altogether a different matter; it requires analytic notions such as convergence.

Example 1.3 In the following, the sets along with the specified addition and scalar multiplication are vector spaces. (Verify.)

(1) Consider the set $\mathbb{F}^n$ of all n-tuples of scalars, that is,

$$\mathbb{F}^n := \{(a_1, \ldots, a_n) : a_1, \ldots, a_n \in \mathbb{F}\}.$$

We assume that two elements in $\mathbb{F}^n$ are equal when their respective components are equal. For $x = (a_1, \ldots, a_n)$, $y = (b_1, \ldots, b_n) \in \mathbb{F}^n$, and $\alpha \in \mathbb{F}$, define the addition and scalar multiplication component-wise, that is,

$$x + y := (a_1 + b_1, \ldots, a_n + b_n), \quad \alpha x := (\alpha a_1, \ldots, \alpha a_n).$$

Then $\mathbb{F}^n$ is a vector space with

$$0 = (0, \ldots, 0) \text{ and } -(a_1, \ldots, a_n) = (-a_1, \ldots, -a_n).$$

(2) We use the notation $\mathbb{F}^{m \times n}$ for the set of all $m \times n$ **matrices** with entries from $\mathbb{F}$. A matrix $A \in \mathbb{F}^{m \times n}$ is usually written as

$$A = \begin{bmatrix} a_{11} & \cdots & a_{1n} \\ \vdots & & \vdots \\ a_{m1} & \cdots & a_{mn} \end{bmatrix},$$

or as $A = [a_{ij}]$ for short, with $a_{ij} \in \mathbb{F}$ for $i = 1, \ldots, m;\ j = 1, \ldots, n$. The number a_{ij} which occurs at the entry in ith row and jth column is referred to as the (i, j)th entry of the matrix $[a_{ij}]$. For $A = [a_{ij}]$ and $B = [b_{ij}]$ in $\mathbb{F}^{m\times n}$, and $\alpha \in \mathbb{F}$, we define

$$A + B = [a_{ij} + b_{ij}] \in \mathbb{F}^{m\times n}, \quad \alpha A = [\alpha a_{ij}] \in \mathbb{F}^{m\times n}.$$

We say that two matrices are equal when their respective entries are equal. That is, for $A = [a_{ij}]$ and $B = [b_{ij}]$, we write $A = B$ if and only if $a_{ij} = b_{ij}$. With these operations of addition and scalar multiplication, $\mathbb{F}^{m\times n}$ becomes a vector space over $\mathbb{F}$. The zero vector in $\mathbb{F}^{m\times n}$ is the zero matrix, i.e. the matrix with all entries 0, and the additive inverse of $A = [a_{ij}] \in \mathbb{F}^{m\times n}$ is the matrix $-A := [-a_{ij}]$.

(3) For $n \in \{0, 1, 2, \ldots\}$, let $\mathcal{P}_n(\mathbb{F})$ denote the set of all polynomials (in the variable t) of degree at most n, with coefficients in $\mathbb{F}$. That is, $x \in \mathcal{P}_n(\mathbb{F})$ if and only if x is of the form

$$x = a_0 + a_1 t + \cdots + a_n t^n$$

for some scalars $a_0, a_1, \ldots, a_n$. Here, we assume that a scalar is a polynomial of degree 0. Further, two polynomials are considered equal when the coefficients of respective powers of t are equal. That is,

$$a_0 + a_1 t + \cdots + a_n t^n = b_0 + b_1 t + \cdots + b_n t^n \text{ if and only if } a_i = b_i \text{ for } i = 0, \ldots, n.$$

Addition and scalar multiplication on $\mathcal{P}_n(\mathbb{F})$ are defined as follows. For $x = a_0 + a_1 t + \cdots + a_n t^n,\ y = b_0 + b_1 t + \cdots + b_n t^n$ in $\mathcal{P}_n(\mathbb{F})$, and $\alpha \in \mathbb{F}$,

$$x + y := (a_0 + b_0) + (a_1 + b_1)t + \cdots + (a_n + b_n)t^n,$$
$$\alpha x := \alpha a_0 + \alpha a_1 t + \cdots + \alpha a_n t^n.$$

The zero vector in $\mathcal{P}_n(\mathbb{F})$ is the polynomial with all its coefficients zero, and

$$-(a_0 + a_1 t + \cdots + a_n t^n) = -a_0 - a_1 t - \cdots - a_n t^n.$$

Then $\mathcal{P}_n(\mathbb{F})$ is a vector space.

(4) Let $\mathcal{P}(\mathbb{F}) = \cup_{n=0}^{\infty} \mathcal{P}_n(\mathbb{F})$, the set of all polynomials with coefficients in $\mathbb{F}$. That is, $x \in \mathcal{P}(\mathbb{F})$ if and only if $x = a_0 + a_1 t + \cdots + a_n t^n$ for some $n \in \{0, 1, 2, \ldots\}$ and for some scalars $a_0, \ldots, a_n \in \mathbb{F}$.

If $x, y \in \mathcal{P}(\mathbb{F})$, then $x \in \mathcal{P}_m(\mathbb{F})$ and $y \in \mathcal{P}_n(\mathbb{F})$ for some m, n. So, $x, y \in \mathcal{P}_k(\mathbb{F})$, with $k = \max\{m, n\}$. The equality relation, addition, and scalar multiplication are defined as in (3). Then $\mathcal{P}(\mathbb{F})$ is a vector space.

(5) Let V be the set of all sequences of scalars. A sequence whose nth term is a_n is written as (a_n). Two sequences are considered equal if and only if their respective terms are equal, that is, $(a_n) = (b_n)$ if and only if $a_n = b_n$ for each $n \in \mathbb{N}$. For $(a_n),\ (b_n) \in V$, and $\alpha \in \mathbb{F}$, define

$$(a_n) + (b_n) := (a_n + b_n), \quad \alpha(a_n) := (\alpha a_n).$$

That is,

$$(a_1, a_2, a_3, \ldots) + (b_1, b_2, b_3, \ldots) = (a_1 + b_1, a_2 + b_2, a_3 + b_3, \ldots),$$

$$\alpha(a_1, a_2, a_3. \ldots) = (\alpha a_1, \alpha a_2, \alpha a_3, \ldots).$$

With this addition and scalar multiplication, V is a vector space, where its zero vector is the sequence with each term as zero, and $-(a_n) = (-a_n)$. This space is called the **sequence space** and is denoted by $\mathbb{F}^\infty$.

(6) Let S be a nonempty set. Let V be a vector space over $\mathbb{F}$. Let $\mathcal{F}(S, V)$ be the set of all functions from S into V. As usual, $x = y$ for $x, y \in \mathcal{F}(S, V)$ when $x(s) = y(s)$ for each $s \in S$. For $x,\ y \in \mathcal{F}(S, V)$ and $\alpha \in \mathbb{F}$, define $x + y$ and αx point-wise; that is,

$$(x + y)(s) := x(s) + y(s), \quad (\alpha x)(s) := \alpha x(s) \quad \text{for } s \in S.$$

Let the functions 0 and $-x$ in $\mathcal{F}(S, V)$ be defined by

$$0(s) = 0, \quad (-x)(s) = -x(s) \quad \text{for } s \in S.$$

Then $\mathcal{F}(S, V)$ is a vector space over $\mathbb{F}$ with the zero vector as 0 and the additive inverse of x as $-x$. We sometimes refer to this space as a **function space**. □

Comments on Notation: $\mathcal{P}_n(\mathbb{R})$ denotes the real vector space of all polynomials of degree at most n with real coefficients. $\mathcal{P}_n(\mathbb{C})$ denotes the complex vector space of all polynomials of degree at most n with complex coefficients. Similarly, $\mathcal{P}(\mathbb{R})$ is the real vector space of all polynomials with real coefficients, and $\mathcal{P}(\mathbb{C})$ is the complex vector space of all polynomials with complex coefficients. Note that $\mathbb{C}$ is also a vector space over $\mathbb{R}$. Similarly, $\mathcal{P}_n(\mathbb{C})$ and $\mathcal{P}(\mathbb{C})$ are vector spaces over $\mathbb{R}$. More generally, if V is a complex vector space, then it is also a real vector space. If at all we require to regard any vector space over $\mathbb{C}$ also as a vector space over $\mathbb{R}$, we will specifically mention it.

As particular cases of Example 1.3(2), (Read: Example 1.3 Part 2) we have the vector spaces $\mathbb{F}^{m\times 1}$, the set of all **column vectors** of size m, and $\mathbb{F}^{1\times n}$, the set of all **row vectors** of size n. To save space, we use the transpose notation in writing a column vector. That is, a column vector v of size n with its entries $a_1, \ldots, a_n$ is written as

$$\begin{bmatrix} a_1 \\ \vdots \\ a_n \end{bmatrix} \quad \text{or as} \quad [a_1 \ \cdots \ a_n]^T.$$

By putting the superscript T over a row vector v we mean that the column vector is obtained by taking *transpose* of the row vector v. When the column vectors are written by the lower case letters u, v, w, x, y, z with or without subscripts (sometimes superscripts), the corresponding row vectors will be written with the transpose notation, that is, as u^T, v^T. Further, we will not distinguish between the square brackets and the parentheses. Usually, we will write a row vector with parentheses. Thus, we will not distinguish between

$$[a_1 \ \cdots \ a_n] \quad \text{and} \quad (a_1, \ \ldots, \ a_n).$$

Thus we regard $\mathbb{F}^n$ same as $\mathbb{F}^{1\times n}$. We may recall that by taking **transpose** of a matrix in $\mathbb{F}^{m\times n}$, we obtain a matrix in $\mathbb{F}^{n\times m}$. That is, if $A = [a_{ij}] \in \mathbb{F}^{m\times n}$, then

$$A^T := [b_{ji}] \in \mathbb{F}^{n\times m} \quad \text{with} \quad b_{ji} = a_{ij} \quad \text{for } i = 1, \ldots, m; \ j = 1, \ldots, n.$$

Many vector spaces can be viewed as function spaces. For example, with $S = \mathbb{N}$ and $V = \mathbb{F}$, we obtain the sequence space of Example 1.3(5). With $S = \{1, \ldots, n\}$ and $V = \mathbb{F}$, each function in $\mathcal{F}(S, \mathbb{F})$ can be specified by an n-tuple of its function values. Therefore, the vector space $\mathcal{F}(\{1, \ldots, n\}, \mathbb{F})$ can be viewed as $\mathbb{F}^n$ and also as $\mathbb{F}^{n\times 1}$. Some more examples of function spaces follow.

Example 1.4 (1) Let I be an interval, and let $C(I, \mathbb{R})$ denote the set of all real-valued continuous functions defined on I. For $x, y \in C(I, \mathbb{R})$ and $\alpha \in \mathbb{R}$, define $x + y$ and αx point-wise as in Example 1.3(6).

The functions $x + y$ and αx are in $C(I, \mathbb{R})$. Then $C(I, \mathbb{R})$ is a real vector space with the zero element as the zero function and the additive inverse of $x \in C(I, \mathbb{R})$ as the function $-x$ defined by $(-x)(t) = -x(t)$ for all $t \in I$.

(2) Let $\mathcal{R}([a, b], \mathbb{R})$ denote the set of all real-valued Riemann integrable functions on $[a, b]$. Define addition and scalar multiplication point-wise, as in Example 1.3(6). From the theory of Riemann integration, it follows that if $x, y \in \mathcal{R}([a, b], \mathbb{R})$ and $\alpha \in \mathbb{R}$, then $x + y, \ \alpha x \in \mathcal{R}([a, b], \mathbb{R})$. It is a real vector space.

(3) For $k \in \mathbb{N}$, let $C^k([a, b], \mathbb{F})$ denote the set of all functions x from $[a, b]$ to $\mathbb{F}$ such that the kth derivative $x^{(k)}$ exists and is continuous on $[a, b]$.

Define addition and scalar multiplication point-wise, as in Example 1.3(6). Then $C^k([a, b], \mathbb{F})$ is a vector space. Notice that

$$C^k([a, b], \mathbb{R}) \subseteq C([a, b], \mathbb{R}) \subseteq \mathcal{R}([a, b], \mathbb{R}) \subseteq \mathcal{F}([a, b], \mathbb{R}). \qquad \square$$

Example 1.5 Let $V_1, \ldots, V_n$ be vector spaces over $\mathbb{F}$. Consider the *Cartesian product*

$$V = V_1 \times \cdots \times V_n = \{(x_1, \ldots, x_n) : x_1 \in V_1, \ldots, x_n \in V_n\}.$$

Define addition and scalar multiplication on V by

$$(x_1, \ldots, x_n) + (y_1, \ldots, y_n) := (x_1 + y_1, \ldots, x_n + y_n),$$

$$\alpha(x_1, \ldots, x_n) := (\alpha x_1, \ldots, \alpha x_n).$$

In V, take the zero vector as $(0, \ldots, 0)$ and $-(x_1, \ldots, x_n) = (-x_1, \ldots, -x_n)$. Here, the addition in the expression $x_i + y_i$ is the addition operation defined in V_i and the scalar multiplication in αx_i is the scalar multiplication defined in V_i. Similarly, in $(0, \ldots, 0)$, the ith component is the zero vector of V_i and $-x_i$ in $(-x_1, \ldots, -x_n)$ is the additive inverse of x_i in V_i.

With these operations, V is a vector space. We call it the **product space** of $V_1, \ldots, V_n$. □

To illustrate Example 1.5, consider $V_1 = \mathbb{R}$ and $V_2 = \mathcal{P}_1(\mathbb{R})$. Then $V_1 \times V_2$ consists of vectors of the form $(a, \alpha + \beta t)$, and the operations of addition and scalar multiplication are defined by

$$(a_1, \alpha_1 + \beta_1 t) + (a_2, \alpha_2 + \beta_2 t) = (a_1 + a_2, (\alpha_1 + \alpha_2) + (\beta_1 + \beta_2)t),$$
$$c(a, \alpha + \beta t) = (ca, c\alpha + c\beta t).$$

Similarly, the space $\mathbb{F}^n$ is a product space with each V_i as $\mathbb{F}$.

Some easy consequences of Definition 1.1 are listed in the following theorem.

Theorem 1.6 *Let V be a vector space over $\mathbb{F}$. Let $x, y, z \in V$ and let $\alpha, \beta \in \mathbb{F}$.*

(1) $0x = 0$.
(2) $\alpha 0 = 0$.
(3) $(-1)x = -x$.
(4) $-(-x) = x$.
(5) *If $x + z = y + z$, then $x = y$.*
(6) *If $\alpha x = 0$, then $\alpha = 0$ or $x = 0$.*
(7) *If $\alpha \neq \beta$ and $x \neq 0$, then $\alpha x \neq \beta x$.*

Proof (1) $0x = (0 + 0)x = 0x + 0x$. Adding $-0x$, we have $0 = 0x$.

(2) $\alpha 0 = \alpha(0 + 0) = \alpha 0 + \alpha 0$. Adding $-\alpha 0$, we obtain $0 = \alpha 0$.

(3) $x + (-1)x = (1 + (-1))x = 0x = 0$, by (1). Adding $-x$ we get $(-1)x = -x$.

(4) From (3), it follows that $-(-x) = (-1)(-1)x = x$.

(5) Suppose that $x + z = y + z$. Then $(x + z) + (-z) = (y + z) + (-z)$. But $(x + z) + (-z) = x + (z + (-z)) = x + 0 = x$. Similarly, $(y + z) + (-z) = y$. Therefore, $x = y$.

(6) Suppose that $\alpha x = 0$. If $\alpha \neq 0$, then $\alpha^{-1} \in \mathbb{F}$. Multiplying α^{-1}, we have $\alpha^{-1}\alpha x = \alpha^{-1}0$. That is, $x = 0$.

(7) Suppose that $\alpha \neq \beta$ and $x \neq 0$. If $\alpha x = \beta x$, then $(\alpha - \beta)x = 0$. By (6), $x = 0$. It leads to a contradiction. Hence, $\alpha x \neq \beta x$. ■

Theorem 1.6 allows us to do algebraic manipulation of vectors like scalars as long as addition is concerned. Thus we abbreviate $x + (-y)$ to $x - y$. However, there is a big difference: vectors cannot be multiplied or raised to powers, in general. Look at the proof of Theorem 1.6(6). We used α^{-1} instead of x^{-1}.

It follows from Theorem 1.6(7) that every nontrivial vector space contains infinitely many vectors.

Exercises for Sect. 1.1

In the following a set V, a field $\mathbb{F}$, which is either $\mathbb{R}$ or C, and the operations of addition and scalar multiplication are given. Check whether V is a vector space over $\mathbb{F}$, with these operations.

1. With addition and scalar multiplication as in $\mathbb{R}^2$,

   (a) $V = \{(a, b) \in \mathbb{R}^2 : 2a + 3b = 0\},\ \mathbb{F} = \mathbb{R}$.
   (b) $V = \{(a, b) \in \mathbb{R}^2 : a + b = 1\},\ \mathbb{F} = \mathbb{R}$.
   (c) $V = \{(a, b) \in \mathbb{R}^2 : ab = 0\},\ \mathbb{F} = \mathbb{R}$.

2. $V = \mathbb{R}^2,\ \mathbb{F} = \mathbb{R}$, with addition as in $\mathbb{R}^2$, and scalar multiplication as given by the following: for $(a, b) \in V,\ \alpha \in \mathbb{R}$,

   (a) $\alpha(a, b) := (a, 0)$.
   (b) $\alpha(a, b) := (b, \alpha a)$.
   (c) $\alpha(a, b) := (\alpha a, -\alpha b)$.
   (d) $\alpha(a, b) := \begin{cases} (0, 0) & \text{if } \alpha = 0 \\ (\alpha a, b/\alpha) & \text{if } \alpha \neq 0. \end{cases}$

3. $V = \{x \in \mathbb{R} : x > 0\},\ \mathbb{F} = \mathbb{R}$, and for $x, y \in V,\ \alpha \in \mathbb{R},\ x + y := xy,\ \alpha x := x^{\alpha}$. The operations on the left are defined by the known operations on the right.
4. $V = \{x \in \mathbb{R} : x \geq 0\},\ \mathbb{F} = \mathbb{R}$, and for $x, y \in V,\ \alpha \in \mathbb{R},\ x + y := xy,\ \alpha x := |\alpha| x$.
5. $V = \mathbb{C}^2, \mathbb{F} = \mathbb{C}$, and for $x = (a, b),\ y = (c, d),\ \alpha \in \mathbb{C}$,

$$x + y := (a + 2c, b + 3d), \quad \alpha x := (\alpha a, \alpha b).$$

6. V is the set of all polynomials of degree 5 with real coefficients, $\mathbb{F} = \mathbb{R}$, and the operations are the addition and scalar multiplication of polynomials.
7. S is a nonempty set, $s \in S$, V is the set of all functions $f : S \to \mathbb{R}$ with $f(s) = 0$, $\mathbb{F} = \mathbb{R}$, and the operations are the addition and scalar multiplication of functions.
8. V is the set of all functions $f : \mathbb{R} \to \mathbb{C}$ satisfying $f(-t) = \overline{f(t)}, \mathbb{F} = \mathbb{R}$, and the operations are the addition and scalar multiplication of functions.
9. $V = \{x\}$, where x is some symbol, and addition and scalar multiplication are defined as $x + x = x,\ \alpha x = x$ for all $\alpha \in \mathbb{F}$.

1.2 Subspaces

A subset of a vector space may or may not be a vector space. It will be interesting if a subset forms a vector space over the same underlying field and with the same operations of addition and scalar multiplication inherited from the given vector space. If U is a subset of a vector space V (over the field $\mathbb{F}$), then the operations of addition and scalar multiplication in U *inherited from* V are defined as follows:

Let $x, y \in U$, $\alpha \in \mathbb{F}$. Consider x, y as elements of V. The vector $x + y$ in V is the result of the inherited addition of x and y in U. Similarly, the vector αx in V is the result of the inherited scalar multiplication of α with x in U.

In order that the operations of addition $(x, y) \mapsto x + y$ and scalar multiplication $(\alpha, x) \mapsto \alpha x$ are well-defined operations on U, we require the vectors $x + y$ and αx to lie in U. This condition is described by asserting that U *is closed under* the inherited operations.

Definition 1.7 Let V be a vector space over $\mathbb{F}$. A nonempty subset U of V is called a **subspace** of V if U is a vector space over $\mathbb{F}$ with respect to the operations of addition and scalar multiplication inherited from V.

To show that a nonempty subset U of a vector space V is a subspace, one must first verify that U is closed under the inherited operations. The closure conditions can be explicitly stated as follows:

For all $x, y \in U$ and for each $\alpha \in \mathbb{F}$, $x + y \in U$ and $\alpha x \in U$.

Here, of course, the operations are the operations in the given vector space V over $\mathbb{F}$. Surprisingly, these closure conditions are enough for establishing that a subset is a subspace, as the following theorem shows.

Theorem 1.8 *Let U be a nonempty subset of a vector space V. Then U is a subspace of V if and only if U is closed under addition and scalar multiplication inherited from V.*

Proof If U is a subspace of V, then $x + y \in U$ and $\alpha x \in U$ for all $x, y \in U$ and for all $\alpha \in \mathbb{F}$.

Conversely, suppose that $x + y \in U$ and $\alpha x \in U$ for all $x, y \in U$ and for all $\alpha \in \mathbb{F}$. Since U is nonempty, let $u \in U$. By Theorem 1.6, $0 = 0\,u \in U$, and for each $x \in U$, $-x = (-1)x \in U$. Since $U \subseteq V$, 0 acts as the zero vector of U, and $-x$ acts as the additive inverse of x in U. Therefore, the conditions (3)–(4) in Definition 1.1 are satisfied for U. The remaining conditions hold since elements of U are elements of V as well. ■

Notice that the closure conditions in Theorem 1.8 can be replaced by the following single condition:

For each scalar $\alpha \in \mathbb{F}$ and for all $x, y \in U$, $x + \alpha y \in U$.

We may also infer from the proof of Theorem 1.8 that

> if U is a subspace of a vector space V, then the zero vector of U is the same as the zero vector of V, and for each $u \in U$, its additive inverse $-u$ as an element of U is the same as $-u$ in V.

Therefore, in order that U is a subspace of V, the zero vector of V must be in U.

Example 1.9 (1) Let $U = \{(a, b) \in \mathbb{R}^2 : b = 0\} \subseteq \mathbb{R}^2$. That is, $U = \{(a, 0) : a \in \mathbb{R}\}$. Addition and scalar multiplication, which are defined component-wise in $\mathbb{R}^2$, are also operations on U since for all $a, b, c, \alpha \in \mathbb{R}$,

$$(a, 0) + (b, 0) = (a + b, 0) \in U, \quad \alpha(c, 0) = (\alpha c, 0) \in U.$$

By Theorem 1.8, U is a subspace of $\mathbb{R}^2$. Notice that the zero vector of U is the same $(0, 0)$ and $-(a, 0) = (-a, 0)$ as in $\mathbb{R}^2$.

(2) The set $U = \{(a, b) \in \mathbb{R}^2 : 2a + 3b = 0\}$ is a subspace of $\mathbb{R}^2$ (Verify).

(3) Let $\mathbb{Q}$ denote the set of all rational numbers. $\mathbb{Q}$ is not a subspace of the real vector space $\mathbb{R}$ since $1 \in \mathbb{Q}$ but $\sqrt{2} \cdot 1 \notin \mathbb{Q}$. Similarly, $\mathbb{Q}^2$ is not a subspace of $\mathbb{R}^2$.

(4) Consider $\mathbb{C}$ as a complex vector space. Let $U = \{a + i\,0 : a \in \mathbb{R}\}$. We see that $1 \in U$ but $i \cdot 1 = i \notin U$. Therefore, U is not a subspace of $\mathbb{C}$.

However, if we consider $\mathbb{C}$ as real vector space, then U is a subspace of $\mathbb{C}$. In this sense, $U = \mathbb{R}$ is a subspace of the real vector space $\mathbb{C}$.

(5) Consider the spaces $\mathcal{P}_m(\mathbb{F})$ and $\mathcal{P}_n(\mathbb{F})$, where $m \le n$. Each polynomial of degree at most m is also a polynomial of degree at most n. Thus, $\mathcal{P}_m(\mathbb{F}) \subseteq \mathcal{P}_n(\mathbb{F})$. Further, $\mathcal{P}_m(\mathbb{F})$ is closed under the operations of addition and scalar multiplication inherited from $\mathcal{P}_n(\mathbb{F})$. So, $\mathcal{P}_m(\mathbb{F})$ is a subspace of $\mathcal{P}_n(\mathbb{F})$ for any $m \le n$.

Also, for each $n \in \mathbb{N}$, $\mathcal{P}_n(\mathbb{F})$ is a subspace of $\mathcal{P}(\mathbb{F})$.

(6) In Examples 1.4(1)–(2), both $\mathcal{C}([a, b], \mathbb{R})$ and $\mathcal{R}([a, b], \mathbb{R})$ are vector spaces. Since $\mathcal{C}([a, b], \mathbb{R}) \subseteq \mathcal{R}([a, b], \mathbb{R})$ and the operations of addition and scalar multiplication in $\mathcal{C}([a, b], \mathbb{R})$ are inherited from $\mathcal{R}([a, b], \mathbb{R})$, we conclude that $\mathcal{C}([a, b], \mathbb{R})$ is a subspace of $\mathcal{R}([a, b], \mathbb{R})$.

(7) Consider $\mathcal{C}^k([a, b], \mathbb{F})$ of Example 1.4(3). For all $\alpha \in \mathbb{F}$, $x, y \in \mathcal{C}^k([a, b], \mathbb{F})$, we have $x + y \in \mathcal{C}^k([a, b], \mathbb{F})$ and $\alpha x \in \mathcal{C}^k([a, b], \mathbb{F})$. By Theorem 1.8, $\mathcal{C}^k([a, b], \mathbb{F})$ is a subspace of $\mathcal{C}([a, b], \mathbb{F})$.

(8) Given $\alpha_1, \ldots, \alpha_n \in \mathbb{F}$, $U = \{(b_1, \ldots, b_n) \in \mathbb{F}^n : \alpha_1 b_1 + \cdots + \alpha_n b_n = 0\}$ is a subspace of $\mathbb{F}^n$. When $(\alpha_1, \ldots, \alpha_n)$ is a nonzero n-tuple, the subspace U is a *hyperplane* passing through the origin in n dimensions. This terminology is partially borrowed from the case of $\mathbb{F} = \mathbb{R}$ and $n = 3$, when the subspace $\{(b_1, b_2, b_3) \in \mathbb{R}^3 : \alpha_1 b_1 + \alpha_2 b_2 + \alpha_3 b_3 = 0,\ \alpha_1, \alpha_2, \alpha_3 \in \mathbb{R}\}$ of $\mathbb{R}^3$ is a plane passing through the origin. However,

$$W = \{(b_1, \ldots, b_n) \in \mathbb{F}^n : \alpha_1 b_1 + \cdots + \alpha_n b_n = 1,\ \alpha_1, \ldots, \alpha_n \in \mathbb{F}\}$$

is not a subspace of $\mathbb{F}^n$, since $(0, \ldots, 0) \notin W$.

(9) Let $\mathcal{P}([a, b], \mathbb{R})$ be the vector space $\mathcal{P}(\mathbb{R})$ where each polynomial is considered as a function from $[a, b]$ to $\mathbb{R}$. Then the space $\mathcal{P}([a, b], \mathbb{R})$ is a subspace of $C^k([a, b], \mathbb{R})$ for each $k \geq 1$. □

Consider two planes passing through the origin in $\mathbb{R}^3$. Their intersection is a straight line passing through the origin. In fact, intersection of two subspaces is a subspace. We prove a more general result.

Theorem 1.10 *Let $\mathcal{C}$ be any collection of subspaces of a vector space V. Let U be the intersection of all subspaces in $\mathcal{C}$. Then U is a subspace of V.*

Proof Let $x, y \in U$ and let $\alpha \in \mathbb{F}$. Then $x, y \in W$ for each $W \in \mathcal{C}$. Since W is a subspace of V, $\alpha x \in W$ and $x + y \in W$ for each $W \in \mathcal{C}$. Then $\alpha x \in U$ and $x + y \in U$. By Theorem 1.8, U is a subspace of V. ■

In contrast, union of two subspaces need not be a subspace. For, consider

$$U := \{(a, b) \in \mathbb{R}^2 : b = a\}, \quad V := \{(a, b) \in \mathbb{R}^2 : b = 2a\}$$

as subspaces of $\mathbb{R}^2$. Here, $(1, 1), (1, 2) \in U \cup V$ but $(1, 1) + (1, 2) = (2, 3) \notin U \cup V$. Hence $U \cup V$ is not a subspace of $\mathbb{R}^2$.

Union of two subspaces can fail to be a subspace since addition of a vector from one with a vector from the other may not be in the union. What happens to the set of all vectors of the form $x + y$, where x is a vector from one subspace and y is from another subspace? To answer this question, we introduce the notion of a (finite) sum of subsets of a vector space.

Definition 1.11 Let $S_1, \ldots, S_n$ be nonempty subsets of a vector space V. Their **sum** is defined by

$$S_1 + \cdots + S_n := \{x_1 + \cdots + x_n \in V : x_i \in S_i, \ i = 1, \ldots, n\}.$$

As expected, sum of two subspaces is a subspace. And, the proof can be generalized easily to any finite sum of subspaces.

Theorem 1.12 *Let $V_1, \ldots, V_n$ be subspaces of a vector space V. Then $V_1 + \cdots + V_n$ is a subspace of V.*

Proof Let $x, y \in V_1 + \cdots + V_n$ and let $\alpha \in \mathbb{F}$. Then, $x = \sum_{i=1}^n x_i$ and $y = \sum_{i=1}^n y_i$ for some $x_i, y_i \in V_i$, $i = 1, \ldots, n$. Since each V_i is a subspace of V, $x_i + y_i \in V_i$ and $\alpha x_i \in V_i$. Then

$$x + y = \sum_{i=1}^n x_i + \sum_{i=1}^n y_i = \sum_{i=1}^n (x_i + y_i) \ \in \ V_1 + \cdots + V_n,$$
$$\alpha x = \alpha \sum_{i=1}^n x_i = \sum_{i=1}^n (\alpha x_i) \ \in \ V_1 + \cdots + V_n.$$

Therefore, $V_1 + \cdots + V_n$ is a subspace of V. ■

Example 1.13 Consider the planes:

$$V_1 = \{(a, b, c) \in \mathbb{R}^3 : a + b + c = 0\}, \quad V_2 = \{(a, b, c) \in \mathbb{R}^3 : a + 2b + 3c = 0\}.$$

Both V_1 and V_2 are subspaces of $\mathbb{R}^3$. Their intersection is the subspace

$$V_1 \cap V_2 = \{(a, b, c) \in \mathbb{R}^3 : a + b + c = 0 = a + 2b + 3c\}.$$

The condition $a + b + c = 0 = a + 2b + 3c$ is equivalent to $b = -2a, \; c = a$. Hence

$$V_1 \cap V_2 = \{(a, -2a, a) : a \in \mathbb{R}\}$$

which is a straight line through the origin. Both $V_1 \cap V_2$ and $V_1 + V_2$ are subspaces of $\mathbb{R}^3$. In this case, we show that $V_1 + V_2 = \mathbb{R}^3$. For this, it is enough to show that $\mathbb{R}^3 \subseteq V_1 + V_2$. It requires to express any $(a, b, c) \in \mathbb{R}^3$ as $(a_1 + a_2, b_1 + b_2, c_1 + c_2)$ for some $(a_1, b_1, c_1) \in V_1$ and $(a_2, b_2, c_2) \in V_2$. This demands determining the six unknowns $a_1, b_1, c_1, a_2, b_2, c_2$ from the five linear equations

$$a_1 + a_2 = a, \quad b_1 + b_2 = b, \quad c_1 + c_2 = c, \quad a_1 + b_1 + c_1 = 0, \quad a_2 + 2b_2 + 3c_2 = 0.$$

It may be verified that with

$$a_1 = -a - 2b - 2c, \quad b_1 = a + 2b + c, \; c_1 = c$$
$$a_2 = 2a + 2b + 2c, \quad b_2 = -a - b - c, \; c_2 = 0,$$

the five equations above are satisfied. Thus, $(a_1, b_1, c_1) \in V_1$, $(a_2, b_2, c_2) \in V_2$, and $(a, b, c) = (a_1 + b_1 + c_1) + (a_2 + b_2 + c_2)$ as desired. □

Exercises for Sect. 1.2

1. In the following, check whether the given subset U is a subspace of V. Assume the usual operations of addition and scalar multiplication along with a suitable field $\mathbb{R}$ or $\mathbb{C}$.

   (a) $V = \mathbb{R}^2$, U is any straight line passing through the origin.
   (b) $V = \mathbb{R}^2$, $U = \{(a, b) : b = 2a - 1\}$.
   (c) $V = \mathbb{R}^3$, $U = \{(a, b, c) : 2a - b - c = 0\}$.
   (d) $V = \mathcal{P}_3(\mathbb{R})$, $U = \{a + bt + ct^2 + dt^3 : c = 0\}$.
   (e) $V = \mathcal{P}_3(\mathbb{C})$, $U = \{a + bt + ct^2 + dt^3 : a = 0\}$.
   (f) $V = \mathcal{P}_3(\mathbb{C})$, $U = \{a + bt + ct^2 + dt^3 : b + c + d = 0\}$.
   (g) $V = \mathcal{P}_3(\mathbb{R})$, $U = \{p \in V : \; p(0) = 0\}$.
   (h) $V = \mathcal{P}_3(\mathbb{C})$, $U = \{p \in V : \; p(1) = 0\}$.
   (i) $V = \mathcal{P}_3(\mathbb{C})$, $U = \{a + bt + ct^2 + dt^3 : a, b, c, d$ integers $\}$.
   (j) $V = C([-1, 1], \mathbb{R})$, $U = \{f \in V : f$ is an odd function$\}$.
   (k) $V = C([0, 1], \mathbb{R})$, $U = \{f \in V : f(t) \geq 0$ for all $t \in [0, 1]\}$.

(l) $V = C^k[a, b]$, for $k \in \mathbb{N}$, $U = \mathcal{P}[a, b]$, the set of all polynomials considered as functions on $[a, b]$.
(m) $V = C([0, 1], \mathbb{R})$, $U = \{f \in V : f \text{ is differentiable}\}$.

2. For $\alpha \in \mathbb{F}$, let $V_\alpha = \{(a, b, c) \in \mathbb{F}^3 : a + b + c = \alpha\}$. Show that V_α is a subspace of $\mathbb{F}^3$ if and only if $\alpha = 0$.
3. Give an example of a nonempty subset of $\mathbb{R}^2$ which is closed under addition and under additive inverse (i.e. if u is in the subset, then so is $-u$), but is not a subspace of $\mathbb{R}^2$.
4. Give an example of a nonempty subset of $\mathbb{R}^2$ which is closed under scalar multiplication but is not a subspace of $\mathbb{R}^2$.
5. Suppose U is a subspace of V and V is a subspace of W. Show that U is a subspace of W.
6. Give an example of subspaces of $\mathbb{C}^3$ whose union is not a subspace of $\mathbb{C}^3$.
7. Show by a counter-example that if $U + W = U + X$ for subspaces U, W, X of V, then W need not be equal to X.
8. Let $m \in \mathbb{N}$. Does the set $\{0\} \cup \{x \in \mathcal{P}(\mathbb{R}) : \text{ degree of } x \text{ is equal to } m\}$ form a subspace of $\mathcal{P}(\mathbb{R})$?
9. Prove that the only nontrivial proper subspaces of $\mathbb{R}^2$ are straight lines passing through the origin.
10. Let $U = \{(a, b) \in \mathbb{R}^2 : a = b\}$. Find a subspace V of $\mathbb{R}^2$ such that $U + V = \mathbb{R}^2$ and $U \cap V = \{(0, 0)\}$. Is such a V unique?
11. Let U be the subspace of $\mathcal{P}(\mathbb{F})$ consisting of all polynomials of the form $at^3 + bt^7$ for $a, b \in \mathbb{F}$. Find a subspace V of $\mathcal{P}(\mathbb{F})$ such that $U + V = \mathcal{P}(\mathbb{F})$ and $U \cap V = \{0\}$.
12. Let U and W be subspaces of a vector space V. Prove the following:

    (a) $U \cup W = V$ if and only if $U = V$ or $W = V$.
    (b) $U \cup W$ is a subspace of V if and only if $U \subseteq W$ or $W \subseteq U$.

13. Let $U = \{A \in \mathbb{F}^{n\times n} : A^T = A\}$ and let $W = \{A \in \mathbb{F}^{n\times n} : A^T = -A\}$. Matrices in U are called *symmetric matrices*, and matrices in W are called *skew-symmetric matrices*. Show that U and W are subspaces of $\mathbb{F}^{n\times n}$, $\mathbb{F}^{n\times n} = U + W$, and $U \cap W = \{0\}$.

1.3 Linear Span

The sum of vectors $x_1, \ldots, x_n$ can be written as $x_1 + \cdots + x_n$, due to Property (2) of vector addition in Definition 1.1. We may also multiply the vectors with scalars and then take their sum. That is, we write the sum

$$(\alpha_1 x_1) + \cdots + (\alpha_n x_n) \text{ as } \alpha_1 x_1 + \cdots + \alpha_n x_n, \text{ and also as } \sum_{i=1}^{n} \alpha_i x_i.$$

If the jth term in $\alpha_1 x_1 + \cdots + \alpha_n x_n$ is absent, that is, if we want to consider the sum $(\alpha_1 x_1 + \cdots + \alpha_n x_n) - \alpha_j x_j$ for some $j \in \{1, \ldots, n\}$, then we write it as $\sum_{i=1,\, i \neq j}^{n} \alpha_i x_i$ and also as

$$\alpha_1 x_1 + \cdots + \alpha_{j-1} x_{j-1} + \alpha_{j+1} x_{j+1} + \cdots \alpha_n x_n,$$

with the understanding that if $j = 1$, then the above sum is $\alpha_2 x_2 + \cdots + \alpha_n x_n$, and if $j = n$, then the sum is equal to $\alpha_1 x_1 + \cdots + \alpha_{n-1} x_{n-1}$.

Definition 1.14 Let V be a vector space.

(a) A vector $v \in V$ is called a **linear combination** of vectors $u_1, \ldots, u_n$ in V if $v = \alpha_1 u_1 + \cdots + \alpha_n u_n$ for some scalars $\alpha_1, \ldots, \alpha_n$.
(b) The **linear span** or the **span** of any nonempty subset S of V is the set of all linear combinations of finite number of vectors from S; it is denoted by $\text{span}(S)$. We define $\text{span}(\varnothing)$ as $\{0\}$.

In view of the above definition, for any nonempty subset S of V, we have

$x \in \text{span}(S)$ if and only if there exist $n \in \mathbb{N}$, vectors $u_1, \ldots, u_n$ in S, and scalars $\alpha_1, \ldots, \alpha_n \in \mathbb{F}$ such that $x = \alpha_1 u_1 + \cdots + \alpha_n u_n$.

It follows that

$$\text{span}(S) = \{\alpha_1 u_1 + \cdots + \alpha_n u_n : n \in \mathbb{N},\ \alpha_1, \ldots, \alpha_n \in \mathbb{F},\ u_1, \ldots, u_n \in S\}.$$

Moreover, for $u_1, \ldots, u_n$ in V,

$$\text{span}(\{u_1, \ldots, u_n\}) = \{\alpha_1 u_1 + \cdots + \alpha_n u_n : \alpha_1, \ldots, \alpha_n \in \mathbb{F}\}.$$

We also write $\text{span}(\{u_1, \ldots, u_n\})$ as $\text{span}\{u_1, \ldots, u_n\}$. Informally, we say that $\text{span}(S)$ is *the set of all linear combinations of elements of* S, remembering the special case that $\text{span}(\varnothing) = 0$.

By a linear combination, we always mean a linear combination of a *finite number* of vectors. As we know an expression of the form $\alpha_1 v_1 + \alpha_2 v_2 + \cdots$ for vectors $v_1, v_2, \ldots$ and scalars $\alpha_1, \alpha_2, \ldots$ has no meaning in a vector space, unless there is some additional structure which may allow infinite sums.

In what follows, we will be freely using *Kronecker's delta* defined as follows:

$$\delta_{ij} = \begin{cases} 1 & \text{if } i = j \\ 0 & \text{if } i \neq j \end{cases} \qquad \text{for } i, j \in \mathbb{N}.$$

Example 1.15 (1) In $\mathbb{R}^3$, consider the set $S = \{(1, 0, 0), (0, 2, 0), (0, 0, 3), (2, 1, 3)\}$. A linear combination of elements of S is a vector in the form

$$\alpha_1(1, 0, 0) + \alpha_2(0, 2, 0) + \alpha_3(0, 0, 3) + \alpha_4(2, 1, 3)$$

for some scalars $\alpha_1, \alpha_2, \alpha_3, \alpha_4$. Since span$(S)$ is the set of all linear combinations of elements of S, it contains all vectors that can be expressed in the above form. For instance, $(1, 2, 3)$, $(4, 2, 9) \in \text{span}(S)$ since

$$(1, 2, 3) = 1(1, 0, 0) + 1(0, 2, 0) + 1(0, 0, 3),$$
$$(4, 2, 9) = 4(1, 0, 0) + 1(0, 2, 0) + 3(0, 0, 3).$$

Also, $(4, 2, 9) = 1(0, 0, 3) + 2(2, 1, 3)$. In fact, span$(S) = \mathbb{R}^3$ since $(\alpha, \beta, \gamma) = \alpha(1, 0, 0) + (\beta/2)(0, 2, 0) + (\gamma/3)(0, 0, 3)$.

(2) In $\mathbb{R}^3$, span$\{(3, 0, 0)\}$ is the set of all scalar multiples of $(3, 0, 0)$. It is the straight line L joining the origin and the point $(3, 0, 0)$. Notice that

$$L = \text{span}\{(1, 0, 0)\} = \text{span}\{(\sqrt{2}, 0, 0),\ (\pi, 0, 0),\ (3, 0, 0)\} = \text{span}(L).$$

It can be seen that span of any two vectors not in a straight line containing the origin is the plane containing those two vectors and the origin.

(3) For each $j \in \{1, \ldots, n\}$, let e_j be the vector in $\mathbb{F}^n$ whose jth coordinate is 1 and all other coordinates are 0, that is, $e_j = (\delta_{1j}, \ldots, \delta_{nj})$. Then for any $(\alpha_1, \ldots, \alpha_n) \in \mathbb{F}^n$, we have

$$(\alpha_1, \ldots, \alpha_n) = \alpha_1 e_1 + \cdots + \alpha_n e_n.$$

Thus, span$\{e_1, \ldots, e_n\} = \mathbb{F}^n$. Also, for any k with $1 \le k < n$,

$$\text{span}\{e_1, \ldots, e_k\} = \{(\alpha_1, \ldots, \alpha_n) \in \mathbb{F}^n : \alpha_j = 0 \text{ for } j > k\}.$$

(4) Consider the vector spaces $\mathcal{P}(\mathbb{F})$ and $\mathcal{P}_n(\mathbb{F})$. Define the polynomials $u_j := t^{j-1}$ for $j \in \mathbb{N}$. Then $\mathcal{P}_n(\mathbb{F})$ is the span of $\{u_1, \ldots, u_{n+1}\}$, and $\mathcal{P}(\mathbb{F})$ is the span of $\{u_1, u_2, \ldots\}$.

(5) Let $V = \mathbb{F}^\infty$, the set of all sequences with scalar entries. For each $n \in \mathbb{N}$, let e_n be the sequence whose nth term is 1 and all other terms are 0, that is,

$$e_n = (\delta_{n1}, \delta_{n2}, \ldots).$$

Then span$\{e_1, e_2, \ldots\}$ is the space of all scalar sequences having only a finite number of nonzero terms. This space is usually denoted by $c_{00}(\mathbb{N}, \mathbb{F})$, also as c_{00}. Notice that $c_{00}(\mathbb{N}, \mathbb{F}) \ne \mathbb{F}^\infty$. □

Clearly, if S is a nonempty subset of a vector space V, then $S \subseteq \text{span}(S)$. For, if $x \in S$ then $x = 1 \cdot x \in \text{span}(S)$. However, span$(S)$ need not be equal to S. For instance, $S = \{(1, 0)\} \subseteq \mathbb{R}^2$ and span$(S) = \{(\alpha, 0) : \alpha \in \mathbb{R}\} \ne S$.

Theorem 1.16 *Let S be a subset of a vector space V. Then the following statements are true:*

(1) $\text{span}(S)$ *is a subspace of V, and it is the smallest subspace containing S.*
(2) $\text{span}(S)$ *is the intersection of all subspaces of V that contain S.*

Proof (1) First, we show that $\text{span}(S)$ is a subspace of V. If $S = \varnothing$, then $\text{span}(\varnothing) = \{0\}$ is a subspace of V that contains $\varnothing$. So, suppose that $S \neq \varnothing$. Trivially, $S \subseteq \text{span}(S)$. To show that $\text{span}(S)$ is a subspace of V, let $x, y \in \text{span}(S)$ and let $\alpha \in \mathbb{F}$. Then

$$x = a_1x_1 + \cdots + a_nx_n, \quad y = b_1y_1 + \cdots + b_my_m$$

for some vectors $x_1, \ldots, x_n, y_1, \ldots, y_m$ in S and scalars $a_1, \ldots, a_n, b_1, \ldots, b_m$ in $\mathbb{F}$. Then

$$x + y = a_1x_1 + \cdots + a_nx_n + b_1y_1 + \cdots + b_my_m \in \text{span}(S),$$
$$\alpha x = \alpha a_1x_1 + \cdots + \alpha a_nx_n \in \text{span}(S).$$

By Theorem 1.8, $\text{span}(S)$ is a subspace of V.

For the second part of the statement in (1), let U be a subspace of V containing S. Let $v \in \text{span}(S)$. There exist vectors $v_1, \ldots, v_n \in S$ and scalars $\alpha_1, \ldots, \alpha_n$ such that $v = \alpha_1v_1 + \cdots + \alpha_nv_n$. Since $v_1, \ldots, v_n \in U$ and U is a subspace of V, $v \in U$. Hence, $\text{span}(S) \subseteq U$. That is, $\text{span}(S)$ is a subset of every subspace that contains S. Therefore, $\text{span}(S)$ is the smallest subspace of V containing S.

(2) Let W be the intersection of all subspaces of V that contain S. Since $\text{span}(S)$ is a subspace of V, $W \subseteq \text{span}(S)$. Also, W is a subspace of V containing S; thus, due to (1), $\text{span}(S) \subseteq W$. ■

Theorem 1.16 implies that taking span of a subset amounts to extending the subset to a subspace in a minimalistic way.

Some useful consequences of the notion of span are contained in the following theorem.

Theorem 1.17 *Let S, S_1 and S_2 be subsets of a vector space V. Then the following are true:*

(1) $S = \text{span}(S)$ *if and only if S is a subspace of V.*
(2) $\text{span}(\text{span}(S)) = \text{span}(S)$.
(3) *If $S_1 \subseteq S_2$, then* $\text{span}(S_1) \subseteq \text{span}(S_2)$.
(4) $\text{span}(S_1) + \text{span}(S_2) = \text{span}(S_1 \cup S_2)$.
(5) *If $x \in S$, then* $\text{span}(S) = \text{span}\{x\} + \text{span}(S \backslash \{x\})$.

Proof (1) Since $\text{span}(S)$ is a subspace of V, the condition $S = \text{span}(S)$ implies that S is a subspace of V. Conversely, if S is a subspace of V, then the minimal subspace containing S is S. By Theorem 1.16, $\text{span}(S) = S$.

(2) As $\text{span}(S)$ is a subspace of V, by (1), $\text{span}(\text{span}(S)) = \text{span}(S)$.

(3) Suppose $S_1 \subseteq S_2$. If $S_1 = \varnothing$, the conclusion is obvious. Suppose $S_1 \neq \varnothing$. Any linear combination of vectors from S_1 is an element of $\text{span}(S_2)$. Hence $\text{span}(S_1) \subseteq \text{span}(S_2)$.

(4) If $S_1 = \varnothing$ or $S_2 = \varnothing$, then the statement is true trivially. Assume that $S_1 \neq \varnothing$ and $S_2 \neq \varnothing$. Let $x \in \text{span}(S_1) + \text{span}(S_2)$. Then x is a linear combination of vectors from S_1 plus a linear combination of vectors from S_2. That is, x is a linear combination of vectors from $S_1 \cup S_2$. Therefore,

$$\text{span}(S_1) + \text{span}(S_2) \subseteq \text{span}(S_1 \cup S_2).$$

Conversely, suppose $x \in \text{span}(S_1 \cup S_2)$. Then x is a linear combination of vectors from $S_1 \cup S_2$. If such a linear combination uses vectors only from S_1, then $x \in \text{span}(S_1) \subseteq \text{span}(S_1) + \text{span}(S_2)$. Similarly, if such a linear combination uses vectors from only S_2, then $x \in \text{span}(S_2) \subseteq \text{span}(S_1) + \text{span}(S_2)$. Otherwise, x is equal to a linear combinations of vectors from S_1 plus a linear combination of vectors from S_2. In that case, $x \in \text{span}(S_1) + \text{span}(S_2)$. Therefore,

$$\text{span}(S_1 \cup S_2) \subseteq \text{span}(S_1) + \text{span}(S_2).$$

(5) This follows from (4) by taking $S_1 = \{x\}$ and $S_2 = S \backslash \{x\}$. ■

As a corollary of Theorem 1.17(4), we obtain the following.

Theorem 1.18 *If V_1 and V_2 are subspaces of a vector space V, then $V_1 + V_2 = \text{span}(V_1 \cup V_2)$.*

Theorems 1.16 and 1.18 show that in extending the union of two subspaces to a subspace could not have been better; the best way is to take their sum.

Definition 1.19 Let S be a subset of a vector space V. If $\text{span}(S) = V$, then we say that S **spans** V, S is a **spanning set** of V, and also V **is spanned by** S.

In case, $S = \{u_1, \ldots, u_n\}$ is a finite spanning set of V, we also say that the vectors $u_1, \ldots, u_n$ span V, and that V is spanned by the vectors $u_1, \ldots, u_n$.

Any vector space spans itself. But there can be much smaller subsets that also span the space as Example 1.15 shows. Note that both $\varnothing$ and $\{0\}$ are spanning sets of the vector space $\{0\}$.

If S is a spanning set of V and $x \in \text{span}(S \backslash \{x\})$, then $S \backslash \{x\}$ is also a spanning set of V. For, in this case, the vector x is a linear combination of some vectors from $S \backslash \{x\}$, and if any vector in V is a linear combination, where x appears, we can replace this x by its linear combination to obtain v as a linear combination of vectors where x does not appear.

Spanning sets are not unique. Given any spanning set, a new spanning set can be obtained by incorporating new vectors into it. Further, new spanning sets can be constructed from old ones by exchanging suitable nonzero vectors. Theorem 1.20 below shows that this is possible.

Let S be a spanning set of a nonzero vector space V. Let $x \in V$ be a nonzero vector. Then $x = \alpha_1 x_1 + \cdots + \alpha_n x_n$ for some scalars $\alpha_1, \ldots, \alpha_n$ and for some distinct nonzero vectors $x_1, \ldots, x_n$ in S. Also, since $x \neq 0$, at least one of $\alpha_1, \ldots, \alpha_n$ is nonzero. If $\alpha_i \neq 0$, then $x = \alpha y + z$, with $\alpha = \alpha_i$, $y = x_i$ and $z \in \text{span}(S \setminus \{y\})$. Indeed, $z = \alpha_1 x_1 + \cdots + \alpha_{i-1} x_{i-1} + \alpha_{i+1} x_{i+1} + \cdots + \alpha_n x_n$.

We use such a writing of a nonzero vector x in the following result.

Theorem 1.20 (Exchange lemma) *Let S be a spanning set of a vector space V and let $x \in V \setminus S$ be a nonzero vector. Let $x = \alpha y + z$ for some nonzero $y \in S$, a nonzero $\alpha \in \mathbb{F}$ and $z \in \text{span}(S \setminus \{y\})$. Then $(S \setminus \{y\}) \cup \{x\}$ is a spanning set of V.*

Proof Let $v \in V$. Since $\text{span}(S) = V$, there exist vectors $u \in \text{span}(S \setminus \{y\})$ and a scalar β such that $v = u + \beta y$. Using $y = \alpha^{-1}(x - z)$ we obtain

$$v = u + \beta y = u + \beta\alpha^{-1}(x - z) = (u - \beta\alpha^{-1}z) + \beta\alpha^{-1}x \in \text{span}\big((S \setminus \{y\}) \cup \{x\}\big).$$

Therefore, $\text{span}\big((S \setminus \{y\}) \cup \{x\}\big) = V$. ■

Example 1.21 To illustrate the proof of the Exchange lemma, consider the set

$$S = \{(1, 0, 1), (0, 1, 1), (1, 1, 0), (1, 2, 3)\}.$$

It can be seen that S spans $\mathbb{R}^3$. Consider the vector $x := (2, 1, 1) \notin S$. Note that

$$(2, 1, 1) = (1, 0, 1) + (1, 1, 0).$$

Thus, $x = \alpha y + z$ with $\alpha = 1$, $y = (1, 0, 1)$ and $z = (1, 1, 0)$. According to the Exchange lemma, the new set

$$\{(2, 1, 1), (0, 1, 1), (1, 1, 0), (1, 2, 3)\}$$

spans $\mathbb{R}^3$. Taking $y = (1, 1, 0)$ and $z = (1, 0, 1)$ we obtain the spanning set

$$\{(1, 0, 1), (0, 1, 1), (2, 1, 1), (1, 2, 3)\}.$$

We could have started with another linear combination of $(2, 1, 1)$ such as

$$(2, 1, 1) = -2(0, 1, 1) + 1(1, 1, 0) + 1(1, 2, 3).$$

Again, by the Exchange lemma, any of the vectors $(0, 1, 1)$, $(1, 1, 0)$, $(1, 2, 3)$ can be replaced by $(2, 1, 1)$ to obtain the following new spanning sets:

$$\{(1, 0, 1), (2, 1, 1), (1, 1, 0), (1, 2, 3)\},$$

$$\{(1, 0, 1), (0, 1, 1), (2, 1, 1), (1, 2, 3)\},$$

$$\{(1, 0, 1), (0, 1, 1), (1, 1, 0), (2, 1, 1)\}.$$ □

Exercises for Sect. 1.3

1. Show that $\text{span}\{e_1 + e_2, e_2 + e_3, e_3 + e_1\} = \mathbb{R}^3$, where $e_1 = (1, 0, 0)$, $e_2 = (0, 1, 0)$, and $e_3 = (0, 0, 1)$.
2. What is $\text{span}\{t^n : n = 0, 2, 4, 6, \ldots\}$?
3. Do the polynomials $t^3 - 2t^2 + 1$, $4t^2 - t + 3$, and $3t - 2$ span $\mathcal{P}_3(\mathbb{C})$?
4. Let $u_1(t) = 1$, and for $j = 2, 3, \ldots$, let $u_j(t) = 1 + t + \ldots + t^{j-1}$. Show that span of $\{u_1, \ldots, u_n\}$ is $\mathcal{P}_n(\mathbb{F})$, and span of $\{u_1, u_2, \ldots\}$ is $\mathcal{P}(\mathbb{F})$.
5. Is it true that $\sin t \in \text{span}\{1, t, t^2, t^3, \ldots\}$?
6. Let x, y, z be nonzero distinct vectors in a vector space V with $x + y + z = 0$. Show that $\text{span}\{x, y\} = \text{span}\{x, z\} = \text{span}\{y, z\}$.
7. Let V be a vector space; U a subspace of V; $x, y \in V$; $X = \text{span}(U \cup \{x\})$; $Y = \text{span}(U \cup \{y\})$; and let $y \in X \backslash U$. Show that $x \in Y$.
8. Let V be a vector space. Let $u, v, w_1, \ldots, w_n$ be distinct vectors in V; $B = \{u, w_1, \ldots, w_n\}$; and let $C = \{v, w_1, \ldots, w_n\}$. Prove that $\text{span}(B) = \text{span}(C)$ if and only if $u \in \text{span}(C)$ if and only if $v \in \text{span}(B)$.
9. Let A and B be subsets of a vector space V.

   (a) Show that $\text{span}(A \cap B) \subseteq \text{span}(A) \cap \text{span}(B)$.
   (b) Give an example where $\text{span}(A) \cap \text{span}(B) \nsubseteq \text{span}(A \cap B)$.

10. Suppose U and W are subspaces of a vector space V. Show that $U + W = U$ if and only if $W \subseteq U$.
11. Let S be a subset of a vector space V. Let $u \in V$ be such that $u \notin \text{span}(S)$. Prove that for each $x \in \text{span}(S \cup \{u\})$, there exists a unique pair (α, v) with $\alpha \in \mathbb{F}$ and $v \in \text{span}(S)$ such that $x = \alpha u + v$.
12. In the following cases, find a finite spanning subset S of U.

    (a) $U = \{(a, b, c) \in \mathbb{F}^4 : a + b + c = 0\}$.
    (b) $U = \{(a, b, c, d) \in \mathbb{F}^4 : 5a + 2b - c = 3a + 2c - d = 0\}$.
    (c) $U = \{a + bt + ct^2 + dt^3 : a - 2b + 3c - 4d = 0\}$.

13. Construct a finite spanning set for $\mathbb{F}^{m \times n}$.
14. Construct a finite spanning set for the subspace $\{A \in \mathbb{R}^{n \times n} : A^T = A\}$ of $\mathbb{R}^{n \times n}$.
15. Let e_i be the sequence of real numbers whose ith term is i and all other terms 0. What is $\text{span}\{e_1, e_2, e_3, \ldots\}$?

1.4 Linear Independence

A spanning set may contain redundancies; a vector in it may be in the span of the rest. In this case, deleting such a vector will result in a smaller spanning set.

Definition 1.22 Let B be a subset of a vector space V.

(a) B is said to be a **linearly dependent set** if there exists a vector $v \in B$ such that $v \in \operatorname{span}(B \setminus \{v\})$.
(b) B is said to be **linearly independent** if B is not linearly dependent.

Example 1.23 (1) In $\mathbb{R}$, the set $\{1, 2\}$ is linearly dependent, since $2 = 2 \times 1$. Notice the double meaning here. The 2 on the left is a vector but the 2 on the right is a scalar. In fact, any set of two vectors in $\mathbb{R}$ is linearly dependent.

(2) In $\mathbb{R}^2$, the set $\{(1, 0), (\pi, 0)\}$ is linearly dependent, since $(\pi, 0) = \pi(1, 0)$. Also, $\{(1, 0), (2, 3), (0, 1)\}$ is linearly dependent, since $(2, 3) = 2(1, 0) + 3(0, 1)$.

(3) In $\mathbb{R}^2$, the set $\{(1, 0), (0, 1)\}$ is linearly independent, since neither $(1, 0)$ is a scalar multiple of $(0, 1)$, nor $(0, 1)$ is a scalar multiple of $(1, 0)$.

(4) In $\mathbb{R}^3$, the set $\{(1, 3, 2), (1, 2, 3), (2, 4, 6)\}$ is linearly dependent since $(2, 4, 6) = 2(1, 2, 3)$.

Notice that $(1, 3, 2)$ is not a linear combination of the other two vectors. That is, a set of vectors to be linearly dependent, it is not necessary that each vector is a linear combination of the others; it is enough if at least one of the vectors is a linear combination of the others.

To illustrate this point further, let u be a nonzero vector. Since the zero vector satisfies $0 = 0 \cdot u$, the set $\{u, 0\}$ is linearly dependent, but u is not scalar multiple of the zero vector.

(5) In $\mathbb{R}^3$, the set $\{(1, 0, 0), (1, 1, 0), (1, 1, 1)\}$ is linearly independent. To show this, we must prove that neither of the vectors is a linear combination of the other two. So, suppose $(1, 0, 0) = \alpha(1, 1, 0) + \beta(1, 1, 1)$. Then comparing the components, we see that $1 = \alpha + \beta,\ 0 = \alpha + \beta,\ 0 = \beta$. This yields $1 = 0$, a contradiction. Similarly, the other two statements can be shown.

(6) In the complex vector space $\mathbb{C}$, the set $\{1, i\}$ is linearly dependent as $i = i \times 1$. The i on the left side is a vector, whereas the i on the right is a scalar.

When $\mathbb{C}$ is considered as a real vector space, the set $\{1, i\}$ is linearly independent, since neither i is equal to any real number times 1, nor 1 is equal to any real number times i.

(7) In the vector space $\mathcal{P}(\mathbb{F})$, the set $\{1, 1 + t, 1 + t^2\}$ is linearly independent. □

The following statements are immediate consequences of Definition 1.22:

1. The empty set $\varnothing$ is linearly independent.
2. The set $\{0\}$ is linearly dependent since $\operatorname{span}(\varnothing) = \{0\}$.
3. A singleton set $\{x\}$, with $x \neq 0$, is linearly independent.

4. The set $\{u, v\}$ is linearly dependent if and only if one of u, v is a scalar multiple of the other.
5. Every proper superset of a spanning set is linearly dependent.
6. Every superset of a linearly dependent set is linearly dependent.
7. Every subset of a linearly independent set is linearly independent.

Since linear dependence involves linear combinations, which are finite sums, the following result should be easy to understand.

Theorem 1.24 *Let B be a subset of a vector space V.*

(1) *B is linearly dependent if and only if some finite subset of B is linearly dependent.*
(2) *B is linearly independent if and only if each finite subset of B is linearly independent.*

Proof (1) Let B be a linearly dependent set. Then there exists a vector $v \in B$ such that $v \in \operatorname{span}(B \setminus \{v\})$. If $B \setminus \{v\} = \varnothing$, then $v = 0$ and hence $\{v\}$ is linearly dependent. If $B \setminus \{v\} \neq \varnothing$, then there exist vectors $v_1, \ldots, v_n$ in $B \setminus \{v\}$ such that v is a linear combination of $v_1, \ldots, v_n$. That is, $v \in \operatorname{span}\{v_1, \ldots, v_n\}$. Then the finite subset $\{v, v_1, \ldots, v_n\}$ of B is linearly dependent.

Conversely, if a finite subset of B is linearly dependent, then as a superset of a linearly dependent set, B is linearly dependent.

(2) It follows from (1). ■

We will have occasions to determine when vectors in a list are linearly dependent or independent. For instance, suppose A is a 2×2 matrix whose rows are $[1\ 2]$ and $[1\ 2]$. Obviously, the two rows of A are linearly dependent, in the sense that one is a linear combination of the other. But the set of the two rows, which is equal to $\{[1\ 2]\}$, is linearly independent. In such a context, Definition 1.25 given below will come of help.

We often abbreviate the phrase "the list of vectors $v_1, \ldots, v_n$" to the phrase "the vectors $v_1, \ldots, v_n$".

Definition 1.25 Let V be a vector space. The vectors $v_1, \ldots, v_n$, from V, are said to be **linearly dependent** if there exists $i \in \{1, \ldots, n\}$ such that $v_i \in \operatorname{span}\{v_j : 1 \leq j \leq n,\ j \neq i\}$.

The vectors $v_1, \ldots, v_n$ are said to be **linearly independent** if they are not linearly dependent.

A finite set of vectors can always be seen as a list of vectors, with some ordering of its elements. As usual, when we write a list of vectors as $u, v, \ldots, z$, we assume that in the ordering, u is the first vector, v is the second vector, and so on, so that z is the last vector.

The following result helps in determining linear dependence or independence of a list of vectors.

Theorem 1.26 *Let $v_1, \ldots, v_n$ be vectors in a vector space V, where $n \geq 2$.*

(1) *The vectors $v_1, \ldots, v_n$ are linearly dependent if and only if there exist scalars $\alpha_1, \ldots, \alpha_n$, with at least one of them nonzero, such that*

$$\alpha_1 v_1 + \cdots + \alpha_n v_n = 0.$$

(2) *The vectors $v_1, \ldots, v_n$ are linearly independent if and only if for scalars $\alpha_1, \ldots, \alpha_n$,*

$$\alpha_1 v_1 + \cdots + \alpha_n v_n = 0 \text{ implies } \alpha_1 = \cdots = \alpha_n = 0.$$

Proof (1) Suppose $v_1, \ldots, v_n$ are linearly dependent. Then there exists j with $1 \leq j \leq n$, and there exist scalars $\beta_1, \ldots, \beta_{j-1}, \beta_{j+1}, \ldots, \beta_n$ such that

$$v_j = \beta_1 v_1 + \cdots + \beta_{j-1} v_{j-1} + \beta_{j+1} v_{j+1} + \cdots + \beta_n v_n.$$

Thus, $\beta_1 v_1 + \cdots + \beta_n v_n = 0$ with $\beta_j = -1$.

Conversely, let $\alpha_1 v_1 + \cdots + \alpha_n v_n = 0$, with at least one of $\alpha_1, \ldots, \alpha_n$ nonzero. Suppose $\alpha_k \neq 0$. Then,

$$v_k = -(\alpha_k)^{-1}(\alpha_1 v_1 + \cdots + \alpha_{k-1} v_{k-1} + \alpha_{k+1} v_{k+1} + \cdots + \alpha_n v_n).$$

Hence $v_1, \ldots, v_n$ are linearly dependent.

(2) It follows from (1). ■

Theorem 1.26 provides the following method:

To show that the vectors $v_1, \ldots v_n$ are linearly independent, we start with a linear combination of these vectors, equate it to 0, and then deduce that each coefficient in that linear combination is 0.

Example 1.27 Let us check whether the vectors $(1, 0, 0), (1, 1, 0), (1, 1, 1)$ are linearly independent in $\mathbb{R}^3$. For this, we start with a linear combination of these vectors and equate it to 0. That is, let $a, b, c \in \mathbb{F}$ be such that

$$a(1, 0, 0) + b(1, 1, 0) + c(1, 1, 1) = 0.$$

This equation gives $(a + b + c, b + c, c) = (0, 0, 0)$. From this, we obtain

$$a + b + c = 0, \; b + c = 0, \; c = 0.$$

It implies that $a = b = c = 0$. Thus, the vectors are linearly independent. □

If a linearly independent set is extended by adding a vector and the extended set is found to be linearly dependent, then the new vector must be in the span of the old set, as the following theorem shows.

Theorem 1.28 *Let S be a linearly independent subset of a vector space V. Let $v \in V \setminus S$. Then $S \cup \{v\}$ is linearly dependent if and only if $v \in \text{span}(S)$. Equivalently, $S \cup \{v\}$ is linearly independent if and only if $v \notin \text{span}(S)$.*

Proof If $v \in \text{span}(S)$, then clearly, $S \cup \{v\}$ is linearly dependent. For the converse, let $S \cup \{v\}$ be linearly dependent. If $S = \varnothing$, then $S \cup \{v\} = \{v\}$ is linearly dependent implies $v = 0 \in \text{span}(S)$.

So, assume that $S \neq \varnothing$. Due to Theorem 1.26, $S \cup \{v\}$ is linearly dependent implies that there exist $v_1, \ldots, v_n \in S$, and $\alpha_1, \ldots, \alpha_n, \alpha \in \mathbb{F}$ with at least one of them nonzero, such that $\alpha_1 v_1 + \cdots + \alpha_n v_n + \alpha v = 0$. Since S is linearly independent, $\alpha \neq 0$. Then $v = \alpha^{-1}(\alpha_1 v_1 + \cdots + \alpha_n v_n) \in \text{span}(S)$. ■

Given a vector occurring in a list, we can speak of vectors preceding it and vectors succeeding it. This helps in streamlining the notion of linear dependence a bit.

Theorem 1.29 *Let $v_1, \ldots, v_n$ be linearly dependent vectors in a vector space V, with $n \geq 2$. Then either $v_1 = 0$ or there exists $k \geq 2$ such that $v_1, \ldots, v_{k-1}$ are linearly independent, and $v_k \in \text{span}\{v_1, \ldots, v_{k-1}\}$.*

Proof Suppose $v_1 \neq 0$. Then the vector v_1 is linearly independent. Let k be the least index such that $v_1, \ldots, v_k$ are linearly dependent. Then $2 \leq k \leq n$ and $v_1, \ldots, v_{k-1}$ are linearly independent. From Theorem 1.28 it follows that $v_k \in \text{span}\{v_1, \ldots, v_{k-1}\}$. ■

A statement similar to Theorem 1.29 holds even for countably infinite subsets of vectors. One may formulate and prove such a statement.

Linear independence is an important notion as to bringing in uniqueness in a linear combination; see the following theorem.

Theorem 1.30 *Let V be a vector space over $\mathbb{F}$. The vectors $v_1, \ldots, v_n \in V$ are linearly independent if and only if for each $v \in \text{span}\{v_1, \ldots, v_n\}$, there exists a unique n-tuple $(\alpha_1, \ldots, \alpha_n) \in \mathbb{F}^n$ such that $v = \alpha_1 v_1 + \cdots + \alpha_n v_n$.*

Proof Let the vectors $v_1, \ldots, v_n$ be linearly independent. Let $v \in \text{span}\{v_1, \ldots, v_n\}$. Then $v = \alpha_1 v_1 + \cdots + \alpha_n v_n$ for some $\alpha_1, \ldots, \alpha_n \in \mathbb{F}$. If there exist $\beta_1, \ldots, \beta_n \in \mathbb{F}$ such that $v = \beta_1 v_1 + \cdots + \beta_n v_n$, then

$$(\alpha_1 - \beta_1)v_1 + \cdots + (\alpha_n - \beta_n)v_n = 0.$$

Since the vectors $v_1, \ldots, v_n$ are linearly independent, $\alpha_1 = \beta_1, \ldots, \alpha_n = \beta_n$. That is, the n-tuple $(\alpha_1, \ldots, \alpha_n)$ is unique.

Conversely, assume that corresponding to each $v \in \text{span}\{v_1, \ldots, v_n\}$, we have a unique n-tuple $(\alpha_1, \ldots, \alpha_n) \in \mathbb{F}^n$ such that $v = \alpha_1 v_1 + \cdots + \alpha_n v_n$. To show linear independence of the vectors $v_1, \ldots, v_n$, suppose that $\beta_1 v_1 + \cdots + \beta_n v_n = 0$. Now, the zero vector is also expressed as $0 = 0 \cdot v_1 + \cdots + 0 \cdot v_n$. Due to uniqueness of the n-tuple $(\beta_1, \ldots, \beta_n)$, it follows that $\beta_1 = \cdots = \beta_n = 0$. Therefore, the vectors $v_1, \ldots, v_n$ are linearly independent. ■

Any proper superset of a spanning set is linearly dependent. A stronger statement holds when a spanning set is finite.

Theorem 1.31 *Let S be a finite spanning set of a vector space V. Then any subset of V having more vectors than those in S is linearly dependent.*

Proof If $S = \varnothing$ or $S = \{0\}$, then $V = \{0\}$. In either case, the result is obvious. If $S = \{v\}$ for a nonzero vector $v \in V$, then $V = \{\alpha v : \alpha \in \mathbb{F}\}$. If $B \subseteq V$ has more than one vector, then any two vectors from B are in the form βv and γv for some scalars β and γ. In that case, one is a scalar multiple of the other, and hence, B is linearly dependent. Next, suppose $S = \{v_1, \ldots, v_n\}$ spans V, where $n \geq 2$. Since supersets of linearly dependent sets are linearly dependent, it is enough to show that any set having $n + 1$ elements is linearly dependent.

So, let $B = \{u_1, \ldots, u_{n+1}\} \subseteq V$. We show that B is linearly dependent. Now, if $B_1 := \{u_1, \ldots, u_n\}$ is linearly dependent, then so is B. Therefore, assume that B_1 is linearly independent. Notice that $u_1 \neq 0, \ldots, u_n \neq 0$. It may happen that B_1 and S have some vectors in common. Suppose there are m vectors in common, where $m \geq 0$. Without loss of generality, write

$$S = \{u_1, \ldots, u_m, v_{m+1}, \ldots, v_n\}, \quad B_1 = \{u_1, \ldots, u_m, u_{m+1}, \ldots, u_n\}.$$

Here, if $m = 0$, then $S = \{v_1, \ldots, v_n\}$ and $B_1 = \{u_1, \ldots, u_n\}$. Since B_1 is linearly independent, u_{m+1} is not a linear combination of $u_1, \ldots, u_m$. As S spans V, u_{m+1} is a linear combination of vectors from S. In such a linear combination, all the coefficients of $v_{m+1}, \ldots, v_n$ cannot be zero. Without loss of generality, suppose the coefficient of v_{m+1} is nonzero. Then by the Exchange lemma, we can replace v_{m+1} in S by u_{m+1} resulting in a new spanning set

$$S_1 = \{u_1, \ldots, u_m, u_{m+1}, v_{m+2}, \ldots, v_n\}$$

In the next stage, we look at the vector u_{m+2} and replace one of the vectors $v_{m+2}, \ldots, v_n$ in S_1 with u_{m+2} to obtain a spanning set S_2. Continuing this process we end up with the spanning set

$$S_{n-m} = \{u_1, \ldots, u_m, u_{m+1}, \ldots, u_n\}.$$

Since S_{n-m} is a spanning set, B is linearly dependent. ■

Theorem 1.31 becomes very helpful when you choose a spanning set having minimum number of elements. For example, $\{(1, 0, 0), (0, 1, 0), (0, 0, 1)\}$ spans $\mathbb{R}^3$; therefore, any set of four or more vectors in $\mathbb{R}^3$ is bound to be linearly dependent. Can a set of two vectors span $\mathbb{R}^3$?

Exercises for Sect. 1.4

1. In each of the following, a vector space V and $S \subseteq V$ are given. Determine whether S is linearly dependent, and if it is, express one of the vectors in S as a linear combination of some or all of the remaining vectors.

   (a) $V = \mathbb{R}^3$, $S = \{(1, -3, -2), (-3, 1, 3), (2, 5, 7)\}$.
   (b) $V = \mathbb{R}^4$, $S = \{(1, 1, 0, 2), (1, 1, 3, 2), (4, 2, 1, 2)\}$.
   (c) $V = \mathbb{C}^3$, $S = \{6, 2, 1), (4, 3, -1), (2, 4, 1)\}$.
   (d) $V = \mathcal{P}_3(\mathbb{F})$, $S = \{t^2 - 3t + 5,\ t^3 + 2t^2 - t + 1,\ t^3 + 3t^2 - 1\}$.
   (e) $V = \mathcal{P}_3(\mathbb{F})$, $S = \{t^2 + 3t + 2,\ t^3 - 2t^2 + 3t + 1,\ 2t^3 + t^2 + 3t - 2\}$.
   (f) $V = \mathcal{P}_3(\mathbb{F}), S = \{6t^3 - 3t^2 + t + 2,\ t^3 - t^2 + 2t + 3,\ 2t^3 + t^2 - 3t + 1\}$.
   (g) $V = \mathcal{F}(\mathbb{R}, \mathbb{R})$, $S = \{1 + t + 3t^2,\ 2 + 4t + t^2,\ 2t + 5t^2\}$.
   (h) $V = \mathcal{F}(\mathbb{R}, \mathbb{R})$, $S = \{2, \sin^2 t, \cos^2 t\}$.
   (i) $V = \mathcal{F}(\mathbb{R}, \mathbb{R})$, $S = \{1, \sin t, \sin 2t\}$.
   (j) $V = \mathcal{F}(\mathbb{R}, \mathbb{R})$, $S = \{e^t,\ te^t,\ t^3 e^t\}$.
   (k) $V = C([-\pi, \pi], \mathbb{R})$, $S = \{\sin t, \sin 2t, \ldots, \sin nt\}$ for some $n \in \mathbb{N}$.

2. For $i, j \in \{1, 2\}$, let E_{ij} be a 2×2 matrix whose (i, j)th entry is 1 and all other entries are 0.

   (a) Show that $E_{11},\ E_{12},\ E_{21}, E_{22}$ are linearly independent in $\mathbb{F}^{2\times 2}$.
   (b) Construct four linearly independent matrices in $\mathbb{F}^{2\times 2}$ none of which is equal to any E_{ij}.

3. Show that the vectors $(1, 0, 0)$, $(0, 2, 0)$, $(0, 0, 3)$, and $(1, 2, 3)$ are linearly dependent, but any three of them are linearly independent.
4. Give three vectors in $\mathbb{R}^2$ such that none of the three is a scalar multiple of another.
5. Let $\{u,\ v,\ w,\ x\}$ be linearly independent in a vector space V. Does it imply that $\{u + v,\ v + w,\ w + x,\ x + u\}$ is linearly independent in V?
6. Let U be a subspace of V, and let $B \subseteq U$. Prove that B is linearly independent in U if and only if B is linearly independent in V.
7. Show that two vectors (a, b) and (c, d) in $\mathbb{R}^2$ are linearly independent if and only if $ad - bc \neq 0$.
8. Answer the following questions with justification:

   (a) Is union of two linearly dependent sets linearly dependent?
   (b) Is union of two linearly independent sets linearly independent?
   (c) Is intersection of two linearly dependent sets linearly dependent?
   (d) Is intersection of two linearly independent sets linearly independent?

9. In the vector space $\mathbb{R}^{3\times 3}$, find six linearly independent vectors.
10. In the real vector space $\mathbb{C}^{2\times 2}$, find six linearly independent vectors.
11. Prove Theorem 1.31 by using Theorem 1.29 instead of Exchange lemma.
12. Prove Theorem 1.31 by using induction.
13. Let $v_1, \ldots, v_n$ be linearly independent vectors in a vector space V. Suppose $w \in V$ is such that the vectors $w + v_1, \ldots, w + v_n$ are linearly dependent. Show that $w \in \text{span}\{v_1, \ldots, v_n\}$.

14. Let V be a vector space. Suppose the vectors $v_1, \ldots, v_n$ span V. Show that the vectors $v_1, v_2 - v_1, \ldots, v_n - v_1$ also span V. Further, show that if $v_1, \ldots, v_n$ are linearly independent, then $v_1, v_2 - v_1, \ldots, v_n - v_1$ are linearly independent.

1.5 Basis and Dimension

For each $j \in \{1, \ldots, n\}$, let $e_j = (\delta_{1j}, \ldots, \delta_{nj})$, that is, the vector in $\mathbb{F}^n$ whose jth component is 1 and all other components are 0. We see that $\mathbb{F}^n$ is spanned by the set $\{e_1, \ldots, e_n\}$; that is, any vector $v \in \mathbb{F}^n$ can be expressed as a linear combination of $e_1, \ldots, e_n$. In any such linear combination the coefficients of $e_1, \ldots, e_n$ are uniquely determined due to Theorem 1.24. In fact, the coefficients give rise to the coordinates of the vector v. We generalize this notion to any vector space and give a name to such sets of vectors.

Definition 1.32 Let B be a subset of a vector space V.

(a) B is called a **basis** of V if B is linearly independent and B spans V.
(b) B is called an **ordered basis** of V if B is a countable ordered set and B is a basis of V.

Usually, when we refer a countable set such as $\{v_1, \ldots, v_n\}$ or $\{v_1, v_2, \ldots\}$ as an ordered set, the order will be taken as the elements are written. That is, v_1 is taken as the first element, v_2 is the second element, and so on.

For instance, the ordered set $\{e_1, \ldots, e_n\}$ is an ordered basis of $\mathbb{F}^n$, and the ordered set $\{1, t, t^2, \ldots, t^n\}$ is an ordered basis of $\mathcal{P}_n(\mathbb{F})$.

Let $E_{ij} \in \mathbb{F}^{m \times n}$ have the (i, j)th entry as 1 and all other entries as 0. Then it is easily verified that $\{E_{ij} : 1 \le i \le m,\ 1 \le j \le n\}$ is a basis of $\mathbb{F}^{m \times n}$.

Example 1.33 The ordered bases of the vector spaces in the following cases are called the **standard bases** of the respective spaces. The reader may verify that they are, indeed, ordered bases:

(a) The basis $\{e_1, \ldots e_n\}$ of $\mathbb{F}^n$ with jth component of e_i as δ_{ij} for $i, j = 1, \ldots, n$.
(b) The basis $\{e_1, \ldots e_n\}$ of $\mathbb{F}^{n \times 1}$ with kth entry of e_i as δ_{ik} for $i, k = 1, \ldots, n$.
(c) The basis $\{u_1, \ldots, u_n\}$ of $\mathcal{P}_{n-1}(\mathbb{F})$ with $u_j(t) = t^{j-1}$ for $j \in \{1, 2, \ldots, n\}$.
(d) The basis $\{u_1, u_2, \ldots\}$ of $\mathcal{P}(\mathbb{F})$ with $u_j(t) = t^{j-1}$ for $j \in \mathbb{N}$.
(e) The basis $\{E_{11}, \ldots, E_{1n}, E_{21}, \ldots, E_{2n}, \ldots, E_{m1}, \ldots, E_{mn}\}$ of $\mathbb{F}^{m \times n}$ where E_{ij} has the (i, j)th entry as 1 and all other entries as 0. □

An important corollary of Theorem 1.30 is the following result.

Theorem 1.34 *An ordered subset $\{u_1, \ldots, u_n\}$ of a vector space V is an ordered basis of V if and only if corresponding to each vector $x \in V$, there exists a unique ordered n-tuple $(\alpha_1, \ldots, \alpha_n)$ of scalars such that $x = \alpha_1 u_1 + \cdots + \alpha_n u_n$.*

Therefore, when a vector space has a finite ordered basis, the basis brings in a coordinate system in the vector space, mapping each vector to the n-tuple of scalars.

Theorem 1.35 *Let V be a vector space with a finite spanning set.*

(1) *Each finite spanning set of V contains a basis of V.*
(2) *Each linearly independent subset of V can be extended to a basis of V.*

Proof (1) Let $S := \{u_1, \ldots, u_n\}$ be a spanning set of the vector space V. Without loss of generality, assume that $u_1 \neq 0$. Consider S as an ordered set. If S is linearly independent, then S is a basis of V. Otherwise, by Theorem 1.29, there exists a vector, say, u_k which is in the span of $\{u_1, \ldots, u_{k-1}\}$. We see that $V = \text{span}(S\backslash\{u_k\})$.

Update S to $S\backslash\{u_k\}$. Continue this process on the new S. The process ends at constructing a basis for V.

(2) Let $S = \{v_1, \ldots, v_n\}$ be a spanning set of V. We know, by Theorem 1.31, that each linearly independent subset has at most n elements. Let $\{u_1, \ldots, u_m\}$ be a linearly independent subset of V. Construct the ordered set $S_1 = \{u_1, \ldots, u_m, v_1, \ldots, v_n\}$. Now, S_1 is a spanning set of V. Using the construction in the proof of (1), we obtain a basis of V containing the vectors $u_1, \ldots, u_m$. ■

Notice that in the proof of Theorem 1.35(1), one may start from u_n and end at u_1 while throwing away a vector which is a linear combination of the earlier ones. This process will also end up in a basis.

It now makes sense to talk of a minimal spanning set and a maximal linearly independent set in the following manner.

Definition 1.36 Let B be a subset of a vector space V.

(a) B is called a **minimal spanning set** of V if $\text{span}(B) = V$, and no proper subset of B spans V.
(b) B is called a **maximal linearly independent set** of V if B is linearly independent, and each proper superset of B contained in V is linearly dependent.

As expected, we have the following theorem.

Theorem 1.37 *Let B be a subset of a vector space V. Then the following are equivalent:*

(1) *B is a basis of V.*
(2) *B is a minimal spanning set of V.*
(3) *B is a maximal linearly independent set in V.*

Proof (1) ⇒ (2): Let B be a basis of V. Let $x \in B$. If $B\backslash\{x\}$ spans V, then $x \in \text{span}(B\backslash\{x\})$. This contradicts the linear independence of B. Thus, no proper subset of B can span V. Therefore, B is a minimal spanning set.

(2) ⇒ (3): Let B be a minimal spanning set of V. Due to minimality, whichever x we choose from B, the set $B\backslash\{x\}$ will never span V. Hence, B is linearly independent. Moreover, since B is a spanning set, any proper superset of it is linearly dependent. Therefore, B is a maximal linearly independent set.

(3) ⇒ (1): Let B be a maximal linearly independent set. If B does not span V, then we have a vector $x \in V$ such that $x \notin \text{span}(B)$. Since B is linearly independent, by

Theorem 1.28, $B \cup \{x\}$ is linearly independent. But this is impossible since every proper superset of B is linearly dependent. Therefore, B spans V. ■

Theorem 1.31 says that if a vector space has a finite spanning set, then no linearly independent subset of the space can have more elements than that in the spanning set. This enables us to compare the number of elements in two bases, provided the bases are finite sets.

Theorem 1.38 *If a vector space has a finite basis, then all bases of it have the same number of elements.*

Proof Let V be a vector space that has a finite basis B with n elements. Let C be any other basis of V. Since B is a spanning set and C is linearly independent, by Theorem 1.31, C has at most n elements. Suppose C has m elements with $m \le n$. Since B is linearly independent and C is a spanning set, again by Theorem 1.31, $n \le m$. ■

Definition 1.39 Let V be a vector space.

(a) V is called **finite dimensional** if it has a finite basis.
(b) If a basis of V has n number of elements, then we say that the **dimension of** V is equal to n and write it as $\dim(V) = n$. In this case, we also say that V is an n-dimensional space.
(c) V is called **infinite dimensional** if it does not have a finite basis. In this case, we write $\dim(V) = \infty$.

In case V is finite dimensional and if $\dim(V)$ is not of any particular interest, we write $\dim(V) < \infty$.

As $\mathrm{span}(\varnothing) = \{0\}$ and $\varnothing$ is linearly independent, $\dim(\{0\}) = 0$. Also,

$$\dim(\mathbb{F}^n) = \dim(\mathbb{F}^{n\times 1}) = \dim(\mathcal{P}_{n-1}(\mathbb{F})) = n \quad \text{and} \quad \dim(\mathcal{P}(\mathbb{F})) = \infty.$$

The real vector space $\mathbb{C}$ has a basis $\{1, i\}$; thus it has dimension 2; while, the complex vector space $\mathbb{C}$ has dimension 1. Similarly, $\mathcal{P}_n(\mathbb{C})$ regarded as a complex vector space has dimension $n+1$. However, the real vector space $\mathcal{P}_n(\mathbb{C})$ has dimension $2(n+1)$.

As a corollary of Theorem 1.31, we obtain the following.

Theorem 1.40 *Let V be a vector space.*

(1) *V is finite dimensional if and only if V has a finite spanning set.*
(2) *If $\dim(V) = n$, then any subset of V having more than n vectors is linearly dependent.*
(3) *V is infinite dimensional if and only if V has an infinite linearly independent subset.*

A linearly independent subset of a subspace remains linearly independent in the parent vector space. Thus, the dimension of a subspace cannot exceed the dimension of the parent space. Therefore, if a subspace is infinite dimensional, then the parent space must also be infinite dimensional. Moreover, dimension helps in proving that certain linearly independent sets or spanning sets are bases. Recall that $|B|$ denotes the number of elements of a finite set B.

Theorem 1.41 *Let B be a finite subset of a finite dimensional vector space V.*

(1) *B is a basis of V if and only if B spans V and $|B| = \dim(V)$.*
(2) *B is a basis of V if and only if B is linearly independent and $|B| = \dim(V)$.*

Proof (1) Suppose $\text{span}(B) = V$ and $|B| = \dim(V)$. By Theorem 1.35, there exists $C \subseteq B$ such that C is a basis of V. But $|C| = \dim(V) = |B|$. Hence $C = B$. Therefore, B is a basis. The converse is trivial.

(2) Suppose B is linearly independent and $|B| = \dim(V)$. By Theorem 1.35, we have a basis C of V with $C \supseteq B$. But $|C| = \dim(V) = |B|$. Hence $C = B$. Therefore, B is a basis. Again, the converse is trivial. ■

Example 1.42 (1) For each $j \in \mathbb{N}$, define the sequence e_j by taking $e_j(i) = \delta_{ij}$, for $i \in \mathbb{N}$. That is, the jth term of the sequence e_j is 1 and every other term of e_j is 0. Recall that $c_{00}(\mathbb{N}, \mathbb{F})$ is the vector space of all sequences having only finite number of nonzero entries. Hence, the set $E := \{e_1, e_2, \ldots\}$ is a linearly independent subset of $c_{00}(\mathbb{N}, \mathbb{F})$. Therefore, $c_{00}(\mathbb{N}, \mathbb{F})$ is infinite dimensional. Each element in $c_{00}(\mathbb{N}, \mathbb{F})$ is a (finite) linear combination of members of E. Thus, E is a basis of $c_{00}(\mathbb{N}, \mathbb{F})$.

The vector space $c_{00}(\mathbb{N}, \mathbb{F})$ is a subspace of $\mathbb{F}^\infty$, the space of all scalar sequences. Therefore, $\mathbb{F}^\infty$ is infinite dimensional.

(2) Let $u_j(t) = t^{j-1}$, $j \in \mathbb{N}$ and $t \in [a, b]$. Then the vector space $\mathcal{P}([a, b], \mathbb{R})$ which is the span of $B := \{u_1, u_2, \ldots\}$ is a subspace of $C^k([a, b], \mathbb{R})$ for any $k \in \mathbb{N}$. Since B is linearly independent it is a basis of $\mathcal{P}([a, b], \mathbb{R})$, and hence $C^k([a, b], \mathbb{R})$ is infinite dimensional.

(3) Suppose S is a finite set consisting of n elements. Then $\mathcal{F}(S, \mathbb{F})$, the space of all functions from S to $\mathbb{F}$, is of dimension n. To see this, let $S = \{s_1, \ldots, s_n\}$. For each $j \in \{1, \ldots, n\}$, define $f_j \in \mathcal{F}(S, \mathbb{F})$ by

$$f_j(s_i) = \delta_{ij} \quad \text{for } i \in \{1, \ldots, n\}.$$

If $\alpha_1, \ldots, \alpha_n$ are scalars such that $\sum_{j=1}^n \alpha_j f_j = 0$, then for each $i \in \{1, \ldots, n\}$, $\sum_{j=1}^n \alpha_j f_j(s_i) = 0$. But,

$$\sum_{j=1}^n \alpha_j f_j(s_i) = \sum_{j=1}^n \alpha_j \delta_{ij} = \alpha_i.$$

Thus, $\alpha_i = 0$ for every $i \in \{1, \ldots, n\}$. Therefore, $\{f_1, \ldots, f_n\}$ is linearly independent. Also, for each $s_i \in S$ and for any $f \in \mathcal{F}(S, \mathbb{F})$,

$$f(s_i) = \sum_{j=1}^n f(s_j)\delta_{ij} = \sum_{j=1}^n f(s_j) f_j(s_i).$$

That is, $f = \sum_{j=1}^{n} f(s_j) f_j$. Thus $\text{span}\{f_1, \ldots, f_n\} = \mathcal{F}(S, \mathbb{F})$. Therefore, the ordered set $\{f_1, \ldots, f_n\}$ is an ordered basis of $\mathcal{F}(S, \mathbb{F})$. □

Exercises for Sect. 1.5

1. Find bases for the following vector spaces:

   (a) $\{(a, b, c) \in \mathbb{R}^3 : a + b + c = 0\}$.
   (b) $\{(a, b, c, d) \in \mathbb{R}^4 : c = b,\ d = -a\}$.
   (c) $\{(a_1, \ldots, a_6) \in \mathbb{R}^6 : a_2 = 2a_1, a_4 = 4a_3, a_6 = 6a_5\}$.
   (d) $\text{span}\{(1, -1, 0, 2, 1), (2, 1, -2, 0, 0), (0, -3, 2, 4, 2), (3, 3, -4, -2, -1), (2, 4, 1, 0, 1), (5, 7, -3, -2, 0)\}$.

2. Are the following bases for $\mathcal{P}_2(\mathbb{F})$?

   (a) $\{1 + 2t + t^2, 3 + t^2, t + t^2\}$.
   (b) $\{1 + 2t + 3t^2, 4 - 5t + 6t^2, 3t + t^2\}$.
   (c) $\{-1 - t - 2t^2, 2 + t - 2t^2, 1 - 2t + 4t^2\}$.

3. Find two bases for $\mathbb{R}^4$ whose intersection is $\{(1, 0, 1, 0), (0, 1, 0, 1)\}$.
4. Construct a basis for $\text{span}\{1 + t^2, -1 + t + t^2, -6 + 3t, 1 + t^2 + t^3, t^3\}$.
5. Extend the set $\{1 + t^2, 1 - t^2\}$ to a basis of $\mathcal{P}_3(\mathbb{F})$.
6. Does there exist a basis for $\mathcal{P}_4(\mathbb{F})$, where no vector is of degree 3?
7. Under what conditions on α, $\{(1, \alpha, 0), (\alpha, 0, 1), (1 + \alpha, \alpha, 1)\}$ is a basis of $\mathbb{R}^3$?
8. Is $\{1 + t^n, t + t^n, \ldots, t^{n-1} + t^n, t^n\}$ a basis for $\mathcal{P}_n(\mathbb{F})$?
9. Is $\text{span}\{e_1 + e_2, e_2 + e_3, e_3 + e_1\}$ a proper subspace of $\mathbb{R}^3$? Why?
10. Let $\{x, y, z\}$ be a basis of a vector space V. Is $\{x + y, y + z, z + x\}$ also a basis of V?
11. Let $u, v, w, x, y_1, y_2, y_3, y_4, y_5 \in \mathbb{C}^9$ satisfy the relations: $y_1 = u + v + w$, $y_2 = 2v + w + x$, $y_3 = u + 3w + x$, $y_4 = -u + v + 4x$, and $y_5 = u + 2v + 3w + 4x$. Are the vectors $y_1, \ldots, y_5$ linearly dependent or independent?
12. Prove that the only nonzero proper subspaces of $\mathbb{R}^3$ are the straight lines and the planes passing through the origin.
13. Give a proof of Theorem 1.35(1) by constructing a basis by visiting the vectors in S in the order $u_n, u_{n-1}, \ldots, u_2, u_1$.
14. Consider each polynomial in $\mathcal{P}(\mathbb{R})$ as a function from the set $\{0, 1, 2\}$ to $\mathbb{R}$. Is the set of vectors $\{t, t^2, t^3, t^4, t^5\}$ linearly independent in $\mathcal{P}(\mathbb{R})$?
15. Consider the set S of all vectors in $\mathbb{R}^4$ whose components are either 0 or 1. How many subsets of S are bases for $\mathbb{R}^4$?
16. Let V be the vector space of all thrice differentiable functions from $\mathbb{R}$ to $\mathbb{R}$. Find a basis and dimension of the following subspaces of V:

    (a) $\{x \in V : x'' + x = 0\}$.
    (b) $\{x \in V : x'' - 4x' + 3x = 0\}$.
    (c) $\{x \in V : x''' - 6x'' + 11x' - 6x = 0\}$.

17. Given real numbers $a_1, \ldots, a_k$, let V be the set of all solutions $x \in C^k[a, b]$ of the differential equation

$$\frac{d^k x}{dt^k} + a_1 \frac{d^{k-1} x}{dt^{k-1}} + \cdots + a_k x = 0.$$

Show that V is a vector space over $\mathbb{R}$. What is $\dim(V)$?
18. Let V be a vector space. Let $B := \{v_i : i \in \mathbb{N}\} \subseteq V$ be such that for each $m \in \mathbb{N}$, the vectors $v_1, \ldots, v_m$ are linearly independent. Show that B is linearly independent. Conclude that $\dim(V) = \infty$.
19. Show that $\dim(\mathcal{F}(\mathbb{N}, \mathbb{F})) = \infty$ and $\dim(C[0, 1]) = \infty$.
20. For $j \in \mathbb{N}$, let e_j be the sequence whose jth term is 1 and all other terms are 0. Why is $\{e_1, e_2, e_3, \ldots\}$ not a basis of $\mathbb{R}^\infty$?

1.6 Basis of Any Vector Space

What happens if a vector space does not have a finite spanning set? Of course, such a vector space spans itself; so, it has an infinite spanning set. Does it have a basis or not, whether finite or infinite? The issue can be settled by using Zorn's lemma, which is equivalent to Axiom of Choice. In order to state this, we introduce some concepts.

A *relation* on a nonempty set X is a subset of $X \times X$. If R is a relation on X and if $(x, y) \in R$, then we say that "x is related to y" under R or xRy. A relation on X is called a *partial order* if the following conditions are met:

(a) Every element of X is related to itself.
(b) For any pair of distinct elements x, y from X, if x is related to y and y related to x, then $x = y$.
(c) For any triple of elements, say x, y, z from X, not necessarily distinct, if x is related to y and y related to z, then x is related to z.

For example, the set $R := \{(x, y) \in \mathbb{R}^2 : x \le y\}$ is a partial order on $\mathbb{R}$. Thus, R, which we also write as $\le$, is a partial order on $\mathbb{R}$. If S is any nonempty set, then the relation of subset, $\subseteq$, is a partial order on the power set of S. A set X with a partial order on it is called a *partially ordered set.*

Suppose a partial order $\preceq$ has been already given on X. A subset Y of X is said to be a *chain* in X if

for every pair of elements $x, y \in Y$, at least one of them is related to the other: $x \preceq y$ or $y \preceq x$.

Any nonempty subset of $\mathbb{R}$ is a chain with the partial order $\le$, since any two real numbers are comparable. However, any subset of the power set of a given nonempty set S need not be a chain with respect to the subset relation. For example, with

$S = \{0, 1\}$, take $Z = \{\{0\}, \{1\}\}$, a subset of the power set of S. Neither $\{0\} \subseteq \{1\}$, nor $\{1\} \subseteq \{0\}$.

If a partially ordered set itself is a chain, then we say that the partial order is a *total order* on the set.

Given a subset Y of a partially ordered set X with partial order $\preceq$, an element $x \in X$ is called an *upper bound of* Y if

every element of Y is related to x, that is, for every $y \in Y$, $y \preceq x$.

Such an upper bound x may or may not be in Y. For example, the interval $(0, 1]$ has an upper bound 1; also 10 is an upper bound. The interval $[0, 1)$ has an upper bound 1; also 5 is an upper bound. The collection $\{\{0\}, \{1\}\}$ of subsets of $\{0, 1, 2\}$ has upper bounds $\{0, 1\}$ and $\{0, 1, 2\}$ in the power set of $\{0, 1, 2\}$ with the partial order as $\subseteq$.

In a partially ordered set X with partial order $\preceq$, an element $x \in X$ is called a *maximal element* of X if

x is related to none other than itself, that is, for every $z \in X$, if $x \preceq z$ then $x = z$.

For example, $\mathbb{R}$ contains no maximal element since every real number is related to some other real number; for instance, if $r \in \mathbb{R}$, then $r \leq r + 1$. The power set of any nonempty set S contains a maximal element, the set S itself.

For a general partially ordered set, we have **Zorn's Lemma**:

If every chain in a partially ordered set has an upper bound, then the set contains a maximal element.

We will apply Zorn's lemma in proving the existence of a basis for any vector space.

Theorem 1.43 *Let E be a linearly independent subset of a vector space V. Then V has a basis B containing E. In particular, each vector space has a basis.*

Proof If V has a finite spanning set, then by Theorem 1.35(2), it has a basis containing E. So, suppose that V does not have a finite spanning set. Denote by $\mathcal{K}$, the collection of all linearly independent subsets of V containing E. The relation $\subseteq$ is a partial order on $\mathcal{K}$.

Consider any chain $\mathcal{C}$ in $\mathcal{K}$. Let A be the union of all subsets in the chain $\mathcal{C}$. We see that $E \subseteq X \subseteq A$ for each subset X in the chain $\mathcal{C}$. Hence A is an upper bound of the chain $\mathcal{C}$.

Thus, every chain in $\mathcal{K}$ has an upper bound. By Zorn's lemma, $\mathcal{K}$ contains a maximal element, say, B. As $B \in \mathcal{K}$, it is a linearly independent subset of V. That is, B is a maximal linearly independent subset of V containing E. Therefore, by Theorem 1.37, B is a basis of V.

In particular, if $V = \{0\}$, then $\varnothing$ is its basis. Otherwise, let u be a nonzero vector in V. The set $\{u\}$ is linearly independent. By what we have just proved, V has a basis containing u. ■

In what follows, we write $|S|$ for the number of elements, more technically called, the *cardinality* of the set S.

We may recall that two sets are of the **same cardinality** if there is a bijection between them. If a set S is a finite set with n elements, then we say that the cardinality of S is n, and if S is in one-to-one correspondence with $\mathbb{N}$, the set of all positive integers, then we say that the cardinality of S is $\aleph_0$, read as *aleph null*, or *aleph naught*. A set with cardinality $\aleph_0$ is called countably infinite, or *denumerable*, and an infinite set which is not in one-to-one correspondence with $\mathbb{N}$ is called an *uncountable set*.

It is clear that for any nonempty set S, the function that takes each $x \in S$ to the singleton set $\{x\}$ is an injective function from S to 2^S, the power set of S. However, it can be shown that there is no surjective function from S to 2^S. Thus, cardinality of 2^S can be considered as strictly bigger than that of S. In particular, there are uncountable sets of different cardinalities.

More generally, if there exists an injective function from a set A to a set B, then we say that the cardinality of A is less than or equal to the cardinality of B and write $|A| \le |B|$. If there is an injective function from A to B, but there is no surjective function from A to B, then we say that the cardinality of A is strictly less than that of B, or the cardinality of B is strictly greater than that of A, and write $|A| < |B|$.

If A and B are two disjoint sets, then $|A| + |B|$ is defined as $|A \cup B|$. In case, A and B are not disjoint, we define $|A| + |B|$ as $|(A \times \{0\}) \cup (B \times \{1\})|$. Further, $|A| \cdot |B|$ is taken as $|A \times B|$. Addition and multiplication of cardinalities of infinite sets are done almost the same way as natural numbers, but with some exceptions. For infinite sets A and B with $|A| \le |B|$, we have $|A| + |B| = |B|$ and $|A| \cdot |B| = |B|$. In particular, $\aleph_0 \cdot |A| = |\mathbb{N}| \cdot |A| = |A|$ for any infinite set A.

If a vector space has an ordered basis having n elements, then each vector is associated with an n-tuple of scalars in a unique way. On the other hand, if V is a vector space having an infinite basis B, then each nonzero vector in V is a unique linear combination of vectors from B, where all coefficients are nonzero. To see this, let $v \in V$ be a nonzero vector having two distinct linear combinations such as

$$v = \alpha_1 u_1 + \cdots + \alpha_m u_m, \quad v = \beta_1 v_1 + \cdots + \beta_n v_n$$

for vectors $u_i, v_j \in B$ and nonzero scalars $\alpha_j,\ \beta_j$. Write

$$B_1 = \{u_1, \ldots, u_m\}, \quad B_2 = \{v_1, \ldots, v_n\}, \quad U = \text{span}(B_1 \cup B_2).$$

Then $B_1 \cup B_2$ is a basis of U. The two linear combinations of v are two distinct linear combinations of the same v that use vectors from the same finite basis $B_1 \cup B_2$. This contradicts Theorem 1.34. Therefore, each nonzero vector $v \in V$ is associated with a unique finite subset $F(v)$ of B such that v is a unique linear combination of vectors from $F(v)$ with nonzero scalars as coefficients.

We use this observation in proving the following theorem.

Theorem 1.44 *Any two bases for the same vector space have equal cardinality.*

Proof If V has a finite basis, then any two bases have the same cardinality due to Theorem 1.38. So, suppose all bases of V are infinite.

Let B and C be infinite bases for V. Each vector in B is nonzero. Since C is a basis, each vector from B can be expressed as a linear combination of finite number of vectors from C. Thus, each $x \in B$ is associated with a unique finite subset of C, write it as $F(x)$, such that x is a linear combination of vectors from $F(x)$ with nonzero coefficients. Write $D := \cup\{F(x) : x \in B\}$.

For each x, $F(x) \subseteq C$. So, $D \subseteq C$. To show that $C \subseteq D$, let $v \in C$. Since B spans V and each vector in B is a linear combination of vectors from D, the set D spans V. In particular, v is a linear combination of some vectors $v_1, \ldots, v_n \in D$. As $D \subseteq C$, we see that v is a linear combination of vectors $v_1, \ldots, v_n$ from C. But C is linearly independent. By Theorem 1.30, the vector v is one of $v_1, \ldots, v_n$. Hence $v \in D$. Therefore, $C \subseteq D$. We then conclude that $C = D$.

Notice that each $F(x)$ is a finite set and B is an infinite set. Hence

$$|C| = |D| = |\cup\{F(x) : x \in B\}| \le |\mathbb{N}| \cdot |B| = |B|.$$

Reversing the roles of B and C results in $|B| \le |C|$. Therefore, $|B| = |C|$. ■

In general, the dimension of a vector space V, denoted by $\dim(V)$, is the number of elements, or the cardinality of any basis of V. Due to Theorem 1.44, dimension of an infinite dimensional vector space is well-defined.

Like an upper bound, an element x of a partially ordered set X is called a *lower bound* of a subset Y if $x \preceq y$ for every $y \in Y$, and an element $x \in X$ is called a *minimal element* of X if for every $z \in X$, $z \preceq x$ implies $z = x$.

A *well-order* on a set is a total order with the property that each nonempty subset of the set has a minimal element. A set is called *well-ordered* if there exists a well-order on it. Like Zorn's lemma, the following result, called the **Well-ordering theorem**, is equivalent to the Axiom of Choice:

Every nonempty set can be well-ordered.

Due to the Well-ordering theorem, the basis B that we obtained in Theorem 1.43 can be well-ordered. Thus every vector space has an ordered basis. For instance, if the basis B of a vector space is a finite set, say with n elements, then we may list the elements of B as $u_1, \ldots u_n$. Similarly, if B is a countably infinite set, then its elements can be listed as $u_1, u_2, \ldots$. However, if a basis has an uncountable number of vectors in it, then it is not clear how to list the vectors.

It is noteworthy that every infinite dimensional vector space which has a basis is equivalent to the Axiom of Choice; see [3]. However, our main interest is in vector spaces that have finite bases. Occasionally we may give examples of vector spaces with infinite bases.

Exercises for Sect. 1.6

1. In each of the following, show that the subset $S = \{u_k : k \in \mathbb{N}\}$ of the given vector space V is linearly independent:

(a) $V = c_{00}(\mathbb{N}, \mathbb{R})$; $u_k = (1, \ldots, 1, 0, 0, \ldots)$, where 1 is repeated k times.
(b) $V = \mathbb{R}^\infty$; $u_k = (1, 1/2^k, 1/3^k, \ldots)$.
(c) $V = C[0, 1]$; $u_k(t) = e^{kt}$ for $0 \le t \le 1$.
(d) $V = C[0, \pi]$; $u_k(t) = \sin(kt)$ for $0 \le t \le 1$.

2. In each of the following, show that the vector space V does not have a countable basis:

(a) $V = \mathbb{R}^\infty$, the real vector space of all real sequences.
(b) $V = c_0(\mathbb{N}, \mathbb{R})$, the real vector space of all real sequences that converge to 0.
(c) $V = \text{span}\{e^{\lambda t} : 0 < \lambda < 1\} \subseteq C[0, 1]$.
(d) V is the subspace of $\mathbb{R}^\infty$ spanned by the set of all sequences (x_n), where each $x_n \in \{0, 1\}$.
(e) V is the subspace of $c_0(\mathbb{N}, \mathbb{R})$ spanned by the set of all sequences $\big(x_n/n\big)$, where (x_n) is a bounded sequence.

3. In the following, check whether S is a basis for the vector space V:

(a) $S = \{t^{k-1} : k \in \mathbb{N}\}$; $V = \mathcal{P}(\mathbb{R})$.
(b) $S = \{e_1, e_2, \ldots\}$, where $e_n = (0, 0, \ldots, 1, 0, 0, \ldots)$ is the sequence whose nth term is 1, and all other terms 0; $V = c_{00}(\mathbb{N}, \mathbb{R})$.
(c) $S = \{e_1, e_2, \ldots\}$; $V = \ell^1(\mathbb{N}, \mathbb{R})$, the vector space of all sequences (x_n) of real numbers such that $\sum_{n=1}^{\infty} |x_n|$ is convergent.
(d) $S = \{\sin(kt) : k \in \mathbb{N}\}$; $V = C[0, \pi]$.

1.7 Sums of Subspaces

Theorem 1.35 reveals more structure of a vector space. For example in $\mathbb{R}^3$, consider the set $C = \{(1, 0, 0),\ (0, 1, 0)\}$. We can extend this linearly independent set to a basis by including another vector v from $\mathbb{R}^3$ which is not in the span of C. For instance, each of the following is a basis for $\mathbb{R}^3$:

$$B_1 := \{(1, 0, 0),\ (0, 1, 0),\ (0, 0, 1)\},$$
$$B_2 := \{(1, 0, 0),\ (0, 1, 0),\ (0, 1, 2)\},$$
$$B_3 := \{(1, 0, 0),\ (0, 1, 0),\ (1, 2, 3)\}.$$

In all these cases, we see that $\text{span}(C) \cap \text{span}\{v\} = \{0\}$. In general, we expect that if E and F are linearly independent sets in a vector space V such that $E \cap F = \varnothing$ and $E \cup F$ spans V, then $V = \text{span}(E) + \text{span}(F)$ and $\text{span}(E) \cap \text{span}(F) = \{0\}$. In view of this, we introduce a definition and prove some related results.

Definition 1.45 Let U and W be subspaces of a vector space V. We say that V is a **direct sum** of U and W, written as $V = U \oplus W$, if $V = U + W$ and $U \cap W = \{0\}$. In such a case, we say that the U and W are **complementary** subspaces.

As an example, we observe that the plane $\mathbb{R}^2$ is a direct sum of the x-axis and the y-axis. It is also the direct sum of the x-axis and the line $\{(a, a) : a \in \mathbb{R}\}$. Once the subspace, namely the x-axis, is given we can always find another subspace W of $\mathbb{R}^2$ such that $\mathbb{R}^2$ is the direct sum of x-axis and W. Of course, there are infinitely many choices for W. Obviously, linear independence has some roles to play with the direct sum.

Theorem 1.46 *Let V be a vector space. Let C and D be disjoint linearly independent subsets of V such that $C \cup D$ is also linearly independent. Then*

$$\text{span}(C \cup D) = \text{span}(C) \oplus \text{span}(D).$$

In particular, each subspace of V has a complementary subspace; that is, for every subspace U of V, there exists a subspace W of V such that $V = U \oplus W$.

Proof If one of the sets C and D is empty, then the result holds trivially. Suppose that $C \neq \varnothing$ and $D \neq \varnothing$. Now, $\text{span}(C \cup D) = \text{span}(C) + \text{span}(D)$. Thus, it is enough to show that $\text{span}(C) \cap \text{span}(D) = \{0\}$. Towards this, let $x \in \text{span}(C) \cap \text{span}(D)$. Then

$$x = \alpha_1 u_1 + \cdots + \alpha_m u_m = \beta_1 v_1 + \cdots + \beta_n v_n$$

for some $u_1, \ldots, u_m \in C,\ v_1, \ldots, v_n \in D$ and $\alpha_1, \ldots, \alpha_m,\ \beta_1, \ldots, \beta_n \in \mathbb{F}$. Then

$$\alpha_1 u_1 + \cdots + \alpha_n u_n - \beta_1 v_1 - \cdots - \beta_m v_m = 0.$$

Since $C \cup D$ is linearly independent, $\alpha_1 = \cdots = \alpha_n = \beta_1 = \cdots = \beta_m = 0$. So, $x = 0$.

For the last statement, let U be a subspace of V, and let C be a basis of U. By Theorem 1.43, there exists $D \subseteq V$ such that $C \cup D$ is a basis for V. Then the subspace $W = \text{span}(D)$ is complementary to U. ∎

We generalize the notion of direct sum to more than two subspaces.

Definition 1.47 Let $U_1, \ldots, U_n$ be subspaces of a vector space V, where $n \geq 2$. We say that V is a **direct sum** of $U_1, \ldots, U_n$ if $V = U_1 + \cdots + U_n$ and for each $i \in \{1, \ldots, n\}$,

$$U_i \cap (U_1 + \cdots + U_{i-1} + U_{i+1} + \cdots + U_n) = \{0\}.$$

In such a case, we write $V = U_1 \oplus \cdots \oplus U_n$.

The direct sum allows writing a sum in a unique way. For example, consider the subspaces $U_1 = \{(\alpha, 0) : \alpha \in \mathbb{R}\}$ and $U_2 = \{(\alpha, \alpha) : \alpha \in \mathbb{R}\}$ of $\mathbb{R}^2$. We see that $\mathbb{R}^2$ is a direct sum of U_1 and U_2. Now, if $(a, b) \in \mathbb{R}^2$, then

$$(a, b) = (a - b, 0) + (b, b),$$

where $(a-b,0)\in U_1,\ (b,b)\in U_2$. Moreover, if $(a,b)=(\beta,0)+(\gamma,\gamma)$, then necessarily $\gamma=b$ and $\beta=a-b$.

Theorem 1.48 *Let $U_1,\ldots,U_n$ be subspaces of a vector space V. Then, $V=U_1\oplus\cdots\oplus U_n$ if and only if for each vector $v\in V$, there exist unique vectors $u_1\in U_1,\ldots,u_n\in U_n$ such that $v=u_1+\cdots+u_n$.*

Proof For any $i\in\{1,\ldots,n\}$, write $W_i=(U_1+\cdots+U_{i-1}+U_{i+1}+\cdots+U_n)$.

Assume that $V=U_1\oplus\cdots\oplus U_n$. Then $U_i\cap W_i=\{0\}$. Let $v\in V$. There exist vectors $u_1\in U_1,\ldots,u_n\in U_n$ such that $v=u_1+\cdots+u_n$.

For uniqueness, suppose that there also exist vectors $v_1\in U_1,\ldots,v_n\in U_n$ such that $v=v_1+\cdots+v_n$. Then, $(u_1-v_1)+\cdots+(u_n-v_n)=0$. That is, for each $i\in\{1,\ldots,n\}$,

$$-(u_i-v_i)=(u_1-v_1)+\cdots+(u_{i-1}-v_{i-1})+(u_{i+1}-v_{i+1})+\cdots+(u_n-v_n).$$

The vector on the left-hand side is in U_i and the vector on the right-hand side is in W_i. As $U_i\cap W_i=\{0\}$, it follows that each is equal to 0. In particular, $u_i=v_i$.

Conversely, suppose that each vector in v can be written as a sum of vectors from $U_1,\ldots,U_n$ in a unique manner. Then $V=U_1+\cdots+U_n$. To show the other condition in the direct sum, let $i\in\{1,\ldots,n\}$. Suppose $v\in U_i\cap W_i$. Then $v\in U_i$ and for some vectors $u_j\in U_j$ for $j=1,\ldots,i-1,i+1,\ldots,n$,

$$v=u_1+\cdots+u_{i-1}+u_{i+1}+\cdots+u_n.$$

It implies that

$$u_1+\cdots+u_{i-1}-v+u_{j+1}+\cdots+u_n=0.$$

But the zero vector is also written as $0=0+\cdots+0$, where the jth 0 in the sum is in U_j. Due to uniqueness in writing, each of the vectors in the above sum is equal to 0. In particular, $v=0$. Therefore, $U_i\cap W_i=\{0\}$. ■

The requirement $U_i\cap(U_1+\cdots+U_{i-1}+U_{i+1}+\cdots+U_n)=\{0\}$ for each i is stronger than $U_i\cap U_j=\{0\}$ for each pair of indices $i,\ j,\ i\neq j$; even in the presence of the sum condition $V=U_1+\cdots+U_n$. For example, take

$$U_1=\{(a,0):a\in\mathbb{R}\},\quad U_2=\{(0,a):a\in\mathbb{R}\},\quad U_3=\{(a,a):a\in\mathbb{R}\}.$$

We see that $U_2+U_3=U_1+U_2+U_3=\mathbb{R}^2$ and $U_1\cap U_2=U_2\cap U_3=U_3\cap U_1=\{0\}$. But $U_1\cap(U_2+U_3)=U_1\neq\{0\}$. Note that

$$(1,1)=0(1,0)+0(0,1)+1(1,1)=1(1,0)+1(0,1)+0(1,1).$$

That is, the same vector $(1,1)\in\mathbb{R}^2$ can be written in two different ways as sum of vectors from $U_1,\ U_2$ and U_3.

In fact, the definition of direct sum of more than two subspaces is motivated by the uniqueness in writing a vector as a sum of vectors from individual subspaces. Many authors define the direct sum using the condition proved in Theorem 1.48. For another equivalent condition on the direct sum, see Problem 22.

As it looks, the intersection of two finite dimensional subspaces has possibly smaller dimension than any of the subspaces involved. Similarly, sum of two subspaces will have possibly larger dimension than any of the two subspaces. An exact formula can be given relating these dimensions.

Theorem 1.49 *Let U and W be subspaces of a vector space V. If $U \cap W = \{0\}$, then*

$$\dim(U + W) = \dim(U) + \dim(W).$$

Proof If one of U and W is the zero space, say $U = \{0\}$, then we have $U + W = W$ and $\dim(U) = 0$ so that the result follows. Also, if one of them is infinite dimensional, say $\dim(U) = \infty$, then U is a subspace of $U + W$, and the result follows. Hence, assume that both U and W are nonzero finite dimensional subspaces. Let B and C be bases of U and W, respectively. To prove the dimension formula, we show that $B \cup C$ is a basis of $U + W$.

By Theorem 1.17(4), $\text{span}(B \cup C) = \text{span}(B) + \text{span}(C) = U + W$. That is, $B \cup C$ spans $U + W$.

Next, to show that $B \cup C$ is linearly independent, let $u_1, \ldots, u_n \in B, v_1, \ldots, v_m \in C$ and let $\alpha_1, \ldots, \alpha_n, \beta_1, \cdots, \beta_m \in \mathbb{F}$ be such that

$$\alpha_1 u_1 + \cdots + \alpha_n u_n + \beta_1 v_1 + \cdots + \beta_m v_m = 0.$$

Then

$$\alpha_1 u_1 + \ldots + \alpha_n u_n = -(\beta_1 v_1 + \cdots + \beta_m v_m).$$

Note that the left-hand side of the above equation is a vector in U while the vector on the right-hand side belongs to W. Since $U \cap W = \{0\}$, we have

$$\alpha_1 u_1 + \cdots + \alpha_n u_n = 0 = \beta_1 v_1 + \cdots + \beta_m v_m.$$

Now, linearly independence of $u_1, \ldots, u_n$ and $v_1, \ldots, v_m$ implies that $\alpha_1, \ldots, \alpha_n, \beta_1, \ldots, \beta_m$ are all zero. Therefore, $B \cup C$ is linearly independent. ■

The converse of Theorem 1.49 is not necessarily true. For example, consider the subspaces of $\mathbb{R}^3$:

$$U = \{(a, b, 0) : a, b \in \mathbb{R}\}, \quad W = \{(a, 0, 0) : a \in \mathbb{R}\}.$$

Here, $\dim(R^3) = 3$, $\dim(U) + \dim(W) = 2 + 1 = 3$ but $U \cap W = W \neq \{0\}$.

However, with $U' = \{(0, a, b) : a, b \in \mathbb{R}\}$, we see that $U' + W = \mathbb{R}^3$. Consequently, $\dim(U') + \dim(W) = \dim(U' + W)$ and also $U' \cap W = \{0\}$.

In fact, a sum of subspaces is a direct sum if and only if the dimensions add up to that of the parent vector space; it is proved below.

Theorem 1.50 *Let $U_1, \ldots, U_n$ be subspaces of a finite dimensional vector space V. Then the following are equivalent:*

(1) $V = U_1 \oplus \cdots \oplus U_n$.
(2) $V = U_1 + \cdots + U_n$ *and* $\dim(V) = \dim(U_1) + \cdots + \dim(U_n)$.

Proof (1) $\Rightarrow$ (2): It follows from Theorem 1.49 by induction on n.
(2) $\Rightarrow$ (1): Suppose $V = U_1 + \cdots + U_n$ and $\dim(V) = \dim(U_1) + \cdots + \dim(U_n)$. For each $i \in \{1, \ldots, n\}$, let B_i be a basis of U_i. Write $B := B_1 \cup \cdots \cup B_n$. Since $V = U_1 + \cdots + U_n$ and $\dim(V) = \dim(U_1) + \cdots + \dim(U_n)$, B spans V. Hence, by Theorem 1.41, B is a basis of V.

For the intersection condition, let $v \in U_1 \cap (U_2 + \cdots + U_n)$. Then $v \in U_1$ and there exist vectors $u_2 \in U_2, \ldots, u_n \in U_n$ such that $v = u_2 + \cdots + u_n$. Consequently,

$$-v + v_2 + \cdots + v_n = 0.$$

Here, the vector v is a linear combination of vectors from B_1, and the vector v_j is a linear combination of vectors from B_j for $j \geq 2$. As B is a basis of V, the scalars in all these linear combinations are 0. So, $v = 0$. Therefore, $U_1 \cap (U_2 + \cdots + U_n) = \{0\}$.

Similar proof holds for any index i instead of 1. ■

When the subspaces U and W have a nonzero intersection, a dimension formula analogous to Theorem 1.49 holds.

Theorem 1.51 *Let U and W be subspaces of a vector space V. Then*

$$\dim(U + W) + \dim(U \cap W) = \dim(U) + \dim(W).$$

Proof By Theorem 1.49, the dimension formula holds when $U \cap W = \{0\}$. Also if $U \cap W = U$ then $U + W = W$, and if $U \cap W = W$, then $U + W = U$. In these cases, the formula holds trivially. So, assume that $U \cap W$ is a nonzero proper subspace of both U and W. Let B_0 be a basis of $U \cap W$. By Theorem 1.43, B_0 can be extended to a basis $B_0 \cup B_1$ of U, and to a basis $B_0 \cup B_2$ of W. To prove the dimension formula, we show that $B_0 \cup B_1 \cup B_2$ is a basis of $U + W$.

Since $\operatorname{span}(B_0 \cup B_1 \cup B_2) = U + W$, it is enough to show that $B_0 \cup B_1 \cup B_2$ is linearly independent. Since $U \cap W$ is a nonzero proper subspace of both U and W, $B_0 \neq \varnothing$, $B_1 \neq \varnothing$, and $B_2 \neq \varnothing$.

So, let $u_1, \ldots, u_m \in B_0$, $v_1, \ldots, v_n \in B_1$, and let $w_1, \ldots, w_k \in B_2$. Let $\alpha_1, \ldots, \alpha_m$, $\beta_1, \ldots, \beta_n$, and $\gamma_1, \ldots, \gamma_k$ be scalars such that

$$\alpha_1 u_1 + \cdots + \alpha_m u_m + \beta_1 v_1 + \cdots + \beta_n v_n + \gamma_1 w_1 + \cdots + \gamma_k w_k = 0.$$

Then

$$\alpha_1 u_1 + \cdots + \alpha_m u_m + \beta_1 v_1 + \cdots + \beta_n v_n = -(\gamma_1 w_1 + \cdots + \gamma_k w_k) \in U \cap W.$$

Hence, there exist vectors $x_1, \ldots, x_r$ in B_0 and scalars $\delta_1, \ldots, \delta_r$ such that

$$\alpha_1 u_1 + \cdots + \alpha_m u_m + \beta_1 v_1 + \cdots + \beta_n v_n = -(\gamma_1 w_1 + \cdots + \gamma_k w_k) = \delta_1 x_1 + \cdots + \delta_r x_r.$$

But, $x_1, \ldots, x_r, w_1, \ldots, w_k \in B_0 \cup B_2$ and $B_0 \cup B_2$ is linearly independent. Hence, the equality

$$-(\gamma_1 w_1 + \cdots + \gamma_k w_k) = \delta_1 x_1 + \cdots + \delta_r x_r$$

implies that $\gamma_1 = \cdots = \gamma_k = \delta_1 = \cdots = \delta_r = 0$. Therefore,

$$\alpha_1 u_1 + \cdots + \alpha_m u_m + \beta_1 v_1 + \cdots + \beta_n v_n = 0.$$

Again, since $u_1, \ldots, u_m, v_1, \ldots, v_n \in B_0 \cup B_1$ and $B_0 \cup B_1$ is linearly independent, we obtain $\alpha_1 = \cdots \alpha_m = \beta_1 = \cdots = \beta_n = 0$. Therefore, $B_0 \cup B_1 \cup B_2$ is linearly independent. ■

Exercises for Sect. 1.7

1. Let $u, v, x, y \in \mathbb{R}^4$. Let $U = \text{span}\{u, v\}$ and let $X = \text{span}\{x, y\}$. In which of the following cases $U \oplus X = \mathbb{R}^4$?

   (a) $u = (0, 1, 0, 1),\ v = (0, 0, 1, 1),\ x = (1, 0, 1, 0),\ y = (1, 1, 0, 0)$.
   (b) $u = (0, 0, 0, 1),\ v = (1, 0, 0, 0),\ x = (1, 1, 1, 0),\ y = (0, 1, 1, 1)$.
   (c) $u = (1, 0, 0, 1),\ v = (0, 1, 1, 0),\ x = (1, 0, 1, 0),\ y = (0, 1, 0, 1)$.
   (d) $u = (1, 1, 1, 0),\ v = (1, 1, 0, 1),\ x = (0, 1, 1, 1),\ y = (0, 0, 1, 1)$.

2. Let $\mathcal{P}_e = \{p(t) \in \mathcal{P}(\mathbb{F}) : p(-t) = p(t)\}$; $\mathcal{P}_o = \{p(t) \in \mathcal{P} : p(-t) = -p(t)\}$. Show that both P_e and P_o are subspaces of $\mathcal{P}(\mathbb{F})$ and that $\mathcal{P}(\mathbb{F}) = \mathcal{P}_e \oplus \mathcal{P}_o$.
3. Consider the subspaces $U = \{(a_1, \ldots, a_{2n}) \in \mathbb{R}^{2n} : a_1 = \cdots = a_n = 0\}$ and $V = \{(a_1, \ldots, a_{2n}) \in \mathbb{R}^{2n} : a_i = a_{n+i} \text{ for } i = 1, \ldots, n\}$ of $\mathbb{R}^{2n}$. Are U and V complementary?
4. Construct three subspaces U, W, X of a vector space V so that $V = U \oplus W$ and $V = U \oplus X$ but $W \neq X$.
5. Find a subspace of $\mathcal{P}(\mathbb{R})$ complementary to $\{\alpha t^3 + \beta t^7 : \alpha, \beta \in \mathbb{R}\}$.
6. Let $U = \{(a, b, c, d) \in \mathbb{R}^4 : b = -a\}$ and let $W = \{(a, b, c, d) : c = -a\}$. Find the dimensions of the subspaces $U,\ W,\ U + W$, and $U \cap W$ of $\mathbb{R}^4$.
7. Show that if U and W are subspace of $\mathbb{R}^9$ such that $\dim(U) = 5 = \dim(W)$, then $U \cap W \neq \{0\}$.
8. Let U and W be subspaces of a vector space of dimension $2n + 1$. Show that if $\dim(U) = \dim(W) \geq n + 1$, then $U \cap W$ contains a nonzero vector.
9. Let U and W be subspaces of $\mathbb{F}^7$ with $\dim(U) = 4$ and $\dim(W) = 3$. Show that $U + W = \mathbb{F}^7$ if and only if $U \cap W = \{0\}$ if and only if $\mathbb{F}^7 = U \oplus W$.

10. Let U be a subspace of a vector space V. Show that $\dim(U) \leq \dim(V)$. Further, if U is a finite dimensional proper subspace of V, then show that $\dim(U) < \dim(V)$.
11. In Theorem 1.50, prove (1) $\Rightarrow$ (2) without using induction.

1.8 Quotient Space

One-dimensional subspaces of $\mathbb{R}^3$ are straight lines passing through the origin. On the other hand, any straight line in $\mathbb{R}^3$ is a translate of such a one-dimensional subspace. Similarly, any plane in $\mathbb{R}^3$ is a translate of a plane passing through the origin which is a two-dimensional subspace of $\mathbb{R}^3$. This notion can be generalized to any vector space.

Definition 1.52 Let U be a subspace of a vector space V. Let $v \in V$. The **sum of** the vector v and the subspace U is the set

$$v + U := \{v\} + U = \{v + x : x \in U\}.$$

Any subset of V which is equal to $v + U$ for some vector $v \in V$ and some subspace U of V is called an **affine space** associated with the subspace U in V. The affine space $v + U$ is also written as $U + v$ or as U_v.

An affine space in $\mathbb{R}^3$ is either a single point, a straight line, a plane, or $\mathbb{R}^3$ itself.

Example 1.53 A plane U through the origin looks like:

$$U = \{(\alpha_1, \alpha_2, \alpha_3) \in \mathbb{R}^3 : a\alpha_1 + b\alpha_2 + c\alpha_3 = 0\}$$

for some $a, b, c \in \mathbb{R}$. Take $v = (\beta_1, \beta_2, \beta_3) \in \mathbb{R}^3$. Then

$$\begin{aligned} v + U &= \{(\beta_1 + \alpha_1, \beta_2 + \alpha_2, \beta_3 + \alpha_3) : a\alpha_1 + b\alpha_2 + c\alpha_3 = 0\} \\ &= \{(\gamma_1, \gamma_2, \gamma_3) : a(\gamma_1 - \beta_1) + b(\gamma_2 - \beta_2) + c(\gamma_3 - \beta_3) = 0\} \\ &= \{(\gamma_1, \gamma_2, \gamma_3) : a\gamma_1 + b\gamma_2 + c\gamma_3 = a\beta_1 + b\beta_2 + c\beta_3\}. \end{aligned}$$

Thus, the affine space $v + U$ is the plane parallel to U passing through the point v.
Also any plane in $\mathbb{R}^3$ is given by

$$S = \{(\gamma_1, \gamma_2, \gamma_3) \in \mathbb{R}^3 : a\gamma_1 + b\gamma_2 + c\gamma_3 = d\}$$

for some $a, b, c, d \in \mathbb{R}$. Now, S is parallel to the plane

$$U = \{(\alpha_1, \alpha_2, \alpha_3) : a\alpha_1 + b\alpha_2 + c\alpha_3 = 0\}$$

that passes through the point $v = (\beta_1, \beta_2, \beta_3)$, where $a\beta_1 + b\beta_2 + c\beta_3 = d$. Hence $S = v + U$. □

In $\mathbb{R}^3$, consider S as a straight line or a plane. Take any two points on S. The line segment joining these two points completely lie on S. The following theorem generalizes this observation and gives a justification as to why affine spaces are so named.

Theorem 1.54 *Let S be a nonempty subset of a vector space V. The following are equivalent:*

(1) *S is an affine space in V.*
(2) *If $x, y \in S$, then $\alpha x + (1 - \alpha)y \in S$ for each scalar α.*
(3) *For any $n \in \mathbb{N}$, if $v_1, \ldots, v_n \in S$ and $\alpha_1, \ldots, \alpha_n \in \mathbb{F}$ with $\sum_{j=1}^{n} \alpha_j = 1$, then $\sum_{j=1}^{n} \alpha_j v_j \in S$.*

Proof (1) ⇒ (3): Assume that S is an affine space in V. Then $S = v + U$ for some $v \in V$ and for some subspace U of V. Let $v_1, \ldots, v_n \in S$ and let $\alpha_1, \ldots, \alpha_n \in \mathbb{F}$ with $\sum_{j=1}^{n} \alpha_j = 1$. Then $v_1 = v + u_1, \ldots, v_n = v + u_n$ for some $u_1, \ldots, u_n \in U$. Hence

$$\sum_{j=1}^{n} \alpha_j v_j = \sum_{j=1}^{n} \alpha_j (v + u_j) = v + \sum_{j=1}^{n} \alpha_j u_j.$$

As U is a subspace of V, $u := \sum_{j=1}^{n} \alpha_j u_j \in U$. Thus, $\sum_{j=1}^{n} \alpha_j v_j = v + u \in S$.
(3) ⇒ (2): Take $n = 2$.

(2) ⇒ (1): Let (2) hold. Fix $v \in S$ and take $U := \{x - v : x \in S\}$. Then $S = v + U$. So, it is enough to show that U is a subspace of V.

Let $u \in U$ and let $\alpha \in \mathbb{F}$. Then $u = x - v$ for some $x \in S$. As $x \in S$ and $v \in S$, $\alpha x + (1 - \alpha)v \in S$. Thus $\alpha u = \alpha(x - v) = \alpha x + (1 - \alpha)v - v \in U$.

Next, let $u_1, u_2 \in U$. Then $u_1 = x_1 - v$ and $u_2 = x_2 - v$ for some $x_1, x_2 \in S$. Now,

$$u_1 + u_2 = (x_1 - v) + (x_2 - v) = (x_1 + x_2) - 2v = 2\Big[\frac{x_1 + x_2}{2} - v\Big].$$

As $(x_1 + x_2)/2 \in S$ we have $u := (x_1 + x_2)/2 - v \in U$, and hence $u_1 + u_2 = 2u \in U$. Therefore, U is a subspace of V. ■

In the above theorem, the fact that (2) ⇒ (3) can also be seen by induction; we leave it as an exercise.

Given a subspace U of a vector space V and a vector $v \in V$, the pair (v, U) corresponds to the affine space $v + U$. Notice that for all $u \in U$, $(x + u) + U = x + U$. That is, unless $U = \{0\}$, the correspondence $(v, U) \mapsto v + U$ is never one-to-one.

Theorem 1.55 *Let U be a subspace of a vector space V, and let $v_1, v_2 \in V$. Then the following are equivalent:*

(1) $(v_1 + U) \cap (v_2 + U) \neq \varnothing$.
(2) $v_1 - v_2 \in U$.
(3) $v_1 + U = v_2 + U$.

Further, the family of all affine spaces associated with the subspace U decomposes the vector space V into disjoint subsets.

Proof (1) $\Rightarrow$ (2): Suppose $(v_1 + U) \cap (v_2 + U) \neq \varnothing$. Then there exist $u_1, u_2 \in U$ such that $v_1 + u_1 = v_2 + u_2$. Consequently, $v_1 - v_2 = u_2 - u_1 \in U$.

(2) $\Rightarrow$ (3): Suppose $v_1 - v_2 \in U$. Now, for any $u \in U$, $v_1 + u = v_2 + (v_1 - v_2) + u \in v_2 + U$. That is, $v_1 + U \subseteq v_2 + U$. Similarly, it follows that $v_2 + U \subseteq v_1 + U$.

(3) $\Rightarrow$ (1): Suppose $v_1 + U = v_2 + U$. Then $(v_1 + U) \cap (v_2 + U) = v_1 + U = v_2 + U$. It implies that $v_1, v_2 \in (v_1 + U) \cap (v_2 + U)$.

From the equivalence of (1) and (3) it follows that for any $v_1, v_2 \in V$, either $(v_1 + U) \cap (v_2 + U) = \varnothing$ or $v_1 + U = v_2 + U$. Moreover, for each $v \in V$, $v \in v + U$. That is, $V = \cup_{v \in V}(v + U)$. Therefore, the set $\{v + U : v \in V\}$ decomposes V into disjoint subsets. ■

For example, let U be a plane passing through the origin, in $\mathbb{R}^3$. The set of all planes parallel to U cover the whole of $\mathbb{R}^3$ and no two such planes intersect. That is, given such a U, the set of all affine spaces $v + U$ decomposes $\mathbb{R}^3$ into disjoint subsets.

Theorem 1.55 implies that for each $v \in V$

$$v + U = \{x \in V : x - v \in U\}.$$

Since affine spaces associated with a subspace U are disjoint subsets of V and their union is V, the subspace U defines an equivalence relation on the vector space V. We call this relation as **congruence modulo** U. It is described as follows:

For $x, y \in V$, x is congruent to y modulo U, written $x \equiv_U y$, if and only if $x - y \in U$.

The equivalence classes of the relation $\equiv_U$ are the affine spaces $v + U$ for $v \in V$. We may then say that the vector v is a *representative* of the class $v + U$. Note that, any vector in the affine space $v + U$ can be considered as its representative.

We have a natural way of adding any two affine spaces and multiplying an affine space with a scalar, namely for $x, y \in V$ and $\alpha \in \mathbb{F}$,

$$(x + U) + (y + U) := (x + y) + U, \quad \alpha(x + U) := (\alpha x) + U.$$

Notice that these definitions of addition and scalar multiplication of affine spaces go along with our earlier notation for the sum of two sets of vectors. That is,

$$(x+U)+(y+U) = \{x+u_1+y+u_2 : u_1, u_2 \in U\} = (x+y)+U.$$
$$\alpha(x+U) = \{\alpha(x+u) : u \in U\} = \{\alpha x + w : w \in U\} = (\alpha x)+U.$$

This happens because U is a subspace of V. It shows that addition and scalar multiplication (of affine spaces) are well-defined.

To see these more directly, suppose $x+U = x'+U$ and $y+U = y'+U$. Then $x-x' \in U$ and $y-y' \in U$. So that $(x+y)-(x'+y') \in U$ and then $x+y+U = x'+y'+U$. Similarly, if $x+U = \hat{x}+U$ and $\alpha \in \mathbb{F}$, then $\alpha(x-\hat{x}) \in U$; consequently, $\alpha x + U = \alpha\hat{x}+U$.

Further, we observe that

$$(x+U)+U = (x+U)+(0+U) = (x+0)+U = x+U,$$
$$(x+U)+((-x)+U) = (x-x)+U = U.$$

That is, the subspace U acts as the zero of the addition of affine spaces associated with U, and $(-x)+U$ acts as the additive inverse of $x+U$. The other conditions in Definition 1.1 can easily be verified. Thus, we have proved the following theorem.

Theorem 1.56 *Let V be a vector space over $\mathbb{F}$. Let U be a subspace of V. Then the family of all affine spaces $\{v+U : v \in V\}$ is a vector space over $\mathbb{F}$ with respect to the operations of addition and scalar multiplication defined by*

$$(x+U)+(y+U) := (x+y)+U, \quad \alpha(x+U) := (\alpha x)+U \quad \textit{for } x, y \in V,\ \alpha \in \mathbb{F}.$$

Definition 1.57 The vector space of all affine spaces of a vector space V associated with a given subspace U with addition and scalar multiplication given by

$$(x+U)+(y+U) := (x+y)+U, \quad \alpha(x+U) := (\alpha x)+U$$

is denoted by V/U and is called a **quotient space**.

In $\mathbb{R}^2$, take a straight line L passing through the origin. Consider all affine spaces which are translates of another straight line U passing through the origin. Now, each such affine space intersects L at exactly one point. In a sense, any such affine space can thus be identified with that single intersection point. Thus the dimension of the quotient space $\mathbb{R}^2/U$ should match with the dimension of L, which happens to be 1.

Similarly, in $\mathbb{R}^3$, each affine space which is a translate of a plane U passing trough the origin will intersect a straight line L passing through the origin exactly at one point. Hence, we expect that the dimension of the quotient space $\mathbb{R}^3/U$ is also 1.

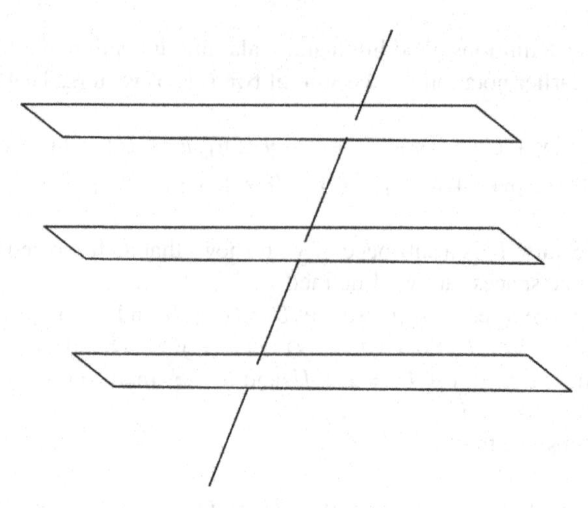

If U is a one-dimensional subspace of $\mathbb{R}^3$, then what would be the dimension of $\mathbb{R}^3/U$? Observe that, in this case, corresponding to each point on a plane passing through the origin, there is exactly one element of $\mathbb{R}^3/U$. Hence, in this case, we expect that dimension of $\mathbb{R}^3/U$ to match with the dimension of the plane, i.e. 2.

The above observations on the dimension of the quotient space can be generalized to yield a dimension formula in a general setting.

Theorem 1.58 *Let U be a subspace of a vector space V. Then*

$$\dim(V) = \dim(U) + \dim(V/U).$$

Proof If $U = V$, then $V/U = \{x + U : x \in V\} = \{U\}$ is the zero space in V/U; so that the result holds. Next, we assume that U is a proper subspace of V.

Let B be a basis of U. Extend this to a basis $B \cup C$ of V. By Theorem 1.46, $V = U \oplus \text{span}(C)$. It is enough to show that $\dim(V/U) = |C|$.

The function that maps any $x \in C$ to $f(C) = x + U \in V/U$ is one-to-one. So, $|C| = |\{x + U : x \in C\}|$. We prove that $\{x + U : x \in C\}$ is a basis of V/U.

Clearly, $\text{span}\{x + U : x \in C\} \subseteq V/U$.

For the other containment, suppose $v + U \in V/U$ for some $v \in V$. Since $V = U \oplus \text{span}(C)$, there exist vectors $u \in U, w_1, \ldots, w_n \in C$ and scalars $\alpha_1, \ldots, \alpha_n$ such that $v = u + \alpha_1 v_1 + \cdots + \alpha_n v_n$. Then

$$v + U = u + (\alpha_1 v_1 + \cdots + \alpha_n v_n) + U = \alpha_1(v_1 + U) + \cdots + \alpha_n(v_n + U).$$

The last vector is in $\text{span}\{x + U : x \in C\}$. So, $V/U \subseteq \text{span}\{x + U : x \in C\}$.

Therefore, $V/U = \text{span}\{x + U : x \in C\}$.

We need to show that $\{x + U : x \in C\}$ is linearly independent. For this, let $v_1, \ldots, v_n \in C$ and let $\beta_1, \ldots, \beta_n$ be scalars such that

$$\beta_1(v_1 + U) + \cdots + \beta_n(v_n + U) = U.$$

Then

$$U = \beta_1(v_1 + U) + \cdots + \beta_n(v_n + U) = (\beta_1 v_1 + \cdots + \beta_n v_n) + U.$$

It follows that $\beta_1 v_1 + \cdots + \beta_n v_n \in U \cap \text{span}(C) = \{0\}$. As C is linearly independent, $\beta_1 = \cdots = \beta_n = 0$. ■

For finite dimensional vector spaces we have proved two-dimensional formulas:

$$\dim(U) + \dim(W) = \dim(U + W) + \dim(U \cap W),$$
$$\dim(V) = \dim(U) + \dim(V/U).$$

Can you relate them?

Exercises for Sect. 1.8

1. Let U be a subspace of a vector space V. Let $u, v, x, y \in V$. Let $\alpha, \beta \in \mathbb{F}$. Show that if $u \equiv_U x$ and $v \equiv_U y$, then $\alpha u + \beta v \equiv_U \alpha x + \beta y$.
2. Describe the following quotient spaces:
   (a) $\mathbb{R}^2/\{0\}$.
   (b) $\mathbb{R}^3/U$, where $U = \{(a, b, c) \in \mathbb{R}^3 : 2a + b - c = 0\}$.
   (c) $\mathbb{F}^\infty/U$, where $U = \{(0, a_2, a_3, \ldots) : a_i \in \mathbb{F}\}$.
3. Determine the quotient space $\mathcal{P}(\mathbb{R})/U$ in each of the following cases:
   (a) $U = \mathcal{P}_n(\mathbb{R})$.
   (b) $U = \{p(t) \in \mathcal{P}(\mathbb{R}) : p(-t) = p(t)\}$.
   (c) $U = \{t^n p(t) : p(t) \in \mathcal{P}(R)\}$ for a fixed n.
4. Consider $\mathbb{C}^n$ as a real vector space. It has a subspace $U := \mathbb{C} \times \{0\}^{n-1}$. What is the dimension of the quotient space $\mathbb{C}^n/U$?

1.9 Problems

1. What is wrong with the following proof of commutativity of addition using other axioms of a vector space?

   $x + x + y + y = 2x + 2y = 2(x + y) = x + y + x + y$. Cancelling x from left and y from right, we have $x + y = y + x$.

2. Prove that a vector space cannot be equal to a finite union of its proper subspaces.
3. Let $V := \mathcal{F}(\mathbb{N}, \mathbb{F})$, the space of all scalar sequences. Let $\ell^1(\mathbb{N}, \mathbb{F})$ be the set of all absolutely summable scalar sequences, that is,

$$\ell^1(\mathbb{N}, \mathbb{F}) := \left\{ x \in V : \sum_{j=1}^{\infty} |x(j)| < \infty \right\}.$$

Show that $\ell^1(\mathbb{N}, \mathbb{F})$ is a subspace of $\mathcal{F}(\mathbb{N}, \mathbb{F})$.

4. For a nonempty set S, let $\ell^\infty(S, \mathbb{F})$ be the set of all bounded functions from S to $\mathbb{F}$, that is,

$$\ell^\infty(S, \mathbb{F}) := \left\{ x \in \mathcal{F}(S, \mathbb{F}) : \sup_{s \in S} |x(s)| < \infty \right\}.$$

Thus, $x \in \ell^\infty(S, \mathbb{F})$ if and only if there exists $M_x > 0$ such that $|x(s)| \le M_x$ for all $s \in S$. In particular, $\ell^\infty(\mathbb{N}, \mathbb{F})$ is the set of all bounded sequences of scalars. Show that $\ell^\infty(S, \mathbb{F})$ is a subspace of $\mathcal{F}(S, \mathbb{F})$.

5. The set $c_{00}(\mathbb{N}, \mathbb{F})$ in Example 1.15 is a subspace of $\ell^1(\mathbb{N}, \mathbb{F})$, and the sets

$$c_0(\mathbb{N}, \mathbb{F}) := \{x \in \mathcal{F}(\mathbb{N}, \mathbb{F}) : x(n) \to 0 \text{ as } n \to \infty\},$$
$$c(\mathbb{N}, \mathbb{F}) := \{x \in \mathcal{F}(\mathbb{N}, \mathbb{F}) : (x(n)) \text{ converges }\}$$

are subspaces of $\ell^\infty(\mathbb{N}, \mathbb{F})$. We observe that

$$c_{00}(\mathbb{N}, \mathbb{F}) \subseteq \ell^1(\mathbb{N}, \mathbb{F}) \subseteq c_0(\mathbb{N}, \mathbb{F}) \subseteq c(\mathbb{N}, \mathbb{F}) \subseteq \ell^\infty(\mathbb{N}, \mathbb{F}).$$

Are these inclusions proper?

6. If one defines $U - W = \{u - v : u \in U, v \in V\}$ for subspaces U, W of a vector space V, then which of the following would hold and which do not?

$$U - U = \{0\}, \quad U - W = U + W, \quad (U - W) + W = U.$$

7. Is sum of subspaces (of a given vector space) commutative and associative? What is the additive identity of the operation of addition of subspaces? What about additive inverse of a subspace? Does every subspace have an additive inverse?
8. Let $S_1 = \{(a, b) \in \mathbb{R}^2 : -1 \le a, b \le 1\}$ and $S_2 = \{(a, b) \in \mathbb{R}^2 : a^2 + b^2 \le 1\}$. What is $S_1 + S_2$?
9. Determine conditions on $\alpha \in \mathbb{C}$ such that in $\mathbb{C}^2$ the vectors $(1 - \alpha, 1 + \alpha)$ and $(1 + \alpha, 1 - \alpha)$ are linearly independent.
10. Determine conditions on $\alpha, \beta \in \mathbb{C}$ such that the vectors $(\alpha, 1), (\beta, 1)$ in $\mathbb{C}^2$ are linearly dependent.
11. Determine conditions on $\alpha, \beta, \gamma \in \mathbb{C}$ such that the vectors $(1, \alpha, \alpha^2), (1, \beta, \beta^2)$, and $(1, \gamma, \gamma^2)$ in $\mathbb{C}^3$ are linearly dependent.
12. Determine conditions on $\alpha \in \mathbb{R}$ such that the vectors $(1, \alpha, 0)$, $(0, 1, \alpha), (\alpha, 1, 1)$ in $\mathbb{R}^3$ are linearly independent.
13. Show that a subset $\{u_1, \ldots, u_n\}$ of V is linearly independent if and only if the function $(\alpha_1, \ldots, \alpha_n) \mapsto \alpha_1 u_1 + \cdots + \alpha_n u_n$ from $\mathbb{F}^n$ into V is injective.

14. Let U be the set of all polynomials $p(t)$ from $\mathcal{P}(\mathbb{C})$, where each $p(t)$ is considered as a function from $S := \{1, \ldots, 2018\}$ to $\mathbb{C}$. Show that U is a subspace of $\mathcal{F}(S, \mathbb{C})$. Determine $\dim(U)$.
15. Let V be an infinite dimensional vector space. Show that there exists a sequence $v_1, v_2, \ldots$, of vectors in V such that for each $n \in \mathbb{N}$, the vectors $v_1, \ldots, v_n$ are linearly independent.
16. Given $a_0, a_1, \ldots, a_n \in \mathbb{R}$, let

$$V = \{f \in C^k[0, 1] : a_n f^{(n)}(t) + \cdots + a_1 f^{(1)}(t) + a_0 f(t) = 0 \text{ for } t \in [0, 1]\}.$$

Show that V is a subspace of $C^k[0, 1]$. Determine $\dim(V)$.
17. In $\mathcal{C}[0, 1]$, consider the function f_n for $n \in \mathbb{N}$, where the graph of f_n in the interval $\left[\frac{1}{n+1}, \frac{1}{n}\right]$ is obtained by joining the point $\left(\frac{1}{n+1}, 0\right)$ to the point $\left(\frac{1}{2}\left(\frac{1}{n+1} + \frac{1}{n}\right), 1\right)$ and then joining $\left(\frac{1}{2}\left(\frac{1}{n+1} + \frac{1}{n}\right), 1\right)$ to the point $\left(\frac{1}{n}, 0\right)$ by straight line segments, and $f_n(x) = 0$ for each $x \notin \left[\frac{1}{n+1}, \frac{1}{n}\right]$. Write $f_n(x)$ as a formula with three cases, and then show that $\{f_n : n \in \mathbb{N}\}$ is linearly independent.
18. For $\lambda \in [a, b]$, let $u_\lambda(t) = \exp(\lambda t)$, $t \in [a, b]$. Show that $\{u_\lambda : \lambda \in [a, b]\}$ is an uncountable linearly independent subset of $\mathcal{C}[a, b]$.
19. Let $p_1(t), \ldots, p_{n+1}(t) \in \mathcal{P}_n(\mathbb{C})$ satisfy $p_1(1) = \cdots = p_{n+1}(1) = 0$. Show that the polynomials $p_1(t), \ldots, p_{n+1}(t)$ are linearly dependent in $\mathcal{P}_n(\mathbb{C})$.
20. Let U, W and X be subspaces of a finite dimensional vector space V. Is it true that if $U \oplus (W \oplus X) = V$, then $(U \oplus W) \oplus X = V$?
21. Let U_1, U_2 and U_3 be subspaces of a vector space V.

    (a) Prove that $U_1 \cap (U_2 + (U_1 \cap U_3)) = (U_1 \cap U_2) + (U_1 \cap U_3)$.
    (b) Give a counter-example for $U_1 \cap (U_2 + U_3) = (U_1 \cap U_2) + (U_1 \cap U_3)$.
    (c) Give a counter-example for
    $\dim(U_1 + U_2 + U_3) + \dim(U_1 \cap U_2) + \dim(U_2 \cap U_3) + \dim(U_3 \cap U_1) =$
    $\dim(U_1) + \dim(U_2) + \dim(U_3) + \dim(U_1 \cap U_2 \cap U_3)$.

22. Let V be a vector space. Show that $V = U_1 \oplus \cdots \oplus U_n$ if and only if $V = U_1 + \cdots + U_n$, and for each $i \in \{1, \ldots, n-1\}$, $U_{i+1} \cap (U_1 + \cdots + U_i) = \{0\}$.
23. Let $U_1, \ldots, U_n$ be finite dimensional subspaces of a vector space V. Prove that $\dim(U_1 + \cdots + U_n) \le \dim(U_1) + \cdots + \dim(U_n)$.
24. Let $U_1, \ldots, U_n$ be subspaces of a finite dimensional vector space V. Prove or disprove: If $\dim(U_1) + \cdots + \dim(U_n) = V$, then $V = U_1 \oplus \cdots \oplus U_n$.
25. Let V be a vector space. Show that $\dim(V) = n \ge 1$ if and only if there exist one-dimensional subspaces $U_1, \ldots, U_n$ such that $V = U_1 \oplus \cdots \oplus U_n$.
26. Let $\mathbb{C}^{n \times n}$ denote the set of all $n \times n$ matrices with complex entries. Show the following:

    (a) $\{B \in \mathbb{C}^{n \times n} : B^T = B\}$ and $\{B \in \mathbb{C}^{n \times n} : B^T = -B\}$ are subspaces of $\mathbb{C}^{n \times n}$.
    (b) $\{B \in \mathbb{C}^{n \times n} : B^T = B\} \oplus \{B \in \mathbb{C}^{n \times n} : B^T = -B\} = \mathbb{C}^{n \times n}$.

27. Let $A \subseteq \mathbb{R}$ be nonempty. Fix $a \in A$. Let $U = \{f \in \mathcal{F}(A, \mathbb{R}) : f(a) = 0\}$. Does there exist a subspace W of $\mathcal{F}(A, \mathbb{R})$ such that $\mathcal{F}(A, \mathbb{R}) = U \oplus W$?
28. Let $t_0, t_1, \ldots, t_n$ be distinct complex numbers. For $i \in \{0, 1, \ldots, n\}$, define $p_i(t) = \Pi_{j \neq i}(t - t_j)$. Show that $\{p_0(t), p_1(t), \ldots, p_n(t)\}$ is a basis for $\mathcal{P}_n(\mathbb{C})$. Further, show that a polynomial $p(t) \in \mathcal{P}_n(\mathbb{C})$ is uniquely determined by the $(n + 1)$-tuple $(p(t_0), \ldots, p(t_n))$.
29. In Theorem 1.54, derive (3) from (2) using induction.
30. Let U and W be subspaces of a finite dimensional vector space V. Prove that any basis of $(U + W)/U$ is in one-to-one correspondence with any basis of $W/(U \cap W)$.

Chapter 2
Linear Transformations

2.1 Linearity

In mathematics one studies structures and the maps between structures. The maps that preserve structures are of fundamental interest. In real analysis, the interesting maps are continuous functions; they map closed intervals onto closed intervals. In group theory, homomorphisms map subgroups onto subgroups. In vector spaces, similarly, the important maps are those which map subspaces onto subspaces. Since the vector spaces get their structure from the two operations of addition and scalar multiplication, the structure preserving maps must preserve these two operations.

Definition 2.1 Let V and W be vector spaces over the same field $\mathbb{F}$.

(a) A function $T : V \to W$ is said to be a **linear transformation from** V **to** W if it satisfies the following two conditions:
Additivity: For all $x, y \in V,\ T(x + y) = T(x) + T(y)$.
Homogeneity: For all $x \in V$ and for all $\alpha \in \mathbb{F},\ T(\alpha x) = \alpha T(x)$.
(b) A linear transformation from V to $\mathbb{F}$ is called a **linear functional on** V.
(c) A linear transformation from V to itself is called a **linear operator on** V.

In Definition 2.1, the $+$ in $T(x + y)$ is the addition operation of V, whereas the $+$ in $T(x) + T(y)$ is the addition of W. Similarly, the scalar multiplication in $T(\alpha x)$ is that of V and that in $\alpha T(x)$ is the scalar multiplication of W. It can be easily seen that the two conditions in Definition 2.1 together are equivalent to either of the following conditions:

$$T(x + \alpha y) = T(x) + \alpha T(y) \quad \text{for all } x, y \in V \text{ and for each } \alpha \in \mathbb{F}$$
$$T(\alpha x + \beta y) = \alpha T(x) + \beta T(y) \quad \text{for all } x, y \in V \text{ and for all } \alpha, \beta \in \mathbb{F}.$$

When we say that $T : V \to W$ is a linear transformation, it is assumed that both V and W are vector spaces, and that too, over the same field. For a linear transformation T and a vector x, the value $T(x)$ is also written as Tx.

M. T. Nair and A. Singh, *Linear Algebra*,
https://doi.org/10.1007/978-981-13-0926-7_2

Linear transformations, in general, are denoted by capital letters such as T, S, A, B, while linear functionals are denoted by small letters f, g, etc.

Linear transformations are also called as *linear maps, linear mappings, linear operators,* and also *homomorphisms*. A linear operator on a vector space is also called an *endomorphism*.

Example 2.2 Let V and W be vector spaces. Define two maps $0 : V \to W$ and $I_V : V \to V$ by

$$0(x) = 0, \quad I_V(x) = x \quad \text{for } x \in V.$$

Notice that the 0 on the left side is a function and the 0 on the right side is the zero vector of W. We see that

$$0(x + y) = 0 = 0(x) + 0(y) \quad 0(\alpha x) = 0 = \alpha\, 0 = \alpha 0(x),$$
$$I_V(x + y) = x + y = I_V(x) + I_V(y), \quad I_V(\alpha x) = \alpha x = \alpha I_V(x).$$

Hence both 0 and I_V are linear transformations; I_V is a linear operator on V.

The linear transformation $0 : V \to W$ is called the **zero map** or the **zero operator**. Sometimes the zero operator is denoted as O instead of 0. The linear transformation $I_V : V \to V$ is called the **identity map**, or the **identity operator**. If there is no confusion, I_V is abbreviated to I. □

Example 2.3 (1) Let V be a vector space. Let λ be a scalar. Define $T : V \to V$ by

$$T(x) = \lambda x \quad \text{for } x \in V.$$

Then T is a linear operator on V, since for every $x, y \in V$ and $\alpha \in \mathbb{F}$,

$$T(x + \alpha y) = \lambda(x + \alpha y) = \lambda x + \alpha\lambda y = Tx + \alpha Ty.$$

This operator T is written as $T = \lambda I$, where I is the identity operator on V, and it is called a **scalar operator**.

(2) The map $T : \mathcal{P}(\mathbb{R}) \to \mathcal{P}(\mathbb{R})$ defined by

$$T(p(t)) = tp(t) \quad \text{for } p(t) \in \mathcal{P}(\mathbb{R})$$

is a linear operator, since for every $p(t), q(t) \in \mathcal{P}(\mathbb{R})$ and $\alpha \in \mathbb{F}$,

$$T(p(t) + \alpha q(t)) = t(p(t) + \alpha q(t)) = tp(t) + \alpha tq(t) = T(p(t)) + \alpha T(q(t)).$$

(3) With $\mathbb{F}^\infty = \mathcal{F}(\mathbb{N}, \mathbb{F})$ as the set of all sequences of scalars, the map $T : \mathbb{F}^\infty \to \mathbb{F}^\infty$ defined by

$$T(a_1, a_2, a_3, \ldots) = (a_2, a_3, \ldots) \quad \text{for } (a_1, a_2, a_3, \ldots) \in \mathbb{F}^\infty$$

is a linear operator, since for all $(a_1, a_2, a_3, \ldots)$, $(b_1, b_2, b_3, \ldots)$ in $\mathbb{F}^\infty$ and $\alpha \in \mathbb{F}$, we see that

$$\begin{aligned} T((a_1, a_2, a_3, \ldots) + \alpha(b_1, b_2, b_3, \ldots)) &= T((a_1 + \alpha b_1, a_2 + \alpha b_2, a_3 + \alpha b_3, \ldots) \\ &= (a_2 + \alpha b_2, a_3 + \alpha b_3, \ldots) \\ &= (a_2, a_3, \ldots) + \alpha(b_2, b_3, \ldots) \\ &= T(a_1, a_2, a_3, \ldots) + \alpha T(b_1, b_2, b_3, \ldots). \end{aligned}$$

This operator T is called the **left shift operator** or the **backward shift operator**.

(4) Let $T : \mathbb{F}^\infty \to \mathbb{F}^\infty$ be defined by

$$T(a_1, a_2, a_3, \ldots) = (0, a_1, a_2, \ldots) \quad \text{for } (a_1, a_2, a_3, \ldots) \in \mathbb{F}^\infty.$$

As in (3), it can be verified that T is a linear operator on $\mathbb{F}^\infty$; it is called the **right shift operator** or the **forward shift operator**. □

Example 2.4 (1) Let V be an n-dimensional vector space, and let $B = \{u_1, \ldots, u_n\}$ be an ordered basis of V. For each $x \in V$, there exist unique scalars $\alpha_1, \ldots, \alpha_n \in \mathbb{F}$ such that $x = \alpha_1 u_1 + \cdots + \alpha_n v_n$. For each $j \in \{1, \ldots, n\}$, define $f_j : V \to \mathbb{F}$ by

$$f_j(x) = \alpha_j \quad \text{for } x = \alpha_1 u_1 + \cdots + \alpha_n v_n \in V.$$

Then f_j is a linear functional. We observe that the linear functionals $f_1, \ldots, f_n$ satisfy

$$f_i(u_j) = \delta_{ij} \quad \text{for all } i, j \in \{1, \ldots, n\}.$$

These linear functionals are called the **coordinate functionals** on V with respect to the basis B. The coordinate functionals depend not only on the basis B but also on the order in which the basis vectors $u_1, \ldots, u_n$ appear in B. The coordinate functionals on $\mathbb{F}^n$ with respect to the standard basis satisfy

$$f_i(\alpha_1, \ldots, \alpha_n) = \alpha_i \quad \text{for each } i = 1, \ldots, n.$$

(2) For a given $t \in [a, b]$, let $f_t : C([a, b], \mathbb{F}) \to \mathbb{F}$ be defined by

$$f_t(x) = x(t) \quad \text{for } x \in C([a, b], \mathbb{F}).$$

Then f_t is a linear functional. The functionals f_t for $t \in [a, b]$, are called **evaluation functionals** on $C[a, b]$.

(3) Given $t_1, \ldots, t_n \in [a, b]$ and $w_1, \ldots w_n \in \mathbb{F}$, define $f : C([a, b], \mathbb{F}) \to \mathbb{F}$ by

$$f(x) = \sum_{i=1}^{n} x(t_i) w_i \quad \text{for } x \in C([a, b], \mathbb{F}).$$

Then f is a linear functional.

This functional is a linear combination of evaluation functionals $f_{t_1}, \ldots, f_{t_n}$ defined in (2) above. The functional f is called a **quadrature formula**.

(4) Let $f : C([a, b], \mathbb{F}) \to \mathbb{F}$ be defined by

$$f(x) = \int_a^b x(t)\, dt \quad \text{for } x \in C([a, b], \mathbb{F}).$$

Then f is a linear functional. □

Example 2.5 (1) Define $T : C^1([a, b], \mathbb{R}) \to C([a, b], \mathbb{R})$ by

$$Tx = x' \quad \text{for } x \in C^1([a, b], \mathbb{R}),$$

where x' denotes the derivative of x. Then T is a linear transformation.

(2) Let $\alpha, \beta \in \mathbb{F}$. Define the function $T : C^1([a, b], \mathbb{R}) \to C([a, b], \mathbb{R})$ by

$$Tx = \alpha x + \beta x', \quad \text{for } x \in C^1([a, b], \mathbb{R})$$

Then T is a linear transformation.

(3) Let $T : C([a, b], \mathbb{R}) \to C([a, b], \mathbb{R})$ be defined by

$$(Tx)(s) = \int_a^s x(t)\, dt \quad \text{for } x \in C([a, b], \mathbb{R}),\ s \in [a, b].$$

Then T is a linear transformation.

(4) Let $\{u_1, \ldots, u_n\}$ be an ordered basis of an n-dimensional vector space V. Let

$$Tx = \sum_{j=1}^{n} \alpha_j u_j \quad \text{for each } x = (\alpha_1, \ldots, \alpha_n) \in \mathbb{F}^n.$$

Then $T : \mathbb{F}^n \to V$ is a linear transformation. □

Here are some easy consequences of Definition 2.1.

Theorem 2.6 *Let $T : V \to W$ be a linear transformation. Then the following are true.*

(1) $T(0) = 0$.
(2) *For each $v \in V$, $T(-v) = -T(v)$.*
(3) $T(\alpha_1 v_1 + \cdots + \alpha_n v_n) = \alpha_1 T v_1 + \cdots + \alpha_n T v_n$ *for all vectors $v_1, \ldots, v_n \in V$ and all scalars $\alpha_1, \ldots, \alpha_n \in \mathbb{F}$.*

Proof (1) $T(0) + 0 = T(0) = T(0 + 0) = T(0) + T(0)$. Thus, $T(0) = 0$.

(2) Let $v \in V$. Then $T(-v) = T((-1)v) = (-1)Tv = -Tv$.

(3) For $n = 2$, we have $T(\alpha_1 v_1 + \alpha_2 v_2) = T(\alpha_1 v_1) + T(\alpha_2 v_2) = \alpha_1 T(v_1) + \alpha_2 T(v_2)$. For any general n, the result follows by induction. ■

Example 2.7 (1) The map $T : \mathbb{R} \to \mathbb{R}$ defined by

$$T(x) = 2x + 3, \quad x \in \mathbb{R},$$

is not a linear transformation, since $T(0) = 3$.

(2) The maps $f, g, h : \mathbb{R} \to \mathbb{R}$ defined by

$$f(x) = x^3, \quad g(x) = \cos x, \quad h(x) = \sin x$$

are not linear transformations since

$$f(2 \times 1) = 8 \neq 2f(1) = 2, \quad g(2\pi) = 1 \neq 2g(\pi) = -2, \quad h(2\pi) = 0 \neq 2h(\pi) = 2.$$

Again, we notice that

$$f(1+1) \neq f(1) + f(1), \quad g(\pi + \pi) \neq g(\pi) + g(\pi), \quad h(\pi) \neq h(\pi) + h(\pi).$$

That is, these functions do not satisfy the additivity requirement also.

(3) Let $T_1, T_2 : V \to W$ be linear transformations. Let a and b be scalars. Define $T : V \to W$ by

$$T(x) = aT_1(x) + bT_2(x) \quad \text{for } x \in V.$$

It is easy to see that T is a linear transformation. □

Theorem 2.6(3) has a nice consequence. Let V be a finite dimensional vector space, $B = \{v_1, \ldots, v_n\}$ a basis of V, and let $v \in V$. There exist unique scalars $\alpha_1, \ldots, \alpha_n$ such that $v = \alpha_1 v_1 + \cdots + \alpha_n v_n$. Then $Tv = \alpha_1 Tv_1 + \cdots + \alpha_n Tv_n$. That is, Tv is completely determined from $Tv_1, \ldots, Tv_n$. Does this hold even if V is not finite dimensional?

Theorem 2.8 *Let V and W be vector spaces over $\mathbb{F}$. Let B be a basis of V, and let $f : B \to W$ be any function. Then there exists a unique linear transformation $T : V \to W$ such that $Tu = f(u)$ for each $u \in B$. In fact, for $v = \alpha_1 v_1 + \cdots + \alpha_k v_k$ with $v_i \in B$ and $\alpha_i \in \mathbb{F}$ for $i = 1, \ldots, k$, $T(v) = \alpha_1 f(v_1) + \cdots + \alpha_k f(v_k)$.*

Proof Let $v \in V$. Since B is a basis of V, there exist unique vectors $v_1 \ldots, v_k \in B$ and correspondingly unique scalars $\alpha_1, \ldots, \alpha_k$ such that $v = \alpha_1 v_1 + \cdots + \alpha_k v_k$. Then the map $T : V \to W$ given by

$$Tv = \alpha_1 f(v_1) + \cdots + \alpha_k f(v_k) \quad \text{for } v = \alpha_1 v_1 + \cdots + \alpha_k v_k$$

is well defined. Notice that this definition takes care of the agreement of T with f on B. That is, $Tu = f(u)$, for each $u \in B$. We show that T is a linear transformation. For this, let $x, y \in V$ and let α be a scalar. Then there exist scalars $\beta_1, \ldots, \beta_m, \gamma_1, \ldots, \gamma_n$ and vectors $u_1, \ldots, u_m, v_1, \ldots, v_n \in B$ such that

$$x = \beta_1 u_1 + \cdots + \beta_m u_m, \quad y = \gamma_1 v_1 + \cdots + \gamma_n v_n.$$

Then

$$\begin{aligned} T(x + \alpha y) &= T(\beta_1 u_1 + \cdots + \beta_m u_m + \alpha\gamma_1 v_1 + \cdots + \alpha\gamma_n v_n) \\ &= \beta_1 f(u_1) + \cdots \beta_m f(u_m) + \alpha\gamma_1 f(v_1) + \cdots + \alpha\gamma_n f(v_n) \\ &= Tx + \alpha Ty. \end{aligned}$$

For uniqueness of such a linear transformation, let $S, T : V \to W$ be linear transformations with $S(u) = T(u)$ for each $u \in B$. Let $z \in V$. Since B is a basis of V, there exist vectors $u_1, \ldots, u_n \in B$ and scalars $\alpha_1, \ldots, \alpha_n$ such that

$$z = \alpha_1 u_1 + \cdots + \alpha_n u_n.$$

As both S and T are linear transformations,

$$S(z) = \alpha_1 S(u_1) + \cdots + \alpha_n S(u_n) = \alpha_1 T(u_1) + \cdots + \alpha_n T(u_n) = T(z). \blacksquare$$

Theorem 2.8 says that a linear transformation is completely determined by its action on a basis. In view of this, we give a definition.

Definition 2.9 Let V and W be vector spaces. Let B be a basis of V. Let $f : B \to W$ be any function. The unique linear transformation $T : V \to W$ defined by

$$T\Big(\sum_{j=1}^{n} \alpha_j v_j\Big) = \sum_{j=1}^{n} \alpha_j f(v_j) \quad \text{for } n \in \mathbb{N},\ \alpha_j \in \mathbb{F},\ v_j \in B$$

is called the **linear extension** of f.

The linear extension of a given map from a basis agrees with the map on the basis; it is defined for other vectors using the linear dependence of the vectors on the basis vectors.

In particular, suppose $B = \{v_1, \ldots, v_n\}$ is a basis for a finite dimensional vector space V. Let $w_1, \ldots, w_n$ be any vectors in any vector space W. If we have a function $f : B \to W$ defined by $f(v_1) = w_1, \ldots, f(v_n) = w_n$, then there exists a unique linear extension of this f to the vector space V. Notice that the vectors $w_1, \ldots, w_n \in W$ need not be linearly independent.

Example 2.10 Let E be the standard basis of $\mathbb{R}^3$, i.e. $E = \{(1, 0, 0), (0, 1, 0), (0, 0, 1)\}$. Let the function $f : E \to \mathbb{R}^3$ be given by

$$f(1, 0, 0) = (0, 1, 1),\ f(0, 1, 0) = (1, 1, 0),\ f(0, 0, 1) = (1, 2, 1).$$

Since $(a, b, c) = a(1, 0, 0) + b(0, 1, 0) + c(0, 0, 1)$, by Theorem 2.8, the linear transformation $T : \mathbb{R}^3 \to \mathbb{R}^3$ is given by

$$T(a, b, c) = a(0, 1, 1) + b(1, 1, 0) + c(1, 2, 1) = (b + c, a + b + 2c, a + c).$$

This is the linear extension of f to $\mathbb{R}^3$. $\square$

In what follows, we will suppress the symbol f for the function that is prescribed on a basis. We will rather start with the symbol T itself and say that T acts on a basis in such and such a way; then denote its linear extension also by T.

Example 2.11 (1) Let $A = [a_{ij}] \in \mathbb{F}^{n \times n}$. Let $\{e_1, \ldots, e_n\}$ be the standard basis of $\mathbb{F}^n$. Suppose T acts on the standard basis in the following way:

$$Te_1 = (a_{11}, \ldots, a_{n1}), \ldots, Te_n = (a_{1n}, \ldots, a_{nn}).$$

Let $x = (\alpha_1, \ldots, \alpha_n) \in \mathbb{F}^n$. Then $x = \alpha_1 e_1 + \cdots + \alpha_n e_n$. The linear extension of T to $\mathbb{F}^n$ is the linear transformation $T : \mathbb{F}^n \to \mathbb{F}^n$ satisfying

$$\begin{aligned} Tx &= \alpha_1 Te_1 + \cdots + \alpha_n Te_n \\ &= \alpha_1 (a_{11}, \ldots, a_{n1}) + \cdots + \alpha_n (a_{1n}, \ldots, a_{nn}) \\ &= \Big(\sum_{j=1}^n a_{1j}\alpha_j, \ldots, \sum_{j=1}^n a_{nj}\alpha_j \Big). \end{aligned}$$

(2) Let $A = [a_{ij}] \in \mathbb{F}^{m \times n}$. Let $\{e_1, \ldots, e_n\}$ be the standard basis for $\mathbb{F}^{n \times 1}$. Let T act on the standard basis in the following manner:

$$Te_j = [a_{1j} \ \cdots \ a_{mj}]^T = j\text{th column of } A.$$

Then its linear extension is the linear transformation $T : \mathbb{F}^{n \times 1} \to \mathbb{F}^{m \times 1}$, where for $x = [\alpha_1 \ \cdots \ \alpha_n]^T \in \mathbb{F}^{n \times 1}$,

$$Tx = \alpha_1 Te_1 + \cdots + \alpha_n Te_n.$$

Note that

$$Tx = \Big[\sum_{j=1}^n a_{1j}\alpha_j \ \cdots \sum_{j=1}^n a_{mj}\alpha_j \Big]^T = Ax \in \mathbb{F}^{m \times 1}.$$

In fact, any matrix $A \in \mathbb{F}^{m \times n}$ is viewed as a linear transformation this way. You can also see the obvious relation between this T and the linear transformation in (1). A more general version of how a matrix gives rise to a linear transformation is described below.

(3) Let $A = [a_{ij}] \in \mathbb{F}^{m\times n}$. Let V and W be vector spaces of dimensions n and m, respectively. Let $B = \{v_1, \ldots, v_n\}$ be an ordered basis of V, and let $C = \{w_1, \ldots, w_m\}$ be an ordered basis of W. Define $T : B \to W$ by

$$T(v_1) = a_{11}w_1 + \cdots + a_{m1}w_m = \sum_{i=1}^{m} a_{i1}w_i$$

$$\vdots$$

$$T(v_n) = a_{1n}w_1 + \cdots + a_{mn}w_m = \sum_{i=1}^{m} a_{in}w_i.$$

Notice that the coefficients of the basis vectors $w_1, \ldots, w_m$ in Tv_j are exactly the entries in the jth column of the matrix A, occurring in that order. Due to Theorem 2.8, this map has a unique linear extension to the linear transformation $T : V \to W$. To see how T works on any vector in V, let $x \in V$. We have a unique n-tuple of scalars $(\alpha_1, \ldots, \alpha_n)$ such that $x = \alpha_1 v_1 + \cdots + \alpha_n v_n$. Then Tx is given by

$$\begin{aligned} Tx &= \alpha_1 \sum_{i=1}^{m} a_{i1}w_i + \cdots + \alpha_n \sum_{i=1}^{m} a_{in}w_i \\ &= \Big(\sum_{j=1}^{n} a_{1j}\alpha_j\Big)w_1 + \cdots + \Big(\sum_{j=1}^{n} a_{nj}\alpha_j\Big)w_n. \end{aligned}$$

Observe that if $V = \mathbb{F}^{n\times 1}$ and $W = \mathbb{F}^{m\times 1}$ with standard bases B and C, then T maps the kth standard basis element in $\mathbb{F}^{n\times 1}$ to the kth column of A; then we obtain the linear transformation in (2) as a particular case.

For any matrix A, the linear transformation $T : V \to W$ defined above is called the linear transformation **induced by** the matrix A; we write this T as T_A. It shows that if V and W are vector spaces with dimensions n and m, and with ordered bases as B and C, respectively, then every $m \times n$ matrix with entries from $\mathbb{F}$ gives rise to a unique linear transformation from V to W. □

Example 2.12 Let $T_1 : U \to V$ and $T_2 : V \to W$ be linear transformations. Let $T : U \to W$ be the *composition map* $T_2 \circ T_1$. That is,

$$T(x) = T_2(T_1(x)) \quad \text{for } x \in V.$$

If $u, v \in U$ and $\alpha \in \mathbb{F}$, then

$$\begin{aligned} T(u + v) &= T_2(T_1(u + v)) = T_2(T_1(u)) + T_2(T_1(v)) = T(u) + T(v), \\ T(\alpha u) &= T_2(T_1(\alpha u)) = T_2(\alpha T_1(u)) = \alpha T_2(T_1(u)) = \alpha T(u). \end{aligned}$$

Hence T is a linear transformation. □

Example 2.13 Suppose $A = [a_{ij}] \in \mathbb{F}^{m \times n}$ and $B = [b_{jk}] \in \mathbb{F}^{r \times m}$. Let U, V, W be vector spaces of dimensions n, m, r, respectively. Let $E_1 = \{u_1, \ldots, u_n\}$, $E_2 = \{v_1, \ldots, v_m\}$ and $E_3 = \{w_1, \ldots, w_r\}$ be ordered bases of U, V, W, respectively. Let $T_A : U \to V$ and $T_B : V \to W$ be the linear transformations induced by A and B, respectively, corresponding to the given ordered bases. As in Example 2.12, the composition map $T_B \circ T_A : U \to W$ is a linear transformation defined by

$$(T_B \circ T_A)(x) = T_B(T_A x), \quad x \in U.$$

Using the properties of the linear transformations T_A and T_B described as in Example 2.11(3), we have

$$\begin{aligned}(T_B \circ T_A)(u_j) = T_B(T_A u_j) &= T_B\Big(\sum_{k=1}^{m} a_{kj} v_k\Big) = \sum_{k=1}^{m} a_{kj} T_B(v_k) \\ &= \sum_{k=1}^{m} a_{kj}\Big[\sum_{i=1}^{r} b_{ik} w_i\Big] = \sum_{i=1}^{r}\Big[\sum_{k=1}^{m} b_{ik} a_{kj}\Big] w_i\end{aligned}$$

for each $j = 1, \ldots, n$. Recall that the product BA is the matrix $[c_{ij}]$ in $\mathbb{F}^{r \times n}$, where

$$c_{ij} = \sum_{k=1}^{m} b_{ik} a_{kj}.$$

Thus, $T_B \circ T_A = T_{BA}$, the linear transformation induced by the matrix BA. □

Example 2.12 shows that composition of linear transformations is a linear transformation. Example 2.13 shows that if A and B are matrices with the product BA well defined, then the induced linear transformation of BA is the composition of the induced linear transformations T_B and T_A, taken in that order. Thus the following definition makes sense.

Definition 2.14 Let $T : U \to V$ and $S : V \to W$ be linear transformations. The composition map $S \circ T$ is also called the **product** of S and T and is denoted by ST. It is defined by

$$ST(x) := (S \circ T)(x) = S(T(x)) \quad \text{for } x \in U.$$

As we have seen, for linear transformations $T : U \to V$ and $S : V \to W$, the product ST is a linear transformation. If $U \neq W$, then the product TS is not well defined. In case, $U = W$, both ST and TS are well defined. If it so happens that $ST = TS$, we say that the linear operators S and T **commute**. In general, two linear operators need not commute; i.e. ST need not be equal to TS.

Example 2.15 Let $V = \mathbb{R}^{2\times 1}$ and $S,\ T : V \to V$ be defined by

$$Sx = \begin{bmatrix} 0 & 1 \\ 0 & 0 \end{bmatrix} \begin{bmatrix} a \\ b \end{bmatrix} := \begin{bmatrix} b \\ 0 \end{bmatrix}, \quad \text{for } x = \begin{bmatrix} a \\ b \end{bmatrix} \in V$$

$$Tx = \begin{bmatrix} 0 & 0 \\ 1 & 0 \end{bmatrix} \begin{bmatrix} a \\ b \end{bmatrix} := \begin{bmatrix} 0 \\ a \end{bmatrix}, \quad \text{for } x = \begin{bmatrix} a \\ b \end{bmatrix} \in V.$$

Then S and T are linear operators on V (verify). Now, for each $x = [a,\ b]^T \in V$,

$$STx = \begin{bmatrix} 0 & 1 \\ 0 & 0 \end{bmatrix} \begin{bmatrix} 0 & 0 \\ 1 & 0 \end{bmatrix} \begin{bmatrix} a \\ b \end{bmatrix} = \begin{bmatrix} 0 & 1 \\ 0 & 0 \end{bmatrix} \begin{bmatrix} 0 \\ a \end{bmatrix} = \begin{bmatrix} a \\ 0 \end{bmatrix}$$

$$TSx = \begin{bmatrix} 0 & 0 \\ 1 & 0 \end{bmatrix} \begin{bmatrix} 0 & 1 \\ 0 & 0 \end{bmatrix} \begin{bmatrix} a \\ b \end{bmatrix} = \begin{bmatrix} 0 & 0 \\ 1 & 0 \end{bmatrix} \begin{bmatrix} b \\ 0 \end{bmatrix} = \begin{bmatrix} 0 \\ b \end{bmatrix}.$$

In particular, if $x = \begin{bmatrix} 1 \\ 0 \end{bmatrix}$, then $STx = x$ but $TSx = 0$. That is, $ST \neq TS$. Thus, S and T do not commute. □

It can be easily seen that a scalar operator on a vector space V commutes with every linear operator on V, that is, corresponding to a given $\alpha \in \mathbb{F}$ if $T : V \to V$ is defined by

$$Tx = \alpha x \quad \text{for } x \in V,$$

and if S is any linear operator on V, then $ST = TS$.

Also, if V is a one-dimensional vector space, then any two linear operators on V commute. This follows from the fact that, if $\dim(V) = 1$, then every linear operator on V is a scalar operator; verify it!

It raises the question that given a vector space V, whether scalar operators are the only linear operators on V that commute with every operator on V? The answer is in the affirmative; prove it!

Product of linear operators on a vector space satisfies associativity and distributivity (over addition) properties: let $T_1,\ T_2,\ T_3$ be linear operators on V. Then we have

Associativity: $T_1(T_2T_3) = (T_1T_2)T_3$.
Distributivity: $(T_1 + T_2)T_3 = T_1T_3 + T_2T_3,\quad T_1(T_2 + T_3) = T_1T_2 + T_1T_3$.

The above properties are satisfied for linear transformations between different vector spaces provided the expressions involved are well defined (verify).

Product of linear transformations enables us to define the powers of linear operators. If T is a linear operator on V, define the **powers of** T, namely T^n for any $n \in \{0, 1, 2, \ldots, \}$ inductively as follows.

$$T^0 := I \quad \text{and} \quad T^n := T\,T^{n-1} \quad \text{for } n > 0.$$

Here, I is the identity operator on V.

We observe that, for any nonnegative integers m, n,

$$T^m T^n = T^n T^m = T^{m+n}.$$

We can also define **polynomials in** T as follows. Let $p \in \mathcal{P}(\mathbb{F})$, say $p(t) = a_1 + a_1 t + \cdots + a_n t^n$. For a linear operator T on V, we define $p(T) : V \to V$ by

$$p(T) = a_1 I + a_1 T + \cdots + a_n T^n.$$

Example 2.16 Consider the *differentiation operator* on $\mathcal{P}_n(\mathbb{F})$ defined by

$$D(a_0 + a_1 t + a_2 t^2 + \cdots + a_n t^n) = a_1 + 2a_2 t + \cdots + n a_n t^{n-1}.$$

The powers of D that is, $D,\ D^2,\ \ldots$ are well defined. We take $D^0 = I$. Then $D^{n+1}(t^k) = 0$ for each $k \in \{0, 1, \ldots, n\}$. That is, D^{n+1} is the zero operator. This gives an example where product of (composition of) two nonzero linear operators can be a zero operator. □

It gives rise to the following notion.

Definition 2.17 A linear operator T on a vector space is called **nilpotent** if for some natural number n, $T^n = 0$. The least natural number k such that $T^k = 0$ is called the **index of nilpotency** of T.

The differentiation operator in Example 2.16 is nilpotent and its index of nilpotency is $n + 1$. We will study properties of nilpotent linear operators later.

If $T : V \to W$ and $S \subseteq W$, then $T^{-1}(S) = \{x \in V : Tx \in S\}$ is the inverse image of S. The notation $T^{-1}(S)$ neither presupposes that T is invertible nor that T^{-1} is a linear operator. In this notation, $T^{-1}(\{0\})$ is the set of all vectors in V which are mapped to 0 by T. Also, for a subset G of a vector space V, we write $T(G) := \{Tx : x \in G\}$.

We see that the structure of a subspace is preserved by a linear transformation.

Theorem 2.18 *Let $T : V \to W$ be a linear transformation. Then the following are true.*

(1) *If V_0 is a subspace of V, then $T(V_0) := \{Tx : x \in V_0\}$ is a subspace of W.*
(2) *If W_0 is a subspace of W, then $T^{-1}(W_0) := \{x \in V : Tx \in W_0\}$ is a subspace of V. In particular, $T^{-1}(\{0\})$ is a subspace of V.*

Proof (1) Let V_0 be a subspace of V. Let $x, y \in T(V_0)$; $\alpha \in \mathbb{F}$. We then have $u, v \in V_0$ such that $x = Tu$ and $y = Tv$. Further, $u + \alpha v \in V_0$. Then

$$x + \alpha y = Tu + \alpha Tv = T(\alpha u + v) \in T(V_0).$$

Therefore, $T(V_0)$ is a subspace of W.

(2) Let W_0 be a subspace of W. Let $x, y \in T^{-1}(W_0)$ and let $\alpha \in \mathbb{F}$. Then $Tx, Ty \in W_0$. It follows that $Tx + \alpha Ty \in W_0$. Since $Tx + \alpha Ty = T(x + \alpha y)$, $x + \alpha y \in T^{-1}(W_0)$. Therefore, $T^{-1}(W_0)$ is a subspace of V. ■

Properties preserved by linear transformations are called *linear properties.* Theorem 2.18(1) says that being a subspace is a linear property. The *line segment* joining u and v in a vector space V is the set

$$S := \{(1 - \lambda)u + \lambda v : 0 \le \lambda \le 1\}.$$

If $T : V \to W$ is a linear transformation, then

$$T(S) = \{(1 - \lambda)T(u) + \lambda T(v) : 0 \le \lambda \le 1\}$$

is a line segment joining $T(u)$ and $T(v)$ in W. We see that *being a triangle* in the plane is a linear property.

This means that a triangle remains a triangle under any linear transformation on $\mathbb{R}^2$. This includes the degenerate cases of a triangle becoming a single point or a line segment. Notice that a linear transformation may change the shape of a triangle; an equilateral triangle may not remain equilateral.

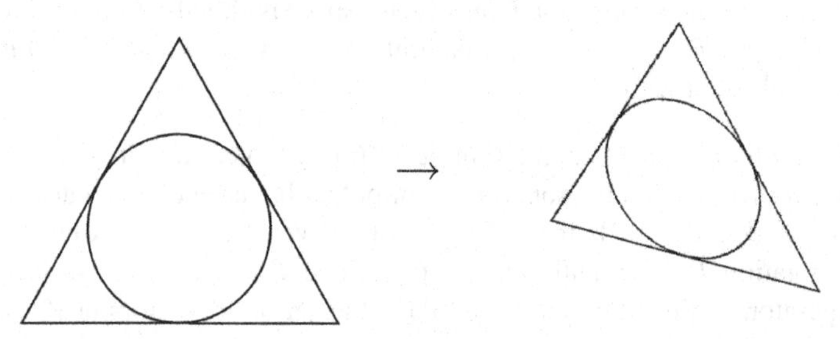

However, *being a circle* is not a linear property. To see this, consider the linear transformation $T : \mathbb{R}^2 \to \mathbb{R}^2$ defined by $T(a, b) = (a, 2b)$ for $a, b \in \mathbb{R}$. Then

$$T(\{(a, b) : a^2 + b^2 = 1\}) = \{(a, 2b) : a^2 + b^2 = 1\} = \{(a, b) : a^2 + b^2/4 = 1\}.$$

Thus, a circle may become an ellipse under a linear transformation.

Exercises for Sect. 2.1

1. Which of the following functions $T : \mathbb{R}^2 \to \mathbb{R}^2$ are linear operators?

   (a) $T(a, b) = (1, b)$.
   (b) $T(a, b) = (a, a^2)$.
   (c) $T(a, b) = (\sin a, 0)$.
   (d) $T(a, b) = (|a|, b)$.
   (e) $T(a, b) = (a + 1, b)$.
   (f) $T(a, b) = (2a + b, a + b^2)$.

2. Consider $\mathbb{C}$ as a vector space over $\mathbb{R}$. Which of the following functions $f : \mathbb{C} \to \mathbb{R}$ are linear functionals?

   (a) $f(a+ib) = a$.
   (b) $f(a+ib) = b$.
   (c) $f(a+ib) = a^2$.
   (d) $f(a+ib) = a - ib$.
   (e) $f(a+ib) = \sqrt{a^2+b^2}$.

   What happens if you consider $\mathbb{C}$ as a vector space over $\mathbb{C}$?
3. Which of the following functions $f : \mathbb{C}^3 \to \mathbb{C}$ are linear functionals?

   (a) $f(a,b,c) = a+b$.
   (b) $f(a,b,c) = b - c^2$.
   (c) $f(a,b,c) = a + 2b - 3c$.

4. Which of the following functions $f : \mathcal{P}(\mathbb{R}) \to \mathbb{R}$ are linear functionals?

   (a) $f(p) = \int_{-1}^{1} p(t)\,dt$.
   (b) $f(p) = \int_0^1 (p(t))^2\,dt$.
   (c) $f(p) = \int_0^1 p(t^2)\,dt$.
   (d) $f(p) = \int_{-1}^{1} t^2 p(t)\,dt$.
   (e) $f(p) = dp/dt$.
   (f) $f(p) = dp/dt$ evaluated at $t = 0$.
   (g) $f(p) = d^2p/dt^2$ evaluated at $t = 1$.

5. Which of the following functions are linear transformations?

   (a) $T : C^1([0,1],\mathbb{R}) \to \mathbb{R}$ with $T(u) = \int_0^1 (u(t))^2\,dt$.
   (b) $T : C^1([0,1],\mathbb{R}) \to \mathbb{R}^2$ with $T(u) = (\int_0^1 u(t)\,dt, u'(0))$.
   (c) $T : \mathcal{P}_n(\mathbb{R}) \to \mathbb{R}$ with $T(p(x)) = p(\alpha)$, for a fixed $\alpha \in \mathbb{R}$.
   (d) $T : \mathbb{C} \to \mathbb{C}$ with $T(z) = \overline{z}$.

6. Let $T : \mathbb{R}^2 \to \mathbb{R}^2$ be a linear transformation with $T(1,0) = (1,4)$ and $T(1,1) = (2,5)$. What is $T(2,3)$?
7. In each of the following, determine whether a linear transformation T with the given conditions exists:

   (a) $T : \mathbb{R}^2 \to \mathbb{R}^3$ with $T(1,1) = (1,0,2)$ and $T(2,3) = (1,-1,4)$.
   (b) $T : \mathbb{R}^3 \to \mathbb{R}^2$ with $T(1,0,3) = (1,1)$ and $T(-2,0,-6) = (2,1)$.
   (c) $T : \mathbb{R}^3 \to \mathbb{R}^2$ with $T(1,1,0) = (0,0)$, $T(0,1,1) = (1,1)$ and $T(1,0,1) = (1,0)$.
   (d) $T : \mathcal{P}_3(\mathbb{R}) \to \mathbb{R}$ with $T(a+bt^2) = 0$ for all $a, b \in \mathbb{R}$.
   (e) $T : \mathbb{C}^3 \to \mathbb{C}^3$ with $T(1,i,-i) = (3i,2i,-i)$, $T(i,2i,-i) = (5,i,1+i)$ and $T(-1,2i-2,1-2i) = (11i,4i-1,1-2i)$.

8. In the following cases, find ST and TS, and determine whether they are linear transformations:

(a) $S,\ T : \mathcal{P}(\mathbb{F}) \to \mathcal{P}(\mathbb{F})$ defined by $S(p(t)) = tp(t),\ T(p(t)) = p'(t)$.
(b) $S,\ T : \mathcal{P}_n(\mathbb{F}) \to \mathcal{P}_n(\mathbb{F})$ defined by $S(p(t)) = tp'(t),\ T(p(t)) = p'(t)$.
(c) $S : C^1([0, 1], \mathbb{R}) \to C([0, 1], \mathbb{R})$ and $T : C([0, 1], \mathbb{R}) \to \mathbb{R}$ defined by $S(u) = u'$ and $T(v) = \int_0^1 v(t)\,dt$.

9. Let $f : \mathbb{R}^2 \to \mathbb{R}$ be defined by $f(x, y) = (x^2 + y^2)^{1/2}$. Show that $f(\alpha(x, y)) = \alpha f(x, y)$, but f is not a linear transformation.
10. Determine all linear operators on $\mathbb{R}^2$ which map the line $y = x$ into the line $y = 3x$.
11. Let $T : V \to V$ be a linear operator, where $\dim(V) = 1$. Show that there exists a scalar α such that for each $v \in V,\ T(v) = \alpha v$.
12. Let $U_1, \ldots, U_n$ be subspaces of a vector space V such that $V = U_1 \oplus \cdots \oplus U_n$. For each $i \in \{1, \ldots, n\}$, let T_i be a linear operator on U_i and let $S_i : V \to V$ be given by
$$S_i(x) = T_i(x) \text{ if } x \in U_i, \quad \text{else } S_i(x) = 0.$$
Show that each S_i is a linear operator on V, and $T = S_1 + \cdots + S_n$.
13. Let $B = \{v_1, \ldots, v_n\}$ be a basis for a vector space V. For each $i \in \{1, \ldots, n\}$, define the linear functional $f_i : V \to \mathbb{F}$ that satisfies
$$f_i(v_j) = \delta_{ij} \quad \text{for } j = 1, \ldots, n.$$
Show that if $f : V \to \mathbb{F}$ is any linear functional then there exist unique scalars $\alpha_1, \ldots, \alpha_n \in \mathbb{F}$ such that $f = \sum_{i=1}^{n} \alpha_i f_i$.
14. Does a linear transformation always map a circle onto an ellipse?
15. Let $\theta \in (0, \pi)$. Let $A, B \in \mathbb{R}^{2\times 2}$ be given by
$$A = \begin{bmatrix} \cos\theta & -\sin\theta \\ \sin\theta & \cos\theta \end{bmatrix}, \quad B = \begin{bmatrix} \cos\theta & \sin\theta \\ \sin\theta & -\cos\theta \end{bmatrix}.$$
Explain why A represents rotation and B represents reflection. By using matrix products, show that
(a) a rotation following a rotation is a rotation,
(b) a rotation following a reflection is a reflection,
(c) a reflection following a rotation is a reflection, and
(d) a reflection following a reflection is a rotation.

2.2 Rank and Nullity

Given a linear transformation $T : V \to W$, Theorem 2.18 asserts that the range of T is a subspace of W, and the inverse image of 0 is a subspace of V. We single out these two special subspaces and their dimensions.

Definition 2.19 Let $T : V \to W$ be a linear transformation. Then

(a) $R(T) := T(V) = \{Tx : x \in V\}$ is called the **range space** of T.
(b) $\dim(R(T))$ is called the **rank** of T, and it is denoted by rank (T).
(c) $N(T) := T^{-1}(\{0\}) = \{x \in V : Tx = 0\}$ is called the **null space** of T.
(d) $\dim(N(T))$ is called the **nullity** of T, and it is denoted by null (T).

Thus rank (T) is the cardinality of a basis of $R(T)$. If $R(T)$ is infinite dimensional, then it is customary to write rank $(T) = \infty$. Curious things can happen for infinite dimensional spaces. Notice that if $R(T) = W$, then rank $(T) = \dim(W)$. Its converse is true for finite dimensional spaces since $R(T)$ is a subspace of W. But for infinite dimensional spaces, rank $(T) = \dim(W)$ does not necessarily imply that $R(T) = W$. For example, consider the right shift operator on $\mathbb{F}^\infty$, that is, $T : \mathbb{F}^\infty \to \mathbb{F}^\infty$ defined by

$$T(a_1, a_2, \ldots) = (0, a_1, a_2, \ldots).$$

Here, rank $(T) = \dim(W) = \infty$, but $R(T) \neq W$, as $(1, 0, 0, \ldots) \notin R(T)$.

For all vector spaces, the map $T : V \to W$ is surjective when the range of T coincides with the co-domain W. Similarly, the null space characterizes whether T is injective or not.

Theorem 2.20 *Let $T : V \to W$ be a linear transformation. Then the following are true:*

(1) *T is surjective if and only if $R(T) = W$; and in that case* rank $(T) = \dim(W)$.
(2) *T injective if and only if $N(T) = \{0\}$ if and only if* null $(T) = 0$.
(3) *For all $v \in V$ and all $w \in W$, $T(v) = w$ if and only if $T^{-1}(\{w\}) = v + N(T)$.*

Proof (1) The first equivalence follows from the definition of surjective maps. Then rank $(T) = \dim(W)$ follows from $R(T) = W$.

(2) Let T be injective. Then 0 has only one inverse image. Since $T(0) = 0$, we conclude that $N(T) = \{0\}$. Conversely, suppose $N(T) = \{0\}$. Let $Tu = Tv$. Then $T(u - v) = 0$ implies that $u - v = 0$. That is, $u = v$. Therefore, T is injective. The second equivalence follows since $N(T)$ is a subspace of V.

(3) Let $v \in V$ and let $w \in W$. Suppose $Tv = w$. Then, for any $x \in V$,

$$\begin{aligned}&x \in T^{-1}(w) \text{ if and only if } Tx = w \text{ if and only if } T(x - v) = 0\\ &\text{if and only if } x - v \in N(T) \text{ if and only if } x = v + N(T).\end{aligned}$$

Therefore, $T^{-1}(\{w\}) = v + N(T)$.

Conversely, let $T^{-1}(\{w\}) = v + N(T)$. Then $v = v + 0 \in v + N(T) = T^{-1}(\{w\})$. Therefore, $T(v) = w$. ∎

Example 2.21 Consider the operator T of Example 2.10 on $\mathbb{R}^3$:

$$T(a, b, c) = (b + c, a + b + 2c, a + c).$$

What are the vectors having image as $(1, 2, 3)$? If (a, b, c) is such a vector, then $T(a, b, c) = (1, 2, 3)$ leads to the system of linear equations

$$b + c = 1, \ a + b + 2c = 2, \ a + c = 3.$$

From the first and the third equations we obtain $a + b + 2c = 4$. It is inconsistent with the second. Therefore, the vector $(1, 2, 3) \notin R(T)$.

Similarly, to find out all vectors whose image under T is $(1, 3, 2)$, we are led to solving the system

$$b + c = 1, \ a + b + 2c = 3, \ a + c = 2.$$

This ends up in all vectors of the form $(a, a - 1, 2 - a)$, where $a \in \mathbb{R}$, that is,

$$T(a, a - 1, 2 - a) = (1, 3, 2) \quad \text{for each } a \in \mathbb{R}.$$

For instance, with $a = 0$, $T(0, -1, 2) = (1, 3, 2)$. It says that

$$T^{-1}(\{(1, 3, 2)\}) = \{(a, a - 1, 2 - a) : a \in \mathbb{R}\}.$$

Since only the vectors of the form $(a, a, -a)$ map to $(0, 0, 0)$,

$$N(T) = T^{-1}(\{(0, 0, 0)\}) = \{(a, a, -a) : a \in \mathbb{R}\}.$$

We see that

$$T^{-1}(\{(1, 3, 2)\}) = (0, -1, 2) + \{(a, a, -a) : a \in \mathbb{R}\} = (0, -1, 2) + N(T).$$

That is, the set of all inverse images of $(1, 3, 2)$ are obtained by taking the sum of a particular inverse image with all inverse images of the origin. □

Surjective linear transformations preserve the spanning sets, and injective linear transformations preserve linear independence. Also, linear transformations generated out of injective maps on a basis are injective, as we see in the following theorem.

Theorem 2.22 *Let $T : V \to W$ be a linear transformation. Let $H \subseteq V$. The following are true.*

(1) $T(\text{span}(H)) = \text{span}(T(H))$. *In particular, if* $\text{span}(H) = V$ *then* $\text{span}(T(H)) = R(T)$; *in addition, if T is surjective, then* $\text{span}(T(H)) = W$.
(2) *If $T(H)$ is linearly independent and T is injective on H, then H is linearly independent. In particular, if $v_1, \ldots, v_n \in V$ are such that $Tv_1, \ldots, Tv_n$ are linearly independent, then $v_1, \ldots, v_n$ are linearly independent.*
(3) *Let T be injective. Then, H is linearly independent if and only if $T(H)$ is linearly independent.*

(4) *Let H be a basis of V. Let T be injective on H. Then T is injective if and only if $T(H)$ is linearly independent.*

Proof (1) Let $H \subseteq V$. Suppose y is any vector in W. Then
$y \in T(\text{span}(H))$
if and only if $y = T(\alpha_1 u_1 + \cdots + \alpha_n u_n)$ for some $\alpha_i \in \mathbb{F}$, and some $u_i \in H$
if and only if $y = \alpha_1 T u_1 + \cdots + \alpha_n T u_n$
if and only if $y \in \text{span}(T(H))$.
In particular, if $\text{span}(H) = V$ then $\text{span}(T(H)) = \text{span}(R(T)) = R(T)$. In addition, if T is surjective, then $W = R(T) = \text{span}(T(H))$.

(2) Let T be injective on H; that is, if $u, v \in H$ and $u \neq v$, then $Tu \neq Tv$. Let $T(H)$ be linearly independent. Then the vectors $Tu_1, \ldots, Tu_m$ are linearly independent for any distinct vectors $u_1, \ldots, u_m \in H$. To prove the linear independence of H, let $v_1, \ldots, v_n \in H$ be distinct vectors, and let $\alpha_1, \ldots, \alpha_n$ be scalars such that

$$\alpha_1 v_1 + \cdots + \alpha_n v_n = 0.$$

Then

$$\alpha_1 T v_1 + \cdots + \alpha_n T v_n = T(\alpha_1 v_1 + \cdots + \alpha_n v_n) = T(0) = 0.$$

Since the vectors $Tv_1, \ldots, Tv_n$ are linearly independent, $\alpha_1 = \cdots = \alpha_n = 0$. Therefore, H is linearly independent.

(3) Suppose that T is injective and that H is linearly independent. Let $v_1, \ldots, v_n \in H$ be such that

$$\alpha_1 T v_1 + \cdots + \alpha_n T v_n = 0.$$

Then $T(\alpha_1 v_1 + \cdots + \alpha_n v_n) = 0$. Since T is injective, $\alpha_1 v_1 + \cdots + \alpha_n v_n = 0$. Since H is linearly independent, $\alpha_1 = 0, \ldots, \alpha_n = 0$. Therefore $T(H)$ is linearly independent. The converse statement follows from (2).

(4) Let H be a basis of V. If T is injective then linear independence of $T(H)$ follows from (3). Conversely suppose that T is injective on H and that $T(H)$ is linearly independent. Let $v \in N(T)$. There are scalars $\alpha_1, \ldots, \alpha_n$ and distinct vectors $v_1, \ldots, v_n \in H$ such that $v = \alpha_1 v_1 + \cdots + \alpha_n v_n$. Now, $Tv = 0$ implies

$$\alpha_1 T(v_1) + \cdots + \alpha_n T(v_n) = 0.$$

Since T is injective on H, there are no repetitions in the list $Tv_1, \ldots, Tv_n$. Then linear independence of $T(H)$ implies that each $\alpha_i = 0$. That is, $v = 0$. So, $N(T) = \{0\}$. Therefore, T is injective. ■

In Theorem 2.22(3), the linear transformation T is assumed to be injective. This condition cannot be replaced with the assumption that T is injective on H. For instance consider the linear transformation $T : \mathbb{R}^3 \to \mathbb{R}^2$ given by

$$T(a, b, c) = (a + c, b + c).$$

Let $H = \{(1,0,0),\ (0,1,0),\ (0,0,1)\}$. We see that

$$T(1,0,0) = (1,0),\ T(0,1,0) = (0,1),\ T(0,0,1) = (1,1).$$

Thus T is injective on H, but $T(H) = \{(1,0),\ (0,1),\ (1,1)\}$ is linearly dependent. In fact, Theorem 2.22(4) now guarantees that T is not an injective linear transformation. This is verified by noting that

$$T(0,0,0) = (0,0) = T(1,1,-1).$$

As a consequence of Theorem 2.22, we have the following theorem.

Theorem 2.23 *Let* $T : V \to W$ *be a linear transformation. Then* $\text{rank}\,(T) \le \dim(V)$. *Moreover, if* T *injective then* $\text{rank}\,(T) = \dim(V)$.

Proof Let B be a basis of V. By Theorem 2.22(1), $T(B)$ spans $R(T)$. Thus $T(B)$ contains a basis of $R(T)$. Let $C \subseteq T(B)$ be such a basis of $R(T)$. Then

$$\text{rank}\,(T) = |C| \le |T(B)| \le |B| = \dim(V).$$

Suppose T is injective. By Theorem 2.22(3), $T(B)$ is a linearly independent subset of $R(T)$. Hence $\dim(V) = |B| = |T(B)| \le \text{rank}\,(T)$. Using the earlier result that $\text{rank}\,(T) \le \dim(V)$, we conclude that $\text{rank}\,(T) = \dim(V)$. ■

Suppose V is a finite dimensional vector space, say, with a basis $\{v_1, \ldots, v_n\}$. If $T : V \to W$ is a linear transformation, then the image of the basis vectors; that is, $Tv_1, \ldots, Tv_n$ span the range space of T. In general, these vectors in $R(T)$ need not be linearly independent; thus $\text{rank}\,(T) \le \dim(V)$. Exactly how much $\text{rank}\,(T)$ falls behind $\dim(V)$?

It follows from Theorem 2.20(3) that the affine space $v + N(T)$ is mapped to the single vector Tv. This means that the quotient space $V/N(T)$ is in one-to-one correspondence with $R(T)$. In that case, due to Theorem 1.58, dimension of V should be equal to the sum of dimensions of $N(T)$ and of $R(T)$. We give a direct proof of this fact below. The proof essentially depends on the following result.

Theorem 2.24 *Let* $T : V \to W$ *be a linear transformation of finite rank. Suppose* $\text{rank}\,(T) = r$. *For vectors* $u_1, \ldots, u_r \in V$ *let* $\{Tu_1, \ldots, Tu_r\}$ *be a basis of* $R(T)$. *If* B *is a basis of* $N(T)$, *then* $B \cup \{u_1, \ldots, u_r\}$ *is a basis of* V.

Proof By Theorem 2.22(2), $\{u_1, \ldots, u_r\}$ is linearly independent. Let $x \in V$, and let B be a basis of $N(T)$. Since $\{Tu_1, \ldots, Tu_r\}$ is a basis of $R(T)$, there exist scalars $\alpha_1, \ldots, \alpha_r$ such that

$$Tx = \alpha_1 Tu_1 + \cdots + \alpha_r Tu_r.$$

Thus, $x - (\alpha_1 u_1 + \cdots + \alpha_r u_r) \in N(T)$. Now, if $B = \varnothing$, then $N(T) = \{0\}$ so that $x = \alpha_1 u_1 + \cdots + \alpha_r u_r$, proving that $\{u_1, \ldots, u_r\}$ spans V. By Theorem 2.22(2),

$\{u_1, \ldots, u_r\}$ is a basis of V. So, suppose $B \neq \varnothing$. Then there exist $v_1, \ldots, v_k \in B$ and $\beta_1, \ldots, \beta_k \in \mathbb{F}$ such that

$$x - (\alpha_1 u_1 + \cdots + \alpha_r u_r) = \beta_1 v_1 + \cdots + \beta_k v_k.$$

Thus,

$$x = (\alpha_1 u_1 + \cdots + \alpha_r u_r) + (\beta_1 v_1 + \cdots + \beta_k v_k).$$

It shows that $B \cup \{u_1, \ldots, u_r\}$ spans V.

For linear independence of $B \cup \{u_1, \ldots, u_r\}$, let $v_1, \ldots, v_k \in B$, $\alpha_1, \ldots, \alpha_r \in \mathbb{F}$ and $\beta_1, \ldots, \beta_k \in \mathbb{F}$ satisfy

$$\alpha_1 u_1 + \cdots + \alpha_r u_r + \beta_1 v_1 + \cdots + \beta_k v_k = 0.$$

Then

$$\alpha_1 T u_1 + \cdots + \alpha_r T u_r + \beta_1 T v_1 + \cdots + \beta_k T v_k = 0.$$

The conditions $Tv_1 = \cdots = Tv_k = 0$ imply that $\alpha_1 T u_1 + \cdots + \alpha_r T u_r = 0$. Linear independence of $Tu_1, \ldots, Tu_r$ implies that

$$\alpha_1 = \cdots = \alpha_r = 0.$$

Using this, we find that $\beta_1 v_1 + \cdots + \beta_k v_k = 0$. Since B is linearly independent,

$$\beta_1 = \cdots = \beta_k = 0.$$

Hence $B \cup \{u_1, \ldots, u_r\}$ is linearly independent. ■

Theorem 2.25 (Rank-nullity theorem) *Let $T : V \to W$ be a linear transformation. Then* rank (T) + null $(T) = \dim(V)$.

Proof If T is of finite rank, then the proof follows from Theorem 2.24. On the other hand, if rank $(T) = \infty$, then by Theorem 2.23, $\dim(V) = \infty$. This completes the proof. ■

Some immediate consequences of Theorems 2.20 and 2.25 are listed in the following theorem.

Theorem 2.26 *Let $T : V \to W$ be a linear transformation, where V is a finite dimensional vector space and W is any vector space. The following statements are true:*

(1) *T is injective if and only if* rank $(T) = \dim(V)$.
(2) *T is surjective if and only if* rank $(T) = \dim(W)$.
(3) *If* $\dim(V) = \dim(W)$, *then T is injective if and only if T is surjective.*
(4) *If* $\dim(V) > \dim(W)$, *then T is not injective.*
(5) *If* $\dim(V) < \dim(W)$, *then T is not surjective.*

Exercises for Sect. 2.2

1. Let T be the linear operator on $\mathbb{R}^4$ defined by
$$T(a, b, c, d) = (a - b, b + c, c - d, b + d).$$
What is nullity of T? Is T surjective?
2. Find the rank and nullity of the linear transformation $T : \mathbb{C}^4 \to \mathbb{C}^4$ given by
$$T(a, b, c, d) = (a + b - 3c + d, a - b - c - d, 2a + b + 2c + d, a + 3b + 3d).$$
Give a basis for $R(T)$; determine $\dim(R(T) \cap \text{span}\{(1, 1, 2, 2), (2, 0, 0, 5)\})$.
3. Determine the rank and nullity of the linear operator T on $\mathcal{P}_n(\mathbb{F})$ given by $T(p(t)) = p'(t)$.
4. Let $T : \mathbb{F}^6 \to \mathbb{F}^3$ be the linear transformation such that
$$N(T) = \{(a_1, \ldots, a_6) \in \mathbb{F}^6 : a_2 = 2a_1, a_4 = 4a_3, a_6 = 6a_5\}.$$
Show that T is surjective.
5. There does not exist a linear transformation from $\mathbb{F}^5$ to $\mathbb{F}^2$ whose null space is equal to $\{(a_1, \ldots, a_5) \in \mathbb{F}^5 : a_2 = 2a_1, a_4 = 4a_3, a_5 = 6a_4\}$. Why?
6. Find a nonzero linear functional f on $\mathbb{C}^3$ with $f(1, -1, 1) = f(1, 1, -1) = 0$.
7. Let f be a linear functional on an n-dimensional vector space V. What could be the nullity of f?
8. Let $T : V \to W$ be a linear transformation such that rank $(T) = \dim(W)$. Does it follow that T is surjective?
9. Let T be a linear operator on a finite dimensional vector space V. Prove that the following are equivalent:
   (a) For each $u \in V$, if $Tu = 0$ then $u = 0$.
   (b) For each $y \in V$, there exists a unique vector $v \in V$ such that $Tv = y$.
10. Let T be a linear operator on a finite dimensional vector space V. Is it true that $V = R(T) \oplus N(T)$?
11. Find linear operators S and T on $\mathbb{R}^3$ such that $ST = 0$ but $TS \neq 0$.
12. If T is a linear operator on a vector space V such that $R(T) = N(T)$, then what can you say about the dimension of V?
13. Let $T : U \to V$ and $S : V \to W$ be linear transformations, where U, V, W are finite dimensional vector spaces. Show the following:
    (a) If S and T are injective, then ST is injective.
    (b) If ST is injective, then T is injective.
    (c) If ST is surjective, then S is surjective.
14. Let U be a subspace of a vector space V. Show that there exists a surjective linear transformation from V to U.

15. Let T be the linear operator on $C([0, 1], \mathbb{R})$ given by

$$(Tf)(t) = \int_0^1 \sin(s+t)f(s)ds \quad \text{for } t \in [0, 1].$$

Determine rank (T).

2.3 Isomorphisms

Recall from set theory that if A, B are nonempty sets then any function $f : A \to B$ is bijective if and only if there exists a function $g : B \to A$ such that $g \circ f = I_A$ and $f \circ g = I_B$, where I_A and I_B denote the identity functions on A and B, respectively. A bijective function is also called invertible, and such a function g is called the inverse of f. It may be noted that if $f : A \to B$ is invertible, then its inverse is unique. Indeed, if $g_1 : B \to A$ and $g_2 : B \to A$ are inverses of f, then we have

$$g_1 = g_1 \circ I_B = g_1 \circ (f \circ g_2) = (g_1 \circ f) \circ g_2 = I_A \circ g_2 = g_2.$$

We prove an analogous result about linear transformations.

Theorem 2.27 *Let $T : V \to W$ be a linear transformation. Then T is bijective if and only if there exists a linear transformation $S : W \to V$ such that $ST = I_V$ and $TS = I_W$. Further, such a linear transformation S is unique.*

Proof Suppose that $T : V \to W$ is bijective. For every $y \in W$, there exists a unique $x \in V$ such that $Tx = y$. This correspondence of y to x defines a function $S : W \to V$ with $Sy = x$.

To show that S is a linear transformation, suppose $y_1, y_2 \in W$ and $\alpha \in \mathbb{F}$. There exist unique vectors $x_1, x_2 \in V$ such that $Tx_1 = y_1$ and $Tx_2 = y_2$. Then $Sy_1 = x_1$ and $Sy_2 = x_2$. Hence

$$S(y_1 + \alpha y_2) = S(Tx_1 + \alpha Tx_2) = S\big(T(x_1 + \alpha x_2)\big) = x_1 + \alpha x_2 = S(y_1) + \alpha S(y_2).$$

Therefore, S is a linear transformation. Further,

$$S(Tx) = Sy = x = I_V(x) \quad \text{and} \quad T(Sy) = Tx = y = I_W(y).$$

Conversely, suppose there exists a linear transformation $S : W \to V$ such that $ST = I_V$ and $TS = I_W$. Now, for any $x \in V$ if $Tx = 0$, then $S(Tx) = 0$. As $ST = I_V$, $x = 0$. Hence, $N(T) = \{0\}$; consequently, T is one-to-one. Further, for any $y \in W$, $y = I_W(y) = TSy = T(Sy)$ shows that T is onto. Therefore, T is bijective.

For the uniqueness of S, let $S_1 : W \to V$ and $S_2 : W \to V$ be linear transformations such that $S_1T = I_V = S_2T$ and $TS_1 = I_W = TS_2$. Then for every $y \in W$, we have $S_1y = S_1(TS_2)y = (S_1T)S_2y = S_2y$. Hence, $S_1 = S_2$. ■

In view of Theorem 2.27, the following definition makes sense.

Definition 2.28 A linear transformation $T : V \to W$ is said to be **invertible** if it is bijective. In that case, the unique linear transformation $S : W \to V$ satisfying $ST = I_V$ and $TS = I_W$ is called the **inverse** of T; it is denoted by T^{-1}.

Recall that a matrix $A \in \mathbb{F}^{n\times n}$ can be considered as a linear operator on $\mathbb{F}^{n\times 1}$. We thus say that a square matrix $A \in \mathbb{F}^{n\times n}$ is **invertible** if there exists a matrix $B \in \mathbb{F}^{n\times n}$ such that

$$AB = BA = I.$$

In this case, the matrix B is called the **inverse of** A and is denoted by A^{-1}.

Example 2.29 (1) The natural correspondence $T : \mathbb{F}^{1\times n} \to \mathbb{F}^{n\times 1}$ defined by

$$T\big([a_1, \ldots, a_n]\big) := [a_1, \ldots, a_n]^T \quad \text{for } [a_1, \ldots, a_n] \in \mathbb{F}^{1\times n}$$

is an invertible linear transformation, and its inverse is $S : \mathbb{F}^{n\times 1} \to \mathbb{F}^{1\times n}$ given by

$$S\big([a_1, \ldots, a_n]^T\big) := [a_1, \ldots, a_n] \quad \text{for } [a_1, \ldots, a_n]^T \in \mathbb{F}^{n\times 1}.$$

(2) The linear transformation $T : \mathcal{P}_n(\mathbb{F}) \to \mathbb{F}^{n+1}$ defined by

$$T(a_0 + a_1 t + \cdots a_n t^n) = (a_0, a_1, \ldots, a_n) \quad \text{for } a_0, a_1, \ldots, a_n \in \mathbb{F}$$

is invertible, and its inverse is the linear transformation $S : \mathbb{F}^{n+1} \to \mathcal{P}_n(\mathbb{F})$ defined by

$$S(a_0, a_1, \ldots, a_n) = a_0 + a_1 t + \cdots a_n t^n \quad \text{for } a_0, a_1, \ldots, a_n \in \mathbb{F}.$$

(3) Consider $\mathcal{F}_n(\mathbb{F}) := \mathcal{F}(\{1, \ldots, n\}, \mathbb{F})$, the space of all functions from $\{1, \ldots, n\}$ to $\mathbb{F}$. Define $T : \mathcal{F}_n(\mathbb{F}) \to \mathbb{F}^n$ by

$$T(x) = \big(x(1), \ldots, x(n)\big) \quad \text{for } x \in \mathcal{F}_n(\mathbb{F}).$$

Then, T is an invertible linear transformation, and its inverse is the linear transformation $S : \mathbb{F}^n \to \mathcal{F}_n(\mathbb{F})$ given by

$$S(a_1, \ldots, a_n) = x, \quad (a_1, \ldots, a_n) \in \mathbb{F}^n,$$

where $x(k) := a_k$, for $k = 1, \ldots, n$. This is why the ith component of a vector $x \in \mathbb{F}^n$ is also written as $x(i)$. □

Definition 2.30 A bijective linear transformation is called an **isomorphism.** A vector space V is said to be **isomorphic** to a vector space W, written as $V \simeq W$, if there exists an isomorphism from V to W.

We observe that the linear transformations in Example 2.29 are isomorphisms. Therefore,

$$\mathbb{F}^{1\times n} \simeq \mathbb{F}^{n\times 1}, \quad \mathcal{P}_n(\mathbb{F}) \simeq \mathbb{F}^{n+1} \quad \text{and} \quad \mathcal{F}_n(\mathbb{F}) \simeq \mathbb{F}^n.$$

Here is another example.

Example 2.31 Let $U = \{(a, b, c) \in \mathbb{R}^3 : 3a + 2b + c = 0\}$. It is a subspace of $\mathbb{R}^3$. We show that $U \simeq \mathbb{R}^2$. For this, let $T : \mathbb{R}^2 \to U$ be defined by

$$T(a, b) = (a, \ b, \ -3a - 2b) \quad \text{for } (a, b) \in \mathbb{R}^2.$$

It can be easily seen that T is a linear transformation. Note that $a, b, c \in U$ implies $c = -3a - 2b$ so that T is onto. Also, if $(a, b) \in \mathbb{R}^2$ such that $T(a, b) = (0, 0)$, then $(a, b, -3a - 2b) = (0, 0, 0)$. It follows that $(a, b) = (0, 0)$. Thus, T is one-to-one. Consequently, T is an isomorphism. Of course, T^{-1} can be defined easily. □

If a linear transformation $T : V \to W$ is injective, then restricting the co-domain space to $R(T)$, we obtain a bijective linear transformation. That is, the function

$$T_0 : V \to R(T) \quad \text{defined by} \quad T_0(x) = T(x) \quad \text{for } x \in V$$

is a bijection. Thus, Theorem 2.27 can be restated in a more general form as in the following.

Theorem 2.32 *A linear transformation $T : V \to W$ is injective if and only if there exists a unique linear transformation $S : R(T) \to V$ such that $ST = I_V$ and $TS = I_{R(T)}$.*

Notice that the conditions $ST = I_V$ and $TS = I_{R(T)}$ in Theorem 2.32 say that $ST(x) = x$ for each $x \in V$, and $TS(y) = y$ for each $y \in R(T)$.

Like linear transformations, when we say that $T : V \to W$ is an isomorphism, we implicitly assume that V and W are vector spaces over some (the same) field $\mathbb{F}$; T is a bijective linear transformation.

Theorem 2.33 *Let $T : V \to W$ be an isomorphism.*

(1) *If $\alpha \neq 0$ is a scalar, then αT is an isomorphism and $(\alpha T)^{-1} = (1/\alpha)T^{-1}$.*
(2) *$T^{-1} : W \to V$ is an isomorphism and $(T^{-1})^{-1} = T$.*
(3) *If $S : U \to V$ is also an isomorphism, then $TS : U \to W$ is an isomorphism and $(TS)^{-1} = S^{-1}T^{-1}$.*

Proof Since both T and T^{-1} are linear transformations, we have

$$(1/\alpha)T^{-1}(\alpha T)(v) = (1/\alpha)\alpha T^{-1}T(v) = v \quad \text{for each } v \in V,$$
$$(\alpha T)(1/\alpha)T^{-1}(w) = \alpha(1/\alpha)TT^{-1}(w) = w \quad \text{for each } w \in W.$$

That is, $(1/\alpha)T^{-1}(\alpha T) = I_V$ and $(\alpha T)(1/\alpha)T^{-1} = I_W$. This proves (1). Similarly, (2) and (3) are proved. ■

Since inverse of an isomorphism is an isomorphism, we use symmetric terminology such as "T is an isomorphism between V and W" and also "V and W are isomorphic".

Theorem 2.22 implies the following.

Theorem 2.34 *Let $T : V \to W$ be a linear transformation, where V and W are finite dimensional vector spaces.*

(1) *Let T be an isomorphism and let $v_1, \ldots, v_n \in V$.*

  (a) *If $\{v_1, \ldots, v_n\}$ is linearly independent in V, then $\{Tv_1, \ldots, Tv_n\}$ is linearly independent in W.*
  (b) *If $\{v_1, \ldots, v_n\}$ spans V, then $\{Tv_1, \ldots, Tv_n\}$ spans W.*
  (c) *If $\{v_1, \ldots, v_n\}$ is a basis of V, then $\{Tv_1, \ldots, Tv_n\}$ is a basis of W.*

(2) *If $\{v_1, \ldots, v_n\}$ is a basis of V such that $\{Tv_1, \ldots, Tv_n\}$ has n vectors and is a basis of W, then T is an isomorphism.*

Notice that composition of isomorphisms is an isomorphism. Thus 'is isomorphic to' is an equivalence relation on the collection of vector spaces. This relation is completely characterized by the dimensions of the spaces, in case the spaces are finite dimensional.

Theorem 2.35 *Let V and W be finite dimensional vector spaces over the same field $\mathbb{F}$. Then $V \simeq W$ if and only if $\dim(V) = \dim(W)$.*

Proof If there exists an isomorphism T from V to W, then Theorem 2.34(1c) implies that $\dim(V) = \dim(W)$.

Conversely, if $\dim(V) = \dim(W)$, then define a bijection f from the basis of V onto the basis of W. Due to Theorem 2.8, there exists a unique linear transformation T from V to W that agrees with f on the basis of V. By Theorem 2.34(2), this T is an isomorphism. Therefore, $V \simeq W$. ■

Let $T : V \to W$ be a linear transformation. If $P : X \to V$ and $Q : W \to Y$ are isomorphisms, can we relate the ranges and null spaces of TP and QT with the range and null space of T?

In the following theorem we abbreviate "if and only if" to "iff" for improving legibility.

Theorem 2.36 *Let $T : V \to W$ be a linear transformation. Let $P : X \to V$ and $Q : W \to Y$ be isomorphisms. Then the following are true:*

(1) $R(TP) = R(T), \quad N(QT) = N(T), \quad R(QT) \simeq R(T), \quad N(TP) \simeq N(T).$
(2) *T is injective iff TP is injective iff QT is injective.*
(3) *T is surjective iff TP is surjective iff QT is surjective.*

Proof (1) For each $v \in V$, $Tv = (TP)(P^{-1}v)$. Thus $R(T) \subseteq R(TP)$. Also, for each $x \in X$, $TPx \in R(T)$. Thus $R(TP) \subseteq R(T)$. That is, $R(TP) = R(T)$.

For each $v \in V$, $Tv = 0$ if and only if $QTv = 0$. It shows that $N(QT) = N(T)$.

Next, define the function $f : R(T) \to R(QT)$ by $f(u) = Qu$ for $u \in R(T)$. As f is a restriction of Q to $R(T) \subseteq W$, it is an injective linear transformation. Now, if $y \in R(QT)$, then $Q^{-1}(y) \in R(T)$ and $f(Q^{-1}y) = QQ^{-1}y = y$. Thus f is surjective. That is, f is an isomorphism. Therefore, $R(QT) \simeq R(T)$.

Similarly, define the function $g : N(T) \to N(TP)$ by $g(v) = P^{-1}v$ for $v \in N(T)$. Again, as a restriction of P^{-1}, g is an injective linear transformation. If $x \in N(TP)$, then $TPx = 0$. That is, $Px \in N(T)$ and $g(Px) = P^{-1}Px = x$. It shows that g is surjective. As earlier g is an isomorphism. Consequently, $N(TP) \simeq N(T)$.

(2) From (1) we have $N(QT) = N(T) \simeq N(TP)$. Hence $N(T) = \{0\}$ if and only if $N(TP) = \{0\}$ if and only if $N(QT) = \{0\}$.

(3) From (1) we have $R(QT) \simeq R(T) = R(TP)$. Since both Q and P are surjective, it follows that $R(T) = W$ if and only if $R(TP) = W$ if and only if $R(QT) = Y$. ■

Exercises for Sect. 2.3

1. Let $m < n$. Show that the subspace $\{(a_1, \ldots, a_m, 0, \ldots, 0) \in \mathbb{F}^n : a_i \in \mathbb{F}\}$ of $\mathbb{F}^n$ is isomorphic to $\mathbb{F}^m$.
2. Is $T : \mathbb{R}^{2\times 2} \to \mathbb{R}^{2\times 2}$ given by $T(A) = \begin{bmatrix} 2 & 3 \\ 5 & 7 \end{bmatrix} A - A \begin{bmatrix} 2 & 3 \\ 5 & 7 \end{bmatrix}$ an isomorphism?
3. Prove that the set of all double sequences of the form $(\ldots, a_{-1}, a_0, a_1, \ldots)$ for $a_i \in \mathbb{F}$ is a vector space over $\mathbb{F}$, with usual addition and scalar multiplication. Also prove that this vector space is isomorphic to $\mathbb{F}^\infty$.
4. Let $\{v_1, \ldots, v_n\}$ be an ordered basis of a vector space V. For each $v \in V$ we have a unique n-tuple $(a_1, \ldots, a_n) \in \mathbb{F}^n$ such that $v = a_1 v_1 + \cdots + a_n v_n$. Define $T(v) = (a_1, \ldots, a_n) \in \mathbb{F}^n$. Show that the function $T : V \to \mathbb{F}^n$ defined this way is an isomorphism.
5. Let $T : V \to W$ be a linear transformation. Let $\{v_1, \ldots, v_n\}$ be a basis of V. If $\{Tv_1, \ldots, Tv_n\}$ is a basis of W, does it follow that T is an isomorphism?
6. Let U and V be subspaces of a finite dimensional vector space W. Prove that the quotient spaces $(U + V)/U$ and $V/(U \cap V)$ are isomorphic. Then deduce the dimension formula $\dim(U + V) = \dim(U) + \dim(V) - \dim(U \cap V)$.
7. Let $T : V \to W$ be a linear transformation, where V is a finite dimensional vector space. Prove that the quotient space $V/N(T)$ is isomorphic to $R(T)$. Then deduce the rank-nullity theorem.
8. Let $T : V \to W$ be a linear transformation. Prove the following:

   (a) Let V be finite dimensional. Then T is surjective if and only if there exists a linear transformation $S : W \to V$ such that $TS = I_W$.
   (b) Let W be finite dimensional. Then T is injective if and only if there exists a linear transformation $S : W \to V$ such that $ST = I_V$.

   What could go wrong if the spaces are infinite dimensional?
9. Find all 2×2 matrices A such that $A^2 = I$. Describe their actions on $\mathbb{F}^{2\times 1}$ as linear transformations. Which ones among them are isomorphisms?
10. Let $T : U \to V$ and $S : V \to W$ be linear transformations, where U, V, W are finite dimensional vector spaces. Prove the following:

(a) rank $(ST) \leq \min\{\text{rank}\,(S), \text{rank}\,(T)\}$.
(b) null $(ST) \leq$ null $(S) +$ null (T).
(c) If T is an isomorphism, then rank $(ST) =$ rank (S).

11. Let $T : U \to V$ and $S : V \to W$ be an isomorphism, where U, V, W are finite dimensional vector spaces. Does it follow that rank $(ST) =$ rank (T)?
12. Let $S,\ T : U \to V$ be linear transformations, where U and V are finite dimensional vector spaces. Prove that rank $(S + T) \leq$ rank $(S) +$ rank (T).

2.4 Matrix Representation

Theorem 2.35 implies that if V is an n-dimensional vector space over $\mathbb{F}$, then there exists an isomorphism from V to $\mathbb{F}^n$. The isomorphism that maps the basis vectors of V to the standard basis vectors of $\mathbb{F}^{n\times 1}$ is one such natural isomorphism.

Definition 2.37 Let $B = \{v_1, \ldots, v_n\}$ be an ordered basis of a vector space V. Let $\{e_1, \ldots, e_n\}$ be the standard basis of $\mathbb{F}^{n\times 1}$. The isomorphism $\phi_B : V \to \mathbb{F}^{n\times 1}$ satisfying

$$\phi_B(v_1) = e_1, \quad \ldots, \quad \phi_B(v_n) = e_n$$

is called the **canonical basis isomorphism** from V to $\mathbb{F}^{n\times 1}$.

For a vector $x \in V$, the column vector $\phi_B(x)$ in $\mathbb{F}^{n\times 1}$ is also written as $[x]_B$ and is read as the **coordinate vector of** x with respect to the ordered basis B. That is, for $x = a_1v_1 + \cdots + a_nv_n \in V$,

$$\phi_B(x) = [x]_B = \begin{bmatrix} a_1 \\ \vdots \\ a_n \end{bmatrix} = (a_1, \ \ldots, \ a_n)^T.$$

Notice that once an ordered basis has been chosen, the canonical basis isomorphism is uniquely determined. Under this isomorphism, the image of a vector $x \in V$ is the vector in $\mathbb{F}^{n\times 1}$ whose components are the coordinates of x with respect to the basis B. The coordinate vector of the zero vector is the zero column vector, whatever be the basis. That is, for $x \in V$, and for any basis B of V,

$$x = 0 \ \text{ if and only if } \ [x]_B = 0 \text{ in } \mathbb{F}^{n\times 1}.$$

Since $\mathbb{F}^{n\times 1} \simeq \mathbb{F}^n$, sometimes canonical basis isomorphisms are defined from V to $\mathbb{F}^n$. We prefer to keep $\mathbb{F}^{n\times 1}$; the reason will be clear soon.

Vectors in a vector space of dimension n can be represented by their coordinate vectors in $\mathbb{F}^{n\times 1}$ by fixing an ordered basis. Similarly, a linear transformation from a finite dimensional space to another can be represented by a matrix by fixing ordered bases in both the spaces. We deliberate on this issue.

Let V and W be vector spaces over $\mathbb{F}$, having dimensions n and m, respectively. Suppose $B = \{v_1, \ldots, v_n\}$ is an ordered basis for V and $C = \{w_1, \ldots, w_m\}$ is an ordered basis for W. Consider the canonical basis isomorphisms $\phi_B : V \to \mathbb{F}^{n\times 1}$ and $\psi_C : W \to \mathbb{F}^{m\times 1}$. For a matrix $A = [a_{ij}] \in \mathbb{F}^{m\times n}$, the function given by

$$T_A := \psi_C^{-1} A \phi_B$$

defines a linear transformation from V to W. As in Example 2.11(3), the linear transformation T_A may be given explicitly as follows. For $x = \sum_{j=1}^{n} \alpha_j v_j$,

$$T_A(x) = \sum_{i=1}^{m} \beta_i w_i \quad \text{with} \quad \beta_i = \sum_{j=1}^{n} a_{ij}\alpha_j \quad \text{for each } i \in \{1, \ldots, m\}.$$

In particular, when $x = v_j$, we have $\alpha_j = 1$ and all other αs are 0; it results in $\beta_i = a_{ij}$. Thus the action of T_A on the basis vectors may be given by

$$T_A(v_j) = a_{1j}w_1 + \cdots + a_{mj}w_m \quad \text{for } j = 1, \ldots, n.$$

Conversely, given ordered bases $\{v_1, \ldots, v_n\}$ for V, $\{w_1, \ldots, w_m\}$ for W, and a linear transformation $T : V \to W$, there exist unique scalars a_{ij} such that the above equation is satisfied with T in place of T_A. Thus a matrix $A = [a_{ij}]$ in $\mathbb{F}^{m\times n}$ can be constructed so that $T_A = T$. It leads to the following definition.

Definition 2.38 Let $B = \{v_1, \ldots, v_n\}$ and $C = \{w_1, \ldots, w_m\}$ be ordered bases for vector spaces V and W, respectively. Let $T : V \to W$ be a linear transformation. For $1 \le i \le m$ and $1 \le j \le n$, let a_{ij} be scalars such that

$$\begin{aligned} T(v_1) &= a_{11}w_1 + \cdots + a_{m1}w_m \\ &\vdots \\ T(v_n) &= a_{1n}w_1 + \cdots + a_{mn}w_m. \end{aligned}$$

Then, the matrix $A \in \mathbb{F}^{m\times n}$ given by

$$A = \begin{bmatrix} a_{11} & \cdots & a_{1n} \\ \vdots & & \vdots \\ a_{m1} & \cdots & a_{mn} \end{bmatrix}$$

is called the **matrix representation** of the linear transformation T with respect to the ordered bases B and C; it is denoted by $[T]_{C,B}$.

Care is needed while constructing the matrix representation. The coefficients of $w_1, \ldots, w_m$ in Tv_j appear in the jth column and not in the jth row. We must remember that when $\dim(V) = n$ and $\dim(W) = m$, that is, when the basis C for

W has m vectors and the basis B for V has n vectors, the matrix $[T]_{C,B}$ is in $\mathbb{F}^{m \times n}$. This is the reason we prefer the notation $[T]_{C,B}$ over the alternate notations $[T]_{B,C}$ and $[T]_C^B$. If the bases B and C for the spaces V and W are fixed in a context, then we write $[T]_{C,B}$ as $[T]$.

Example 2.39 Choose standard bases $B = \{e_1, e_2\}$ for $\mathbb{R}^2$ and $E = \{\tilde{e}_1, \tilde{e}_2, \tilde{e}_3\}$ for $\mathbb{R}^3$. Consider the linear transformation $T : \mathbb{R}^2 \to \mathbb{R}^3$ given by

$$T(a, b) = (2a - b, a + b, b - a).$$

Then

$$\begin{aligned} T(e_1) &= (2, 1, -1) = 2\tilde{e}_1 + 1\tilde{e}_2 + (-1)\tilde{e}_3 \\ T(e_2) &= (-1, 1, 1) = (-1)\tilde{e}_1 + 1\tilde{e}_2 + 1\tilde{e}_3. \end{aligned}$$

Therefore

$$[T]_{E,B} = \begin{bmatrix} 2 & -1 \\ 1 & 1 \\ -1 & 1 \end{bmatrix}.$$

Notice that $[T(a, b)]_E = [T]_{E,B}\,[a\ b]^T = [T]_{E,B}\,[(a, b)]_B$. □

If the ordering of vectors in the basis of the domain space changes, then a corresponding permutation of columns occurs in the matrix representation of the linear transformation; if the change is in the ordering of vectors in the basis of the co-domain space, then the matrix representation will have permutation of rows.

Example 2.40 Consider the polynomials in $\mathcal{P}_3(\mathbb{R})$ and in $\mathcal{P}_2(\mathbb{R})$ as functions from $\mathbb{R}$ to $\mathbb{R}$. Then the function $T : \mathcal{P}_3(\mathbb{R}) \to \mathcal{P}_2(\mathbb{R})$ given by

$$T(p(t)) = \frac{dp(t)}{dt}$$

is a linear transformation. With the standard bases $B = \{1, t, t^2, t^3\}$ and $E = \{1, t, t^2\}$ for the spaces, we have

$$\begin{aligned} T(1) &= \ 0 \ = (0)\,1 + 0\,t + 0\,t^2 \\ T(t) &= \ 1 \ = (1)\,1 + 0\,t + 0\,t^2 \\ T(t^2) &= 2\,t = (0)\,1 + 2\,t + 0\,t^2 \\ T(t^3) &= 3\,t^2 = (0)\,1 + 0\,t + 3\,t^2. \end{aligned}$$

The matrix representation of T with respect to the given ordered bases B and E is

$$[T]_{E,B} = \begin{bmatrix} 0 & 1 & 0 & 0 \\ 0 & 0 & 2 & 0 \\ 0 & 0 & 0 & 3 \end{bmatrix}.$$

With the ordered basis $C = \{t^2, t, 1\}$ for $\mathcal{P}_2(\mathbb{R})$, we have

$$[T]_{C,B} = \begin{bmatrix} 0 & 0 & 0 & 3 \\ 0 & 0 & 2 & 0 \\ 0 & 1 & 0 & 0 \end{bmatrix}.$$

Notice that $[T(a + bt + ct^2 + dt^3)]_C = [T]_{C,B}[(a + bt + ct^2 + dt^3)]_B$. □

Let $B = \{v_1, \ldots, v_n\}$ and let $C = \{w_1, \ldots, w_m\}$ be ordered bases of V and W, respectively. With the standard bases for $\mathbb{F}^{n\times 1}$ and $\mathbb{F}^{m\times 1}$, let the canonical basis isomorphisms be denoted by $\phi_B : V \to \mathbb{F}^{n\times 1}$ and $\psi_C : W \to \mathbb{F}^{m\times 1}$, respectively. The coordinate vectors of $x \in V$ and $y \in W$ can then be written as

$$[x]_B = \phi_B(x) \quad \text{and} \quad [y]_C = \psi_C(y).$$

The discussion prior to Definition 2.38 can then be summarized as the commutative diagram below.

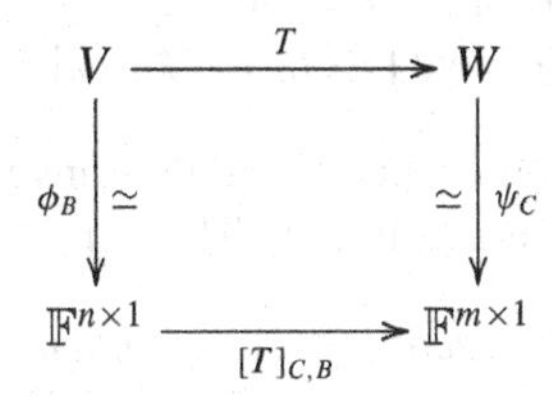

This means

$$[T]_{C,B} = \psi_C T \phi_B^{-1} \quad \text{and} \quad \psi_C T = [T]_{C,B}\phi_B.$$

We prove it formally.

Theorem 2.41 *Let $B = \{v_1, \ldots, v_n\}$ and $C = \{w_1, \ldots, w_m\}$ be ordered bases of vector spaces V and W, respectively. Let $T : V \to W$ be a linear transformation. Then*

$$[T]_{C,B} = \psi_C T \phi_B^{-1} \quad \textit{and} \quad [Tx]_C = [T]_{C,B}[x]_B \ \textit{for each}\ x \in V.$$

Proof Let $x \in V$. Let $x = \beta_1 v_1 + \cdots + \beta_n v_n$ for some $\beta_1, \ldots, \beta_n \in \mathbb{F}$. Suppose $a_{ij} \in \mathbb{F}$ are such that

$$T(v_1) = a_{11}w_1 + \cdots + a_{m1}w_m$$
$$\vdots$$
$$T(v_n) = a_{1n}w_1 + \cdots + a_{mn}w_m.$$

Then

$$\begin{aligned} Tx &= \beta_1 T(v_1) + \cdots + \beta_n T(v_n) \\ &= \beta_1(a_{11}w_1 + \cdots + a_{m1}w_m) + \cdots + \beta_n(a_{1n}w_1 + \cdots + a_{mn}w_m) \\ &= (\beta_1 a_{11} + \cdots + \beta_n a_{1n})w_1 + \cdots + (\beta_1 a_{m1} + \cdots + \beta_n a_{mn})w_m. \end{aligned}$$

Consequently,

$$\psi_C(Tx) = [(\beta_1 a_{11} + \cdots + \beta_n a_{1n}), \ldots, (\beta_1 a_{m1} + \cdots + \beta_n a_{mn})]^T.$$

On the other hand, $\phi_B(x) = [\beta_1, \ldots, \beta_n]^T$. And then

$$[T]_{C,B}\,\phi_B(x) = \begin{bmatrix} a_{11} & \cdots & a_{1n} \\ & \vdots & \\ a_{m1} & \cdots & a_{mn} \end{bmatrix} \begin{bmatrix} \beta_1 \\ \vdots \\ \beta_n \end{bmatrix} = \psi_C(Tx).$$

This shows that $[T]_{C,B}\phi_B = \psi_C T$, and hence $[T]_{C,B} = \psi_C T \phi_B^{-1}$. Since $\phi_B(x) = [x]_B$ and $\psi_C(Tx) = [Tx]_C$, we have $[Tx]_C = [T]_{C,B}[x]_B$. ■

Since the canonical basis isomorphisms and their inverses are isomorphisms, the following result can be seen as a corollary to Theorem 2.34.

Theorem 2.42 *Let V and W be vector spaces of dimensions n and m, with ordered bases B and C, respectively. Let $T : V \to W$ be a linear transformation and let $H \subseteq V$.*

(1) *H is linearly independent in V if and only if $\{[x]_B : x \in H\}$ is linearly independent in $\mathbb{F}^{n\times 1}$.*
(2) *H spans V if and only if $\{[x]_B : x \in H\}$ spans $\mathbb{F}^{n\times 1}$.*
(3) *T is injective if and only if $[T]_{C,B}$ is injective if and only if the columns of $[T]_{C,B}$ are linearly independent.*
(4) *T is surjective if and only if $[T]_{C,B}$ is surjective if and only if the columns of $[T]_{C,B}$ span $\mathbb{F}^{m\times 1}$.*

In fact, Theorem 2.41 justifies why the product of a matrix and a column vector is defined in such a clumsy way. From Definition 2.38, it is clear that

> the jth column of the matrix representation of T is the coordinate vector of the image of the jth basis vector under T, that is,
> if $B = \{v_1, \ldots, v_n\}$ is an ordered basis of V, C is an ordered basis of W, and $T : V \to W$, then the jth column of $[T]_{C,B}$ is $[Tv_j]_C$.

Thus, for a linear transformation $T : \mathbb{F}^{n\times 1} \to \mathbb{F}^{m\times 1}$, its matrix representation can easily be computed by putting together the images of the standard basis vectors as columns. Therefore, T and $[T]_{C,B}$ coincide when B and C are the standard bases for the spaces $\mathbb{F}^{n\times 1}$ and $\mathbb{F}^{m\times 1}$, respectively.

This gives rise to a geometrical interpretation of a matrix. When B and C are the standard bases for $\mathbb{F}^{n\times 1}$ and $\mathbb{F}^{m\times 1}$, respectively, the matrix $[T]_{C,B}$ is a matrix representation of the linear transformation T and also it is a linear transformation from $\mathbb{F}^{n\times 1}$ to $\mathbb{F}^{m\times 1}$. Thus

$$[T]_{C,B}(e_j) = \text{ the } j\text{th column of the matrix } [T]_{C,B}.$$

In general, for any matrix $A \in \mathbb{F}^{m\times n}$, we have

$$A(e_j) = Ae_j = \text{ the } j\text{th column of } A.$$

Since the range of a linear transformation is the span of the image of a basis of the domain space, the range space of a matrix (as a linear transformation) is simply the span of its columns. So, the rank of a matrix is simply the maximum number of linearly independent columns in it.

In Example 2.11(3), we had seen how a linear transformation is induced by a matrix. Definition 2.38 says how a linear transformation is represented by a matrix. Of course, the keys in passing from one to the other is achieved by fixing two ordered bases for the spaces involved.

Let $B = \{v_1, \ldots, v_n\}$ and $C = \{w_1, \ldots, w_n\}$ be ordered bases for the vector spaces V and W, respectively. Let $A = [a_{ij}] \in \mathbb{F}^{m\times n}$. Suppose $T : V \to W$ is a linear transformation. If T is induced by A, then as in Example 2.11(3),

$$Tv_j = a_{1j}w_1 + \cdots + a_{mj}w_m \quad \text{for } j \in \{1, \ldots, n\}.$$

From Definition 2.38, we see that the matrix representation of T is also A. Similarly, if T has the matrix representation as A, then T is also induced by A.

It thus follows that the matrix representation of the composition of two linear transformations must be the product of the matrices that represent them. This is the reason we abbreviate the composition map $T \circ S$ to TS.

Theorem 2.43 *Let U, V, and W be finite dimensional vector spaces having bases B, C, and D, respectively. Let $S : U \to V$ and $T : V \to W$ be linear transformations. Then*

$$[TS]_{D,B} = [T]_{D,C}[S]_{C,B}.$$

Proof We already know that TS is a linear transformation. Let $z \in U$. By Theorem 2.41,

$$[TS]_{D,B}[z]_B = [(TS)(z)]_D = [T(Sz)]_D = [T]_{D,C}[Sz]_C = [T]_{D,C}[S]_{C,B}[z]_B.$$

Therefore, $[TS]_{D,B} = [T]_{D,C}[S]_{C,B}$. ■

Example 2.44 Let the linear transformations $S : \mathbb{R}^3 \to \mathbb{R}^2$ and $T : \mathbb{R}^2 \to \mathbb{R}^2$ be defined by

$$S(a_1, a_2, a_3) = (a_1 + a_2 + 2a_3,\ 2a_1 - a_2 + a_3),$$
$$T(b_1, b_2) = (b_1 + b_2,\ b_1 - b_2).$$

Then the composition map TS is given by

$$\begin{aligned}(TS)(a_1, a_2, a_2) &= T(a_1 + a_2 + 2a_3,\ 2a_1 - a_2 + a_3)\\ &= (3a_1 + 3a_3,\ -a_1 + 2a_2 + a_3).\end{aligned}$$

With the standard bases B and E for $\mathbb{R}^3$ and $\mathbb{R}^2$, respectively, we see that

$$[S]_{E,B} = \begin{bmatrix} 1 & 1 & 2 \\ 2 & -1 & 1 \end{bmatrix},\quad [T]_{E,E} = \begin{bmatrix} 1 & 1 \\ 1 & -1 \end{bmatrix},\quad [TS]_{E,B} = \begin{bmatrix} 3 & 0 & 3 \\ -1 & 2 & 1 \end{bmatrix}.$$

Notice that $[TS]_{E,B} = [T]_{E,E}\,[S]_{E,B}$. □

If $\dim(U) = n$ with basis B, $\dim(V) = m$ with basis C, and $\dim(W) = k$ with basis D, then the size of the matrices, and the domain and co-domain spaces of the associated maps can be summarized as follows:

$S : U \to V$	$[S]_{C,B} : \mathbb{F}^{n\times 1} \to \mathbb{F}^{m\times 1}$	$[S]_{C,B} \in \mathbb{F}^{m\times n}$
$T : V \to W$	$[T]_{D,C} : \mathbb{F}^{m\times 1} \to \mathbb{F}^{k\times 1}$	$[T]_{D,C} \in \mathbb{F}^{k\times m}$
$TS : U \to W$	$[TS]_{D,B} = [T]_{D,C}[S]_{C,B} : \mathbb{F}^{n\times 1} \to \mathbb{F}^{k\times 1}$	$[TS]_{D,B} \in \mathbb{F}^{k\times n}$

You may prove associativity of matrix multiplication by using the composition maps. Also, think about how to determine a linear transformation if its matrix representation with respect to a known basis is given.

Example 2.45 Let e_1, e_2, e_3 be the standard basis vectors in $\mathbb{R}^3$, and let $\tilde{e}_1, \tilde{e}_2$ be the standard basis vectors in $\mathbb{R}^2$. Consider bases $B = \{e_1, e_1 + e_2, e_1 + e_2 + e_3\}$ and $C = \{\tilde{e}_1, \tilde{e}_1 + \tilde{e}_2\}$ for $\mathbb{R}^3$ and $\mathbb{R}^2$, respectively. Let $T : \mathbb{R}^3 \to \mathbb{R}^2$ be the linear transformation such that its matrix representation with respect to the bases B and C is given by

$$[T]_{C,B} = \begin{bmatrix} 4 & 2 & 1 \\ 0 & 1 & 3 \end{bmatrix}.$$

How does T act on a typical vector $(a, b, c) \in \mathbb{R}^3$? We solve the problem in three ways, though essentially, they are one and the same.
(1) Denote by $D = \{e_1, e_2, e_3\}$ and $E = \{\tilde{e}_1, \tilde{e}_2\}$ the standard bases for $\mathbb{R}^3$ and $\mathbb{R}^2$, respectively. The composition maps formula says that

$$[T]_{E,D} = [I]_{E,C}[T]_{C,B}[I]_{B,D}.$$

The matrix $[I]_{E,C}$ is simply the matrix obtained by expressing vectors in C by those in E:

$$\tilde{e}_1 = 1.\tilde{e}_1 + 0.\tilde{e}_2,\quad \tilde{e}_1 + \tilde{e}_2 = 1.\tilde{e}_1 + 1.\tilde{e}_2.$$

For $[I]_{B,D}$, we express the vectors in D in terms of those in B :

$$e_1 = e_1,\ e_2 = -e_1 + (e_1 + e_2),\ e_3 = -(e_1 + e_2) + (e_1 + e_2 + e_3).$$

Hence

$$[I]_{E,C} = \begin{bmatrix} 1 & 1 \\ 0 & 1 \end{bmatrix},\quad [I]_{B,D} = \begin{bmatrix} 1 & -1 & 0 \\ 0 & 1 & -1 \\ 0 & 0 & 1 \end{bmatrix},$$

$$[T]_{E,D} = \begin{bmatrix} 1 & 1 \\ 0 & 1 \end{bmatrix} \begin{bmatrix} 4 & 2 & 1 \\ 0 & 1 & 3 \end{bmatrix} \begin{bmatrix} 1 & -1 & 0 \\ 0 & 1 & -1 \\ 0 & 0 & 1 \end{bmatrix} = \begin{bmatrix} 4 & -1 & 1 \\ 0 & 1 & 2 \end{bmatrix}.$$

Therefore, $T(a, b, c) = (4a - b + c,\ b + 2c)$.

(2) T is given with respect to the bases B and C. We must use these bases to obtain $T(a, b, c)$, which is understood to be expressed in standard bases. We see that

$$(a, b, c) = (a - b)e_1 + (b - c)(e_1 + e_2) + c(e_1 + e_2 + e_3).$$

Thus

$$\begin{aligned} T(a, b, c) &= T\big((a - b)e_1 + (b - c)(e_1 + e_2) + c(e_1 + e_2 + e_3)\big) \\ &= \big(4(a - b) + 2(b - c) + 1(c)\big)\tilde{e}_1 + \big(0(a - b) + 1(b - c) + 3(c)\big)(\tilde{e}_1 + \tilde{e}_2) \\ &= \big((4a - 2b - c) + (b + 2c)\big)\tilde{e}_1 + (b + 2c)\tilde{e}_2 \\ &= (4a - b + c,\ b + 2c). \end{aligned}$$

(3) The matrix $[T]_{C,B}$, as given, means the following:

$$T(e_1) = 4\tilde{e}_1,\ T(e_1 + e_2) = 2\tilde{e}_1 + (\tilde{e}_1 + \tilde{e}_2),\ T(e_1 + e_2 + e_3) = \tilde{e}_1 + 3(\tilde{e}_1 + \tilde{e}_2).$$

This gives

$$\begin{aligned} T(e_1) &= 4\tilde{e}_1 \\ T(e_2) &= -2\tilde{e}_1 + (\tilde{e}_1 + \tilde{e}_2) = -\tilde{e}_1 + \tilde{e}_2 \\ T(e_3) &= -\tilde{e}_1 + 2(\tilde{e}_1 + \tilde{e}_2)\ = \tilde{e}_1 + 2\tilde{e}_2 \end{aligned}$$

Therefore, $T(a, b, c) = aT(e_1) + bT(e_2) + cT(e_3) = (4a - b + c,\ b + 2c)$. □

Exercises for Sect. 2.4

In the following exercises, a basis means an ordered basis.

1. Let $T : \mathbb{R}^3 \to \mathbb{R}^3$ be defined by $T(a, b, c) = (b + c,\ c + a,\ a + b)$. Find $[T]_{C,B}$ in each of the following cases:

(a) $B = \{(1,0,0), (0,1,0), (0,0,1)\}$, $C = \{(1,0,0), (1,1,0), (1,1,1)\}$.
(b) $B = \{(1,0,0), (1,1,0), (1,1,1)\}$, $C = \{(1,0,0), (0,1,0), (0,0,1)\}$.
(c) $B = \{(1,1,2), (2,1,1), (1,2,1)\}$, $C = \{(2,1,1), (1,2,1), (1,1,2)\}$.

2. Let $T : \mathcal{P}^3 \to \mathcal{P}^2$ be defined by $T(a_0 + a_1 t + a_2 t^2 + a_3 t^3) = a_1 + 2a_2 t + 3a_3 t^2$. Find $[T]_{C,B}$ in each of the following cases:

(a) $B = \{1, t, t^2, t^3\}$, $C = \{1+t, 1-t, t^2\}$.
(b) $B = \{1, 1+t, 1+t+t^2, t^3\}$, $C = \{1, 1+t, 1+t+t^2\}$.
(c) $B = \{1, 1+t, 1+t+t^2, 1+t+t^2+t^3\}$, $C = \{t^2, t, 1\}$.

3. Determine the matrix of the linear transformation $T : \mathcal{P}_3 \to \mathcal{P}_4$ defined by $T(p(t)) = (2-t)p(t)$, with respect to the standard bases of $\mathcal{P}_3$ and $\mathcal{P}_4$.
4. Let $T : \mathcal{P}_2 \to \mathcal{P}_3$ be defined by $T(a + bt + ct^2) = at + bt^2 + ct^3$. If $B = \{1+t, 1-t, t^2\}$ and $C = \{1, 1+t, 1+t+t^2, t^3\}$, then what is $[T]_{C,B}$?
5. Let $T : \mathcal{P}_2 \to \mathcal{P}_3$ be defined by $T(a_0 + a_1 t + a_2 t^2) = a_0 t + \frac{a_1}{2} t^2 + \frac{a_2}{3} t^3$. Find $[T]_{C,B}$ in each of the following cases:

(a) $B = \{t^2, t, 1\}$, $C = \{1, 1+t, 1+t+t^2, 1+t+t^2+t^3\}$.
(b) $B = \{1+t, 1-t, t^2\}$, $C = \{1, t, t^2, t^3\}$.
(c) $B = \{1, 1+t, 1+t+t^2\}$, $C = \{1, 1+t, 1+t+t^2, t^3\}$.

6. Define $T : \mathcal{P}_2(\mathbb{R}) \to \mathbb{R}$ by $T(f) = f(2)$. Compute $[T]$ using the standard bases of the spaces.
7. Define $T : \mathbb{R}^2 \to \mathbb{R}^3$ by $T(a,b) = (a-b, a, 2b+b)$. Suppose E is the standard basis for $\mathbb{R}^2$, $B = \{(1,2), (2,3)\}$, and $C = \{(1,1,0), (0,1,1), (2,2,3)\}$. Compute $[T]_{C,E}$ and $[T]_{C,B}$.
8. Let T be the linear operator on $\mathbb{C}^2$ defined by $T(a,b) = (a,0)$. Let E be the standard basis, and let $B = \{(1,i), (-i,2)\}$ be another basis of $\mathbb{C}^2$. Determine the matrices $[T]_{E,E}$, $[T]_{B,B}$, $[T]_{E,B}$, and $[T]_{B,E}$.
9. Let $(a,b), (c,d) \in \mathbb{R}^2$ satisfy $ac + bd = 0$ and $a^2 + b^2 = c^2 + d^2 = 1$. Show that $B = \{(a,b), (c,d)\}$ is a basis of $\mathbb{R}^2$. What is $[(\alpha, \beta)]_B$?
10. Denote by $\mathbb{C}_R$ the vector space of all complex numbers over the field $\mathbb{R}$. Show that $Tz = \overline{z}$ is a linear operator on $\mathbb{C}_R$. What is $[T]_{B,B}$ with respect to the basis $B = \{1, i\}$?
11. Let $S,\ T : \mathbb{R}^3 \to \mathbb{R}^3$ be given by $S(a,b,c) = (a+2b+c, a-b-c, b+3c)$ and $T(a,b,c) = (c,b,a)$. Consider bases $B = \{(1,2,3), (1,0,1), (1,1,0)\}$ and $E = \{e_1, e_2, e_3\}$ for $\mathbb{R}^3$. Determine the matrices $[ST]_{E,E}$, $[ST]_{E,B}$, $[TS]_{B,E}$, and $[TS]_{B,B}$.
12. Given bases $B = \{1+t,\ 1-t,\ t^2\}$ and $C = \{1,\ 1+t,\ 1+t+t^2,\ t^3\}$ for $\mathcal{P}_2(\mathbb{R})$ and $\mathcal{P}_3(\mathbb{R})$, respectively, and the linear transformation $S : \mathcal{P}_2(\mathbb{R}) \to \mathcal{P}_3(\mathbb{R})$ with $S(p(t)) = t\,p(t)$, find the matrix $[\,S\,]_{C,B}$.
13. Let $T : \mathbb{R}^3 \to \mathbb{R}^3$ be given by $T(a,b,c) = (a+b, 2a-b-c, a+b+c)$. Consider $B = \{(1,0,0), (1,1,0), (1,1,1)\}$ and $C = \{(1,2,1), (2,1,0), (3,2,1)\}$ as bases for $\mathbb{R}^3$. Determine the matrices $[T]_{B,B}$, $[T]_{B,C}$, $[T]_{C,B}$, and $[T]_{C,C}$. Also, find the rank(s) of all these matrices.

14. Let $T : \mathcal{P}_n(\mathbb{F}) \to \mathcal{P}_n(\mathbb{F})$ be given by $T(p(t)) = p(t+1)$ for $p(t) \in \mathcal{P}_n(\mathbb{F})$. Determine $[T]_{E,E}$ with respect to the standard basis E of $\mathcal{P}_n(\mathbb{F})$.
15. Let V be the vector space of all functions from $\mathbb{R}$ to $\mathbb{C}$ over the field $\mathbb{C}$. Let $f_1(t) = 1$, $f_2(t) = e^{it}$, $f_3(t) = e^{-it}$, $g_1(t) = 1$, $g_2(t) = \cos t$, and $g_3(t) = \sin t$. Find $A = [a_{ij}] \in \mathbb{C}^{3\times 3}$ such that $g_j(t) = \sum_{i=1}^{3} a_{ij} f_i$ for each $j \in \{1, 2, 3\}$.
16. Define $T : \mathcal{P}_2(\mathbb{R}) \to \mathbb{R}^{2\times 2}$ by $T(f(t)) = \begin{bmatrix} f'(0) & 2f(1) \\ 0 & f'(3) \end{bmatrix}$. Compute $[T]_{E,B}$, where B and E are standard bases for $\mathcal{P}_2(\mathbb{R})$ and $\mathbb{R}^{2\times 2}$, respectively.
17. Let π be a permutation, i.e. $\pi : \{1, \ldots, n\} \to \{1, \ldots, n\}$ is a bijection. Define a function $T_\pi : \mathbb{F}^n \to \mathbb{F}^n$ by $T_\pi(\alpha_1, \ldots, \alpha_n) = (\alpha_{\pi(1)}, \ldots, \alpha_{\pi(n)})$. Show that T_π is a linear transformation. Describe the matrix representation of T_π with respect to the standard basis of $\mathbb{F}^n$.
18. Let $B = \left\{ \begin{bmatrix} 1 & 0 \\ 0 & 0 \end{bmatrix}, \begin{bmatrix} 0 & 1 \\ 0 & 0 \end{bmatrix}, \begin{bmatrix} 0 & 0 \\ 1 & 0 \end{bmatrix}, \begin{bmatrix} 0 & 0 \\ 0 & 1 \end{bmatrix} \right\}$ and let $E = \{1\}$.

    (a) Compute $[T]_{B,B}$, where $T : \mathbb{R}^{2\times 2} \to \mathbb{R}^{2\times 2}$ is given by $T(A) = A^T$.
    (b) Compute $[T]_{E,B}$, where $T : \mathbb{R}^{2\times 2} \to \mathbb{R}$ is given by $T(A) = tr(A)$.

19. Let $B = \{u_1, \ldots, u_n\}$ be an ordered basis of a vector space V. Let f be a linear functional on V. Prove that there exists a unique $(\beta_1, \ldots, \beta_n) \in \mathbb{F}^n$ such that $f(\alpha_1 u_1 + \cdots + \alpha_n u_n) = \alpha_1\beta_1 + \cdots + \alpha_n\beta_n$. Conclude that the matrix representation $[f]_{\{1\},B}$ is the row vector $[\beta_1 \ \cdots \ \beta_n]$.
20. Let $B = \{v_1, \ldots, v_n\}$ be an ordered basis for a vector space V. Let T be a linear operator on V. Show that T is an isomorphism if and only if $[T]_{B,B}$ is an invertible matrix.
21. Let $A, B \in \mathbb{F}^{n\times n}$. Does it follow that if $AB = 0$, then $BA = 0$?

2.5 Change of Basis

If the bases in the spaces change, the matrix representation of a linear transformation changes. It is a boon because by choosing suitable bases for the spaces, a nice form for the matrix representation can be obtained. We will discuss this issue later. It is also a curse, since even sparse matrices can loose sparseness.

Example 2.46 Consider the identity map $I : \mathbb{R}^3 \to \mathbb{R}^3$, i.e. $I(x) = x$. Take the ordered basis for the domain space as $B = \{(1, 1, 1), (1, 0, 1), (1, 1, 0)\}$ and for the co-domain space as the standard basis E. In order to get the matrix representation of I, we express the basis vectors in B in terms of those in E. It leads to

$$[I]_{E,B} = \begin{bmatrix} 1 & 1 & 1 \\ 1 & 0 & 1 \\ 1 & 1 & 0 \end{bmatrix}.$$

Notice that $[I]_{E,B}$ is not the identity matrix. □

It raises the questions as to how does a matrix representation of a linear transformation change and how do the coordinate vectors change when bases are changed. We consider the second question first. Let V be a vector space of dimension n. Let $B = \{u_1, \ldots, u_n\}$ and $C = \{v_1, \ldots, v_n\}$ be ordered bases of V. Let $x \in V$. Suppose

$$x = \alpha_1 u_1 + \cdots + \alpha_n u_n \quad \text{and} \quad x = \beta_1 v_1 + \cdots + \beta_n v_n.$$

Then what is the connection between the scalars α_i's and the β_j's? Here, the same vector x has two different representations in terms of the bases B and C. So the linear transformation which takes x to itself, the identity linear transformation, corresponds to the linear transformation which takes $[\alpha_1, \ldots, \alpha_n]^T$ to $[\beta_1, \ldots, \beta_n]^T$ from $\mathbb{F}^{n\times 1}$ to itself. Let us write V_B for the vector space V where we consider the basis B; similarly V_C. Then the commutative diagram considered after Example 2.40 looks as follows:

$$\begin{array}{ccc} V_B & \xrightarrow{\quad I \quad} & V_C \\ \phi_B \downarrow \simeq & & \simeq \downarrow \psi_C \\ \mathbb{F}^{n\times 1} & \xrightarrow[{[I]_{C,B}}]{} & \mathbb{F}^{n\times 1} \end{array}$$

This means that $\psi_C(x) = [I]_{C,B}\phi_B I^{-1}(x) = [I]_{C,B}\phi_B(x)$. Since the canonical basis isomorphisms give simply the coordinate vectors, we have:

$$[x]_C = [I(x)]_C = [I]_{C,B}[x]_B. \tag{2.1}$$

We thus call the matrix $[I]_{C,B}$ as the **change of basis matrix**. This is the same matrix that you get by expressing the basis vectors in B as linear combination of basis vectors in C due to Definition 2.38.

However, if bases are the same in both the domain space and the co-domain space, then the matrix representation of the identity transformation remains as the identity matrix. That is, for each basis B, $[I]_{B,B} = I$.

Example 2.47 We continue Example 2.46, where

$$B = \{(1, 1, 1), (1, 0, 1), (1, 1, 0)\} \text{ and } E = \{(1, 0, 0), (0, 1, 0), (0, 0, 1)\}.$$

Let $x = (3, 2, 2) \in \mathbb{R}^3$. We see that

$$(3, 2, 2) = 3\,(1, 0, 0) + 2\,(0, 1, 0) + 2\,(0, 0, 1) = 1\,(1, 1, 1) + 1\,(1, 0, 1) + 1\,(1, 1, 0).$$

Therefore

$$[x]_E = \begin{bmatrix} 3 \\ 2 \\ 2 \end{bmatrix} = \begin{bmatrix} 1 & 1 & 1 \\ 1 & 0 & 1 \\ 1 & 1 & 0 \end{bmatrix} \begin{bmatrix} 1 \\ 1 \\ 1 \end{bmatrix} = [I]_{E,B}[x]_B. \qquad \square$$

In Example 2.47, the change of basis matrix $[I]_{E,B}$ has the columns as the basis vectors in B. In general, let $E = \{e_1, \ldots, e_n\}$ be the standard basis of $\mathbb{F}^{n\times 1}$ and let $B = \{v_1, \ldots, v_n\}$ be an ordered basis of $\mathbb{F}^{n\times 1}$. Let $i \in \{1, \ldots, n\}$. Then the coordinate vector of v_i with respect to B is the ith standard basis vector, i.e. $[v_i]_B = e_i$. Also, $[v_i]_E = v_i$. Thus, $[I]_{E,B}(e_i) = [I]_{E,B}\,[v_i]_B = [I(v_i)]_E = [v_i]_E = v_i$. Therefore, the ith column of $[I]_{E,B}$ is equal to v_i. We summarize:

> If $B = \{v_1, \ldots, v_n\}$ is any basis for $\mathbb{F}^{n\times 1}$ and E is the standard basis of $\mathbb{F}^{n\times 1}$, then $[I]_{E,B} = [v_1 \;\cdots\; v_n]$, i.e. the ith column of $[I]_{E,B}$ is the ith basis vector in B.

In fact, if P is any invertible $n \times n$ matrix, then its columns form a basis of $\mathbb{F}^{n\times 1}$. Then such a matrix P is simply the change of basis matrix $[I]_{E,B}$, where B is the basis consisting of the columns of P and E is the standard basis of $\mathbb{F}^{n\times 1}$. We obtain the following result.

Theorem 2.48 *Change of basis matrices are precisely the invertible matrices.*

The inverses of change of basis matrices are also change of basis matrices. Further, $[I]_{E,B}[I]_{B,E} = [I]_{E,E}$ is the identity matrix. Therefore, the change of basis matrices $[I]_{E,B}$ and $[I]_{B,E}$ are inverses of each other.

Thus, in addition, if $C = \{u_1, \ldots, u_n\}$ is another ordered basis for $\mathbb{F}^{n\times 1}$, then

$$[I]_{C,B} = [I]_{C,E}[I]_{E,B} = [I]_{E,C}^{-1}[I]_{E,B}. \tag{2.2}$$

Similarly, $[I]_{B,C} = [I]_{E,B}^{-1}[I]_{E,C}$. We then obtain

$$[I]_{C,B} = [I]_{B,C}^{-1}. \tag{2.3}$$

Example 2.49 Consider the ordered bases $B = \left\{\begin{bmatrix}1\\1\end{bmatrix}, \begin{bmatrix}1\\-1\end{bmatrix}\right\}$ and $C = \left\{\begin{bmatrix}1\\0\end{bmatrix}, \begin{bmatrix}1\\1\end{bmatrix}\right\}$ of $\mathbb{F}^{2\times 1}$. Let E be the standard basis of $\mathbb{F}^{2\times 1}$. The change of basis matrices are as follows.

$$[I]_{E,B} = \begin{bmatrix}1 & 1\\1 & -1\end{bmatrix}, \quad [I]_{B,E} = [I]_{E,B}^{-1} = \tfrac{1}{2}\begin{bmatrix}1 & 1\\1 & -1\end{bmatrix}.$$

$$[I]_{E,C} = \begin{bmatrix}1 & 1\\0 & 1\end{bmatrix}, \quad [I]_{C,E} = [I]_{E,C}^{-1} = \begin{bmatrix}1 & -1\\0 & 1\end{bmatrix}.$$

$$[I]_{B,C} = [I]_{B,E}[I]_{E,C} = \tfrac{1}{2}\begin{bmatrix}1 & 1\\1 & -1\end{bmatrix}\begin{bmatrix}1 & 1\\0 & 1\end{bmatrix} = \tfrac{1}{2}\begin{bmatrix}1 & 2\\1 & 0\end{bmatrix}.$$

$$[I]_{C,B} = [I]_{C,E}[I]_{E,B} = \begin{bmatrix}1 & -1\\0 & 1\end{bmatrix}\begin{bmatrix}1 & 1\\1 & -1\end{bmatrix} = \begin{bmatrix}0 & 2\\1 & -1\end{bmatrix} = [I]_{C,B}^{-1}.$$

Equation 2.1 says that $[x]_C = [I]_{C,B}[x]_B$ and $[x]_B = [I]_{B,C}[x]_C$.

In order to verify this, let $x = \begin{bmatrix} a \\ b \end{bmatrix}$. Suppose $[x]_B = \begin{bmatrix} \alpha \\ \beta \end{bmatrix}$ and $[x]_C = \begin{bmatrix} \gamma \\ \delta \end{bmatrix}$. Then

$$\begin{bmatrix} a \\ b \end{bmatrix} = \alpha \begin{bmatrix} 1 \\ 1 \end{bmatrix} + \beta \begin{bmatrix} 1 \\ -1 \end{bmatrix}, \quad \begin{bmatrix} a \\ b \end{bmatrix} = \gamma \begin{bmatrix} 1 \\ 0 \end{bmatrix} + \delta \begin{bmatrix} 1 \\ 1 \end{bmatrix}.$$

The first equation gives $a = \alpha + \beta,\ b = \alpha - \beta$; that is, $\alpha = \frac{a+b}{2}$ and $\beta = \frac{a-b}{2}$. The second equation leads to $\gamma = a - b$ and $\delta = b$. Therefore,

$$[x]_B = \begin{bmatrix} \alpha \\ \beta \end{bmatrix} = \tfrac{1}{2} \begin{bmatrix} a+b \\ a-b \end{bmatrix}, \quad [x]_C = \begin{bmatrix} a-b \\ b \end{bmatrix}.$$

We then find that

$$[I]_{C,B}[x]_B = \begin{bmatrix} 0 & 2 \\ 1 & -1 \end{bmatrix} \tfrac{1}{2} \begin{bmatrix} a+b \\ a-b \end{bmatrix} = \tfrac{1}{2} \begin{bmatrix} 2(a-b) \\ 2b \end{bmatrix} = [x]_C.$$

$$[I]_{B,C}[x]_C = \frac{1}{2} \begin{bmatrix} 1 & 2 \\ 1 & 0 \end{bmatrix} \begin{bmatrix} a-b \\ b \end{bmatrix} = \frac{1}{2} \begin{bmatrix} a+b \\ a-b \end{bmatrix} = [x]_B.$$

□

We now consider the first question as to how does a change of bases affects the matrix representation of a linear transformation.

Theorem 2.50 *Let V and W be finite dimensional vector spaces, and let $T : V \to W$ be a linear transformation. Let $B,\ B'$ be ordered bases for V, and let $C,\ C'$ be ordered bases for W. Write $P := [I_V]_{B,B'}$ and $Q := [I_W]_{C,C'}$. Then $[T]_{C',B'} = Q^{-1}[T]_{C,B}P$.*

Proof Let $T : V \to W$ be a linear transformation. Then $T = I_V T I_W$. By Theorem 2.43 and (2.3), we obtain

$$\begin{aligned}[T]_{C',B'} &= [I_W T I_V]_{C',B'}, = [I_W\, T]_{C',B}[I_V]_{B,B'} = [I_W]_{C',C}[T]_{C,B}[I_V]_{B,B'} \\ &= [I_W]^{-1}_{C,C'}[T]_{C,B}[I_V]_{B,B'} = Q^{-1}[T]_{C,B}P.\end{aligned}$$

■

To understand Theorem 2.50, a diagram may be helpful. We write the linear transformation and its matrix representation on different sides of the arrow.

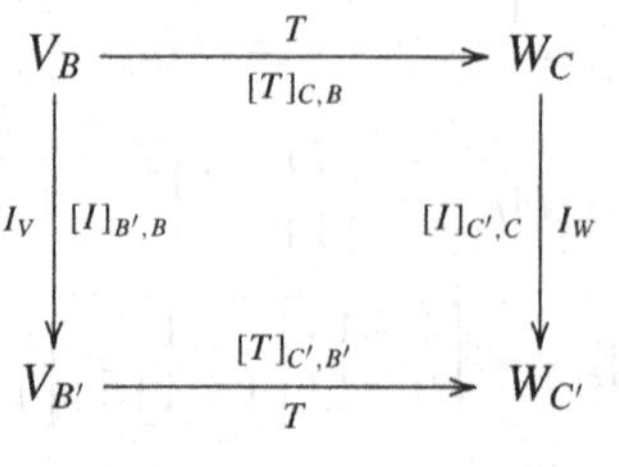

$$\begin{array}{ccc} V_B & \xrightarrow[\ [T]_{C,B}\ ]{T} & W_C \\ I_V \Big\downarrow [I]_{B',B} & & [I]_{C',C} \Big\downarrow I_W \\ V_{B'} & \xrightarrow[\ T\ ]{[T]_{C',B'}} & W_{C'} \end{array}$$

It shows that $[T]_{C',B'} = [I]_{C',C}\,[T]_{C,B}\,[I]_{B',B}^{-1} = [I]_{C,C'}^{-1}[T]_{C,B}\,[I]_{B,B'}$.

In Theorem 2.50, take $C = B$ and $C' = B' = E$ to obtain the following result.

Theorem 2.51 *Let T be a linear operator on a finite dimensional vector space V. Let B and E be ordered bases of V. Write $P := [I_V]_{B,E}$. Then $[T]_{E,E} = P^{-1}[T]_{B,B}P$.*

To interpret Theorem 2.50 for matrices, let $V = \mathbb{F}^{n\times 1}$ and $W = \mathbb{F}^{m\times 1}$ have their standard bases, say, B and C, respectively. Look at a given matrix $A \in \mathbb{F}^{m\times n}$ as a linear transformation from $\mathbb{F}^{n\times 1}$ to $\mathbb{F}^{m\times 1}$. The columns of A are the images of the standard basis vectors $Ae_j \in \mathbb{F}^{n\times 1}$. Suppose we change the bases of both the spaces, say, $B' = \{v_1, \ldots, v_n\}$ for $\mathbb{F}^{n\times 1}$ and $C' = \{u_1, \ldots, u_m\}$ for $\mathbb{F}^{m\times 1}$ are the new bases. Since B and C are the standard bases, we have

$$[I]_{B,B'} = [v_1 \ \cdots \ v_n], \quad [I]_{C,C'} = [w_1 \ \cdots \ w_m].$$

The effect of change of bases is as follows:

$$[A]_{C',B'} = [I]_{C',C}\,[A]_{C,B}\,([I]_{B',B})^{-1} = [w_1 \ \cdots \ w_m]^{-1} A\,[v_1 \ \cdots \ v_n].$$

Observe that this result is simply a matrix interpretation of Theorem 2.50. We note down this result for future use. In the following statement, we write the new bases for $\mathbb{F}^{n\times 1}$ and $\mathbb{F}^{m\times 1}$ as B and C. The standard bases are used by default when a matrix is viewed as a linear transformation.

Theorem 2.52 *Let $A \in \mathbb{F}^{m\times n}$. Let $B = \{v_1, \ldots, v_n\}$ and $C = \{w_1, \ldots, w_m\}$ be ordered bases for $\mathbb{F}^{n\times 1}$ and $\mathbb{F}^{m\times 1}$, respectively. Write $P := [v_1 \ \cdots \ v_n]$ and $Q := [w_1 \ \cdots \ w_m]$. Then $[A]_{C,B} = Q^{-1} A\, P$.*

Notice that if $T = I$ with $m = n$, then $[I]_{C,B} = Q^{-1}P$ as obtained earlier.

Converse of Theorem 2.52 also holds. That is, for a given matrix $A \in \mathbb{F}^{m\times n}$, and given invertible matrices $P \in \mathbb{F}^{n\times n}$ and $Q \in \mathbb{F}^{m\times m}$ there exist bases B for $\mathbb{F}^{n\times 1}$ and C for $\mathbb{F}^{m\times 1}$ such that $[A]_{C,B} = Q^{-1}AP$. To prove this, take B as the ordered set of columns of P and C as the ordered set of columns of Q.

We interpret Theorem 2.51 for square matrices by taking B as the standard basis for $\mathbb{F}^{n\times 1}$ and E as another basis C. It yields the following result.

Theorem 2.53 *Let $A \in \mathbb{F}^{n\times n}$. Let $C = \{v_1, \ldots, v_n\}$ be an ordered basis for $\mathbb{F}^{n\times 1}$. Construct the matrix $P := [v_1 \ \cdots \ v_n]$. Then $[A]_{C,C} = P^{-1} A\, P$.*

In view of Theorems 2.52 and 2.53, we give the following definition.

Definition 2.54 Let A and M be matrices in $\mathbb{F}^{m\times n}$.

(a) A and M are called **equivalent** if there exist invertible matrices $P \in \mathbb{F}^{n\times n}$ and $Q \in \mathbb{F}^{m\times m}$ such that $M = Q^{-1}AP$.
(b) For $m = n$, the matrices A and M are called **similar** if there exists an invertible matrix $P \in \mathbb{F}^{n\times n}$ such that $M = P^{-1}AP$.

The symmetric terminology in the definition is justified since when $M = Q^{-1}AP$, we see that $A = QMP^{-1}$; if $M = P^{-1}AP$ then $A = PMP^{-1}$. Theorems 2.52 and 2.53 say that change of bases in the domain and co-domain spaces gives rise to equivalent matrices. Thus equivalent matrices represent the same linear transformation with respect to different pairs of ordered bases; similar matrices represent the same linear operator on a vector space with respect to different ordered bases.

Notice that both the relations of equivalence on $\mathbb{F}^{m\times n}$ and similarity of matrices on $\mathbb{F}^{n\times n}$ are equivalence relations (reflexive, symmetric, and transitive).

It is obvious that if two square matrices are similar, then they are equivalent. However, two equivalent square matrices need not be similar. For, the identity matrix is equivalent to any invertible matrix whereas it is similar to only itself; see the following example.

Example 2.55 Consider the 2×2 matrices

$$A = \begin{bmatrix} 1 & 0 \\ 0 & 1 \end{bmatrix} = I, \; B = \begin{bmatrix} 1 & 1 \\ 0 & 1 \end{bmatrix}, \; Q = \begin{bmatrix} 1 & -1 \\ 0 & 1 \end{bmatrix}, \text{ and } P = \begin{bmatrix} 1 & 0 \\ 0 & 1 \end{bmatrix} = I.$$

Since $BQ = QB = I$, $B = Q^{-1} = Q^{-1}AP$. That is B is equivalent to A.

If B is similar to A, then there exists a 2×2 invertible matrix R such that $B = R^{-1}AR = R^{-1}R = I$. But $B \neq I$. Hence B is not similar to A. □

Since two equivalent matrices represent the same linear transformation, they must have the same rank. Does the rank criterion alone suffice for equivalence?

Theorem 2.56 (Rank theorem) *Let V and W be finite dimensional vector spaces, and let $S, T : V \to W$ be linear transformations. Then* $\text{rank}\,(S) = \text{rank}\,(T)$ *if and only if there exist isomorphisms $P : V \to V$ and $Q : W \to W$ such that $T = Q^{-1}SP$.*

Proof Let $T = Q^{-1}SP$, where $Q : V \to V$ and $P : W \to W$ are isomorphisms. Then $QT = SP$. Due to Theorem 2.36(1)

$$\text{rank}\,(T) = \text{rank}\,(QT) = \text{rank}\,(SP) = \text{rank}\,(S).$$

Conversely, let $\text{rank}\,(T) = \text{rank}\,(S) = r$. Suppose $\dim(V) = n$ and $\dim(W) = m$. Necessarily, $n \geq r$ and $m \geq r$. Let $\{u_1, \ldots u_r\}$ and $\{v_1, \ldots, v_r\}$ be bases of $R(T)$ and $R(S)$, respectively. Let $x_i, y_i \in V$ be such that $Tx_i = u_i$ and $Sy_i = v_i$ for $i = 1, \ldots, r$.

Notice that $\text{null}\,(T) = \text{null}\,(S) = n - r$. So, let $\{x_{r+1}, \ldots, x_n\}$ and $\{y_{r+1}, \ldots, y_n\}$ be bases of $N(T)$ and $N(S)$, respectively. By Theorem 2.24, $\{x_1, \ldots x_r, x_{r+1}, \ldots, x_n\}$ and $\{y_1, \ldots, y_r, y_{r+1}, \ldots, y_n\}$ are bases of V. Since $R(T)$ is a subspace of W, there exist vectors $u_{r+1}, \ldots u_m, \; v_{r+1}, \ldots, v_m \in W$ such that $\{u_1, \ldots, u_r, u_{r+1}, \ldots u_m\}$ and $\{v_1, \ldots, v_r, v_{r+1}, \ldots, v_m\}$ are bases of W.

Using Theorem 2.8, define linear transformations $P : V \to V$ and $Q : W \to W$ by

$$Px_i = y_i \quad \text{and} \quad Q(u_j) = v_j$$

for $i = 1, \ldots, n$ and $j = 1, \ldots, m$. Both P and Q map bases onto bases; thus they are isomorphisms. We look at the actions of $Q^{-1}SP$ and T on the basis vectors x_i.

If $i \in \{1, \ldots, r\}$, then $Q^{-1}SPx_i = Q^{-1}Sy_i = Q^{-1}v_i = u_i = Tx_i$.

If $i \in \{r+1, \ldots, n\}$, then $Q^{-1}SPx_i = Q^{-1}Sy_i = Q^{-1}(0) = 0 = Tx_i$.

Since $\{x_1, \ldots, x_n\}$ is a basis of V, due to Theorem 2.8, $Q^{-1}SP = T$. ■

We interpret the rank theorem for matrices. Suppose $A, M \in \mathbb{F}^{m\times n}$. Then both are linear transformations from $\mathbb{F}^{n\times 1}$ to $\mathbb{F}^{m\times 1}$. If they are equivalent, then their ranks are equal. Conversely, if their ranks are equal, then there exist invertible matrices $P \in \mathbb{F}^{n\times n}$ and $Q \in \mathbb{F}^{m\times m}$ such that $M = Q^{-1}AP$. That is, they are equivalent. The rank theorem for matrices is stated as follows:

> Two matrices of the same size are equivalent if and only if they have the same rank.

It characterizes the equivalence of two matrices. Thus, any two invertible matrices of the same order are equivalent; they are equivalent to the identity matrix. However, two invertible matrices of the same order need not be similar; see Example 2.55.

It is relatively easy to construct an $m \times n$ matrix of rank r, since $r \le \min\{m, n\}$. We take the first r columns of the matrix as $e_1, \ldots, e_r$, the first r standard basis vectors of $\mathbb{F}^{m\times 1}$. And then take the other $n - r$ columns as zero vectors in $\mathbb{F}^{m\times 1}$. That is, define the matrix $E_r = [e_{ij}] \in \mathbb{F}^{m\times n}$ with

$$e_{11} = \cdots = e_{rr} = 1, \quad \text{and} \quad e_{ij} = 0, \text{ otherwise.}$$

We will call such a matrix E_r as the **rank echelon matrix** of size $m \times n$ and rank r. If the size $m \times n$ of E_r is not obvious from the context, we write it as $E_r^{m,n}$. For example, the rank echelon matrix E_2 of size 3×4 and its transpose $(E_2)^T$ of size 4×3 look like

$$E_2^{3,4} = \begin{bmatrix} 1 & 0 & 0 & 0 \\ 0 & 1 & 0 & 0 \\ 0 & 0 & 0 & 0 \end{bmatrix}, \quad (E_2^{3,4})^T = \begin{bmatrix} 1 & 0 & 0 \\ 0 & 1 & 0 \\ 0 & 0 & 0 \\ 0 & 0 & 0 \end{bmatrix}.$$

We observe that $\operatorname{rank}(E_r^T) = \operatorname{rank}(E_r) = r$.

By the rank theorem, every matrix of rank r is equivalent to the rank echelon matrix E_r of the same size. We will use rank echelon matrices for proving that the **row rank** of a matrix, that is, the dimension of the vector space spanned by the rows of the matrix, is same as its (column) rank.

Theorem 2.57 *The row rank of each matrix is equal to its rank. Further, a square matrix and its transpose are equivalent.*

Proof Let $A \in \mathbb{F}^{m\times n}$ have rank r. By the rank theorem, A is equivalent to E_r. That is, there exist invertible matrices $P \in \mathbb{F}^{n\times n}$ and $Q \in \mathbb{F}^{m\times m}$ such that $A = Q^{-1}E_rP$. Then $A^T = P^TE_r^T(Q^T)^{-1}$. Again, by the rank theorem,

$$\text{row-rank of } A = \text{rank}\,(A^T) = \text{rank}\,(E_r^T) = \text{rank}\,(E_r) = \text{rank}\,(A).$$

When $m = n$, the rank theorem implies that A^T and A are equivalent. ■

In fact, the transpose of a square matrix is similar to the matrix; but its proof uses a characterization of similarity, which is a difficult task worthy of being postponed for a while.

Exercises for Sect. 2.5

In the following exercises, consider any given basis as an ordered basis.

1. Let T be a linear operator on $\mathbb{R}^2$ such that all entries of $[T]$ with respect to the basis $\{(1,0),(0,1)\}$ are 1. What is $[T]_{B,B}$, where $B = \{(1,-1),(1,1)\}$?
2. Let T be a linear operator on $\mathbb{R}^3$ such that $Te_1 = (0,1,-1)$, $T(e_2) = (1,0,-1)$ and $T(e_3) = (-1,-1,0)$. Let $B = \{(0,1,-1),(1,0,-1),(-1,-1,0)\}$. What is $[T]_{B,B}$?
3. Show that $B = \{(1,0,i),(1+i,1,-1)\}$ and $C = \{(1,1,0),(1,i,1+i)\}$ are bases for span(B) as a subspace of $\mathbb{C}^3$. What are the coordinate vectors $[(1,0,i)]_C$ and $[(1,i,1+i)]_B$?
4. Determine the change of basis matrix in each of the following cases, considering the vector space as $\mathbb{R}^n$:

   (a) Old basis is $\{e_1,\ldots,e_n\}$ and new basis is $\{e_n,\ldots,e_1\}$.
   (b) Old basis is $\{e_1,\ldots,e_n\}$ and new basis is $\{e_1+e_2,\ldots,e_{n-1}+e_n\}$.
   (c) Old basis is $\{e_1-e_2,\ldots,e_{n-1}-e_n\}$ and new basis is $\{e_1,\ldots,e_n\}$.
   (d) For $n = 2$, old basis is $\{e_1, v_2\}$, where v_2 is a unit vector making an angle of $2\pi/3$ with e_1, and new basis is $\{e_2, e_1\}$.

5. Consider the standard basis E, and the basis $B = \{(1,2,3),(3,2,1),(0,0,1)\}$ for $\mathbb{R}^3$.

   (a) Compute the change of basis matrices $[I]_{B,E}$ and $[I]_{E,B}$.
   (b) Determine the matrices $[T]_{B,B}$, $[T]_{B,E}$, $[T]_{E,B}$, and $[T]_{E,E}$ for the linear operator T on $\mathbb{R}^3$ given by $T(a,b,c) = (6a+b, a-b-c, 2a-b+3c)$.

6. Let $B := \{u_1,\ldots,u_n\}$ and $C := \{v_1,\ldots,v_n\}$ be ordered bases for a vector space V. Let T be a linear operator on V defined by $Tu_1 = v_1,\ldots,Tu_n = v_n$. Show that $[T]_{B,B} = [I]_{C,B}$.
7. Given any invertible matrix $A \in \mathbb{F}^{n\times n}$, show that ordered bases B and C can be chosen for $\mathbb{F}^{n\times 1}$ such that $A = [I]_{C,B}$.
8. Give infinitely many matrices $A \in \mathbb{F}^{n\times n}$ where A is similar to only A.
9. Prove that the matrices $\begin{bmatrix} a & 0 \\ 0 & d \end{bmatrix}$ and $\begin{bmatrix} a & b \\ 0 & d \end{bmatrix}$ are similar if and only if $a \neq d$.

10. Let $\theta \in \mathbb{R}$. Show that the matrices $\begin{bmatrix} \cos\theta & -\sin\theta \\ \sin\theta & \cos\theta \end{bmatrix}$ and $\begin{bmatrix} e^{i\theta} & 0 \\ 0 & e^{-i\theta} \end{bmatrix}$ are similar in $\mathbb{C}^{2\times 2}$.
11. Let $\theta \in \mathbb{R}$. Let $A = \begin{bmatrix} \cos\theta & -\sin\theta \\ \sin\theta & \cos\theta \end{bmatrix}$ and $B = \begin{bmatrix} \cos\theta & \sin\theta \\ \sin\theta & -\cos\theta \end{bmatrix}$.

    (a) Are A and B similar in $\mathbb{R}^{2\times 2}$?
    (b) Are A and B similar in $\mathbb{C}^{2\times 2}$?

2.6 Space of Linear Transformations

Let V and W be vector spaces over $\mathbb{F}$, having dimensions n and m, respectively. Each linear transformation $T : V \to W$ has a matrix representation. Conversely each matrix in $\mathbb{F}^{m\times n}$ induces a linear transformation from V to W, provided we fix ordered bases for V and W a priori. It suggests that the *matrix representation* itself is some sort of isomorphism. To explore it further we need to see the set of all linear transformations as a vector space.

In general, if V and W are vector spaces over the same field $\mathbb{F}$, the set of all linear transformations from V to W, denoted by $\mathcal{L}(V, W)$, is a vector space over $\mathbb{F}$. See Example 2.7(3). We shall denote $\mathcal{L}(V, V)$ by $\mathcal{L}(V)$.

In fact, $\mathcal{L}(V)$ has more structure than a vector space, since a multiplication is well defined here. The multiplication of two maps in $\mathcal{L}(V)$ is their composition. This multiplication (composition of maps here) is associative, distributive over addition, and satisfies

$$\alpha(ST) = (\alpha S)T = S(\alpha T) \quad \text{for each } \alpha \in \mathbb{F}.$$

Moreover, the identity map I serves as the multiplicative identity in $\mathcal{L}(V)$, since $TI = IT = T$. Such a structure is called an *algebra*.

We show that if V and W are vector spaces of dimensions n and m, respectively, then $\mathcal{L}(V, W)$ and $\mathbb{F}^{m\times n}$ are isomorphic.

Theorem 2.58 *Let V and W be finite dimensional vector spaces. Let B and C be bases for V and W, respectively. Then the map $\chi : \mathcal{L}(V, W) \to \mathbb{F}^{m\times n}$ defined by $\chi(T) = [T]_{C,B}$ is an isomorphism; consequently, $\mathcal{L}(V, W) \simeq \mathbb{F}^{m\times n}$.*

Proof Let $B = \{v_1, \ldots, v_n\}$ and $C = \{w_1, \ldots, w_m\}$ be ordered bases for the vector spaces V and W, respectively. To keep notation simple, write $[T]$ in place of $[T]_{C,B}$ and $[x]$ instead of $[x]_B$ for any $x \in V$. We verify that $\chi : \mathcal{L}(V, W) \to \mathbb{F}^{m\times n}$ defined by $\chi(T) = [T]$ is a bijective linear transformation. For this, let $\alpha \in \mathbb{F}$, $S, T \in \mathcal{L}(V, W)$ and let $x \in V$. Now,

$$\begin{aligned}
[S+T][x] &= [(S+T)(x)] = [Sx + Tx] = [Sx] + [Tx] \\
&= [S][x] + [T][x] = ([S] + [T])[x], \\
[\alpha S][x] &= [\alpha Sx] = \alpha[Sx] = \alpha[S][x].
\end{aligned}$$

Hence, we have $[S+T]=[S]+[T]$ and $[\alpha S]=\alpha[S]$. That is,

$$\mathcal{X}(S+T)=\mathcal{X}(S)+\mathcal{X}(T), \quad \mathcal{X}(\alpha S)=\alpha\mathcal{X}(S).$$

Therefore, $\mathcal{X}$ is a linear transformation.

Let $T\in\mathcal{L}(V,W)$ be such that $\mathcal{X}(T)=0$, the zero matrix in $\mathbb{F}^{m\times n}$. Then $T=\psi_C^{-1}\,0\,\phi_B=0$, the zero operator, where ϕ_B and ψ_C are the canonical basis isomorphisms. Thus, $N(\mathcal{X})=\{0\}$. This shows that $\mathcal{X}$ is injective.

Now, to show that $\mathcal{X}$ is surjective, let $A\in\mathbb{F}^{m\times n}$. Consider the linear operator T_A induced by the matrix A as defined in Example 2.11(3). Then we see that $\mathcal{X}(T_A)=A$. Hence, $\mathcal{X}$ is onto.

Therefore, $\mathcal{X}$ is an isomorphism; consequently, $\mathcal{L}(V,W)\simeq\mathbb{F}^{m\times n}$. ■

Given ordered bases $\{v_1,\ldots,v_n\}$ for V and $\{w_1,\ldots,w_m\}$ for W, the isomorphism $\mathcal{X}$ can also be used to construct a basis for $\mathcal{L}(V,W)$. Since isomorphisms map bases onto bases, $\mathcal{X}^{-1}$ of the standard basis of $\mathbb{F}^{m\times n}$ does the job. This gives rise to the basis

$$\{T_{ij}: i=1,\ldots,m,\ j=1,\ldots,n\},$$

of $\mathcal{L}(V,W)$, where T_{ij} is the unique linear transformation from V to W defined by

$$T_{ij}(v_j)=w_i, \quad \text{and} \quad T_{ij}(v_k)=0 \text{ for } k\neq j.$$

This basis is called the **standard basis** of $\mathcal{L}(V,W)$ with respect to the ordered bases $\{v_1,\ldots,v_n\}$ for V, and $\{w_1,\ldots,w_m\}$ for W. Since $\mathcal{L}(V,W)\simeq\mathbb{F}^{m\times n}$, we see that $\dim(\mathcal{L}(V,W))=\dim(\mathbb{F}^{m\times n})=mn$. You can also prove this fact independently by showing that the set of linear maps $\{T_{ij}: 1\le i\le m, 1\le j\le m\}$ is a basis of $\mathcal{L}(V,W)$. This will be an alternative proof of $\mathcal{L}(V,W)\simeq\mathbb{F}^{m\times n}$.

The particular case of $W=\mathbb{F}$ leads to the space of linear functionals on V.

Definition 2.59 The space $\mathcal{L}(V,\mathbb{F})$ of all linear functionals on V is called the **dual space** of V; this space is denoted by V'.

Due to Theorem 2.58, if $\dim(V)=n$, then the dual space V' of V is isomorphic to $\mathbb{F}^n$. But V itself is isomorphic to $\mathbb{F}^n$. In this case, $V'\simeq V$. By taking help from the coordinate functionals (see Example 2.4), we can have alternate ways of proving the results about $\mathcal{L}(V,W)$. Recall that, given an ordered basis $\{v_1,\ldots,v_n\}$ of V, the corresponding coordinate functionals $f_i\in V'$ are defined by $f_i(v_j)=\delta_{ij}$ for $i,j=1\ldots,n$.

Theorem 2.60 *Let $B=\{v_1,\ldots,v_n\}$ be an ordered basis of a vector space V. Let $f_1,\ldots,f_n$ be the coordinate functionals on V with respect to B. Then the following are true:*

(1) *Every vector $x\in V$ can be written as $x=\sum_{j=1}^n f_j(x)v_j$.*
(2) *Let $v\in V$ be such that $f(v)=0$ for each $f\in V'$. Then $v=0$.*

(3) *The set* $\{f_1, \ldots, f_n\}$ *is a basis of* V'.

Proof (1) Let $x \in V$. Since B is a basis of V, there exist unique scalars $\alpha_1, \ldots \alpha_n$ such that $x = \sum_{j=1}^n \alpha_j v_j$. Due to the relation $f_i(v_j) = \delta_{ij}$,

$$f_i(x) = \sum_{j=1}^n \alpha_j f_i(v_j) = \alpha_i, \quad i = 1, \ldots, n.$$

Therefore, $x = \sum_{j=1}^n f_j(x) v_j$.
(2) It follows from (1).
(3) Let $f \in V'$. For each $x \in V$, using (1), we have

$$f(x) = \sum_{j=1}^n f_j(x) f(v_j) = \big(\sum_{j=1}^n f(v_j) f_j\big)(x).$$

That is, $f = \sum_{j=1}^n f(v_j) f_j$. Thus, $f \in \text{span}\{f_1, \ldots, f_n\}$.
Next, suppose that $\sum_{i=1}^n \alpha_i f_i = 0$. Then

$$0 = \sum_{i=1}^n \alpha_i f_i(v_j) = \sum_{i=1}^n \alpha_i \delta_{ij} = \alpha_j \quad \text{for } j = 1, \ldots, n.$$

Hence, $\{f_1, \ldots, f_n\}$ is linearly independent in V'.
Therefore, $\{f_1, \ldots, f_n\}$ is a basis of V'. ■

In view of Theorem 2.60, we have the following definition.

Definition 2.61 Let $B = \{v_1, \ldots, v_n\}$ be an ordered basis of the vector space V. Let $f_1, \ldots, f_n \in V'$ be such that $f_i(v_j) = \delta_{ij}$ for $i, j \in \{1, \ldots, n\}$. The ordered basis $B' := \{f_1, \ldots, f_n\}$ of V' is called the **dual basis** of V', corresponding to the basis B of V.

Theorem 2.60 implies that $\dim(V') = n = \dim(V)$. By Theorem 2.35, $V' \simeq V$.

Example 2.62 Consider the standard basis $\{e_1, \ldots, e_n\}$ for the vector space $\mathbb{F}^{n\times 1}$. The ith dual basis vector in the dual of $\mathbb{F}^{n\times 1}$ is that f_i which maps e_i to 1 and every other e_j to 0. Following the convention that $f_i(e_j)$ is just the product of f_i and e_j, we write f_j as the row vector, in which the jth component alone is 1 and all others are 0. That is, $f_j = e_j^T$. So that

$$f_j(e_i) = e_j^T e_i = \delta_{ji}.$$

It shows that the dual space of $\mathbb{F}^{n\times 1}$ is isomorphic to $\mathbb{F}^{1\times n}$. Notice that the coordinate vector of f_j in this dual basis is again e_j. Further, the dual space of the dual space of $\mathbb{F}^{n\times 1}$ can be identified with $\mathbb{F}^{n\times 1}$ and the dual basis of this space is again the standard basis of $\mathbb{F}^{n\times 1}$. Also, the coordinate vector of each e_j in the dual basis is e_j itself. □

Remember that the dual basis is constructed from the coordinate functionals. Therefore, the dual basis can be used to construct a basis for $\mathcal{L}(V, W)$.

Theorem 2.63 *Let $B = \{v_1, \ldots, v_n\}$ and $C = \{w_1, \ldots, w_m\}$ be ordered bases of the vector spaces V and W, respectively. Let $B' = \{f_1, \ldots, f_n\}$ and $C' = \{g_1, \ldots, g_m\}$ be the dual bases of V' and W' corresponding to B and C. For $i \in \{1, \ldots, m\}$ and $j \in \{1, \ldots, n\}$, define linear transformations $T_{ij} : V \to W$ by*

$$T_{ij}(x) := f_j(x)w_i, \quad \text{for } x \in V.$$

Then the following are true:

(1) *Every $T \in \mathcal{L}(V, W)$ can be written as $T = \sum_{i=1}^{m} \sum_{j=1}^{n} g_i(Tv_j)T_{ij}$.*
(2) *The set $\{T_{ij} : i = 1, \ldots, m, \ j = 1, \ldots, n\}$ is a basis of $\mathcal{L}(V, W)$.*

Proof (1) Let $T \in \mathcal{L}(V, W)$. Let $x \in V$. By Theorem 2.60, $x = \sum_{j=1}^{n} f_j(x)v_j$. Then

$$T(x) = \sum_{j=1}^{n} f_j(x)T(v_j) = \sum_{j=1}^{n} f_j(x)\Big(\sum_{i=1}^{m} g_i(Tv_j)w_i\Big)$$
$$= \sum_{i=1}^{m}\sum_{j=1}^{n} g_i(Tv_j)f_j(x)w_i = \sum_{i=1}^{m}\sum_{j=1}^{n} g_i(Tv_j)T_{ij}(x).$$

(2) Let $\sum_{i=1}^{m} \sum_{j=1}^{n} \alpha_{ij}T_{ij} = 0$ for scalars α_{ij}. Then, for each $x \in V$,

$$\sum_{i=1}^{m}\sum_{j=1}^{n} \alpha_{ij}T_{ij}(x) = \sum_{i=1}^{m}\sum_{j=1}^{n} \alpha_{ij}f_j(x)w_i = 0.$$

In particular, for each $k \in \{1, \ldots, n\}$,

$$\sum_{i=1}^{m}\sum_{j=1}^{n} \alpha_{ij}f_j(v_k)w_i = \sum_{i=1}^{m} \alpha_{ik}w_i = 0.$$

as $f_j(v_k) = \delta_{jk}$. Since $w_1, \ldots, w_m$ are linearly independent, we obtain $\alpha_{ik} = 0$ for $i = 1, \ldots, m$. This is true for each $k \in \{1, \ldots, n\}$. Thus T_{ij}s are linearly independent. Due to (1), they form a basis for $\mathcal{L}(V, W)$. ■

Once again, Theorem 2.63 implies that if $\dim(V) = n$ and $\dim(W) = m$, then $\dim(\mathcal{L}(V, W)) = mn$, and that $\mathcal{L}(V, W) \simeq \mathbb{F}^{m\times n}$.

Suppose V is an n-dimensional vector space. Then its dual V' has dimension n. Also, the dual of V' has the same dimension n. That is,

$$V \simeq V' \simeq V''.$$

Example 2.62 shows that the isomorphism between V and V'' is more natural than that of V and V' in the particular case of $V = \mathbb{F}^{n\times 1}$. To see it for any finite dimensional vector space V, suppose $v \in V$ and $f \in V'$. Then $f(v)$ is a scalar. If

$\phi \in V''$, then $\phi(f)$ is also a scalar. The natural isomorphism between V and V'' comes from equating these two scalars. That is, define $\chi : V \to V''$ by

$$\chi(v)(f) = f(v) \quad \text{for } f \in V', \ v \in V.$$

We must show that χ is an isomorphism. Towards this, let $v_1, v_2 \in V$ and $\alpha \in \mathbb{F}$. Then, for every $f \in V'$,

$$\begin{aligned}\chi(v_1 + \alpha v_2)(f) &= f(v_1 + \alpha v_2) = f(v_1) + \alpha f(v_2) \\ &= \chi(v_1)(f) + \alpha \chi(v_2)(f) = \big(\chi(v_1) + \alpha \chi(v_2)\big)(f)\end{aligned}$$

Thus, χ is a linear transformation.

Let $v \in N(\chi)$. That is, $v \in V$ with $\chi(v) = 0$. Then, for every $f \in V'$, we have

$$f(v) = \chi(v)(f) = 0.$$

That is, $f(v) = 0$ for all $f \in V'$. By Theorem 2.60(2), we conclude that $v = 0$. Therefore, $N(\chi) = \{0\}$ so that χ is injective. Since the dimensions of V and V'' are finite and equal, the injective linear map χ is an isomorphism.

We remark that the problem of identifying V and V'' for an infinite dimensional space is not easy, and it is known that the map $\chi : V \to V''$ need not be onto. Discussion on such issues is beyond the scope of this book.

Exercises for Sect. 2.6

1. Let $T \in \mathcal{L}(\mathbb{F}^{n\times 1}, \mathbb{F}^{m\times 1})$. Show that there exists a matrix $A \in \mathbb{F}^{m\times n}$ such that for each $x \in \mathbb{F}^{n\times 1}$, $T(x) = Ax$.
2. Consider the basis $\{(-1, -1, 1), (-1, 1, 1), (1, 1, 1)\}$ of $\mathbb{C}^3$. Let $\{f_1, f_2, f_3\}$ be the corresponding dual basis. Compute $f_i(0, 1, 0)$ for $i = 1, 2, 3$.
3. Let $f : \mathbb{C}^3 \to \mathbb{F}$ be defined by $f(a, b, c) = a + b + c$. Show that $f \in (\mathbb{C}^3)'$. Find a basis for $N(f)$.
4. Let f be a nonzero functional on a vector space V. Let $\alpha \in \mathbb{F}$. Does there exist a vector $v \in V$ such that $f(v) = \alpha$?
5. Establish a one-to-one correspondence between the set of all invertible $n \times n$ matrices and the set of all ordered bases for $\mathbb{F}^n$.
6. Let V be a vector space with $1 < \dim(V) < \infty$. Prove that neither the set of all invertible linear operators nor the set of all noninvertible linear operators on V is a subspace of $\mathcal{L}(V, V)$. What happens if $\dim(V) = 1$?
7. For $p(t) = \sum_{j=0}^{n} \alpha_j t^j \in \mathcal{P}_n(\mathbb{F})$, and any $(n + 1)$-tuples of scalars $(\beta_0, \beta_1, \ldots, \beta_n)$ define $f(p) = \sum_{j=0}^{n} \alpha_j \beta_j$. Prove that f is a linear functional on $\mathcal{P}_n(\mathbb{F})$. Conversely, show that each linear functional on $\mathcal{P}(\mathbb{F})$ can be obtained this way by a suitable choice of an $(n + 1)$-tuple of scalars.
8. Let $B = \{v_1, \ldots, v_n\}$ and $C = \{w_1, \ldots, w_m\}$ be ordered bases of the vector spaces V and W, respectively. Let $B' = \{f_1, \ldots, f_n\}$ and $C' = \{g_1, \ldots, g_m\}$ be the corresponding dual bases for V' and W'. Show the following:

(a) If $T \in \mathcal{L}(V, W)$, then $[T]_{C,B} = [(g_i(Tv_j)]$ for $i = 1, \ldots, m,\ j = 1, \ldots, n$.
(b) Let $\{A_{ij} : i = 1 \ldots, m;\ j = 1, \ldots, n\}$ be any basis of $\mathbb{F}^{m \times n}$. If $T_{ij} \in \mathcal{L}(V, W)$ is such that $[T_{ij}]_{C,B} = A_{ij}$, then $\{T_{ij} : i = 1 \ldots, m,\ j = 1, \ldots, n\}$ is a basis of $\mathcal{L}(V, W)$.

9. Let V be a finite dimensional vector space. Prove the following:

   (a) For each nonzero $v \in V$, there exists $f \in V'$ such that $f(v) \neq 0$.
   (b) For every pair of distinct vectors $u, v \in V$, there exists $f \in V'$ such that $f(u) \neq f(v)$.

2.7 Problems

1. Let $S,\ T : C[a, b] \to C[a, b]$ be defined by

$$[S(x)](t) = \int_a^t x(s)ds, \quad [(T(x)](t) = tx(t) \quad \text{for } x \in C[a, b],\ t \in [a, b].$$

   Show that the map $x \mapsto S(x) \circ T(x)$ is not a linear transformation.
2. Let T be a linear operator on a vector space V. If $T^2 = 0$, what can you say about $R(T)$ and $N(T)$?
3. Let S and T be linear operators on a finite dimensional vector space. Prove that rank $(ST) =$ rank $(T) - \dim(N(S) \cap R(T))$.
4. Let T be a nilpotent operator on a finite dimensional vector space V. Let $m \in \mathbb{N}$ and $v \in V$ be such that $T^{m-1}v \neq 0$ but $T^m v = 0$. Show that $\{v, Tv, \ldots, T^{m-1}v\}$ is linearly independent.
5. Let U be a subspace of a finite dimensional vector space V over $\mathbb{F}$. Let W be any vector space over $\mathbb{F}$. Let $T : U \to W$ be a linear transformation. Show that there exists a linear transformation $S : V \to W$ such that $Tu = Su$ for each $u \in U$. Does the conclusion hold when $\dim(V) = \infty$?
6. Do there exist square matrices A and B such that A is equivalent to B but A^2 is not equivalent to B^2?
7. Let T be a linear operator on a vector space V of dimension $n \geq 2$. Suppose $TS = ST$ for each linear operator S on V. Prove the following:

   (a) Let v be a nonzero vector in V. Then there exists a scalar α such that $Tv = \alpha v$. (Hint: If not, then extend $\{v, Tv\}$ to a basis of V. Show that the map $S : V \to V$ defined by $S(\alpha_1 v + \alpha_2 Tv + y) = \alpha_2 v$ is a linear operator satisfying $ST(v) = v$ but $TS(v) = 0$.)
   (b) Let v and w be nonzero vectors in V such that $Tv = \alpha v$ and $Tw = \beta w$ for some scalars α and β. Then $\alpha = \beta$. (Hint: Consider (i) $\{v, w\}$ is linearly independent; (ii) $\{v, w\}$ is linearly dependent.)

8. Let T be a linear operator on a finite dimensional vector space V. Prove that if $TS = ST$ for each linear operator S on V, then there exists a unique scalar α such that $T = \alpha I$. (Hint: Use the results of Exercise 7.)
9. Prove that if $B \in \mathbb{F}^{m\times m}$ is such that $AB = BA$ for each invertible $A \in \mathbb{F}^{m\times m}$, then B is a scalar matrix.
10. Let T be a linear operator on a vector space V. Prove that $N(T) \cap R(T) = \{0\}$ if and only if $T^2 v = 0$ implies that $Tv = 0$ for each $v \in V$.
11. Let T be a linear operator on a vector space V. Prove that if rank $(T^2) =$ rank (T) then $N(T) \cap R(T) = \{0\}$. Does the converse hold?
12. $\mathbb{C}$ is a vector space over $\mathbb{C}$, as usual. Denote by $\mathbb{C}_R$ the vector space $\mathbb{C}$ over the field $\mathbb{R}$. Give an example of a linear operator on $\mathbb{C}_R$ which is not a linear operator on $\mathbb{C}$.
13. Let $\mathbb{C}_{\mathbb{R}}$ be the vector space of all complex numbers over the field $\mathbb{R}$. Define $T : \mathbb{C}_R \to \mathbb{R}^{2\times 2}$ by $T(a+ib) = \begin{bmatrix} a+7b & 5b \\ -10b & a-7b \end{bmatrix}$. Answer the following:

    (a) Is T injective?
    (b) Is it true that $T(xy) = T(x)T(y)$ for all $x, y \in \mathbb{C}_R$?
    (c) How do you describe $R(T)$?

14. Let A and B be nonzero matrices in $\mathbb{R}^{n\times n}$. For $x = (x_1, \ldots, x_n)^T$ and $y = (y_1, \ldots, y_n)^T$ in $\mathbb{R}^{n\times 1}$, define $xy = (x_1 y_1, \ldots, x_n y_n)^T$. Prove that there exists $z \in \mathbb{R}^{n\times 1}$ such that $(BA)(x) \neq (Bx)(Ax)$.
15. Let $\mathbb{R}^\infty$ be the vector space of all sequences of real numbers. Let S_ℓ and S_r be the left shift and right shift linear operators on $\mathbb{R}^\infty$, i.e.

$$S_\ell(a_1, a_2, \ldots) = (a_2, a_3, \ldots), \quad S_r(a_1, a_2, \ldots) = (0, a_1, a_2, \ldots).$$

    Prove that null $(S_\ell) > 0$, $R(S_\ell) = \mathbb{R}^\infty$, null $(S_r) = 0$ and $R(S_r) \neq \mathbb{R}^\infty$.
16. Let $A_t = \begin{bmatrix} \sin 2\pi t & \sin(\pi t/6) \\ \cos 2\pi t & \cos(\pi t/6) \end{bmatrix}$ for $0 \le t \le 12$. Determine rank (A_t) for each t. Determine the values of t for which rank $(A_t) = 1$.
17. Show that if $M = AB - iBA$, then $M^3 + M^2 + M = 0$, where

$$A = \begin{bmatrix} 0 & 1 & 0 & 0 \\ 0 & 0 & 1 & 0 \\ 0 & 0 & 0 & 1 \\ 1 & 0 & 0 & 0 \end{bmatrix} \quad \text{and} \quad B = \begin{bmatrix} i & 0 & 0 & 0 \\ 0 & -1 & 0 & 0 \\ 0 & 0 & -i & 0 \\ 0 & 0 & 0 & 1 \end{bmatrix}.$$

18. Let E_{ij} denote the matrix in $\mathbb{F}^{n\times n}$ whose (i, j)th entry is 1 and all other entries are 0. Show that $E_{ij}E_{k\ell} = 0$ if $j \neq k$, and $E_{ij}E_{j\ell} = E_{i\ell}$.
19. Let $A, E \in \mathbb{F}^{m\times m}$, $B, F \in \mathbb{F}^{m\times n}$, $C, G \in \mathbb{F}^{n\times m}$, and $D, H \in \mathbb{F}^{n\times n}$. Show that

$$\begin{bmatrix} A & B \\ C & D \end{bmatrix} \begin{bmatrix} E & F \\ G & H \end{bmatrix} = \begin{bmatrix} AE+BG & AF+BH \\ CE+DG & CF+DH \end{bmatrix}.$$

    Describe how this formula can be used to multiply two matrices in $\mathbb{F}^{r\times r}$.

20. Let V and W be finite dimensional vector spaces. Prove the following:

    (a) There exists an injective linear transformation from V to W if and only if $\dim(V) \le \dim(W)$.
    (b) There exists a surjective linear transformation from V to W if and only if $\dim(V) \ge \dim(W)$.

21. Let $A \in \mathbb{F}^{m\times n}$, where $m < n$. Show that there exist nonzero scalars $\alpha_1, \ldots, \alpha_n$ such that $A[\alpha_1, \ldots, \alpha_n]^T = 0$.
22. Using Exercise 21, give an alternate proof for Theorem 1.38.
23. Let T be a linear operator on a finite dimensional vector space. Suppose for an $m \in \mathbb{N}$, $R(T^{m-1}) = R(T^m)$. Show that $R(T^{m-1}) = R(T^k)$ for each $k \ge m$.
24. Let $S : U \to V$ and $T : V \to W$ be linear transformations such that TS is an isomorphism. Prove that T is injective and S is surjective.
25. Let T be a linear operator on a vector space V. Prove that if there exists a unique linear operator S on V such that $ST = I$, then T is an isomorphism.
26. Let T be a linear operator on a finite dimensional vector space V. Show that if there exist distinct linear operators S_1, S_2 on V such that $S_1 T = I = S_2 T$, then T need not be an isomorphism.
27. Let U be a subspace of an infinite dimensional vector space V. Show that even if $\dim(U) = \dim(V)$, U and V may not be isomorphic.
28. Let V be a vector space of dimension n. Let S and T be linear operators on V. Prove that there exist bases B and C for V such that $[S]_B = [T]_C$ if and only if there exists an isomorphism P on V such that $T = PSP^{-1}$.
29. Let S and T be linear operators on a finite dimensional vector space V. Prove the following:

    (a) ST is an isomorphism if and only if both S and T are isomorphisms.
    (b) $ST = I$ if and only if $TS = I$.

30. Let $A, B \in \mathbb{F}^{n\times n}$. Prove that if $AB = I$, then $BA = I$.
31. Let S and T be linear operators on a vector space V. Is it true that both S and T are invertible if and only if both ST and TS are invertible?
32. Let S and T be isomorphisms between finite dimensional vector spaces U and V. Is $S + T$ an isomorphism?
33. Let S and T be linear transformations from a finite dimensional vector space U to a finite dimensional vector space V. If S and T are not isomorphisms, is it possible that $S + T$ is an isomorphism?
34. Let $T : V \to W$ be a linear transformation where $\dim(V) < \infty$. Show that there exists a subspace U of V such that $R(T) = T(U)$ and $U \cap N(T) = \{0\}$.
35. Let V be a finite dimensional vector space, U be a subspace of V, and let T be a linear operator on V. Prove that $\dim(T(U)) \ge \dim(U) - \text{null}\,(T)$.
36. Let V and W be finite dimensional vector spaces. Let U be a subspace of V. Show that $\dim(U) + \dim(W) \ge \dim(V)$ if and only if there exists a linear transformation $T : V \to W$ such that $U = N(T)$.

37. Let $T : V \to W$ be a linear transformation, where both rank (T) and null (T) are finite. Show that V is finite dimensional. (Note: Since $\dim(V) < \infty$ is not assumed, you cannot use the formula $\dim(V) = \text{rank}\,(T) + \text{null}\,(T)$.)
38. Prove Theorem 2.42.
39. Let T be a linear operator on a finite dimensional vector space V. Prove that there exists an $n \in \mathbb{N}$ such that $N(T^n) \cap R(T^n) = \{0\}$.
40. Let V be a vector space of dimension n. Prove or give counter-examples:

    (a) If S and T are linear operators on V with $ST = 0$, then $\text{rank}\,(S) + \text{rank}\,(T) \le n$.
    (b) For each linear operator T on V, there exists a linear operator S on V such that $ST = 0$ and $\text{rank}\,(S) + \text{rank}\,(T) = n$.

41. Let S and T be linear operators on $\mathcal{P}$, where S is invertible. Show that for any polynomial $p(t) \in \mathcal{P}$, $S^{-1}p(T)S = p(S^{-1}TS)$.
42. Let $\{u_1, \ldots, u_m\}$ and $\{v_1, \ldots, v_m\}$ be linearly independent sets of vectors in an n-dimensional vector space V. Does there exist an invertible operator on V with $Au_j = v_j$ for each $j \in \{1, \ldots, m\}$?
43. Let T be a linear operator on a vector space. Show that if $T^2 - T + I = 0$, then T is invertible.
44. Let $\{u_1, \ldots, u_n\}$ be a basis for a vector space U. Let $v_1, \ldots, v_n$ be distinct vectors in any vector space V. We know that there exists a linear transformation $T : U \to V$ with $T(u_i) = v_i$ for $i = 1, 2, \ldots, n$. Prove or disprove:

    (a) T is one-to-one if and only if $\{v_1, \ldots, v_n\}$ is linearly independent.
    (b) T is onto if and only if $\text{span}\{v_1, \ldots, v_n\} = V$.
    (c) T is unique if and only if $\{v_1, \ldots, v_n\}$ is a basis of V.

45. If a linear operator T on a finite dimensional vector space V is invertible, then show that $T^{-1} = p(T)$ for some polynomial $p(t)$.
46. Let $V = U_1 \oplus \cdots \oplus U_k$. For each $i \in \{1, \ldots, k\}$, let B_i be a basis for U_i, and let T_i be a linear operator on U_i. Define the *direct sum of the maps* T_i by $T : V \to V$, where $T(u_1 + \cdots + u_k) = T_1(u_1) + \cdots T_k(u_k)$. Write $A_i = [T_i]_{B_i,B_i}$ and $B = B_1 \cup \cdots \cup B_k$. Show that $[T]_{B,B} = \text{diag}(A_1, \ldots, A_k)$.
47. Let $A, B \in \mathbb{F}^{n\times n}$ satisfy $A + B = 2B^T$ and $3A + 2B = I$. Determine A and B.
48. Let P be a linear operator on a finite dimensional vector space V such that $P^2 = P$. Prove that $\text{tr}(P) = \text{rank}\,(P)$.
49. A linear operator T is called *idempotent* when $T^2 = T$.

    (a) Give an example to show that $T^2(T - I) = 0$ but T is not idempotent.
    (b) Give an example to show that $(T - I)T^2 = 0$ but T is not idempotent.
    (c) Prove that $T^2(T - I) = 0 = (T - I)T^2$ implies that T is idempotent.

50. Show that if $A, B \in \mathbb{C}^{m\times m}$ with at least one of them invertible, then AB and BA are similar matrices. Is the result still valid when neither A nor B is invertible?
51. Let $A \in \mathbb{R}^{n\times n}$. Prove that $\text{rank}\,(A) \le 1$ if and only if there exist vectors $x, y \in \mathbb{R}^{n\times 1}$ such that $A = xy^T$.

52. Let $A, B \in \mathbb{R}^{n\times n}$. Suppose that there exists an invertible matrix $P \in \mathbb{C}^{n\times n}$ such that $P^{-1}AP = B$. Does there exist an invertible matrix $Q \in \mathbb{R}^{n\times n}$ such that $Q^{-1}AQ = B$? In other words, is it true that if two real matrices are similar over $\mathbb{C}$, then they are similar over $\mathbb{R}$?
53. Prove or disprove: If T is a linear operator of rank 1 then there exists a unique scalar β such that $A^2 = \beta A$.
54. Let S and T be linear operators on a finite dimensional vector space V. Prove or disprove:

    (a) Suppose $S^2 = S$ and $T^2 = T$. Then S and T are similar if and only if $\text{rank}\,(S) = \text{rank}\,(T)$.
    (b) Suppose $S \neq 0,\ T \neq 0$ and $S^2 = T^2 = 0$. Then S and T are similar if and only if $\text{rank}\,(S) = \text{rank}\,(T)$.

55. Let $B = \{v_1, \ldots, v_n\}$ be a basis for a vector space V. Let T be a linear operator on V satisfying $T(v_1) = v_2, \ldots, T(v_{n-1}) = v_n$ and $T(v_n) = 0$.

    (a) What is $[T]_B$?
    (b) Prove that T is nilpotent with its index of nilpotency n.
    (c) Prove that if S is a nilpotent operator on V with index of nilpotency as n, then there exists a basis C for V such that $[S]_C = [T]_B$.
    (d) Deduce that if $A, B \in \mathbb{F}^{n\times n}$ are nilpotent matrices with index of nilpotency as n, then A and B are similar matrices.

56. Let S and T be linear operators on a finite dimensional vector space. Show that if $S(ST - TS) = (ST - TS)S$, then $S^kT - TS^k = kS^{k-1}(ST - TS)$ for each $k \in \mathbb{N}$.
57. Let U be a subspace of a vector space V. Define the map $P : V \to V/U$ by $P(v) = v + U$ for $v \in V$. Prove the following:

    (a) The map P is a surjective linear transformation.
    (b) Let W be a vector space and let $T : V \to W$ be a linear transformation such that $N(T) = U$. Then there exists a unique linear transformation $S : V/U \to W$ such that $T = S \circ P$.

58. Let D be the differentiation operator on $\mathcal{P}_n$. Let T be a linear operator on $\mathcal{P}_n$ satisfying $T(p(t)) = p(t+1)$. Then show that $T = I + D + \frac{D^2}{2!} + \cdots + \frac{D^{n-1}}{(n-1)!}$.
59. Let V be a finite dimensional vector space. Fix a vector $u \in V$ and a linear functional f on V. Define the operator T on V by $Tx = f(x)u$. Find a polynomial $p(t)$ such that $p(T) = 0$.
60. Let $\{v_1, \ldots, v_n\}$ be a basis for a vector space V. Let $\alpha_1, \ldots, \alpha_n$ be distinct scalars. Let A be a linear operator on V with $Av_j = \alpha_j v_j$ for $j \in \{1, \ldots, n\}$. Suppose B is a linear operator on V such that $AB = BA$. Does it guarantee the existence of scalars β_j such that $Bv_j = \beta_j v_j$ for $j \in \{1, \ldots, n\}$?

61. Let $k, m, n \in \mathbb{N}$. Let $A \in \mathbb{F}^{k \times m}$. Define $T : \mathbb{F}^{m \times n} \to \mathbb{F}^{k \times n}$ by $T(X) = AX$. Show the following:

    (a) If $m < n$, then T can be surjective but not injective.
    (b) If $m > n$, then T can be injective but not surjective.
    (c) T is invertible if and only if $k = m$ and A is invertible.

62. Let A, B, C, D be linear operators on a finite dimensional vector space V. Prove or disprove: If both $A + B$ and $A - B$ are invertible, then there exist linear operators S and T on V such that $AS + BT = C$ and $BS + AT = D$.
63. Let A be a linear operator on a finite dimensional vector space V. Define a map $\phi : \mathcal{L}(V, V) \to \mathcal{L}(V, V)$ by $\phi(X) = AX$.

    (a) Prove that ϕ is a linear operator on $\mathcal{L}(V, V)$.
    (b) Under what conditions on A, is ϕ invertible?
    (c) Can each operator on $\mathcal{L}(V, V)$ be obtained this way by fixing A?

64. Let f and g be linear functionals on a vector space V such that $f(v) = 0$ if and only if $g(v) = 0$ for every $v \in V$. Prove that $g = \alpha f$ for some scalar α.
65. Let $f : V \to \mathbb{F}$ be a linear functional and let $v \in V \setminus N(f)$. Show that $V = N(f) \oplus \{\alpha v : \alpha \in \mathbb{F}\}$.
66. Let V be a vector space of dimension n and let $f_1, \ldots, f_m \in V'$, where $m < n$. Determine the conditions on the scalars $\alpha_1, \ldots, \alpha_n$ to guarantee the existence of a vector $x \in V$ such that $f_j(x) = \alpha_j$ for $j \in \{1, \ldots, m\}$. Interpret the result for linear systems.
67. Let V be a vector space of dimension n. Call an $(n - 1)$-dimensional subspace of V a *hyperspace*. Prove that if U is a k-dimensional subspace of V, then U is the intersection of $n - k$ hyperspaces.
68. In general, a *hyperspace* in a vector space v is a maximal proper subspace of V. Prove the following:

    (a) If f is a nonzero linear functional on V, then $N(f)$ is a hyperspace in V.
    (b) Each hyperspace in V is the null space of some linear functional on V.
    (c) Such a linear functional in (b) need not be unique.

69. Let $t_1, \ldots, t_n \in \mathbb{R}$ be distinct. For any $p(t) \in \mathcal{P}_{n-1}(\mathbb{R})$, let $L_i(p) = p(t_i)$ for each $i \in \{1, \ldots, n\}$. Let

$$p_j(t) := \frac{(t - t_1) \cdots (t - t_{j-1})(t - t_{j+1}) \cdots (t - t_n)}{(t_j - t_1) \cdots (t_j - t_{j-1})(t_j - t_{j+1}) \cdots (t_j - t_n)} \quad \text{for } j \in \{1, \ldots, n\},$$

    Prove the following:

    (a) $\{p_1, \ldots, p_n\}$ is a basis of $\mathcal{P}_{n-1}(\mathbb{R})$.
    (b) $\{L_1, \ldots, L_n\}$ is a basis of the dual space of $\mathcal{P}_{n-1}(\mathbb{R})$.
    (c) Given $a_1, \ldots, a_n \in \mathbb{R}$, there exists a unique polynomial $p(t) \in \mathcal{P}_{n-1}(\mathbb{R})$ such that $p(t_1) = a_1, \ldots, p(t_n) = a_n$.

The polynomials $p_j(t)$ are called the *Lagrange polynomials*. By doing this exercise you have solved the interpolation problem which asks for constructing a polynomial that takes prescribed values at prescribed points.

70. Let U be a subspace of a vector space V with $\dim(U) = m \leq n = \dim(V)$. Show that $W := \{A \in \mathcal{L}(V, V) : Ax = 0 \text{ for all } x \in U\}$ is a subspace of $\mathcal{L}(V, V)$. What is $\dim(W)$?
71. Let V be a vector space. Let $f_1, \ldots, f_n$ and g be linear functionals on V. Then $g \in \text{span}\{f_1, \ldots, f_n\}$ if and only if $N(f_1) \cap \cdots \cap N(f_n) \subseteq N(g)$.
72. Let V be a vector space of dimension 2. Let $A, B, T \in \mathcal{L}(V, V)$. Show that $(AB - BA)^2 T = T(AB - BA)^2$. Does it happen if $\dim(V) > 2$?
73. Let $T \in \mathcal{L}(\mathbb{C}^2, \mathbb{C}^2)$ be given by $T(a, b) = (a + b, b)$. Show that if $S \in \mathcal{L}(\mathbb{C}^2, \mathbb{C}^2)$ commutes with T, then $S = p(T)$ for some polynomial $p(t)$.
74. Let $T : U \to V$ be an isomorphism. Define $\phi : \mathcal{L}(U, U) \to \mathcal{L}(V, V)$ by $\phi(S) = T^{-1}ST$. Prove that ϕ is an isomorphism.
75. Suppose $A \in \mathbb{C}^{n \times n}$ has all diagonal entries 0. Do there exist matrices M, $P \in \mathbb{C}^{n \times n}$ such that $A = MP - PM$?
76. Let $T : V \to W$ be a linear transformation. Show that rank (T) is finite if and only if there exist $n \in \mathbb{N}$, $\{v_1, \ldots, v_n\} \subseteq V$ and $\{f_1, \ldots, f_n\} \subseteq V'$ such that $Tx = \sum_{j=1}^{n} f_j(x) v_j$ for each $x \in V$. Such a linear transformation is said to be of *finite rank*.
77. Let f and g be linear functionals on a vector space V. If h is a linear functional on V with $h(v) = f(v)g(v)$ for each $v \in V$, then prove that either $f = 0$ or $g = 0$.
78. Let S be a subset of a vector space V. The *annihilator* S^0 of S is defined by $S^0 := \{f \in V' : f(v) = 0 \text{ for each } v \in S\}$. Thus $\{0\}^0 = V'$ and $V^0 = \{0\} \subseteq V'$.

    (a) Prove that S^0 is a subspace of V'.
    (b) Let U be a subspace of a finite dimensional vector space V. Prove that $\dim(V) = \dim(U) + \dim(U^0)$.
    (c) If $\dim(V) < \infty$, then prove that $\text{span}(S) = S^{00}$. (The identification uses the natural isomorphism between V and V''.)
    (d) Let A, B be subsets of a finite dimensional vector space V. Prove that if $A \subseteq B$, then $B^0 \subseteq A^0$.
    (e) Let U and W be subspaces of a finite dimensional vector space V. Prove that $(U \cap W)^0 = U^0 + W^0$ and $(U + W)^0 = U^0 \cap W^0$.
    (f) Let U and W be subspaces of a finite dimensional vector space V. Prove that if $V = U \oplus W$, then $U' \simeq W^0$, $W' \simeq U^0$, and $V' = U^0 \oplus W^0$.
    (g) The restriction of the isomorphism $T : V \to V''$ to U, which is defined by $(Tv)(g) = g(v)$, is an isomorphism from U to U^{00}.
    (h) If U and W are subspaces of a finite dimensional vector space V, then $U = W$ if and only if $U^0 = W^0$.

79. Let $U_1, \ldots, U_k$ be subspaces of a vector space V. Write

$$V_i := U_1 \oplus \cdots \oplus U_{i-1} \oplus U_{i+1} \cdots \oplus U_k.$$

Suppose that $V = U_1 \oplus \cdots \oplus U_k$. Prove that $V' = V_1^0 \oplus \cdots \oplus V_k^0$.

80. Let U be a subspace of a vector space V. Corresponding to each linear functional F on V/U, define a linear functional f on V by $f(x) = F(x + U)$ for each $x \in V$. Prove that the correspondence $F \mapsto f$ is an isomorphism from $(V/U)'$ to U^0.

Chapter 3
Elementary Operations

3.1 Elementary Row Operations

In Sect. 2.5, we have seen how the rank of a matrix determines its equivalence class. If two $m \times n$ matrices have the same rank, then they represent the same linear transformation from $\mathbb{F}^{n\times 1}$ to $\mathbb{F}^{m\times 1}$, with respect to (possibly) different pairs of bases. The question is, how do we determine the rank of a matrix?

A rank echelon matrix E_r has from top left the first r diagonal entries as 1 and all other entries as 0. That is, E_r of order n looks like

$$E_r = \begin{bmatrix} I_r & 0 \\ 0 & 0 \end{bmatrix},$$

where I_r is the identity matrix of order r and the 0s are the matrices of suitable size with all entries as 0. Clearly, E_r is of rank r. Thus to compute the rank of a matrix, a *rank preserving reduction* to a rank echelon matrix will do the job. We will see that such a reduction requires operating with both rows and columns of a given matrix, in general. The column operations are nothing but row operations on the transpose of a matrix. Therefore, we will consider the row operations in detail.

Consider the 3×4 matrix

$$A = \begin{bmatrix} 1 & 1 & 2 & 3 \\ 1 & 2 & 3 & 4 \\ 1 & 4 & 5 & 6 \end{bmatrix}.$$

Its rows are $u_1 = (1, 1, 2, 3)$, $u_2 = (1, 2, 3, 4)$, and $u_3 = (1, 4, 5, 6)$. To determine the *row rank* of A, that is, the dimension of the space spanned by the rows of A, (which is also equal to the maximum number of linearly independent vectors out of u_1, u_2, u_3), we must find out whether any one of them is a linear combination of the others. If that happens, then surely that vector can safely be omitted for computing the row rank. Here, we see that

M. T. Nair and A. Singh, *Linear Algebra*,
https://doi.org/10.1007/978-981-13-0926-7_3

$$u_3 = 3u_2 - 2u_1.$$

Also, u_1 and u_2 are linearly independent. Hence, the row rank of A is 2.

To implement the changes in the entries of the matrix A we replace the third row by this row minus thrice the second row plus twice the first row. Then the new matrix will have the third row as a zero row. Now, going a bit further on the same line of computation, we replace the second row of the new matrix, which is u_2 by $u_2 - u_1$. This results in the new second row as $(0, 1, 1, 1)$. The main goal here is to bring in as many zero entries as possible by using linear combinations so that we progress towards a rank echelon matrix. See the following computation.

$$\begin{bmatrix} 1 & 1 & 2 & 3 \\ 1 & 2 & 3 & 4 \\ 1 & 4 & 5 & 6 \end{bmatrix} \xrightarrow{R1} \begin{bmatrix} 1 & 1 & 2 & 3 \\ 0 & 1 & 1 & 1 \\ 0 & 0 & 0 & 0 \end{bmatrix} \xrightarrow{R2} \begin{bmatrix} 1 & 0 & 1 & 2 \\ 0 & 1 & 1 & 1 \\ 0 & 0 & 0 & 0 \end{bmatrix}.$$

Here, the row operation $R1$ replaces $row(2)$ by $row(2) - row(1)$, and $row(3)$ by $row(3) - 3row(2) + 2row(1)$ simultaneously. Similarly, $R2$ stands for replacing $row(1)$ with $row(1) - row(2)$.

Such row operations are combinations of the three kinds of operations on the rows of a matrix. We define these operations.

Definition 3.1 The following are called **elementary row operations** applied on a matrix:

(a) *Type 1:* Exchanging two rows;
(b) *Type 2:* Multiplying a row by a nonzero scalar;
(c) *Type 3:* Adding to a row a nonzero scalar multiple of another row.

The elementary row operations can be seen as matrix products. For this purpose, we introduce the so-called *elementary matrices*, which are obtained by performing analogous elementary row operations on the identity matrix.

Definition 3.2 Let I be the identity matrix of order m. Let $i,\ j \in \{1, \ldots, m\}$ be the row indices and let α be a nonzero scalar. The matrices obtained as in the following are called the **elementary matrices** of order m:

(a) $E[i, j]$: by exchanging the ith and the jth rows in I *(Type 1)*;
(b) $E_\alpha[i]$: by multiplying α with the ith row in I *(Type 2)*;
(c) $E_\alpha[i, j]$: by adding to the ith row α times the jth row, in I *(Type 3)*.

To avoid heavy overload on symbolism, our notation for elementary matrices do not reflect their orders. The orders or sizes of the matrices will either be specified separately or will be clear from the context.

Example 3.3 Some of the elementary matrices of order 4 are as follows:

$$E[2,4] = \begin{bmatrix} 1 & 0 & 0 & 0 \\ 0 & 0 & 0 & 1 \\ 0 & 0 & 1 & 0 \\ 0 & 1 & 0 & 0 \end{bmatrix}, \quad E_6[3] = \begin{bmatrix} 1 & 0 & 0 & 0 \\ 0 & 1 & 0 & 0 \\ 0 & 0 & 6 & 0 \\ 0 & 0 & 0 & 1 \end{bmatrix}, \quad E_3[2,3] = \begin{bmatrix} 1 & 0 & 0 & 0 \\ 0 & 1 & 3 & 0 \\ 0 & 0 & 1 & 0 \\ 0 & 0 & 0 & 1 \end{bmatrix}.$$

Of course, there are other elementary matrices of order 4 such as $E[1,2]$, $E[1,3]$, $E[1,4]$, $E[2,3]$, $E[3,4]$; $E_\alpha[1]$, $E_\alpha[2]$, $E_\alpha[3]$, $E_\alpha[4]$; $E_\alpha[1,2]$, $E_\alpha[1,3]$, $E_\alpha[1,4]$; and $E_\alpha[2,3]$, $E_\alpha[2,4]$, $E_\alpha[3,4]$ for any nonzero $\alpha \in \mathbb{F}$. □

As it turns out, the products $E[i,j]A$, $E_\alpha[i]A$ and $E_\alpha[i,j]A$ are the matrices obtained by performing the corresponding elementary row operations of Types 1, 2, 3 on A. They are as follows. (See Exercises 4–6.)

1. $E[i,j]A$ is obtained from A by exchanging the ith and the jth rows in A.
2. $E_\alpha[i]A$ is obtained from A by replacing the ith row of A by α times the ith row of A.
3. $E_\alpha[i,j]A$ is obtained from A by replacing the ith row by ith row plus α times the jth row of A.

We specify an elementary row operation by its corresponding elementary matrix. In showing the elementary row operations in numerical examples, we will write

$$A \xrightarrow{E} B,$$

where $B = EA$ has been obtained from A by using the elementary row operation corresponding to the elementary matrix E. See the following example.

Example 3.4 Using our notation for elementary row operations, the sequence of elementary row operations

replace $row(2)$ with $row(2) - row(1)$;
replace $row(3)$ with $row(3) - 3row(2)$;
replace $row(3)$ with $row(3) - row(1)$

applied on the matrix $\begin{bmatrix} 1 & 1 & 2 & 3 \\ 1 & 2 & 3 & 4 \\ 1 & 4 & 5 & 6 \end{bmatrix}$ can now be written as follows:

$$\begin{bmatrix} 1 & 1 & 2 & 3 \\ 1 & 2 & 3 & 4 \\ 1 & 4 & 5 & 6 \end{bmatrix} \xrightarrow{E_{-1}[2,1]} \begin{bmatrix} 1 & 1 & 2 & 3 \\ 0 & 1 & 1 & 1 \\ 1 & 4 & 5 & 6 \end{bmatrix} \xrightarrow{E_{-3}[3,2]} \begin{bmatrix} 1 & 1 & 2 & 3 \\ 0 & 1 & 1 & 1 \\ 1 & 1 & 2 & 3 \end{bmatrix} \xrightarrow{E_{-1}[3,1]} \begin{bmatrix} 1 & 1 & 2 & 3 \\ 0 & 1 & 1 & 1 \\ 0 & 0 & 0 & 0 \end{bmatrix}.$$

The operations above say that we first apply $E_{-1}[2,1]$. Then on the resulting matrix we apply $E_{-3}[3,2]$. Again, on the resulting matrix, $E_{-1}[3,1]$ is applied. The computation can be written more compactly as

$$E_{-1}[3,1]\,E_{-3}[3,2]\,E_{-1}[2,1]\begin{bmatrix}1&1&2&3\\1&2&3&4\\1&4&5&6\end{bmatrix}=\begin{bmatrix}1&1&2&3\\0&1&1&1\\0&0&0&0\end{bmatrix}.$$

Notice the reverse order in multiplying the elementary matrices. □

Exercises for Sect. 3.1

1. Compute $E_{-1/5}[2,3]\,E[1,3]\,E_{1/2}[2]A$, where $A=\begin{bmatrix}0&1&1&1\\0&2&3&1\\1&0&2&-3\end{bmatrix}$.
2. Determine the elementary row operations and the corresponding elementary matrices in the following reduction:

$$\begin{bmatrix}5&4&1&0\\6&5&0&1\end{bmatrix}\longrightarrow\begin{bmatrix}5&4&1&0\\1&1&-1&1\end{bmatrix}\longrightarrow\begin{bmatrix}1&0&5&-4\\1&1&-1&1\end{bmatrix}\longrightarrow\begin{bmatrix}1&0&5&-4\\0&1&-6&5\end{bmatrix}.$$

   Further, verify that the product of those elementary matrices is $\begin{bmatrix}5&-4\\-6&5\end{bmatrix}$.
3. Let $A\in\mathbb{F}^{3\times 3}$. Determine three matrices $B, C, D\in\mathbb{F}^{3\times 3}$ such that

   (a) BA is obtained from A by adding the third row to the first row in A.
   (b) CA is obtained from A by adding the third row to the second row in A, and simultaneously by adding the second row to the first row in A.
   (c) DA is obtained from A by adding the third row to the second row in A, and then by adding the second row to the first row in A.

4. Let $e_1,\ldots,e_m$ be the standard basis vectors of $\mathbb{F}^{m\times 1}$. For $i,j\in\{1,\ldots,m\}$, let $E_{ij}:=e_ie_j^T$ and let I be the identity matrix of order m. Show that the three kinds of elementary matrices $E[i,j],\ E_\alpha[i],\ E_\alpha[i,j]$ of order m can be given as in the following:

   (a) $E[i,j]:=I-E_{ii}-E_{jj}+E_{ij}+E_{ji}$ for $i\neq j$.
   (b) $E_\alpha[i]:=I-E_{ii}+\alpha E_{ii}$ for a nonzero scalar α.
   (c) $E_\alpha[i,j]:=I+\alpha E_{ij}$ for a nonzero scalar α and $i\neq j$.

5. Let $A\in\mathbb{F}^{m\times n}$ and let $E_{ij}:=e_ie_j^T$ be as given in Exercise 4. Show that the ith row of $E_{ij}A$ is the jth row of A; all other rows of $E_{ij}A$ are zero rows.
6. Let $A\in\mathbb{F}^{m\times n}$. Let $E[i,j],\ E_\alpha[i]$ and $E_\alpha[i,j]$ be the matrices as given in Exercise 4. Show that the matrices $E[i,j]A,\ E_\alpha[i]A$, and $E_\alpha[i,j]A$ are equal to those obtained by applying the corresponding the elementary row operations on A.
7. Show that the elementary matrices are invertible and their inverses are given by $(E[i,j])^{-1}=E[j,i],\ (E_\alpha[i])^{-1}=E_{1/\alpha}[i]$, and $(E_\alpha[i,j])^{-1}=E_{-\alpha}[i,j]$.

8. Show that if a matrix B is obtained from another matrix A by applying a sequence of elementary row operations, then A can also be obtained from B by applying a sequence of elementary row operations.
9. Take a 3×3 matrix A and explore what happens to the rows or columns of A in the products $AE[i, j]$, $AE_\alpha[i]$, $AE_\alpha[i, j]$, $AE[i, j]^T$, $AE_\alpha[i]^T$, and $AE_\alpha[i, j]^T$.
10. Is it true that the transpose of an elementary matrix is also an elementary matrix?

3.2 Row Echelon Form

Row operations can be used to obtain a matrix with fewer nonzero entries. Given a matrix, our goal is to obtain a matrix from the given one, all of whose nonzero rows are linearly independent. Moreover, the linear independence should be visible without any further computation. Towards this, we give the following definition.

Definition 3.5 A **row operation** is a finite sequence of elementary row operations. If a matrix B is obtained from a matrix A by applying a row operation, we say that B is **row equivalent to** A.

In Example 3.4, $\begin{bmatrix} 1 & 1 & 2 & 3 \\ 0 & 1 & 1 & 1 \\ 0 & 0 & 0 & 0 \end{bmatrix}$ is row equivalent to $\begin{bmatrix} 1 & 1 & 2 & 3 \\ 1 & 2 & 3 & 4 \\ 1 & 4 & 5 & 6 \end{bmatrix}$.

The elementary matrices are invertible, and their inverses are given by

$$(E[i, j])^{-1} = E[j, i], \quad (E_\alpha[i])^{-1} = E_{1/\alpha}[i], \quad (E_\alpha[i, j])^{-1} = E_{-\alpha}[i, j],$$

which are also elementary matrices. Using this it is easily seen that row equivalence is an equivalence relation on $\mathbb{F}^{m \times n}$. We are in search of a canonical representative which would somehow *show us* the linearly independent rows as in the first matrix above.

Towards this, we first define such a form. Recall that the (i, j)th entry of a matrix has row index i and column index j. We also say that the row index of the ith row is i, and the column index of the jth column is j.

Definition 3.6 In a nonzero row of a matrix, the first nonzero entry from the left is called a **pivot**.
A matrix is said to be in **row echelon form** if the following are satisfied:

(1) The row index of each nonzero row is smaller than the row index of each zero row.
(2) Among any two pivots, the pivot with larger row index also has larger column index.

We mark a pivot in a row by putting a box around it. All entries to the left of a pivot in that row are zero.

For example, the following matrices are in row echelon form:

$$\begin{bmatrix} \boxed{1} & 2 & 0 & 0 \\ 0 & 0 & \boxed{1} & 0 \\ 0 & 0 & 0 & \boxed{1} \end{bmatrix}, \begin{bmatrix} 0 & \boxed{2} & 3 & 0 \\ 0 & 0 & 0 & \boxed{1} \\ 0 & 0 & 0 & 0 \\ 0 & 0 & 0 & 0 \end{bmatrix}, \begin{bmatrix} 0 & \boxed{1} & 3 & 1 \\ 0 & 0 & 0 & \boxed{1} \\ 0 & 0 & 0 & 0 \\ 0 & 0 & 0 & 0 \end{bmatrix}, \begin{bmatrix} 0 & \boxed{1} & 0 & 0 \\ 0 & 0 & \boxed{1} & 0 \\ 0 & 0 & 0 & 0 \\ 0 & 0 & 0 & \boxed{1} \end{bmatrix}.$$

And, the following matrices are not in row echelon form:

$$\begin{bmatrix} 0 & 0 & \boxed{1} & 2 \\ 0 & \boxed{1} & 1 & 0 \\ 0 & 0 & 0 & \boxed{1} \end{bmatrix}, \begin{bmatrix} 0 & \boxed{2} & 3 & 0 \\ \boxed{1} & 0 & 0 & 0 \\ 0 & 0 & 0 & 0 \\ 0 & 0 & 0 & 0 \end{bmatrix}, \begin{bmatrix} 0 & \boxed{1} & 3 & 1 \\ 0 & 0 & 0 & \boxed{1} \\ 0 & 0 & 0 & 0 \\ 0 & \boxed{1} & 0 & 0 \end{bmatrix}, \begin{bmatrix} 0 & \boxed{1} & 3 & 0 \\ 0 & 0 & 0 & \boxed{1} \\ 0 & 0 & 0 & \boxed{1} \\ 0 & 0 & 0 & 0 \end{bmatrix}.$$

The first condition in Definition 3.6 says that all zero rows are at the bottom. The second condition is sometimes stated informally as

The pivot in the $(i + 1)$th row is to the right of the pivot in the ith row.

This condition implies the following:

In a pivotal column of a matrix in row echelon form, all entries below the pivot are zero.

We will see that satisfying such a goal systematically will result in converting a matrix to a row echelon form.

Theorem 3.7 (Row echelon form) *Each matrix is row equivalent to one in row echelon form.*

Proof Let $A \in \mathbb{F}^{m\times n}$. We use induction on n, the number of columns in A. For $n = 1$, A is a column vector. If all entries in this single column are 0, then this itself is the row echelon form. Otherwise, let j be the minimum index out of $1, \ldots, m$ such that $a_{j1} \neq 0$. Exchange $row(1)$ and $row(j)$. Now, the first row of this new column vector has the first entry as $a_{j1} \neq 0$. Write $\alpha = a_{j1}$. Replace each other row, $row(i)$ by $row(i) - \frac{a_{i1}}{\alpha}\, row(1)$. This brings the column vector to row echelon form. The result is the column vector

$$E_{-\frac{a_{m1}}{\alpha}}[m, 1] \cdots E_{-\frac{a_{21}}{\alpha}}[2, 1]\, E[j, 1]\, A,$$

which is equal to αe_1.

Assume that all matrices with $n - 1$ columns can be brought to row echelon form by pre-multiplying it with elementary matrices. Let A have n columns. Look at the first column $col(1)$ of A. Use the same elementary matrices for reducing $col(1)$ to row reduced form as in the basis case. These elementary row operations are applied on A and not only on $col(1)$. That is, there are elementary matrices $E_1', \ldots, E_m'$ such that

$$E_m' \cdots E_1' A = \begin{bmatrix} \alpha & * \\ 0 & B \end{bmatrix}.$$

The entries in the first row marked $*$ are some numbers, and the matrix $B \in \mathbb{F}^{(m-1)\times(n-1)}$. By induction hypothesis, there exist elementary matrices $E_1, \ldots, E_r$ such that $E_r \cdots E_1 B$ is in row echelon form. Then

$$\begin{bmatrix} 1 & 0 \\ 0 & E_r \end{bmatrix} \cdots \begin{bmatrix} 1 & 0 \\ 0 & E_1 \end{bmatrix} E'_m \cdots E'_1 A$$

is the required row echelon form of the matrix A. Notice that any matrix $\begin{bmatrix} 1 & 0 \\ 0 & E_k \end{bmatrix}$ used in the above product is also an elementary matrix, whenever E_k is an elementary matrix. ■

The proof of Theorem 3.7 gives the following algorithm for reducing a matrix to row echelon form. We use a pivot in a row to zero-out the entries below the pivot systematically. Initially, we work with the whole matrix, and subsequently, with its submatrices. In the following algorithm we write such submatrices as R, and call them as *search regions*.

Reduction to Row Echelon Form

1. Set the search region R to the whole matrix A.
2. If all entries in R are 0, then stop.
3. Else, find the leftmost nonzero column in R. Mark this column in A as a pivotal column.
4. Find the topmost nonzero entry in the pivotal column that occurs in R; box it; it is a pivot.
5. If the top row of R does not contain the pivot, then exchange the row of A that contains the top row of R with the row that contains the pivot.
6. Mark the row in A, which is the top row of R, as the pivotal row.
7. Zero-out all entries below the pivot, in the pivotal column, by replacing each row below the pivotal row using the third type of elementary row operations with that row and the pivotal row.
8. Find the submatrix of R to the right and below the current pivot. If no such submatrix exists, then stop. Else, reset the search region R to this submatrix, and go to 2.

Clearly, the output of the algorithm on an input matrix A is a matrix in row echelon form. We refer to the resulting matrix as *the* row echelon form of A.

Example 3.8 Check how A is converted to B, which is in row echelon form:

$$A = \begin{bmatrix} \boxed{1} & 1 & 2 & 0 \\ 3 & 5 & 5 & 1 \\ 1 & 5 & 4 & 5 \\ 2 & 8 & 7 & 9 \end{bmatrix} \xrightarrow{R1} \begin{bmatrix} \boxed{1} & 1 & 2 & 0 \\ 0 & \boxed{2} & -1 & 1 \\ 0 & 4 & 2 & 5 \\ 0 & 6 & 3 & 9 \end{bmatrix} \xrightarrow{R2} \begin{bmatrix} \boxed{1} & 1 & 2 & 0 \\ 0 & \boxed{2} & -1 & 1 \\ 0 & 0 & \boxed{4} & 3 \\ 0 & 0 & 6 & 6 \end{bmatrix}$$

$$\xrightarrow{R3} \begin{bmatrix} \boxed{1} & 1 & 2 & 0 \\ 0 & \boxed{2} & 1 & 1 \\ 0 & 0 & \boxed{4} & 3 \\ 0 & 0 & 0 & \boxed{3/2} \end{bmatrix} = B.$$

Here, the row operations are

$$R1 = E_{-3}[2,1],\ E_{-1}[3,1],\ E_{-2}[4,1];\ R2 = E_{-2}[3,2],\ E_{-3}[4,2];\ R3 = E_{-3/2}[4,3].$$

The matrix B can be written as a product of elementary matrices with A:

$$B = E_{-3/2}[4,3]\, E_{-3}[4,2]\, E_{-2}[3,2]\, E_{-2}[4,1]\, E_{-1}[3,1]\, E_{-3}[2,1]\, A. \qquad \square$$

In the row echelon form, the nonzero rows are linearly independent. Thus, the dimension of the space spanned by all the rows is the number of nonzero rows in the row echelon form. That is, the row rank of a matrix in row echelon form is the number of pivots. Moreover, the row rank is equal to the rank of a matrix. Therefore, rank (A) is the number of pivots in the row echelon form of A.

Example 3.9 To compute the rank of the matrix $A = \begin{bmatrix} 0 & 2 & 0 & 1 \\ 1 & 2 & 2 & 1 \\ 2 & 2 & 4 & 1 \\ 1 & -2 & 2 & -1 \end{bmatrix}$, we reduce it to its row echelon form as follows:

$$\begin{bmatrix} 0 & 2 & 0 & 1 \\ 1 & 2 & 2 & 1 \\ 2 & 2 & 4 & 1 \\ 1 & -2 & 2 & -1 \end{bmatrix} \xrightarrow{E[1,2]} \begin{bmatrix} \boxed{1} & 2 & 2 & 1 \\ 0 & 2 & 0 & 1 \\ 2 & 2 & 4 & 1 \\ 1 & -2 & 2 & -1 \end{bmatrix} \xrightarrow{R} \begin{bmatrix} \boxed{1} & 2 & 2 & 1 \\ 0 & \boxed{2} & 0 & 1 \\ 0 & 0 & 0 & 0 \\ 0 & 0 & 0 & 0 \end{bmatrix}.$$

Here, R is the row operation $E_{-2}[3,1], E_{-1}[4,1];\ E_1[3,2], E_2[4,2]$. The number of pivots shows that rank $(A) = 2$. $\square$

Looking at the reduction to row echelon form, we end up at a useful matrix factorization. For convenience, we will use the following terminology.

Definition 3.10 A **permutation matrix** of order m is a matrix obtained by permuting the rows of the identity matrix of order m.

Observe that a permutation matrix is a product of elementary matrices of Type 1. Moreover each permutation matrix is invertible.

Theorem 3.11 (LU-factorization) *Let $A \in \mathbb{F}^{m \times n}$. Then there exist a permutation matrix P of order m, a lower triangular matrix L of order m with all its diagonal entries as 1, and a matrix $U \in \mathbb{F}^{m \times n}$ in row echelon form such that $PA = LU$.*

Proof Let rank $(A) = r$. If $r = 0$, then $A = 0$, and we take $P = L = I$ and $U = A$. So, let $r \geq 1$. Let U be the row echelon form of A. In U, there are exactly r number of pivots. If no row exchanges have been used for reducing A to its row echelon form U, then we take $P = I$. Otherwise, suppose the row exchanges applied in the reduction process are $E[i_1, j_1], \ldots, E[i_\ell, j_\ell]$, in that order. These row exchanges have been carried out in the reduction process so that starting from the first row of A to the last, exactly r number of linearly independent rows come up as the first r rows in U. If we apply the same row exchanges on A in the beginning, then exactly the same r linearly independent rows of A will come up as the first r rows in PA, where

$$P := E[i_\ell, j_\ell] \cdots E[i_1, j_1].$$

Notice that P is a permutation matrix; it is invertible. Thus rank $(PA) = r$. In PA, the first r rows are linearly independent. While reducing PA to its row echelon form, no row exchanges are required. Moreover, the pivots in this reduction process are the same entries as in U. Thus U is also the row echelon form of the matrix PA. Observe that the elementary row operations of Type 3 used in the reduction of PA to U are the same as in the reduction of A to U. However, the row indices i and j in each such $E_\alpha[i, j]$ may change due to initial row exchanges carried out in pre-multiplying A with P.

We look at how the reduction process works on $PA = [b_{ij}]$ to reach at U. Initially, the $(1, 1)$th entry b_{11} of PA is a pivot. Any other entry b_{j1} for $j > 1$, in the first column is zeroed-out by using the elementary row operation $E_{-b_{j1}/b_{11}}[j, 1]$. In general, while operating with the kth pivot, we see that all entries in the first $k - 1$ columns below the diagonal are already 0. Next, if the kth column is

$$[c_1, \ldots, c_{k-1}, c_k, c_{k+1}, \ldots, c_n]^T,$$

then to zero-out the jth entry in this column, for $j > k$, we use the elementary row operation $E_{-c_j/c_k}[j, k]$. This process stops when k becomes equal to r. And, in that case, we reach at the matrix U. Taking the product of all these elementary matrices as L_0, we obtain

$$U = L_0 P A,$$

where L_0 is the product of elementary matrices of each of the form $E_\alpha[j, k]$ with $j > k$. Notice that each such matrix $E_\alpha[j, k]$ is a lower triangular matrix with all diagonal entries as 1. Therefore, L_0 is invertible, and L_0^{-1} is also a lower triangular matrix with all its diagonal entries as 1. With $L = L_0^{-1}$, we have $PA = LU$. ∎

Notice that if A is a square matrix, then its row echelon form U is upper triangular.

Example 3.12 To illustrate the proof of LU-factorization, consider the matrix A and its reduction to row echelon form as given below:

$$A = \begin{bmatrix} \boxed{1} & 2 & 1 & 2 & 1 \\ 2 & 4 & 2 & 5 & 2 \\ 3 & 7 & 4 & 6 & 4 \\ 1 & 4 & 5 & 4 & 5 \\ 2 & 6 & 4 & 6 & 4 \end{bmatrix} \xrightarrow{R1} \begin{bmatrix} \boxed{1} & 2 & 1 & 2 & 1 \\ 0 & 0 & 0 & 1 & 0 \\ 0 & 1 & 1 & 0 & 1 \\ 0 & 2 & 4 & 2 & 4 \\ 0 & 2 & 2 & 2 & 2 \end{bmatrix} \xrightarrow{E[2,3]} \begin{bmatrix} \boxed{1} & 2 & 1 & 2 & 1 \\ 0 & \boxed{1} & 1 & 0 & 1 \\ 0 & 0 & 0 & 1 & 0 \\ 0 & 2 & 4 & 2 & 4 \\ 0 & 2 & 2 & 2 & 2 \end{bmatrix}$$

$$\xrightarrow{R2} \begin{bmatrix} \boxed{1} & 2 & 1 & 2 & 1 \\ 0 & \boxed{1} & 1 & 0 & 1 \\ 0 & 0 & 0 & 1 & 0 \\ 0 & 0 & 2 & 2 & 2 \\ 0 & 0 & 0 & 2 & 0 \end{bmatrix} \xrightarrow{E[3,4]} \begin{bmatrix} \boxed{1} & 2 & 1 & 2 & 1 \\ 0 & \boxed{1} & 1 & 0 & 1 \\ 0 & 0 & \boxed{2} & 2 & 2 \\ 0 & 0 & 0 & 1 & 0 \\ 0 & 0 & 0 & 2 & 0 \end{bmatrix}$$

$$\xrightarrow{E_{-2}[5,4]} \begin{bmatrix} \boxed{1} & 2 & 1 & 2 & 1 \\ 0 & \boxed{1} & 1 & 0 & 1 \\ 0 & 0 & \boxed{2} & 2 & 2 \\ 0 & 0 & 0 & \boxed{1} & 0 \\ 0 & 0 & 0 & 0 & 0 \end{bmatrix} = U.$$

Here, $R1 = E_{-2}[2,1],\ E_{-3}[3,1],\ E_{-1}[4,1],\ E_{-2}[5,1]$, and $R2 = E_{-2}[4,2]$, $E_{-2}[5,2]$.

We look at the row exchanges. They are $E[2,3]$ and $E[3,4]$ in that order. That is, first $E[2,3]$ and then $E[3,4]$ have been multiplied with A from the left. Thus

$$P = E[3,4]\, E[2,3] = \begin{bmatrix} 1 & 0 & 0 & 0 & 0 \\ 0 & 0 & 1 & 0 & 0 \\ 0 & 0 & 0 & 1 & 0 \\ 0 & 1 & 0 & 0 & 0 \\ 0 & 0 & 0 & 0 & 1 \end{bmatrix}, \quad PA = \begin{bmatrix} 1 & 2 & 1 & 2 & 1 \\ 3 & 7 & 4 & 6 & 4 \\ 1 & 4 & 5 & 4 & 5 \\ 2 & 4 & 2 & 5 & 2 \\ 2 & 6 & 4 & 6 & 4 \end{bmatrix}.$$

Next, PA is reduced to its row echelon form as follows:

$$\begin{bmatrix} 1 & 2 & 1 & 2 & 1 \\ 3 & 7 & 4 & 6 & 4 \\ 1 & 4 & 5 & 4 & 5 \\ 2 & 4 & 2 & 5 & 2 \\ 2 & 6 & 4 & 6 & 4 \end{bmatrix} \xrightarrow{R3} \begin{bmatrix} \boxed{1} & 2 & 1 & 2 & 1 \\ 0 & \boxed{1} & 1 & 0 & 1 \\ 0 & 2 & 4 & 2 & 4 \\ 0 & 0 & 0 & 1 & 0 \\ 0 & 2 & 2 & 2 & 2 \end{bmatrix} \xrightarrow{R4} \begin{bmatrix} \boxed{1} & 2 & 1 & 2 & 1 \\ 0 & \boxed{1} & 1 & 0 & 1 \\ 0 & 0 & \boxed{2} & 2 & 2 \\ 0 & 0 & 0 & \boxed{1} & 0 \\ 0 & 0 & 0 & 2 & 0 \end{bmatrix} \xrightarrow{E_{-2}[5,4]} U.$$

Here, $R3 = E_{-3}[2,1],\ E_{-1}[3,1],\ E_{-2}[4,1],\ E_{-2}[5,1]$, and $R4 = E_{-2}[3,2]$, $E_{-2}[5,2]$. Then

$$\begin{aligned} L &= \left(E_{-2}[5,4]\, E_{-2}[5,2]\, E_{-2}[3,2]\, E_{-2}[5,1]\, E_{-2}[4,1]\, E_{-1}[3,1]\, E_{-3}[2,1]\right)^{-1} \\ &= E_3[2,1]\, E_1[3,1]\, E_2[4,1]\, E_2[5,1]\, E_2[3,2]\, E_2[5,2]\, E_2[5,4] \\ &= \begin{bmatrix} 1 & 0 & 0 & 0 & 0 \\ 3 & 1 & 0 & 0 & 0 \\ 1 & 2 & 1 & 0 & 0 \\ 2 & 0 & 0 & 1 & 0 \\ 2 & 2 & 0 & 2 & 1 \end{bmatrix}. \end{aligned}$$

It is easy to verify that $PA = LU$. □

Though the factorization is called LU-factorization, it is not possible to express every square matrix as a product of a lower triangular matrix with nonzero diagonal entries, and an upper triangular matrix; see Exercise 5.

Exercises for Sect. 3.2

1. List out all 3×4 matrices in row echelon form by writing $\star$ for any nonzero entry, and 0 for a zero entry.
2. Reduce the following matrices to row echelon form:

$$\text{(a)} \begin{bmatrix} 1 & 1 & 1 \\ 1 & 1 & 0 \\ 0 & 0 & 1 \end{bmatrix} \quad \text{(b)} \begin{bmatrix} 0 & 0 & 4 & 0 \\ 2 & 2 & -2 & 5 \\ 5 & 5 & -5 & 5 \end{bmatrix} \quad \text{(c)} \begin{bmatrix} 0 & 2 & 3 & 7 \\ 1 & 1 & 1 & 1 \\ 1 & 3 & 4 & 8 \\ 0 & 0 & 0 & 1 \end{bmatrix}.$$

3. Prove that row equivalence is an equivalence relation on $\mathbb{F}^{m \times n}$.
4. Give examples of two unequal $m \times n$ matrices in row echelon form, which are row equivalent.
5. Show that there do not exist a lower triangular matrix $L \in \mathbb{F}^{2\times 2}$ with nonzero diagonal entries, and an upper triangular matrix $U \in \mathbb{F}^{2\times 2}$ such that

$$LU = \begin{bmatrix} 0 & 1 \\ 1 & 0 \end{bmatrix}.$$

6. In each of the following cases, determine a lower triangular matrix L and an upper triangular matrix U so that the given matrix equals LU. If not possible, determine also a permutation matrix P so that the given matrix equals PLU.

$$\text{(a)} \begin{bmatrix} 0 & -6 & 4 \\ 2 & 1 & 2 \\ 1 & 4 & 1 \end{bmatrix} \quad \text{(b)} \begin{bmatrix} 1 & 1 & 1 \\ 1 & 1 & 3 \\ 2 & 5 & 8 \end{bmatrix} \quad \text{(c)} \begin{bmatrix} 1 & 2 & 4 \\ -1 & -3 & 3 \\ 4 & 9 & 14 \end{bmatrix}.$$

7. Let $A \in \mathbb{F}^{m\times n}$. Suppose $P \in \mathbb{F}^{m\times m}$ and $Q \in \mathbb{F}^{n\times n}$ are permutation matrices. Show that PA permutes the rows of A and AQ permutes the columns of A.
8. Prove that a permutation matrix is invertible without using elementary matrices.

3.3 Row Reduced Echelon Form

Elementary operations can be used to reduce a matrix to a still simpler form, where all pivots are 1 and all nonpivotal entries in a pivotal column are 0. This will also answer our question regarding a canonical representative for the equivalence relation of row equivalence. Such a form is defined below.

Definition 3.13 A matrix $A \in \mathbb{F}^{m\times n}$ is said to be in **row reduced echelon form** if the following conditions are satisfied:

(1) Each pivot is equal to 1.
(2) In a pivotal column, all entries other than the pivot are zero.
(3) The row index of each nonzero row is smaller than the row index of each zero row.
(4) Among any two pivots, the pivot with larger row index also has larger column index.

For example, the matrix $\begin{bmatrix} \boxed{1} & 2 & 0 & 0 \\ 0 & 0 & \boxed{1} & 0 \\ 0 & 0 & 0 & 0 \end{bmatrix}$ is in row reduced echelon form whereas the following matrices are not in row reduced echelon form:

$$\begin{bmatrix} 0 & \boxed{2} & 3 & 0 \\ 0 & 0 & 0 & \boxed{1} \\ 0 & 0 & 0 & 0 \\ 0 & 0 & 0 & 0 \end{bmatrix}, \begin{bmatrix} 0 & \boxed{1} & 3 & 1 \\ 0 & 0 & 0 & \boxed{1} \\ 0 & 0 & 0 & 0 \\ 0 & 0 & 0 & 0 \end{bmatrix}, \begin{bmatrix} 0 & \boxed{1} & 3 & 0 \\ 0 & 0 & 0 & \boxed{1} \\ 0 & 0 & 0 & \boxed{1} \\ 0 & 0 & 0 & 0 \end{bmatrix}, \begin{bmatrix} 0 & \boxed{1} & 0 & 0 \\ 0 & 0 & \boxed{1} & 0 \\ 0 & 0 & 0 & 0 \\ 0 & 0 & 0 & \boxed{1} \end{bmatrix}.$$

Clearly, a matrix which is in row reduced echelon form is also in row echelon form. There are row echelon matrices which are not in row reduced echelon form. A matrix in row echelon form with r number of pivots is also in row reduced echelon form if and only if the pivotal columns are the first r standard basis vectors $e_1, \ldots, e_r$, appearing in that order from left to right. Therefore, a column vector (an $m \times 1$ matrix) is in row reduced echelon form if and only if it is either a zero vector or e_1.

Theorem 3.14 (Row reduced echelon form) *Each matrix is row equivalent to one in row reduced echelon form.*

Proof Let $A \in \mathbb{F}^{m\times n}$. By Theorem 3.7, there exists a row echelon matrix B which is row equivalent to A. If a pivot in the ith row is α, then $E_\alpha[i]\, B$ brings that pivot to 1. Thus B is reduced to a row equivalent matrix C where all pivots are 1.

Next, we look at any nonpivotal entry in a pivotal column. If it is the (i, j)th entry and it is $\alpha \neq 0$, and the pivot is the (k, j)th entry, then we compute $E_{-\alpha}[i, k]\, C$. In this new matrix, the (i, j)th entry has been reduced to 0. We continue doing this looking at all such nonpivotal nonzero entries. Finally, the obtained matrix is in row reduced echelon form. ■

The proof of Theorem 3.14 says that the algorithm for reducing a matrix to its row echelon form can be modified for reducing it to a row reduced echelon form. The modifications are in Step 7 for *Reduction to Row Echelon Form.*

Reduction to Row Reduced Echelon Form

All steps except Step 7 are as in *Reduction to Row Echelon Form.*

7A. Divide each entry in the pivotal row by the pivot, using the second type of elementary row operation.
7B. Zero-out all entries below and above the pivot, in the pivotal column, by replacing each nonpivotal row using the third type of elementary row operations with that row and the pivotal row.

As earlier, we will refer to the output of the above reduction algorithm as *the* row reduced echelon form of a given matrix.

Example 3.15 In the following the matrix A is brought to its row reduced echelon form B.

$$A = \begin{bmatrix} \boxed{1} & 1 & 2 & 0 \\ 3 & 5 & 7 & 1 \\ 1 & 5 & 4 & 5 \\ 2 & 8 & 7 & 9 \end{bmatrix} \xrightarrow{R1} \begin{bmatrix} \boxed{1} & 1 & 2 & 0 \\ 0 & 2 & 1 & 1 \\ 0 & 4 & 2 & 5 \\ 0 & 6 & 3 & 9 \end{bmatrix} \xrightarrow{E_{1/2}[2]} \begin{bmatrix} \boxed{1} & 1 & 2 & 0 \\ 0 & \boxed{1} & 1/2 & 1/2 \\ 0 & 4 & 2 & 5 \\ 0 & 6 & 3 & 9 \end{bmatrix}$$

$$\xrightarrow{R2} \begin{bmatrix} \boxed{1} & 0 & 3/2 & -1/2 \\ 0 & \boxed{1} & 1/2 & 1/2 \\ 0 & 0 & 0 & 3 \\ 0 & 0 & 0 & 6 \end{bmatrix} \xrightarrow{E_{1/3}[3]} \begin{bmatrix} \boxed{1} & 0 & 3/2 & -1/2 \\ 0 & \boxed{1} & 1/2 & 1/2 \\ 0 & 0 & 0 & \boxed{1} \\ 0 & 0 & 0 & 6 \end{bmatrix} \xrightarrow{R3} \begin{bmatrix} \boxed{1} & 0 & 3/2 & 0 \\ 0 & \boxed{1} & 1/2 & 0 \\ 0 & 0 & 0 & \boxed{1} \\ 0 & 0 & 0 & 0 \end{bmatrix} = B$$

with $R1 = E_{-3}[2, 1],\ E_{-1}[3, 1], E_{-2}[4, 1]$; $R2 = E_{-1}[1, 2],\ E_{-4}[3, 2],\ E_{-6}[4, 2]$; and $R3 = E_{1/2}[1, 3],\ E_{-1/2}[2, 3],\ E_{-6}[4, 3]$. Notice that

$$B = E_{-2}[4, 1]\, E_{-1}[3, 1]\, E_{-3}[2, 1]\, E_{1/2}[2]\, E_{-6}[4, 2]\, E_{-4}[3, 2]\, E_{-1}[1, 2]$$
$$E_{1/3}[3]\, E_{-6}[4, 3]\, E_{-1/2}[2, 3]\, E_{1/2}[1, 3]\, A.$$ □

We observe that in the row reduced echelon form of $A \in \mathbb{F}^{m\times n}$, $r :=$ rank (A) is the number of pivots. The r pivotal rows are the nonzero rows; the $m - r$ nonpivotal rows are the zero rows, which occur at the bottom. The r pivotal columns are $e_1, \ldots, e_r$, the first r standard basis vectors of $\mathbb{F}^{m\times 1}$; the $n - r$ nonpivotal columns are linear combinations of these pivotal columns.

Theorem 3.16 *A square matrix is invertible if and only if it is a product of elementary matrices.*

Proof Each elementary matrix is invertible. Therefore, a product of elementary matrices is invertible. Conversely, let $A \in \mathbb{F}^{n \times n}$ be an invertible matrix. Let E be a product of elementary matrices such that EA is in row reduced echelon form. As $\operatorname{rank}(A) = n$, the matrix EA has exactly n pivots. The entries in each pivotal column above and below the pivot are 0. The pivots are each equal to 1. Therefore, $EA = I$. Consequently, $A = E^{-1}$, a product of elementary matrices. ■

Notice that row equivalence of matrices can now be stated in an easier way. If A and B are $m \times n$ matrices, then A is row equivalent to B if and only if there exists an invertible matrix P such that $B = PA$.

When a sequence of elementary row operations reduce an invertible matrix A to I, the product of the corresponding elementary matrices is equal to A^{-1}. In order to use this in computation, the following writing aid will be helpful.

Definition 3.17 Let A be an $m \times n$ matrix and let B be an $m \times k$ matrix. The $m \times (n+k)$ matrix obtained by writing first all the columns of A and then the columns of B, in that order, is called the **augmented matrix** corresponding to the matrices A and B, and it is denoted by $[A|B]$.

Given a matrix $A \in \mathbb{F}^{n \times n}$, we start with the augmented matrix $[A|I]$; then apply elementary row operations for reducing A to its row reduced echelon form, while simultaneously applying the operations on the entries of I. This will reduce the augmented matrix $[A|I]$ to $[B|C]$.

If $B = I$, then $C = A^{-1}$. If A is not invertible, then B will have zero row(s) at the bottom. As $\operatorname{rank}([A|I]) = n$, this will result in at least one pivot in C. Therefore, in the reduction process if a pivot appears in C, we may stop and report that A is not invertible.

Example 3.18 Consider the following square matrices:

$$A = \begin{bmatrix} 1 & -1 & 2 & 0 \\ -1 & 0 & 0 & 2 \\ 2 & 1 & -1 & -2 \\ 1 & -2 & 4 & 2 \end{bmatrix}, \quad B = \begin{bmatrix} 1 & -1 & 2 & 0 \\ -1 & 0 & 0 & 2 \\ 2 & 1 & -1 & -2 \\ 0 & -2 & 0 & 2 \end{bmatrix}.$$

We want to find the inverses of the matrices, if at all they are invertible.

Augment A with an identity matrix to get

$$\left[\begin{array}{cccc|cccc} \boxed{1} & -1 & 2 & 0 & 1 & 0 & 0 & 0 \\ -1 & 0 & 0 & 2 & 0 & 1 & 0 & 0 \\ 2 & 1 & -1 & -2 & 0 & 0 & 1 & 0 \\ 1 & -2 & 4 & 2 & 0 & 0 & 0 & 1 \end{array}\right].$$

Use elementary row operations towards reducing A to its row reduced echelon form. Since $a_{11} = 1$, we leave $row(1)$ untouched. To zero-out the other entries in the first column, use the sequence of elementary row operations $E_1[2, 1]$; $E_{-2}[3, 1]$; $E_{-1}[4, 1]$ to obtain

$$\left[\begin{array}{rrrr|rrrr} \boxed{1} & -1 & 2 & 0 & 1 & 0 & 0 & 0 \\ 0 & -1 & 2 & 2 & 1 & 1 & 0 & 0 \\ 0 & 3 & -5 & -2 & -2 & 0 & 1 & 0 \\ 0 & -1 & 2 & 2 & -1 & 0 & 0 & 1 \end{array}\right].$$

The pivot is -1 in (2, 2) position. Use $E_{-1}[2]$ to make the pivot 1.

$$\left[\begin{array}{rrrr|rrrr} 1 & -1 & 2 & 0 & 1 & 0 & 0 & 0 \\ 0 & \boxed{1} & -2 & -2 & -1 & -1 & 0 & 0 \\ 0 & 3 & -5 & -2 & -2 & 0 & 1 & 0 \\ 0 & -1 & 2 & 2 & -1 & 0 & 0 & 1 \end{array}\right].$$

Use $E_1[1, 2],\ E_{-3}[3, 2],\ E_1[4, 2]$ to zero-out all nonpivot entries in the pivotal column to 0:

$$\left[\begin{array}{rrrr|rrrr} \boxed{1} & 0 & 0 & -2 & 0 & -1 & 0 & 0 \\ 0 & \boxed{1} & -2 & -2 & -1 & -1 & 0 & 0 \\ 0 & 0 & \boxed{1} & 4 & 1 & 3 & 1 & 0 \\ 0 & 0 & 0 & 0 & \boxed{-2} & -1 & 0 & 1 \end{array}\right].$$

Since a zero row has appeared, we find that A is not invertible. Also, observe that a pivot has appeared in the I portion of the augmented matrix. Though the reduced matrix is not in row reduced form, it can be brought to this form with further row operations. And rank $(A) = 3$; A is not invertible.

The second portion of the augmented matrix has no meaning now. However, it records the elementary row operations which were carried out in the reduction process. Verify that this matrix is equal to

$$E_1[4, 2]\, E_{-3}[3, 2]\, E_1[1, 2]\, E_{-1}[2]\, E_{-1}[4, 1]\, E_{-2}[3, 1]\, E_1[2, 1]$$

and that the A portion is equal to this matrix times A.

For B, we proceed similarly. The augmented matrix $[B|I]$ with the first pivot looks like:

$$\left[\begin{array}{rrrr|rrrr} \boxed{1} & -1 & 2 & 0 & 1 & 0 & 0 & 0 \\ -1 & 0 & 0 & 2 & 0 & 1 & 0 & 0 \\ 2 & 1 & -1 & -2 & 0 & 0 & 1 & 0 \\ 0 & -2 & 0 & 2 & 0 & 0 & 0 & 1 \end{array}\right].$$

The sequence of elementary row operations $E_1[2, 1],\ E_{-2}[3, 1]$ yields

$$\left[\begin{array}{rrrr|rrrr} \boxed{1} & -1 & 2 & 0 & 1 & 0 & 0 & 0 \\ 0 & -1 & 2 & 2 & 1 & 1 & 0 & 0 \\ 0 & 3 & -5 & -2 & -2 & 0 & 1 & 0 \\ 0 & -2 & 0 & 2 & 0 & 0 & 0 & 1 \end{array}\right].$$

Next, the pivot is -1 in $(2, 2)$ position. Use $E_{-1}[2]$ to get the pivot as 1.

$$\left[\begin{array}{cccc|cccc} \boxed{1} & -1 & 2 & 0 & 1 & 0 & 0 & 0 \\ 0 & \boxed{1} & -2 & -2 & -1 & -1 & 0 & 0 \\ 0 & 3 & -5 & -2 & -2 & 0 & 1 & 0 \\ 0 & -2 & 0 & 2 & 0 & 0 & 0 & 1 \end{array}\right].$$

And then $E_1[1, 2],\ E_{-3}[3, 2],\ E_2[4, 2]$ gives

$$\left[\begin{array}{cccc|cccc} \boxed{1} & 0 & 0 & -2 & 0 & -1 & 0 & 0 \\ 0 & \boxed{1} & -2 & -2 & -1 & -1 & 0 & 0 \\ 0 & 0 & 1 & 4 & 1 & 3 & 1 & 0 \\ 0 & 0 & -4 & -2 & -2 & -2 & 0 & 1 \end{array}\right].$$

Next pivot is 1 in $(3, 3)$ position. Now, $E_2[2, 3],\ E_4[4, 3]$ produces

$$\left[\begin{array}{cccc|cccc} \boxed{1} & 0 & 0 & -2 & 0 & -1 & 0 & 0 \\ 0 & \boxed{1} & 0 & 6 & 1 & 5 & 2 & 0 \\ 0 & 0 & \boxed{1} & 4 & 1 & 3 & 1 & 0 \\ 0 & 0 & 0 & 14 & 2 & 10 & 4 & 1 \end{array}\right].$$

Next pivot is 14 in $(4, 4)$ position. Use $E_{1/14}[4]$ to get the pivot as 1:

$$\left[\begin{array}{cccc|cccc} \boxed{1} & 0 & 0 & -2 & 0 & -1 & 0 & 0 \\ 0 & \boxed{1} & 0 & 6 & 1 & 5 & 2 & 0 \\ 0 & 0 & \boxed{1} & 4 & 1 & 3 & 1 & 0 \\ 0 & 0 & 0 & \boxed{1} & 1/7 & 5/7 & 2/7 & 1/14 \end{array}\right].$$

Use $E_2[1, 4],\ E_{-6}[2, 4],\ E_{-4}[3, 4]$ to zero-out the entries in the pivotal column:

$$\left[\begin{array}{cccc|cccc} \boxed{1} & 0 & 0 & 0 & 2/7 & 3/7 & 4/7 & 1/7 \\ 0 & \boxed{1} & 0 & 0 & 1/7 & 5/7 & 2/7 & -3/7 \\ 0 & 0 & \boxed{1} & 0 & 3/7 & 1/7 & -1/7 & -2/7 \\ 0 & 0 & 0 & \boxed{1} & 1/7 & 5/7 & 2/7 & 1/14 \end{array}\right].$$

Thus $B^{-1} = \dfrac{1}{7}\begin{bmatrix} 2 & 3 & 4 & 1 \\ 1 & 5 & 2 & -3 \\ 3 & 1 & -1 & -2 \\ 1 & 5 & 2 & 1/2 \end{bmatrix}$. Verify that $B^{-1}B = BB^{-1} = I$. □

Another alternative for computing the inverse of a matrix is to employ the method of bookkeeping by augmenting a symbolic vector. If we start with the augmented matrix $[A|y]$ for an invertible matrix A, then its row reduction will be the augmented matrix $[I|A^{-1}y]$. From the expression for the vector $A^{-1}y$, the expression for A^{-1} can be recovered with its columns as $A^{-1}e_j$ for $j = 1, \ldots, n$. This procedure can also be used to solve the equation

$$Ax = y.$$

Of course, when A is not invertible, a zero row will appear at the bottom of the first part of the augmented matrix.

Example 3.19 For computing the inverse of $A = \begin{bmatrix} 1 & 0 & -2 \\ 2 & 2 & 0 \\ 2 & 0 & -3 \end{bmatrix}$, we start with the augmented matrix $[A|y]$ and reduce A to its row reduced echelon form while applying the same operation for y.

$$\left[\begin{array}{ccc|c} 1 & 0 & -2 & y_1 \\ 2 & 2 & 0 & y_2 \\ 2 & 0 & -3 & y_3 \end{array}\right] \longrightarrow \left[\begin{array}{ccc|c} \boxed{1} & 0 & 0 & -3y_1 + 2y_3 \\ 0 & \boxed{1} & 0 & 3y_1 + \frac{1}{2}y_2 - 2y_3 \\ 0 & 0 & \boxed{1} & -2y_1 + y_3 \end{array}\right]$$

Looking at the y-entries, which is equal to $A^{-1}(y_1, y_2, y_3)^T$, we find that

$$A^{-1} = \begin{bmatrix} -3 & 0 & 2 \\ 3 & 1/2 & -2 \\ -2 & 0 & 1 \end{bmatrix}.$$

You can decide for yourself which of the two methods such as augmenting I or working with a symbolic vector is easier to work with. □

A matrix in row reduced echelon form is special. Let $A \in \mathbb{F}^{m\times n}$ have rank r. Suppose the columns of A are $u_1, \ldots, u_n$, from $\mathbb{F}^{m\times 1}$. That is,

$$A = [u_1 \; u_2 \; \cdots \; u_n].$$

Let B be the row reduced echelon form of A obtained by applying a sequence of elementary row operations. Let E be the $m \times m$ invertible matrix (which is the product of the corresponding elementary matrices) so that

$$EA = E[u_1 \; u_2 \; \cdots \; u_n] = B.$$

The number of pivots in B is r. Then the standard basis vectors $e_1, \ldots, e_r$ of $\mathbb{F}^{m\times 1}$ occur as the pivotal columns in B in the same order form left to right. Any two consecutive pivotal columns are of the form e_j and e_{j+1}; but they need not be two consecutive columns. Suppose the $n - r$ nonpivotal columns in B are $v_1, \ldots, v_{n-r}$, occurring in that order from left to right. In general, if v_i is a nonzero nonpivotal column, then there exists a maximum j such that $e_1, \ldots, e_j$ are all the pivotal columns that occur to the left of v_i. In this case, we say that e_j is the rightmost pivotal column to the left of v_i. Notice that a nonpivotal column that occurs to the right of some pivotal columns can also be a zero column.

Observation 3.1 In B, if e_j is the rightmost pivotal column to the left of a nonpivotal column v_i, then $v_i = [a_1\ a_2\ \cdots\ a_j\ 0\ 0\ \cdots 0]^T = a_1 e_1 + \cdots + a_j e_j$ for some $a_1, \ldots, a_j \in \mathbb{F}$.

In B, if e_1 occurs as k_1th column, e_2 occurs as k_2th column, and so on, then $v_i = a_1 E u_{k_1} + \cdots + a_j E u_{k_j}$. That is,

$$E^{-1} v_i = a_1 u_{k_1} + \cdots + a_j u_{k_j}.$$

However, $E^{-1} v_i$ is the ith column of A. Thus we observe the following.

Observation 3.2 In B, if e_j is the rightmost pivotal column to the left of a nonpivotal column $v_i = [a_1\ a_2\ \cdots\ a_j\ 0\ 0\ \cdots 0]^T$, and if $e_1, \ldots, e_j, v_i$ occur in the columns $k_1, \ldots, k_j, k$ respectively, then $u_k = a_1 u_{k_1} + \cdots + a_j u_{k_j}$.

It thus follows that each column of A can be written as a linear combination of $u_{k_1}, \ldots, u_{k_r}$, where $k_1, \ldots, k_r$ are all the column indices of pivotal columns in B. Notice that the range space of A is the span of all its columns, and $u_{k_1}, \ldots, u_{k_r}$ are linearly independent. Thus we have the following.

Observation 3.3 If $k_1, \ldots, k_r$ are all the column indices of pivotal columns in B, then $\{u_{k_1}, \ldots, u_{k_r}\}$ is a basis of $R(A)$.

If $v \in R(A)$, then $v = \beta_1 u_{k_1} + \cdots + \beta_r u_{k_r} = \beta_1 E^{-1} e_1 + \cdots + \beta_r E^{-1} e_r$ for some scalars $\beta_1, \ldots, \beta_r$. Since $e_{r+1} \notin \text{span}\{e_1, \ldots, e_r\}$, we see that $E^{-1} e_{r+1}$ is not expressible in the form $\alpha_1 E^{-1} e_1 + \cdots + \alpha_r E^{-1} e_r$. We conclude the following.

Observation 3.4 If $r = \text{rank}\,(A) < m$, then $E^{-1} e_{r+k} \notin R(A)$ for $1 \le k \le m - r$.

In B, the $m - r$ bottom rows are zero rows. They have been obtained from the pivotal rows by elementary row operations. Monitoring the row exchanges that have been applied on A to reach B, we see that the zero rows correspond to some $m - r$ rows of A. Therefore, we find the following.

Observation 3.5 Let $w_{k_1}, \ldots, w_{k_r}$ be the rows of A which have become the pivotal rows in B. Then $\{w_{k_1}, \ldots, w_{k_r}\}$ is a basis for the span of all rows of A.

The subspace spanned by all the rows of a matrix $A \in \mathbb{F}^{m \times n}$ is called the **row space** of A. Since the rows of A are the columns of A^T, we see that the row space of A is the same as the subspace $\{v^T : v \in R(A^T)\}$ of $\mathbb{F}^{1 \times n}$. The dimension of the row space of A is called the *row rank* of A. From Observation 3.5 it follows that the number of pivots in the row reduced echelon form of A is same as the row rank of A. That is,

The rank of A is equal to the row rank of A, which is also equal to the number of pivots in the row reduced form of A.

If B and C are two matrices in row reduced echelon form and they have been obtained from another matrix $A \in \mathbb{F}^{m\times n}$ by elementary row operations, then $B = E_1 A$ and $C = E_2 A$ for some invertible matrices E_1 and E_2 in $\mathbb{F}^{m\times m}$. We see that $B = E_1(E_2)^{-1}C$. Therefore, B and C are row equivalent. To show that the row reduced echelon form of a matrix is unique, we prove the following theorem.

Theorem 3.20 *Two matrices in row reduced echelon form are row equivalent if and only if they are equal.*

Proof Any matrix is row equivalent to itself. We thus prove only the "only if" part in the theorem. Let B, C be in row reduced echelon form with $B = EC$ for some invertible matrix E. We prove that $B = C$ by induction on the number of columns in B and C. In the basis step, suppose both B and C are column vectors. Since both B and C are in row reduced echelon form, either $B = 0$ or $B = e_1$; similarly, $C = 0$ or $C = e_1$. Notice that $B = EC$ and $C = E^{-1}B$. Thus $B = 0$ if and only if $C = 0$. This implies that $B = C = 0$ or $B = C = e_1$. In either case, $B = C$.

Lay out the induction hypothesis that the statement of the theorem holds true for all matrices with $k \geq 1$ number of columns. Let B and C be matrices having $k+1$ columns. Write $B = [B' \mid u]$, $C = [C' \mid v]$ in augmented form, where u and v are the last columns of B and C, respectively. Clearly, B' and C' are in row reduced echelon form, $B' = EC'$, and $u = Ev$.

By the induction hypothesis, $B' = C'$. Suppose that $e_1, \ldots, e_j$ are all the pivotal columns in C'. Then $e_1, \ldots, e_j$ are also all the pivotal columns in B'. Now, $B' = EC'$ implies that

$$e_1 = Ee_1, \ldots, Ee_j = e_j, \quad e_1 = E^{-1}e_1, \ldots, e_j = E^{-1}e_j.$$

In B, we see that either $u = e_{j+1}$ or $u = \alpha_1 e_1 + \cdots + \alpha_j e_j$ for some scalars $\alpha_1, \ldots, \alpha_j$. The latter case includes the possibility that $u = 0$. (If none of the first k columns in B is a pivotal column, we have $u = 0$.) Similarly, $v = e_{j+1}$ or $v = \beta_1 e_1 + \cdots + \beta_j e_j$ for some scalars $\beta_1, \ldots, \beta_j$. We consider the following exhaustive cases.

If $u = e_{k+1}$ and $v = e_{k+1}$, then $u = v$.

If $u = \alpha_1 e_1 + \cdots + \alpha_j e_j$, then $u = \alpha_1 E^{-1}e_1 + \cdots + \alpha_j E^{-1}e_j = E^{-1}u = v$.

If $v = \beta_1 e_1 + \cdots + \beta_j e_j$, then $u = Ev = \beta_1 Ee_1 + \cdots + \beta_j Ee_j = v$.

In all cases, $u = v$. Since $B' = C'$, we conclude that $B = C$. ■

Theorem 3.20 justifies our use of the term *the* row reduced echelon form of a matrix. Given a matrix, it does not matter whether you compute its row reduced echelon form by following our algorithm or any other algorithm; the end result is the same matrix in row reduced echelon form.

Given $u_1, \ldots, u_m \in \mathbb{F}^{1\times n}$, a basis for $U = \text{span}\{u_1, \ldots, u_n\}$ can be constructed by using elementary operations. We construct a matrix A with its rows as the given vectors:

$$A = \begin{bmatrix} u_1 \\ \vdots \\ u_m \end{bmatrix}.$$

We then convert A to its row echelon form or row reduced echelon form. The pivotal rows form a basis for U; the zero rows (nonpivotal rows) are linear combinations of the pivoted ones. Notice that the rows of A that correspond to the pivotal rows (monitoring row exchanges) also form a basis for U.

On the other hand, if $v_1, \ldots, v_m \in \mathbb{F}^{n\times 1}$ are column vectors, then we form the matrix A by taking these as columns. That is,

$$A = \begin{bmatrix} v_1 & \cdots & v_m \end{bmatrix}.$$

Then, we bring A to its row reduced echelon form, say, B. The columns in A corresponding to the pivotal columns in B form a basis for $V = \text{span}\{v_1, \ldots, v_m\} = R(A)$. The other columns of A are linear combinations of the basis vectors. The coefficients of such a linear combination is given by the entries in the corresponding nonpivotal column in B.

In fact, we can use any of the above methods by taking the transpose of the given vectors as required. This is illustrated in the following example.

Example 3.21 The problem is to determine a basis for the subspace U of $\mathbb{R}^4$ spanned by the vectors $u_1 = (0, 2, 0, 1)$, $u_2 = (1, 2, 2, 1)$, $u_3 = (2, 2, 4, 1)$ and $u_4 = (1, -2, 2, -1)$.

By taking the vectors as rows of a matrix A, and then bringing it to row echelon form, we obtain

$$A = \begin{bmatrix} 0 & 2 & 0 & 1 \\ 1 & 2 & 2 & 1 \\ 2 & 2 & 4 & 1 \\ 1 & -2 & 2 & -1 \end{bmatrix} \xrightarrow{E[1,2]} \begin{bmatrix} \boxed{1} & 2 & 2 & 1 \\ 0 & 2 & 0 & 1 \\ 2 & 2 & 4 & 1 \\ 1 & -2 & 2 & -1 \end{bmatrix} \xrightarrow{R} \begin{bmatrix} \boxed{1} & 2 & 2 & 1 \\ 0 & \boxed{2} & 0 & 1 \\ 0 & 0 & 0 & 0 \\ 0 & 0 & 0 & 0 \end{bmatrix}.$$

Here, R is the row operation $E_{-2}[3, 1]$, $E_{-1}[4, 1]$; $E_1[3, 2]$, $E_2[4, 2]$. The row echelon form shows that the third row and the fourth row in the row echelon form are linear combinations of the first and the second. In the row operation, the first and the second rows were exchanged. Taking the corresponding rows in A, we conclude that u_3 and u_4 are linear combinations of the pivotal rows u_2 and u_1; a basis for U is $\{u_2, u_1\}$. Same conclusion is drawn by bringing A to its row reduced echelon form.

Alternatively, we take the transposes of the given row vectors and solve the problem in $\mathbb{R}^{4\times 1}$; that is, let $v_1 = u_1^T$, $v_2 = u_2^T$, $v_3 = u_3^T$ and $v_4 = u_4^T$. Our aim is to extract a basis for $V = \{u^T : u \in U\}$. We then form the 4×4 matrix A with columns as these transposed vectors:

$$A = \begin{bmatrix} v_1 & v_2 & v_3 & v_4 \end{bmatrix} = \begin{bmatrix} 1 & 0 & 2 & 1 \\ 2 & 2 & 2 & -2 \\ 2 & 0 & 4 & 2 \\ 1 & 1 & 1 & -1 \end{bmatrix}.$$

Next, we bring A to its row reduced echelon form as follows:

$$\begin{bmatrix} \boxed{1} & 0 & 2 & 1 \\ 2 & 2 & 2 & -2 \\ 2 & 0 & 4 & 2 \\ 1 & 1 & 1 & -1 \end{bmatrix} \xrightarrow{R1} \begin{bmatrix} \boxed{1} & 0 & 2 & 1 \\ 0 & \boxed{2} & -2 & -4 \\ 0 & 0 & 0 & 0 \\ 0 & 1 & -1 & -2 \end{bmatrix} \xrightarrow{R2} \begin{bmatrix} \boxed{1} & 0 & 2 & 1 \\ 0 & \boxed{1} & -1 & -2 \\ 0 & 0 & 0 & 0 \\ 0 & 0 & 0 & 0 \end{bmatrix}.$$

Here, $R1 = E_{-2}[2,1],\ E_{-2}[3,1],\ E_{-1}[4,1]$, and $R2 = E_{1/2}[2],\ E_{-1}[4,2]$. The pivotal columns are $col(1)$ and $col(2)$. Thus v_1, v_2 is a basis of V. The entries in $col(3)$ and $col(4)$ say that $v_3 = 2v_1 - v_2$ and $v_4 = v_1 - 2v_2$. Therefore,

$$u_3 = 2u_1 - u_2, \quad u_4 = u_1 - 2u_2.$$

And $\{u_1, u_2\}$ is a basis for U. □

Let $A \in \mathbb{F}^{m\times n}$. Suppose E is an invertible matrix such that EA is in row reduced echelon form. If A has only $k < n$ number of pivotal columns, then the pivotal columns in EA are $e_1, \ldots, e_k$. The nonpivotal columns in EA are linear combinations of these pivotal columns. So, $R(EA) = \text{span}\{e_1, \ldots, e_k\}$. Thus, $e_{k+1} \notin R(EA)$. It follows that $E^{-1}e_{k+1} \notin R(A)$. Therefore, if a vector in $\mathbb{F}^{n\times 1}$ exists outside the span of the given vectors $v_1, \ldots, v_m \in \mathbb{F}^{n\times 1}$, then elementary row operations can be used to construct such a vector.

Exercises for Sect. 3.3

1. Reduce the following matrices to row reduced echelon form:

   (a) $\begin{bmatrix} 0 & 2 & 3 & 7 \\ 1 & 1 & 1 & 1 \\ 1 & 3 & 4 & 8 \\ 0 & 0 & 0 & 1 \end{bmatrix}$ (b) $\begin{bmatrix} 0 & 0 & 4 & 0 \\ 2 & 2 & -2 & 5 \\ 5 & 5 & -5 & 5 \end{bmatrix}$ (c) $\begin{bmatrix} 1 & 1 & 1 \\ 1 & 1 & 0 \\ 0 & 0 & 1 \end{bmatrix}$.
2. Construct three different matrices in $\mathbb{F}^{4\times 3}$, each of which is of rank
   (a) 0 (b) 1 (c) 2 (d) 3.
3. Determine whether the given pair of matrices are row equivalent.

   (a) $\begin{bmatrix} 1 & 2 \\ 4 & 8 \end{bmatrix}$, $\begin{bmatrix} 0 & 1 \\ 1 & 2 \end{bmatrix}$ (b) $\begin{bmatrix} 1 & 0 & 2 \\ 0 & 2 & 10 \\ 2 & 0 & 4 \end{bmatrix}$, $\begin{bmatrix} 1 & 0 & 2 \\ 3 & -1 & 1 \\ 5 & -1 & 5 \end{bmatrix}$.
4. Are the matrices $\begin{bmatrix} 1 & 0 & 0 \\ 0 & 1 & 0 \\ 0 & 0 & 1 \end{bmatrix}$ and $\begin{bmatrix} 1 & 0 & 1 \\ 0 & 1 & 0 \\ 0 & 0 & 1 \end{bmatrix}$ row equivalent? What are their row reduced echelon forms?
5. Let $A \in \mathbb{F}^{6\times 7}$ be in row reduced echelon form.

(a) If the fourth column of A is the second (from left) pivotal column, then how does this column look like?
(b) Delete the fifth column of A. Is the resulting matrix in $\mathbb{F}^{6\times 6}$ in row reduced echelon form?

6. Determine the inverse of the matrix $\begin{bmatrix} 1 & 1 & 1 \\ 1 & 2 & 3 \\ 1 & 4 & 9 \end{bmatrix}$.

7. In each of the following subspaces U and W of a vector space V, determine the bases and dimensions of U, W, $U \cap W$ and of $U + W$.

(a) $V = \mathbb{R}^3$, $U = \text{span}\{(1, 2, 3), (2, 1, 1)\}$, $W = \text{span}\{(1, 0, 1), (3, 0, -1)\}$.
(b) $V = \mathbb{R}^4$, $U = \text{span}\{(1, 0, 2, 0), (1, 0, 3, 0)\}$,
$W = \text{span}\{(1, 0, 0, 0), (0, 1, 0, 0), (0, 0, 1, 1)\}$.
(c) $V = \mathbb{C}^4$, $U = \text{span}\{(1, 0, 3, 2), (10, 4, 14, 8), (1, 1, -1, -1)\}$,
$W = \text{span}\{(1, 0, 0, 2), (3, 1, 0, 2), (7, 0, 5, 2)\}$.

8. Let $A \in \mathbb{F}^{m\times n}$. Show the following:

(a) If $m > n$, then there exists an invertible matrix $E \in \mathbb{F}^{m\times m}$ such that the last row of EA is a zero row. Conclude that there exists no matrix $B \in \mathbb{F}^{n\times m}$ such that $AB = I_m$.
(b) If $m < n$, then there exists an invertible matrix $D \in \mathbb{F}^{n\times n}$ such that the last column of AD is a zero column. Conclude that there exists no matrix $C \in \mathbb{F}^{n\times m}$ such that $CA = I_n$.

3.4 Reduction to Rank Echelon Form

We remarked earlier that reducing a matrix to its rank echelon form requires column operations. The column operations work with columns instead of rows. Once again, we break down column operations to simpler ones.

Definition 3.22 The following three kinds of operations on a matrix are called **elementary column operations**:

(a) Exchanging two columns;
(b) Multiplying a nonzero scalar to one of the columns;
(c) Adding to a column a nonzero scalar multiple of another column.

A **column operation** is a finite sequence of elementary column operations. We say that a matrix B is **column equivalent to** a matrix A if B has been obtained from A by using column operations.

Elementary matrices can be used for expressing elementary column operations as matrix products. In this regard we observe the following:

Exchanging ith and jth columns of I results in $E[i, j]$.
Multiplying α with the ith column of I results in $E_\alpha[i]$.
Adding to the ith column α times the jth column results in $E_\alpha[j, i]$.

Notice that $(E[i, j])^T = E[i, j]$, $(E_\alpha[i])^T = E_\alpha[i]$, and $(E_\alpha[i, j])^T = E_\alpha[j, i]$. Therefore, applying an elementary column operation on a matrix A is equivalent to the following:

Apply the corresponding row operation on A^T.
And then, take the transpose of the result.

We summarize the discussion in the following observation.

Observation 3.6 Let $A \in \mathbb{F}^{m \times n}$. For $i, j \in \{1, \ldots, n\}$, $i \neq j$, and scalar $\alpha \neq 0$, the following matrices are the results of applying the corresponding elementary column operations:

(1) $A\,E[i, j]$: exchanging the ith and the jth columns in A;
(2) $A\,E_\alpha[i]$: multiplying α with the ith column in A;
(3) $A\,E_\alpha[i, j]$: Adding to the jth column α times the ith column, in A.

We will use the following notation to specify an elementary column operation in a computation:

$$A \xrightarrow{E'} B,$$

where $B = A\,E$ and E is an elementary matrix. We also refer to the corresponding elementary column operation as E. We remember to use the "prime" with E to say that it is a column operation; thus E is to be post-multiplied.

Example 3.23 We continue with the matrix obtained in Example 3.4 for illustrating elementary column operations.

$$\begin{bmatrix} 1 & 1 & 2 & 3 \\ 0 & 1 & 1 & 1 \\ 0 & 0 & 0 & 0 \end{bmatrix} \xrightarrow{C1'} \begin{bmatrix} 1 & 0 & 0 & 0 \\ 0 & 1 & 1 & 1 \\ 0 & 0 & 0 & 0 \end{bmatrix} \xrightarrow{C2'} \begin{bmatrix} 1 & 0 & 0 & 0 \\ 0 & 1 & 0 & 0 \\ 0 & 0 & 0 & 0 \end{bmatrix},$$

where $C1 = E_{-1}[1, 2],\ E_{-2}[1, 3],\ E_{-3}[1, 4]$, and $C2 = E_{-1}[2, 3],\ E_{-1}[2, 4]$ are the column operations. As a matrix product the computation can be written as

$$\begin{bmatrix} 1 & 1 & 2 & 3 \\ 0 & 1 & 1 & 1 \\ 0 & 0 & 0 & 0 \end{bmatrix} E_{-1}[1, 2]\,E_{-2}[1, 3]\,E_{-3}[1, 4]E_{-1}[2, 3]\,E_{-1}[2, 4] = \begin{bmatrix} 1 & 0 & 0 & 0 \\ 0 & 1 & 0 & 0 \\ 0 & 0 & 0 & 0 \end{bmatrix}.$$

Notice that the order of (the transposes of) the elementary matrices in the product appear as they are, whereas in elementary row operations, they appeared in reverse order. □

Therefore, a matrix B is column equivalent to a matrix A if and only if $B = AE$, where E is a finite product of elementary matrices if and only if there exists an invertible matrix E such that $B = AE$.

From Theorem 2.36, it follows that if $A \in \mathbb{F}^{m\times n}$, and B, C are invertible matrices of size $m \times m$ and $n \times n$, respectively, then

$$\text{rank}\,(BA) = \text{rank}\,(A) = \text{rank}\,(AC).$$

That is, multiplying a matrix with an invertible matrix preserves the rank. This implies that both row operations and column operations preserve (column) rank of a matrix. Since the row rank of A is same as the rank of A^T, it also follows that both row and column operations preserve the row rank. It provides another proof of Theorem 2.57.

In fact, starting from the row reduced echelon form of a given matrix, we may use column operations to zero-out the nonpivotal columns. Then the result is the rank echelon matrix E_r of rank r. It gives an algorithm for constructing the required matrices P and Q in the following theorem.

Theorem 3.24 (Rank factorization) *Let $A \in \mathbb{F}^{m\times n}$. Then rank $(A) = r$ if and only if there exist invertible matrices $P \in \mathbb{F}^{m\times m}$ and $Q \in \mathbb{F}^{n\times n}$ such that $A = P \begin{bmatrix} I_r & 0 \\ 0 & 0 \end{bmatrix} Q$.*

The row and column operations provide a proof of rank factorization without depending on the rank theorem. Further, the rank theorem can be proved using rank factorization.

Notice that the matrix E_r can be written as the following matrix product:

$$E_r = \begin{bmatrix} I_r & 0 \\ 0 & 0 \end{bmatrix} = \begin{bmatrix} I_r \\ 0 \end{bmatrix} \begin{bmatrix} I_r & 0 \end{bmatrix}.$$

Then, in the rank factorization of A, write

$$C = P \begin{bmatrix} I_r \\ 0 \end{bmatrix}, \quad F = \begin{bmatrix} I_r & 0 \end{bmatrix} Q.$$

Since both P and Q are invertible, we see that rank $(C) = r = $ rank (F). We thus obtain the following result.

Theorem 3.25 (Rank decomposition) *Let $A \in \mathbb{F}^{m\times n}$ be of rank r. Then there exist matrices $C \in \mathbb{F}^{m\times r}$ and $F \in \mathbb{F}^{r\times n}$ each of rank r, such that $A = CF$.*

Rank decomposition is also referred to as the *full rank factorization*, since both the matrices C and F have maximum possible ranks.

Elementary column operations can also be used for computing the inverse. We can start with an augmented matrix A by writing I below it and apply elementary column operations. Once the part that contains A is reduced to I, the other block which originally contained I must give A^{-1}.

In yet another variation, augment A with I on the right and also I below A. Then perform row and column operations on the augmented matrix. The I portion on the right will record the row operations, whereas the I portion below A will record the column operations. Finally, when A is reduced to I, if the I portion on the right is the matrix P and the I portion below is the matrix Q, then $A^{-1} = PQ$.

Exercises for Sect. 3.4

1. Prove that a matrix $\begin{bmatrix} a & b \\ c & d \end{bmatrix}$ can be reduced to $\begin{bmatrix} 0 & \alpha \\ 1 & \beta \end{bmatrix}$ for some α, β by row and column operations unless $b = 0 = c$ and $a = d$. Explain the cases $b = 0$ and $c = 0$.
2. Compute the rank echelon form of $\begin{bmatrix} 1 & 2 & 1 & 1 \\ 3 & 0 & 0 & 4 \\ 1 & 4 & 2 & 2 \end{bmatrix}$ in two different ways:
   (a) by using elementary row and column operations in any order you like;
   (b) by reducing first to its row reduced echelon form, and then converting the row reduced echelon matrix to its column reduced echelon form.
3. Let $A = \begin{bmatrix} 11 & 21 & 31 & 41 \\ 12 & 22 & 32 & 42 \\ 13 & 23 & 33 & 43 \\ 14 & 24 & 34 & 44 \end{bmatrix}$.
   Compute the rank factorization and a rank decomposition of A.
4. Let $A = [a_{ij}]$ and $B = [b_{ij}]$ be $n \times n$ matrices with $a_{ii} = i$, $b_{i(n-i)} = 1$ and all other entries in both of them are 0. Using elementary row operations, determine whether A and B are equivalent.
5. Let $A \in \mathbb{F}^{m \times n}$. Show the following:
   (a) If A has a zero row, then there does not exist a matrix $B \in \mathbb{F}^{n \times m}$ such that $AB = I_m$.
   (b) If A has a zero column, then there does not exist a matrix $C \in \mathbb{F}^{n \times m}$ such that $CA = I_n$.
6. Compute the inverse of the matrix $\begin{bmatrix} 1 & 1 & 1 \\ 2 & 1 & 1 \\ 1 & 2 & 1 \end{bmatrix}$ using row operations, column operations, and also both row and column operations.
7. Use the rank theorem to prove the rank factorization theorem. Also, derive the rank theorem from rank factorization.
8. Using the rank theorem, prove that the row rank of a matrix is equal to its columns rank.
9. Using rank factorization show that $\operatorname{rank}(AB) \le \min\{\operatorname{rank}(A), \operatorname{rank}(B)\}$ for matrices $A \in \mathbb{F}^{m \times n}$ and $B \in \mathbb{F}^{n \times k}$. Construct an example where equality is achieved.

3.5 Determinant

There are two important quantities associated with a square matrix. One is the trace and the other is the determinant.

Definition 3.26 The **trace** of a square matrix is the sum of its diagonal entries. We write the trace of a matrix A by $\mathrm{tr}(A)$.

Thus, if $A = [a_{ij}] \in \mathbb{F}^{n\times n}$, then $\mathrm{tr}(A) = a_{11} + \cdots + a_{nn} = \sum_{k=1}^{n} a_{kk}$. Clearly, $\mathrm{tr}(I_n) = n$ and $\mathrm{tr}(0) = 0$. In addition, the trace satisfies the following properties:

Let $A, B \in \mathbb{F}^{n\times n}$.

1. $\mathrm{tr}(\alpha A) = \alpha \mathrm{tr}(A)$ for each $\alpha \in \mathbb{F}$.
2. $\mathrm{tr}(A^T) = \mathrm{tr}(A)$ and $\mathrm{tr}(A^*) = \overline{\mathrm{tr}(A)}$.
3. $\mathrm{tr}(A + B) = \mathrm{tr}(A) + \mathrm{tr}(B)$.
4. $\mathrm{tr}(AB) = \mathrm{tr}(BA)$.
5. $\mathrm{tr}(A^*A) = 0$ if and only if $\mathrm{tr}(AA^*) = 0$ if and only if $A = 0$.

The determinant of a square matrix is a bit involved.

Definition 3.27 The **determinant** of a square matrix $A = [a_{ij}] \in \mathbb{F}^{n\times n}$, written as $\det(A)$, is defined inductively as follows:

If $n = 1$, then $\det(A) = a_{11}$.

If $n > 1$, then $\det(A) = \sum_{j=1}^{n} (-1)^{1+j}\, a_{1j}\, \det(A_{1j})$

where the matrix $A_{1j} \in \mathbb{F}^{(n-1)\times(n-1)}$ is obtained from A by deleting the first row and the jth column of A.

When $A = [a_{ij}]$ is written showing all its entries, we also write $\det(A)$ by replacing the two big closing brackets [and] by two vertical bars | and |. Thus, for a 2×2 matrix $[a_{ij}]$, its determinant is seen as follows:

$$\begin{vmatrix} a_{11} & a_{12} \\ a_{21} & a_{22} \end{vmatrix} = (-1)^{1+1} a_{11} \det[a_{22}] + (-1)^{1+2} a_{12} \det[a_{21}] = a_{11}a_{22} - a_{12}a_{21}.$$

Similarly, for a 3×3 matrix, we need to compute three 2×2 determinants. For example,

$$\begin{aligned}
&\begin{vmatrix} 1 & 2 & 3 \\ 2 & 3 & 1 \\ 3 & 1 & 2 \end{vmatrix} \\
&= (-1)^{1+1} \times 1 \times \begin{vmatrix} 3 & 1 \\ 1 & 2 \end{vmatrix} + (-1)^{1+2} \times 2 \times \begin{vmatrix} 2 & 1 \\ 3 & 2 \end{vmatrix} + (-1)^{1+3} \times 3 \times \begin{vmatrix} 2 & 3 \\ 3 & 1 \end{vmatrix}. \\
&= 1 \times \begin{vmatrix} 3 & 1 \\ 1 & 2 \end{vmatrix} - 2 \times \begin{vmatrix} 2 & 1 \\ 3 & 2 \end{vmatrix} + 3 \times \begin{vmatrix} 2 & 3 \\ 3 & 1 \end{vmatrix} \\
&= (3 \times 2 - 1 \times 1) - 2 \times (2 \times 2 - 1 \times 3) + 3 \times (2 \times 1 - 3 \times 3) \\
&= 5 - 2 \times 1 + 3 \times (-7) = -18.
\end{aligned}$$

To see the determinant geometrically, consider a 2×2 matrix $A = [a_{ij}]$ with real entries. Let u be the vector with initial point at $(0, 0)$ and end point at (a_{11}, a_{12}). Similarly, let v be the vector starting from the origin and ending at the point (a_{21}, a_{22}). Their sum $u + v$ is the vector whose initial point is the origin and end point is $(a_{11} + a_{21}, a_{21} + a_{22})$. Denote by Δ, the area of the parallelogram with one vertex at the origin, and other vertices at the end points of vectors u, v, and $u + v$. Writing the acute angle between the vectors u and v as θ, we have

$$\begin{aligned}\Delta^2 &= |u|^2|v|^2 \sin^2\theta = |u|^2|v|^2(1 - \cos^2\theta)\\ &= |u|^2|v|^2\left(1 - \frac{(u \cdot v)^2}{|u|^2|v|^2}\right) = |u|^2|v|^2 - (u \cdot v)^2\\ &= (a_{11}^2 + a_{12}^2)(a_{21}^2 + a_{22}^2) - (a_{11}a_{21} + a_{12}a_{22})^2\\ &= (a_{11}a_{22} - a_{12}a_{21})^2 = (\det(A))^2.\end{aligned}$$

That is, the absolute value of $\det(A)$ is the area of the parallelogram whose sides are represented by the row vectors of A. In $\mathbb{R}^3$, similarly, it can be shown that the absolute value of $\det(A)$ is the volume of the parallelepiped whose sides are represented by the row vectors of A.

It is easy to see that if a matrix is either lower triangular, or upper triangular, or diagonal, then its determinant is the product of its diagonal entries. In particular, $\det(I_n) = 1$ and $\det(-I_n) = (-1)^n$.

Notation: Let $A \in \mathbb{F}^{n\times n}$, $v \in \mathbb{F}^{n\times 1}$, $i, j \in \{1. \ldots, n\}$.

1. We write the jth column of A as $\tilde{a_j}$. Thus, $\tilde{a_j} \in \mathbb{F}^{n\times 1}$ and $A = [\tilde{a_1} \quad \cdots \quad \tilde{a_n}]$.
2. A_{ij} denotes the $(n-1) \times (n-1)$ matrix obtained from A by deleting the ith row and the jth column of A.
3. $A_j(v)$ denotes the matrix obtained from A by replacing the jth column of A with v, and keeping all other columns unchanged.
4. We write $v = \begin{bmatrix} v_1 \\ v_1' \end{bmatrix}$, where v_1 is the first component of v and v_1' is the vector in $\mathbb{F}^{(n-1)\times 1}$ obtained from v by deleting its first component.

We consider the determinant as a function from $\mathbb{F}^{n\times n}$ to $\mathbb{F}$, and list its characterizing properties in the following theorem.

Theorem 3.28 *Let f be a function from $\mathbb{F}^{n\times n}$ to $\mathbb{F}$ and let $A \in \mathbb{F}^{n\times n}$. Then $f(A) = \det(A)$ if and only if f satisfies the following properties:*

(1) *For any $A \in \mathbb{F}^{n\times n}$, $x, y \in \mathbb{F}^{n\times 1}$, and $j \in \{1, \ldots, n\}$,*

$$f\big(A_j(x + y)\big) = f\big(A_j(x)\big) + f\big(A_j(y)\big).$$

(2) *For any $A \in \mathbb{F}^{n\times n}$, $x \in \mathbb{F}^{n\times 1}$, $\alpha \in \mathbb{F}$, and $j \in \{1, \ldots, n\}$,*

$$f\big(A_j(\alpha x)\big) = \alpha f\big(A_j(x)\big).$$

(3) *For* $A \in \mathbb{F}^{n\times n}$, *if* $B \in \mathbb{F}^{n\times n}$ *is the matrix obtained from A by exchanging two columns, then* $f(B) = -f(A)$.
(4) $f(I) = 1$.

Proof First, we show that the function the function $f(A) := \det(A)$, $A \in \mathbb{F}^{n\times n}$, satisfies all the four properties. For $n = 1$, the results in (1)–(3) are obvious. Lay out the induction hypothesis that for $n = m - 1$, the determinant function satisfies Properties (1)–(3). We use this to show that all of (1)–(3) hold for $n = m$. Let $x, y \in \mathbb{F}^{n\times 1}$.

$$\begin{aligned}(1) \qquad \Delta_1 &:= \det[\tilde{a}_1, \ldots, \tilde{a}_{k-1}, x + y, \tilde{a}_{k+1}, \ldots, \tilde{a}_m] \\ &= a_{11}\beta_1 - a_{12}\beta_2 + \cdots + (-1)^k(x_1 + y_1)\beta_k + \cdots + (-1)^{1+m} a_{mm}\beta_m,\end{aligned}$$

where for $j \neq k$, β_j is the determinant of the matrix obtained from A_{1j} by replacing the kth column with $x_1' + y_1'$ and $\beta_k = \det(A_{1k})$. By the induction hypothesis,

$$\beta_j = \gamma_j + \delta_j \quad \text{for } j \neq k$$

where γ_j is the determinant of the matrix obtained from A_{1j} by replacing the kth column with x_1', and δ_j is the determinant of the matrix obtained from A_{1j} by replacing the kth column with y_1'. Since $\beta_k = \det(A_{1k})$, we have

$$\begin{aligned}\Delta_1 &= (a_{11}\gamma_1 - a_{12}\gamma_2 + \cdots + (-1)^k x_1 \det(A_{1k}) + \cdots + (-1)^{1+m} a_{1m}\gamma_m) \\ &\quad +(a_{11}\delta_1 - a_{12}\delta_2 + \cdots + (-1)^k y_1 \det(A_{1k}) + \cdots + (-1)^{1+m} a_{1m}\delta_m) \\ &= \det[\tilde{a}_1, \ldots, \tilde{a}_{k-1}, x, \tilde{a}_{k+1}, \ldots, \tilde{a}_m] + \det[\tilde{a}_1, \ldots, \tilde{a}_{k-1}, y, \tilde{a}_{k+1}, \ldots, \tilde{a}_m].\end{aligned}$$

(2) For $j \neq k$, let λ_j be the determinant of the matrix obtained from A_{1j} by replacing the kth column with αu_1. Write $\lambda_k := \det(A_{1k})$. Then

$$\begin{aligned}\Delta_2 &:= \det[\tilde{a}_1, \ldots, \tilde{a}_{k-1}, \alpha x, \tilde{a}_{k+1}, \ldots, \tilde{a}_n] \\ &= a_{11}\lambda_1 - a_{12}\lambda_2 + \cdots + (-1)^k(\alpha x_1)\alpha_k + \cdots + (-1)^{1+m} a_{mm}\lambda_m.\end{aligned}$$

By the induction hypothesis,

$$\lambda_j = \alpha\mu_j \quad \text{for } j \neq k$$

where μ_j is the determinant of the matrix obtained from A_{1j} by replacing the kth column with u. Since $\lambda_k = \det(A_{1k})$, we have

$$\begin{aligned}\Delta_2 &= (a_{11}\alpha\mu_1 - a_{12}\alpha\mu_2 + \cdots + (-1)^k x_1 \det(A_{1k}) + \cdots + (-1)^{1+m} a_{1m}\alpha\mu_m) \\ &= \alpha \; \det[\tilde{a}_1, \ldots, \tilde{a}_{k-1}, x, \tilde{a}_{k+1}, \ldots, \tilde{a}_m].\end{aligned}$$

(3)
$$\Delta_3 := \det[\tilde{a}_1, \ldots, \tilde{a}_j, \ldots, \tilde{a}_i, \ldots, \tilde{a}_m]$$
$$= a_{11}\tau_1 + \cdots + a_{1(i-1)}\tau_{i-1} + a_{1j}\tau_j + \cdots a_{1i}\tau_i + \cdots a_{1m}\tau_m,$$

where τ_k is the determinant of the matrix obtained from A_{1k} by interchanging the ith and jth columns. Then by the induction hypothesis,

$$\tau_k = -\det(A_{ik}) \quad \text{for } k \neq i \text{ and } k \neq j.$$

For $k = i$, we see that $\tau_i = -\det(A_{1j})$ and for $k = j$, $\tau_j = -\det(A_{1i})$; again by the induction hypothesis. Hence,

$$\Delta_3 = -\big(a_{11}\det(A_{11}) + \cdots + (-1)^{1+n}a_{1m}\det(A_{1m})\big) = -\det(A).$$

(4) Since determinant of a diagonal matrix is the product of its diagonal entries, $\det(I) = 1$.

Next, we show that $\det(\cdot)$ is the only such function satisfying the four properties. For this, let $f : \mathbb{F}^{n\times n} \to \mathbb{F}$ be any function that satisfies Properties (1)–(3). We will come back to Property (4) in due course. We use, without mention, the already proved fact that $\det(\cdot)$ satisfies all the four properties.

First, let B be a matrix whose columns are from the standard basis $\{e_1, \ldots, e_n\}$ of $\mathbb{F}^{n\times 1}$. If two columns of B are identical, then by interchanging them we see that $f(B) = -f(B)$, due to Property (3). Hence, $f(B) = 0$. In this case, due to the same reason, $\det(B) = 0$. If no two columns in B are identical, then each basis vector e_k appears in B exactly once. Now, find, where is e_1 occurring. If it is the ith column, then exchange the first column and the ith column to bring e_1 to the first column. Then look for e_2. Continuing this way, we have a certain fixed number of exchanges after which B becomes I. Suppose the number of such exchange of columns is r. Then $f(B) = (-1)^r f(I)$. Again, due to the same reason, $\det(B) = (-1)^r \det(I) = (-1)^r$. Thus, for this particular type of matrices B, if f satisfies Properties(1)–(3), then $f(B) = \det(B)\, f(I)$.

We now consider the general case. Let $A = [a_{ij}] \in \mathbb{F}^{n\times n}$. Then the jth column of A is given by

$$\tilde{a}_j = a_{1j}e_1 + \cdots + a_{nj}e_n.$$

Writing each column of A in this manner and using Properties (1)–(2) for $f(\cdot)$ and $\det(\cdot)$, we see that

$$f(A) = \sum_{k=1}^{m} \alpha_k f(B_k), \quad \det(A) = \sum_{k=1}^{m} \alpha_k \det(B_k), \tag{3.1}$$

where $m = n^n$, each constant α_k is a scalar, and B_k is a matrix whose columns are chosen from the standard basis vectors $e_1, \ldots, e_n$. Notice that the same α_k appears as coefficient of $f(B_k)$ and $\det(B_k)$ in the sums for $f(\cdot)$ and $\det(A)$, since both are obtained using Properties (1)–(2).

For instance, if B_1 is the matrix with each column as e_1, then $\alpha_1 = a_{11}a_{12}\cdots a_{1n}$. Now, if a column is repeated in any B_k, then $f(B_k) = 0 = \det(B_k)$. In particular, $f(B_k) = \det(B_k)\, f(I)$. If no column is repeated in B_k, then each vector $e_1, \ldots, e_n$ occurs exactly once as a column in B_k. That is, B_k is a matrix obtained by permuting the columns of I. In this case, B_k can be obtained from I by some number of column exchanges, as we have shown earlier. Suppose the number of exchanges is s. By Property (3), $f(B_k) = (-1)^s f(I)$, and $\det(B_k) = (-1)^s \det(I) = (-1)^s$. Thus,

$$f(B_k) = \det(B_k)\, f(I) \quad \text{for } k = 1, \ldots, m.$$

Using (3.1), we conclude that $f(A) = \det(A)\, f(I)$. Now, Property (4) implies that $f(A) = \det(A)$. ■

Properties (1) and (2) say that the determinant is a *multi-linear map,* in the sense that it is linear in each column. Property (3) is the *alternating property* of the determinant. Property (4) is the *normalizing property*. The proof of Theorem 3.28 reveals the following fact.

Observation 3.7 If a function $f : \mathbb{F}^{n\times n} \to \mathbb{F}$ satisfies Properties (1)–(3) in Theorem 3.28, then for each $A \in \mathbb{F}^{n\times n}$, $f(A) = \det(A)\, f(I)$

Our definition of determinant expands the determinant in the first row. In fact, the same result may be obtained by expanding it in any other row, or even, in any column. Along with this, some more properties of the determinant are listed in Theorem 3.30 below. We first give a definition and then prove those properties.

Definition 3.29 Let $A = [a_{ij}] \in \mathbb{F}^{n\times n}$.

(a) The quantity $\det(A_{ij})$ is called the **(i, j)th minor** of A; it is denoted by $M_{ij}(A)$.
(b) The signed minor $(-1)^{i+j} \det(A_{ij})$ is called the **(i, j)th co-factor** of A; it is denoted by $C_{ij}(A)$.
(c) The matrix in $\mathbb{F}^{n\times n}$ whose (i, j)th entry is the (j, i)th co-factor $C_{ji}(A)$ is called the **adjugate** of A.

Following the notation in Definition 3.29, we obtain

$$\det(A) = \sum_{j=1}^{n} a_{1j}(-1)^{1+j} \det(A_{1j}) = \sum_{j=1}^{n} a_{1j}(-1)^{1+j} M_{1j}(A) = \sum_{j=1}^{n} a_{1j}\, C_{1j}(A).$$

The adjugate, $\operatorname{adj}(A)$, is the transpose of the matrix whose (i, j)th entry is $C_{ij}(A)$.

Theorem 3.30 *Let $A \in \mathbb{F}^{n\times n}$. Let $i, j, k \in \{1, \ldots, n\}$. Then the following statements are true.*

(1) $\det(A) = \sum_j a_{ij}(-1)^{i+j} \det(A_{ij}) = \sum_j a_{ij} C_{ij}(A)$ *for any fixed i.*
(2) *If some column of A is the zero vector, then* $\det(A) = 0$.
(3) *If a column of A is a scalar multiple of another column, then* $\det(A) = 0$.

(4) *If a column of A is replaced by that column plus a scalar multiple of another column, then determinant does not change.*
(5) *If A is a triangular matrix, then* $\det(A)$ *is equal to the product of the diagonal entries of A.*
(6) $\det(AB) = \det(A)\ \det(B)$ *for any matrix* $B \in \mathbb{F}^{n\times n}$.
(7) *If A is invertible, then* $\det(A) \neq 0$ *and* $\det(A^{-1}) = (\det(A))^{-1}$.
(8) *Columns of A are linearly dependent if and only if* $\det(A) = 0$.
(9) *If A and B are similar matrices, then* $\det(A) = \det(B)$.
(10) $\det(A) = \sum_i a_{ij}(-1)^{i+j}\ \det(A_{ij})$ *for any fixed j.*
(11) *rank* $(A) = n$ *if and only if* $\det(A) \neq 0$.
(12) *All of* (2), (3), (4) *and* (8) *are true for rows instead of columns.*
(13) $\det(A^T) = \det(A)$.
(14) $A\, adj(A) = adj(A) A = \det(A)\, I$.

Proof (1) Construct a matrix $C \in \mathbb{F}^{(n+1)\times(n+1)}$ by taking its first row as $e_1 \in \mathbb{F}^{n+1}$, its first column as e_1^T, and filling up the rest with the entries of A. In block form, it looks like:

$$C = \begin{bmatrix} 1 & 0 \\ 0 & A \end{bmatrix}.$$

Then $\det(C) = \det(A)$. Now, exchange the first row and the $(i+1)$th rows in C. Call the matrix so obtained as D. Then

$$\det(C) = -\det(D) = -\sum_j a_{ij}(-1)^{i+1+j}\det(D_{ij}).$$

The ith row of D_{ij} is $e_i \in \mathbb{F}^{1\times n}$. To compute $\det(D_{ij})$, exchange the first and the ith rows in D_{ij}. Then $\det(D_{ij}) = -\det(A_{ij})$. Therefore,

$$\det(A) = \det(C) = -\sum_j a_{ij}(-1)^{i+1+j}\det(D_{ij}) = \sum_j a_{ij}(-1)^{i+j}\det(A_{ij}).$$

(2)–(4) These properties follow from Properties (1)–(3) in Theorem 3.28.
(5) If A is lower triangular, then expand along the first row. Similarly, if A is upper triangular expand it along the first column.
(6) Let $A \in \mathbb{F}^{n\times n}$. Consider the function $g : \mathbb{F}^{n\times n} \to \mathbb{F}$ defined by

$$g(B) = \det(AB).$$

It is routine to verify that g satisfies Properties (1)–(3) of Theorem 3.28, where we take $g(\cdot)$ instead of $\det(\cdot)$. By Observation 3.7,

$$\det(AB) = g(B) = \det(B)\, g(I) = \det(A)\det(B).$$

(7) Let A be invertible. Due to (6), $\det(A)\det(A^{-1}) = \det(AA^{-1}) = \det(I) = 1$. It shows that $\det(A) \neq 0$ and $\det(A^{-1}) = (\det(A))^{-1}$.
(8) If the columns of A are linearly independent, then A is invertible. Due to (7), $\det(A) \neq 0$. So, suppose the columns of A are linearly dependent. Say, the kth column of A is a linear combination of the columns 1 through $k-1$. Then by the linearity of $\det(\cdot)$ in each of the columns, and by (3) above, $\det(A) = 0$.
(9) Matrices A and B are similar if and only if there exists an invertible matrix P such that $B = P^{-1}AP$. Then, due to (6) and (7), we have

$$\det(B) = \det(P^{-1})\det(A)\det(P) = \det(A).$$

(10) This property asserts that a determinant can be expanded in any column. Due to (1), we show that expansion of a determinant can be done in its first column. For a matrix of size 1×1 or 2×2, the result is obvious. Suppose that determinants of all matrices of order less than or equal to $n-1$ can be expanded in their first column. Let $A \in \mathbb{F}^{n\times n}$. As per our definition of the determinant, expanding in the first row,

$$\det(A) = \sum_j a_{1j}(-1)^{1+j}\det(A_{1j}).$$

The minors $\det(A_{ij})$ can be expanded in their first column, due to the induction assumption. That is,

$$\det(A_{1j}) = \sum_{i=2}^{m}(-1)^{i-1+1}a_{i1}B_{ij},$$

where B_{ij} denotes the determinant of the $(n-2)\times(n-2)$ matrix obtained from A by deleting the first and the ith rows, and deleting the first and the jth columns. Thus the only term in $\det(A)$ involving $a_{1j}a_{i1}$ is $(-1)^{i+j+1}a_{1j}a_{i1}Bij$.

Also, examining the expression

$$\sum_i a_{i1}(-1)^{i+1}\det(A_{i1}),$$

we see that the only term involving $a_{1j}a_{i1}$ is $(-1)^{i+j+1}a_{1j}a_{i1}Bij$.

Therefore, $\det(A) = \sum_i a_{i1}(-1)^{i+1}\det(A_{i1})$.

Alternatively, one can verify all the four properties in Theorem 3.28 for the function $f(A) = \sum_j a_{ij}(-1)^{i+j}\det(A_{ij})$.
(11) It follows from (8), since rank $(A) = n$ if and only if columns of A are linearly independent.
(12) It follows from (10).
(13) Use induction and (10).
(14) The adjugate $\text{adj}(A)$ of A is the matrix whose (i, j)th entry is $C_{ji}(A)$. Consider $A\,\text{adj}(A)$. Due to (1), the jth diagonal entry in this product is

$$\sum_{i=1}^{n} a_{ij} C_{ij}(A) = \det(A).$$

For the nondiagonal entries, let $i \neq j$. Construct a matrix B which is identical to A except at the jth row. The jth row of B is the same as the ith row of A. Now, look at the co-factors $C_{kj}(B)$. Such a co-factor is obtained by deleting the kth row and the jth column of B and then taking the determinant of the resulting $(n-1) \times (n-1)$ matrix. This is same as $C_{kj}(A)$. Since the ith and the kth rows are equal in B, the (i, k)th entry in $A \operatorname{adj}(A)$ is

$$\sum_{k=1}^{n} a_{ik} C_{ik}(A) = \sum_{k=1}^{n} a_{ik} C_{ik}(B) = \det(B) = 0,$$

Therefore, $A \operatorname{adj}(A) = \det(A)\, I$.

The product formula $\operatorname{adj}(A)A = \det(A)\, I$ is proved similarly. ■

In Theorem 3.30, Statement (2) follows from (3); it is an important particular case to be mentioned. Using (1)–(5) and (10)–(11), the *computational complexity* for evaluating a determinant can be reduced drastically. The trick is to bring a matrix to its row echelon form by using elementary row operations. While using row exchanges we must take care of the change of sign of the determinant. Similarly, we must account for α if we use an elementary row operation of the type $E_\alpha[i]$.

Example 3.31

$$\begin{vmatrix} 1 & 0 & 0 & 1 \\ -1 & 1 & 0 & 1 \\ -1 & -1 & 1 & 1 \\ -1 & -1 & -1 & 1 \end{vmatrix} \overset{R1}{=} \begin{vmatrix} 1 & 0 & 0 & 1 \\ 0 & 1 & 0 & 2 \\ 0 & -1 & 1 & 2 \\ 0 & -1 & -1 & 2 \end{vmatrix} \overset{R2}{=} \begin{vmatrix} 1 & 0 & 0 & 1 \\ 0 & 1 & 0 & 2 \\ 0 & 0 & 1 & 4 \\ 0 & 0 & -1 & 4 \end{vmatrix} \overset{R3}{=} \begin{vmatrix} 1 & 0 & 0 & 1 \\ 0 & 1 & 0 & 2 \\ 0 & 0 & 1 & 4 \\ 0 & 0 & 0 & 8 \end{vmatrix}.$$

Here, $R1 = E_1[2, 1],\ E_1[3, 1],\ E_1[4, 1]$; $R2 = E_1[3, 2],\ E_1[4, 2]$; and $R3 = E_1[4, 3]$.

Notice that the last determinant is easily computed, and its value is 8. □

In Theorem 3.30, Statement (12) implies that a square matrix is invertible if and only if its determinant is nonzero. Further, (14) gives a method of computing the inverse of a matrix if it exists, though it is less efficient than using elementary row operations.

From Theorem 3.11(9), we obtain the following theorem.

Theorem 3.32 *Let T be a linear operator on a finite dimensional vector space V, and let B and C be ordered bases of V. Then* $\det([T]_{B,B}) = \det([T]_{C,C})$.

In view of Theorem 3.32, we introduce the following definition.

Definition 3.33 Let T be a linear operator on a finite dimensional vector space V. Let $[T]$ be a matrix representation of T with respect to an ordered basis of V. The **determinant** of T is defined as $\det(T) := \det([T])$.

Note that, by Theorem 3.32, $\det(T)$ is well defined; it is independent of the basis used.

For linear operators T and S on a finite dimensional vector space, Theorem 3.30(6)–(8) entail the following.

Theorem 3.34 *Let T be a linear operator on a finite dimensional vector space V.*

(1) $\det(T) = 0$ *if and only if* rank $(T) < \dim(V)$.
(2) $\det(T) \neq 0$ *if and only if* T *is invertible.*
(3) *For each linear operator S on V,* $\det(ST) = \det(S)\det(T)$.

For a square matrix A, the matrix whose entries are the complex conjugates of the respective entries of A, is denoted by $\overline{A}$. Then the **adjoint**, of A, denoted by A^*, is the transpose of $\overline{A}$. A **hermitian** matrix is one for which its adjoint is the same as itself; a **unitary** matrix is one for which its adjoint coincides with its inverse. An **orthogonal** matrix is a real unitary matrix.

The determinant of a hermitian matrix is a real number, since $A = A^*$ implies

$$\det(A) = \det(A^*) = \det(\overline{A}) = \overline{\det(A)}.$$

For a unitary matrix A,

$$|\det(A)|^2 = \overline{\det(A)}\det(A) = \det(\overline{A})\det(A) = \det(A^*A) = \det(I) = 1.$$

That is, the determinant of a unitary matrix is of absolute value 1. It follows that the determinant of an orthogonal matrix is ± 1.

Execises for Sect. 3.5

1. Compute the determinant of $\begin{bmatrix} 3 & 1 & 1 & 2 \\ 1 & 2 & 0 & 1 \\ 1 & 1 & 2 & -1 \\ -1 & 1 & -1 & 3 \end{bmatrix}$.
2. Compute the inverse of the following matrices in two ways: by using elementary row (or column) operations, and by using the determinant-adjugate formula.
(a) $\begin{bmatrix} 4 & -2 & -2 & 0 \\ -2 & 1 & -1 & 1 \\ 0 & 1 & 1 & -1 \\ -2 & 1 & 1 & 1 \end{bmatrix}$ (b) $\begin{bmatrix} 4 & -7 & -5 \\ -2 & 4 & 3 \\ 3 & -5 & -4 \end{bmatrix}$ (c) $\begin{bmatrix} 1 & 0 & 2 & -2 \\ 2 & 1 & 1 & -1 \\ 1 & 0 & 0 & 2 \\ -1 & 1 & 1 & -1 \end{bmatrix}$.
3. For which values of $\alpha \in \mathbb{C}$ the matrix $\begin{bmatrix} 1 & \alpha & 0 & 0 \\ \alpha & 1 & 0 & 0 \\ 0 & \alpha & 1 & 0 \\ 0 & 0 & \alpha & 1 \end{bmatrix}$ is invertible?

Find the inverse in all such cases.

4. Compute $\det[e_n \ e_{n-1} \ \cdots \ e_2 \ e_1]$.
5. Show that $\operatorname{rank}(A) = \max\{k : \text{there exists a nonzero } k \times k \text{ minor of } A\}$ for any matrix $A \in \mathbb{F}^{n\times n}$. How do you generalize this for an $m \times n$ matrix?
6. Let $T : V \to V$ be a linear transformation, where $\dim(V) < \infty$. Then T is invertible if and only if for every basis E of V, $\det([T]_{E,E}) \neq 0$.
7. Let $u, v, w \in \mathbb{R}^3$. Show that $u \cdot (v \times w) =$ determinant of the matrix $[u \ v \ w]$.
8. Let $A, B \in \mathbb{F}^{n\times n}$. Is it true that $\det(A + B) = \det(A) + \det(B)$?
9. Let $v_1, \ldots, v_n \in \mathbb{F}^n$. Let A be the matrix whose rows are v_1 through v_n and let B be the matrix whose columns are v_1^T through v_n^T. Show that $\{v_1, \ldots, v_n\}$ is a basis for $\mathbb{F}^n$ if and only if $\det(A) \neq 0$ if and only if $\det(B) \neq 0$.
10. Let $A \in \mathbb{F}^{n\times n}$. Show that if $A \neq I$ and $A^2 = A$, then $\det(A) = 0$.
11. Let $A \in \mathbb{F}^{n\times n}$. Show that if $A^k = 0$ for some $k \in \mathbb{N}$, then $\det(A) = 0$.

3.6 Linear Equations

Consider a **linear system** with m equations in n unknowns:

$$\begin{aligned} a_{11}x_1 + a_{12}x_2 + \cdots a_{1n}x_n &= b_1 \\ a_{21}x_1 + a_{22}x_2 + \cdots a_{2n}x_n &= b_2 \\ &\vdots \\ a_{m1}x_1 + a_{m2}x_2 + \cdots a_{mn}x_n &= b_m \end{aligned}$$

Using the abbreviation $x = (x_1, \ldots, x_n)^T$, $b = (b_1, \ldots, b_m)^T$ and $A = [a_{ij}]$, the system can be written in a compact form as

$$Ax = b.$$

Of course, we can now look at the matrix $A \in \mathbb{F}^{m\times n}$ as a linear transformation from $\mathbb{F}^{n\times 1}$ to $\mathbb{F}^{m\times 1}$, where m is the *number of equations* and n is the *number of unknowns* in the system.

Notice that for linear systems, we deviate from our symbolism and write b as a column vector, and x_i for unknown scalars.

Definition 3.35 Let $A \in \mathbb{F}^{m\times n}$ and let $b \in \mathbb{F}^{m\times 1}$.

(a) The **solution set** of the linear system $Ax = b$ is given by

$$\operatorname{Sol}(A, b) := \{x \in \mathbb{F}^{n\times 1} : Ax = b\}.$$

(b) The system $Ax = b$ is said to be **solvable**, if and only if $\operatorname{Sol}(A, b) \neq \varnothing$.
(c) Each vector in $\operatorname{Sol}(A, b)$ is called a **solution** of $Ax = b$.

Also, we say that a system **is consistent** or **has a solution** when it is solvable. It follows that the system $Ax = b$ is solvable if and only if $b \in R(A)$. Further, $Ax = b$ has a unique solution if and only if $b \in R(A)$ and A is injective. These issues are better tackled with the help of the corresponding **homogeneous system** $Ax = 0$.

Notice that $N(A) = \mathrm{Sol}(A, 0)$ is the solution set of the homogeneous system. The homogeneous system always has a solution since $0 \in N(A)$. It has infinitely many solutions when $N(A)$ contains a nonzero vector.

To study the nonhomogeneous system, we use the augmented matrix $[A|b] \in \mathbb{F}^{m\times(n+1)}$ which has its first n columns as those of A in the same order, and its $(n+1)$th column is b.

As earlier, we write $A_j(b)$ to denote the matrix obtained from A by replacing its jth column with the column vector b.

The following theorem lists some important facts about solutions of a system of linear equations.

Theorem 3.36 *Let $A \in \mathbb{F}^{m\times n}$ and $b \in \mathbb{F}^{m\times 1}$. Then the following statements are true.*

(1) *$Ax = b$ is solvable if and only if $\mathrm{rank}\,([A|b]) = \mathrm{rank}\,(A)$.*
(2) *If u is a solution of $Ax = b$, then $\mathrm{Sol}(A, b) = u + N(A)$.*
(3) *$Ax = b$ has a unique solution if and only if $\mathrm{rank}\,([A|b]) = \mathrm{rank}\,(A) = n$.*
(4) *If u is a solution of $Ax = b$ and $\mathrm{rank}\,(A) < n$, then*

$$\mathrm{Sol}(A, b) = \{u + \alpha_1 v_1 + \cdots \alpha_k v_k : \alpha_1, \ldots, \alpha_k \in \mathbb{F}\},$$

where $k = n - \mathrm{rank}\,(A) > 0$ and $\{v_1, \ldots, v_k\}$ is a basis for $N(A)$.
(5) *If $[A'|b']$ is obtained from $[A|b]$ by a finite sequence of elementary row operations, then $\mathrm{Sol}(A', b') = \mathrm{Sol}(A, b)$.*
(6) *If $m = n$, then $Ax = b$ has a unique solution if and only if $\det(A) \neq 0$.*
(7) (Cramer's Rule) *If $m = n$, $\det(A) \neq 0$ and $x = (x_1, \ldots, x_n)^T$ is the solution of $Ax = b$, then $x_j = \det(\,A_j(b)\,)/\det(A)$ for each $j \in \{1, \ldots, n\}$.*

Proof (1) The system $Ax = b$ is solvable if and only if $b \in R(A)$ if and only if $\mathrm{rank}\,([A|b]) = \mathrm{rank}\,(A)$.
(2) Let u be a solution of $Ax = b$. Then $Au = b$. Now, $y \in \mathrm{Sol}(A, b)$ if and only if $Ay = b$ if and only if $Ay = Au$ if and only if $A(y - u) = 0$ if and only if $y - u \in N(A)$ if and only if $y \in u + N(A)$.
(3) Due to (1)–(2), and the rank-nullity theorem,
$Ax = b$ has a unique solution
if and only if $\mathrm{rank}\,([A|b]) = \mathrm{rank}\,(A)$ and $N(A) = \{0\}$
if and only if $\mathrm{rank}\,([A|b]) = \mathrm{rank}\,(A)$ and $\mathrm{rank}\,(A) = n$.

(4) Recall that, by the rank-nullity theorem, $\dim(N(A)) = k = n - \mathrm{rank}\,(A) > 0$. Hence, the result follows from (2).
(5) If $[A'|b']$ has been obtained from $[A|b]$ by a finite sequence of elementary row operations, then $A' = EA$ and $b' = Eb$, where E is the product of corresponding

elementary matrices. Since each elementary matrix is invertible, so is E. Therefore, for a vector u, we have $Au = b$ if and only if $EAu = Eb$, that is, $A'u = b'$.
(6) Suppose $m = n$. Then A is a linear operator on $\mathbb{F}^{n\times 1}$. Due to (3), $Ax = b$ has a unique solution if and only if rank $([A|b]) =$ rank $(A) = n$ if and only if rank $(A) = n$ if and only if $\det(A) \neq 0$ by Theorem 3.30(12).
(7) Since $\det(A) \neq 0$, by (6), $Ax = b$ has a unique solution, say $y \in \mathbb{F}^{n\times 1}$. Write the identity $Ay = b$ in the form:

$$y_1 \begin{bmatrix} a_{11} \\ \vdots \\ a_{n1} \end{bmatrix} + \cdots + y_j \begin{bmatrix} a_{1j} \\ \vdots \\ a_{nj} \end{bmatrix} + \cdots + y_n \begin{bmatrix} a_{1n} \\ \vdots \\ a_{nn} \end{bmatrix} = \begin{bmatrix} b_1 \\ \vdots \\ b_n \end{bmatrix}.$$

This gives

$$y_1 \begin{bmatrix} a_{11} \\ \vdots \\ a_{n1} \end{bmatrix} + \cdots + \begin{bmatrix} y_j a_{1j} - b_1 \\ \vdots \\ y_j a_{nj} - b_n \end{bmatrix} + \cdots + y_n \begin{bmatrix} a_{1n} \\ \cdots \\ a_{nn} \end{bmatrix} = 0.$$

So, the column vectors in this sum are linearly dependent with their coefficients as $y_1, \ldots, 1, \ldots, y_n$. Due to Theorem 3.30(8),

$$\begin{vmatrix} a_{11} & \cdots & (y_j a_{1j} - b_1) & \cdots & a_{1n} \\ & & \vdots & & \\ a_{n1} & \cdots & (y_j a_{nj} - b_n) & \cdots & a_{nn} \end{vmatrix} = 0.$$

From Theorem 3.28 it now follows that

$$y_j \begin{vmatrix} a_{11} & \cdots & a_{1j} & \cdots & a_{1n} \\ & & \vdots & & \\ a_{n1} & \cdots & a_{nj} & \cdots & a_{nn} \end{vmatrix} - \begin{vmatrix} a_{11} & \cdots & b_1 & \cdots & a_{1n} \\ & & \vdots & & \\ a_{n1} & \cdots & b_n & \cdots & a_{nn} \end{vmatrix} = 0.$$

Therefore, $y_j = \det(A_j(b))/\det(A)$. ■

Theorem 3.36(2) asserts that all solutions of the nonhomogeneous system can be obtained by adding a particular solution to a solution of the corresponding homogeneous system.

Example 3.37 (1) The system of linear equations

$$x_1 + x_2 = 3$$
$$x_1 - x_2 = 1$$

is rewritten as $Ax = b$, where

$$A = \begin{bmatrix} 1 & 1 \\ 1 & -1 \end{bmatrix}, \quad b = \begin{bmatrix} 3 \\ 1 \end{bmatrix}.$$

Here, $A \in \mathbb{F}^{2\times 2}$ is a square matrix, and

$$\text{rank}\,([A|b]) = \text{rank}\left[\begin{array}{cc|c} 1 & 1 & 3 \\ 1 & -1 & 1 \end{array}\right] = 2, \quad \text{rank}\,(A) = \text{rank}\begin{bmatrix} 1 & 1 \\ 1 & -1 \end{bmatrix} = 2.$$

Thus the linear system has a unique solution. It is given by $x_1 = 2,\ x_2 = 1$, or

$$x = \begin{bmatrix} 2 \\ 1 \end{bmatrix}.$$

(2) The linear system

$$\begin{aligned} x_1 + x_2 &= 3 \\ x_1 - x_2 &= 1 \\ 2x_1 - x_2 &= 3 \end{aligned}$$

is rewritten as $Ax = b$, where

$$A = \begin{bmatrix} 1 & 1 \\ 1 & -1 \\ 2 & -1 \end{bmatrix}, \quad b = \begin{bmatrix} 3 \\ 1 \\ 3 \end{bmatrix}.$$

Here, $A \in \mathbb{F}^{3\times 2}$. We find that

$$\text{rank}\,([A|b]) = \text{rank}\begin{bmatrix} 1 & 1 & 3 \\ 1 & -1 & 1 \\ 2 & -1 & 3 \end{bmatrix} = 2, \quad \text{rank}\,(A) = \text{rank}\begin{bmatrix} 1 & 1 \\ 1 & -1 \\ 2 & -1 \end{bmatrix} = 2.$$

Thus the linear system has a unique solution. The vector $x = [2,\ 1]^T$ is the solution of this system as in (1). The extra equation does not put any constraint on the solution(s) that we obtained earlier.

(3) Consider the system

$$\begin{aligned} x_1 + x_2 &= 3 \\ x_1 - x_2 &= 1 \\ 2x_1 + x_2 &= 3 \end{aligned}$$

In matrix form, the system looks like $Ax = b$, where

$$A = \begin{bmatrix} 1 & 1 \\ 1 & -1 \\ 2 & 1 \end{bmatrix}, \quad b = \begin{bmatrix} 3 \\ 1 \\ 3 \end{bmatrix}.$$

Here, $A \in \mathbb{F}^{3\times 2}$. We find that

$$\text{rank}\,([A|b]) = \text{rank} \begin{bmatrix} 1 & 1 & 3 \\ 1 & -1 & 1 \\ 2 & -1 & 3 \end{bmatrix} = 3, \quad \text{rank}\,(A) = \text{rank} \begin{bmatrix} 1 & 1 \\ 1 & -1 \\ 2 & -1 \end{bmatrix} = 2.$$

The system has no solution. We see that the first two equations again have the same solution $x_1 = 2, x_2 = 1$. But this time, the third is not satisfied by these values of the unknowns.

(4) Finally, we consider the linear equation

$$x_1 + x_2 = 3.$$

Here, the system is $Ax = b$, where

$$A = \begin{bmatrix} 1 & 1 \end{bmatrix}, \quad b = \begin{bmatrix} 3 \end{bmatrix}.$$

The matrix A has size 1×2. We see that

$$\text{rank}\,([A|b]) = \text{rank} \begin{bmatrix} 1 & 1 & 3 \end{bmatrix} = 1, \quad \text{rank}\,(A) = \text{rank} \begin{bmatrix} 1 & 1 \end{bmatrix} = 1.$$

Thus the system has a solution. The old solution $x = [2, \ 1]^T$ is still a solution of this system. As $k = n - \text{rank}\,(A) = 2 - 1 = 1$, the solution set is given by

$$\text{Sol}(A, b) = \{x + v : v \in N(A)\}.$$

The null space of A is obtained by solving the homogeneous system

$$x_1 + x_2 = 0.$$

Choosing $x_1 = 1, \ x_2 = -1$, we have $v = [1, \ -1]^T$, and

$$N(A) = \text{span}\{[1, \ 1]^T\} = \{[\alpha, \ -\alpha]^T : \alpha \in \mathbb{F}\}.$$

Therefore, the solution set is

$$\text{Sol}(A, b) = \left\{ \begin{bmatrix} 2+\alpha \\ 1-\alpha \end{bmatrix} : \alpha \in \mathbb{F} \right\} = \left\{ \begin{bmatrix} \beta \\ 3-\beta \end{bmatrix} : \beta \in \mathbb{F} \right\}.$$

This describes the "infinitely many solutions" as expected. □

Exercises for Sect. 3.6

1. Let $A \in \mathbb{R}^{n \times n}$. Give brief explanations for the following:

   (a) $\mathrm{Sol}(A, 0)$ is a subspace of $\mathbb{R}^{n \times 1}$.
   (b) A homogeneous system is always consistent.
   (c) For no $k \in \mathbb{N}$, $Ax = b$ has exactly k number of solutions.

2. For which triple (a, b, c) the following system has a solution?

   (a) $3x_1 - x_2 + 2x_3 = a,\ 2x_1 + x_2 + x_3 = b,\ x_1 - 3x_2 = c.$
   (b) $3x_1 - 6x_2 + 2x_3 - x_4 = a,\ \ -2x_1 + 4x_2 + x_3 - 3x_4 = b,\ x_3 + x_4 = c,$ $x_1 - 2x_2 + x_3 = 0.$

3. Determine whether the following system is solvable. In that case, find the solution set.
$$x_1 + 2x_2 + 3x_3 = 1,\ 4x_1 + 5x_2 + 6x_3 = 2,$$
$$7x_1 + 8x_2 + 9x_3 = 3,\ 5x_1 + 7x_2 + 9x_3 = 4.$$

4. Express the solution set of the following system by keeping the unknown x_3 arbitrary: $2x_1 + 2x_2 - x_3 + 3x_4 = 2,\ 2x_1 + 3x_2 + 4x_3 = -2,\ x_2 - 6x_3 = 6.$
5. Determine the value of k for which the following linear system has more than one solution: $6x_1 + x_2 + kx_3 = 11,\ 2x_1 + x_2 + x_3 = 3,\ 3x_1 - x_2 - x_3 = 7.$ Then find the solution set of the system.
6. Can a system of two linear equations in three unknowns be inconsistent?
7. Suppose a system of two linear equations in three unknowns is consistent. How many solutions the system has?
8. Does there exist a matrix B such that $B \begin{bmatrix} 1 & -1 \\ 2 & 2 \\ 1 & 0 \end{bmatrix} = \begin{bmatrix} 3 & 1 \\ -4 & 4 \end{bmatrix}$?
9. Let A be a 4×6 matrix. Show that the space of solutions of the linear system $Ax = 0$ has dimension at least 2.
10. Suppose that $A \in \mathbb{R}^{m \times n}$ and $y \in \mathbb{R}^{m \times 1}$. Show that if the linear system $Ax = y$ has a complex solution, then it has a real solution.
11. Let $A \in \mathbb{F}^{m \times n}$, and let $b \in \mathbb{F}^{m \times 1}$. Show the following:

    (a) The linear system $Ax = 0$ has a nonzero solution $x \in \mathbb{F}^{n \times 1}$ if and only if the n columns of A are linearly dependent.
    (b) The linear system $Ax = b$ has at most one solution $x \in \mathbb{F}^{n \times 1}$ if and only if the n columns of A are linearly independent.

12. Prove that two $m \times n$ matrices A and B are row equivalent if and only if $\mathrm{Sol}(A, 0) = \mathrm{Sol}(B, 0)$.
13. Let $A, B \in \mathbb{F}^{2 \times 3}$ be in row reduced echelon form. Show that if $\mathrm{Sol}(A, 0) = \mathrm{Sol}(B, 0)$, then $A = B$. What happens if $A, B \in \mathbb{F}^{m \times n}$?

14. Let $u_1, \ldots, u_m$ be linearly independent vectors in a vector space V. Let $A = [a_{ij}] \in \mathbb{F}^{m \times n}$. For $j \in \{1, \ldots, n\}$, define

$$v_j := a_{1j}u_1 + a_{2j}u_2 + \cdots + a_{mj}u_m.$$

Show that $v_1, \ldots, v_n$ are linearly independent if and only if the columns of A are linearly independent in $\mathbb{F}^{m \times 1}$.

3.7 Gaussian and Gauss–Jordan Elimination

Cramer's rule in Theorem 3.36(7) gives an explicit formula for determining the unique solution of a linear system, whenever it exists. Though a theoretical tool, it is not very useful for computation, as it involves determinants. A more efficient method is the *Gaussian elimination*, known to us from our school days. This, and its refinement called *Gauss–Jordan Elimination* use the elementary row operations.

The idea is simple. Suppose that A is an $m \times n$ matrix and b is a column vector of size m. Then for every invertible $m \times m$ matrix E,

$$Ax = b \quad \text{if and only if} \quad EAx = Eb.$$

We choose E in such a way that the matrix EA is simpler so that the linear system $EAx = Eb$ may be solved easily. Since E is a product of elementary matrices, we may use elementary row operations on the *system matrix* A and the right-hand side vector b simultaneously to reach a solution.

The method of Gaussian elimination employs such a scheme. It starts with reducing the system matrix A to its row echelon form. To use LU-factorization, the system $Ax = b$ is transformed to $PAx = Pb$; then to

$$LUx = Pb,$$

where L is a lower triangular matrix with each diagonal entry as 1, U is matrix in row echelon form, and P is a permutation matrix.

Writing $Ux = y$, we first solve the system $Ly = Pb$ for y. In the second stage, we solve the system $Ux = y$. Notice that it is easier to solve the systems $Ly = Pb$ and $Ux = y$; the first, by forward substitution, and the second, by backward substitution.

Example 3.38 Consider the following system of linear equations:

$$\begin{array}{rrrrl} x_1 & +x_2 & -4x_3 & & = \ \ 1 \\ 2x_1 & -x_2 & & -x_4 & = \ \ 2 \\ & x_2 & +4x_3 & -x_4 & = -4 \end{array}$$

The system matrix $A = [a_{ij}]$ is

$$A = \begin{bmatrix} 1 & 1 & -4 & 0 \\ 2 & -1 & 0 & -1 \\ 0 & 1 & 4 & -1 \end{bmatrix}.$$

As we have stated above, the first step in Gaussian elimination is to transform the system $Ax = b$ to a system $LUx = Pb$. For this, we first zero-out the entry a_{21} by replacing $row(2)$ with $row(2) - 2 \cdot row(1)$. The corresponding elementary matrix to be pre-multiplied with A is

$$E_{-2}[2, 1] = \begin{bmatrix} 1 & 0 & 0 \\ -2 & 1 & 0 \\ 0 & 0 & 1 \end{bmatrix}.$$

This matrix is obtained from the 3×3 identity matrix by replacing its second row $e_2 = [0 \;\; 1 \;\; 0]$ by $e_2 - 2e_1 = [-2 \;\; 1 \;\; 0]$. Then

$$E_{-2}[2, 1]\, A = \begin{bmatrix} 1 & 1 & -4 & 0 \\ 0 & -3 & 8 & -1 \\ 0 & 1 & 4 & -1 \end{bmatrix}.$$

Next, the entry a_{32} in $E_{-2}[2, 1]\, A$ is zeroed-out by replacing $row(3)$ with $row(3) + {}^1\!/_3 \cdot row(2)$. The corresponding elementary matrix is

$$E_{1/3}[3, 2] = \begin{bmatrix} 1 & 0 & 0 \\ 0 & 1 & 0 \\ 0 & 1/3 & 1 \end{bmatrix}.$$

And then

$$E_{1/3}[3, 2]\, E_{-2}[2, 1]\, A = U = \begin{bmatrix} 1 & 1 & -4 & 0 \\ 0 & -3 & 8 & -1 \\ 0 & 0 & 20/3 & -4/3 \end{bmatrix}.$$

Thus the system $Ax = b$ has now been brought to the form

$$E_{1/3}[3, 2]\, E_{-2}[2, 1]\, Ax = E_{1/3}[3, 2]\, E_{-2}[2, 1]\, b,$$

where $U = E_{1/3}[3, 2]\, E_{-2}[2, 1]\, A$ is in row echelon form; the lower triangular matrix L is given by

$$L = (E_{1/3}[3, 2]\, E_{-2}[2, 1])^{-1} = E_2[2, 1]\, E_{-1/3}[3, 2] = \begin{bmatrix} 1 & 0 & 0 \\ 2 & 1 & 0 \\ 0 & -1/3 & 1 \end{bmatrix}.$$

We verify that

$$A = \begin{bmatrix} 1 & 1 & -4 & 0 \\ 2 & -1 & 0 & -1 \\ 0 & 1 & 4 & -1 \end{bmatrix} = LU = \begin{bmatrix} 1 & 0 & 0 \\ 2 & 1 & 0 \\ 0 & -1/3 & 1 \end{bmatrix} \begin{bmatrix} 1 & 1 & -4 & 0 \\ 0 & -3 & 8 & -1 \\ 0 & 0 & 20/3 & -4/3 \end{bmatrix}.$$

Notice that no permutation of the rows of A has been used. That is, the permutation matrix P is equal to I. For solving the system $Ax = b$ with $b = [1\ 2\ -4]^T$, we first solve $Ly = Pb$, i.e., $Ly = b$, by forward substitution. The system is

$$y_1 = 1, \quad 2y_1 + y_2 = 2, \quad -\tfrac{1}{3}y_2 + y_3 = -4.$$

Its solution is $y = [1\ 0\ -4]^T$. Next, we solve $Ux = y$ by using back substitution. The system $Ux = y$ is

$$\begin{aligned} x_1 + x_2 - 4x_3 &= 1 \\ -3x_2 + 8x_3 - x_4 &= 0 \\ + \tfrac{20}{3}x_3 - \tfrac{4}{3}x_4 &= -4 \end{aligned}$$

The last equation gives $x_3 = -\frac{3}{5} + \frac{1}{5}x_4$. Substituting on the second equation, we have $x_2 = -\frac{8}{5} + \frac{1}{5}x_4$. Substituting both of these on the first equation, we obtain $x_1 = \frac{1}{5} + \frac{3}{5}x_4$. Here, x_4 can take any arbitrary value. □

Since for obtaining the LU-factorization of A, we use elementary row operations, a direct algorithm using the conversion to echelon form can be developed. In this method, we reduce the augmented matrix $[A|b]$ to its row echelon form and compute the solution set directly. This is usually referred to as *Gaussian elimination*. Before giving the algorithm, let us see how the idea works.

Example 3.39 Consider the following system of linear equations:

$$\begin{aligned} x_1 + x_2 + 2x_3 + x_5 &= 1 \\ 3x_1 + 5x_2 + 5x_3 + x_4 + x_5 &= 2 \\ 4x_1 + 6x_2 + 7x_3 + x_4 + 2x_5 &= 3 \\ x_1 + 5x_2 + 5x_4 + x_5 &= 2 \\ 2x_1 + 8x_2 + x_3 + 6x_4 + 0x_5 &= 2 \end{aligned}$$

We reduce the augmented matrix to its row echelon form as in the following:

$$\left[\begin{array}{ccccc|c} 1&1&2&0&1&1 \\ 3&5&5&1&1&2 \\ 4&6&7&1&2&3 \\ 1&5&0&5&1&2 \\ 2&8&1&6&0&2 \end{array}\right] \xrightarrow{R1} \left[\begin{array}{ccccc|c} \boxed{1}&1&2&0&1&1 \\ 0&2&-1&1&-2&-1 \\ 0&2&-1&1&-2&-1 \\ 0&4&-2&5&0&1 \\ 0&6&-3&6&-2&0 \end{array}\right] \xrightarrow{R2} \left[\begin{array}{ccccc|c} \boxed{1}&1&2&0&1&1 \\ 0&\boxed{2}&-1&1&-2&-1 \\ 0&0&0&0&0&0 \\ 0&0&0&3&4&3 \\ 0&0&0&3&4&3 \end{array}\right]$$

$$\xrightarrow{R3} \left[\begin{array}{cccccc|c} \boxed{1} & 1 & 2 & 0 & 1 & 1 \\ 0 & \boxed{2} & -1 & 1 & -2 & -1 \\ 0 & 0 & 0 & \boxed{3} & 4 & 3 \\ 0 & 0 & 0 & 0 & 0 & 0 \\ 0 & 0 & 0 & 3 & 4 & 3 \end{array}\right] \xrightarrow{R4} \left[\begin{array}{ccccc|c} \boxed{1} & 1 & 2 & 0 & 1 & 1 \\ 0 & \boxed{2} & -1 & 1 & -2 & -1 \\ 0 & 0 & 0 & \boxed{3} & 4 & 3 \\ 0 & 0 & 0 & 0 & 0 & 0 \\ 0 & 0 & 0 & 0 & 0 & 0 \end{array}\right].$$

Here, the row operations are given by
$R1 = E_{-3}[2, 1],\ E_{-4}[3, 1],\ E_{-1}[4, 1],\ E_{-2}[5, 1];\ \ R2 = E_{-1}[3, 2],\ E_{-2}[4, 2],$ $E_{-3}[5, 2];\ R3 = E[3, 4]$; and $R4 = E_{-1}[5, 3]$.
The equations now look like

$$\begin{array}{rrrrr} x_1 + x_2 + 2x_3 & & + x_5 &= & 1 \\ 2x_2 & -x_3 + x_4 & -2x_5 &= & -1 \\ & 3x_4 & +4x_5 &= & 3 \end{array}$$

The variables corresponding to the pivots are x_1, x_2, x_4; the other variables are x_3 and x_5. In back substitution, we express the variables corresponding to the pivots in terms of the others. It leads to

$$\begin{aligned} x_4 &= 1 - \tfrac{4}{3}x_5 \\ x_2 &= \tfrac{1}{2}[-1 + x_3 - x_4 + 2x_5] = \tfrac{1}{2}[-1 + x_3 - (1 - \tfrac{4}{3}x_5) + 2x_5] = -1 + \tfrac{1}{2}x_3 + \tfrac{5}{3}x_5 \\ x_1 &= 1 - x_2 - 2x_3 - x_5 = 1 - [-1 + \tfrac{1}{2}x_3 + \tfrac{5}{3}x_5] - 2x_3 - x_5 = 2 - \tfrac{5}{2}x_3 - \tfrac{8}{3}x_5. \end{aligned}$$

The variables other than those corresponding to pivots, that is, x_3 and x_5 can take any arbitrary values. Writing $x_3 = -\alpha$ and $x_5 = -\beta$ for any $\alpha, \beta \in \mathbb{F}$, the solution set is given by

$$\mathrm{Sol}(A, b) = \left\{ \begin{bmatrix} 2 \\ -1 \\ 0 \\ 1 \\ 0 \end{bmatrix} + \alpha \begin{bmatrix} 5/2 \\ -1/2 \\ -1 \\ 0 \\ 0 \end{bmatrix} + \beta \begin{bmatrix} 8/3 \\ -5/3 \\ 0 \\ 4/3 \\ -1 \end{bmatrix} : \alpha, \beta \in \mathbb{F} \right\}.$$

Instead of taking x_3 and x_5 as α and β, we take $-\alpha$ and $-\beta$ for convenience in writing an algorithm for the general case. Look at the algorithm of Gaussian Elimination below. □

The unknowns in the system corresponding to the pivots are marked as **basic variables**, and others are marked as **free variables**. In Example 3.39, the basic variables are x_1, x_2, x_4, and the free variables are x_3, x_5. Using the row echelon form the basic variables are expressed in terms of the free variables to get a solution.

We present Gaussian elimination as an algorithm. It starts from a linear system $Ax = b$ and ends at producing the solution set $\mathrm{Sol}(A, b)$.

Gaussian Elimination

1. Construct the augmented matrix $[A|b]$.
2. Use the algorithm *Reduction to Row Echelon Form* to compute the row echelon form of $[A|b]$. Call the result as $[A'|b']$.
3. If a pivot occurs in b', then the system does not have a solution. Stop.
4. Else, delete zero rows at the bottom of $[A'|b']$.
5. Use the second type of elementary row operations to bring all pivots to 1.
6. Zero-out all nonpivotal entries in each pivotal column by using elementary row operations, starting from the rightmost pivotal column back to the leftmost pivotal column.
7. If in the current augmented matrix $[B|c]$, every column of B is a pivotal column, then $\mathrm{Sol}(A, b) = \{c\}$. Stop.
8. Else, insert required number of zero rows to $[B|c]$ so that in the resulting matrix $[B'|c']$, each pivot is a diagonal entry.
9. Adjoin a required number of zero rows at the bottom of $[B'|c']$ to make the B' portion a square matrix.
10. Change all nonpivotal diagonal entries to -1 to obtain $[\tilde{B}|\tilde{c}]$.
11. Suppose the nonpivotal columns in $[\tilde{B}|\tilde{c}]$ are $v_1, \ldots, v_k$. Then

$$\mathrm{Sol}(A, b) = \{\tilde{c} + \alpha_1 v_1 + \cdots + \alpha_k v_k : \alpha_1, \ldots, \alpha_k \in \mathbb{F}\}.$$

Example 3.40 We continue Example 3.39, starting from the row echelon form of the augmented matrix, as follows.

$$\left[\begin{array}{ccccc|c} \boxed{1} & 1 & 2 & 0 & 1 & 1 \\ 0 & \boxed{2} & -1 & 1 & -2 & -1 \\ 0 & 0 & 0 & \boxed{3} & 4 & 3 \\ 0 & 0 & 0 & 0 & 0 & 0 \\ 0 & 0 & 0 & 0 & 0 & 0 \end{array}\right] \xrightarrow{del-0} \left[\begin{array}{ccccc|c} \boxed{1} & 1 & 2 & 0 & 1 & 1 \\ 0 & \boxed{2} & -1 & 1 & -2 & -1 \\ 0 & 0 & 0 & \boxed{3} & 4 & 3 \end{array}\right]$$

$$\xrightarrow{R5} \left[\begin{array}{ccccc|c} \boxed{1} & 1 & 2 & 0 & 1 & 1 \\ 0 & \boxed{1} & -1/2 & 1/2 & -1 & -1/2 \\ 0 & 0 & 0 & \boxed{1} & 4/3 & 1 \end{array}\right] \xrightarrow{E_{-1/2}[2,3]} \left[\begin{array}{ccccc|c} \boxed{1} & 1 & 2 & 0 & 1 & 1 \\ 0 & \boxed{1} & -1/2 & 0 & -5/3 & -1 \\ 0 & 0 & 0 & \boxed{1} & 4/3 & 1 \end{array}\right]$$

$$\xrightarrow{E_{-1}[1,2]} \left[\begin{array}{ccccc|c} \boxed{1} & 0 & 5/2 & 0 & 8/3 & 2 \\ 0 & \boxed{1} & -1/2 & 0 & -5/3 & -1 \\ 0 & 0 & 0 & \boxed{1} & 4/3 & 1 \end{array}\right].$$

Here, $del - 0$ means deletion of all zero rows at the bottom, and $R5$ is $E_{1/2}[2]$, $E_{1/3}[3]$, which brings all pivots to 1.

This completes Step 5. Now, there are two nonpivotal columns; so we insert a zero rows between the second and the third to bring pivots to the diagonal. Next, we adjoin a zero row at the bottom to make the resulting B' portion in $[B'|c']$ a square matrix. Then we continue changing the nonpivotal entries on the diagonal to -1 as follows.

$$\xrightarrow{ins-0}\left[\begin{array}{ccccc|c} \boxed{1} & 0 & 5/2 & 0 & 8/3 & 2 \\ 0 & \boxed{1} & -1/2 & 0 & -5/3 & -1 \\ 0 & 0 & 0 & 0 & 0 & 0 \\ 0 & 0 & 0 & \boxed{1} & 4/3 & 1 \\ 0 & 0 & 0 & 0 & 0 & 0 \end{array}\right] \xrightarrow{ch(-1)} \left[\begin{array}{ccccc|c} \boxed{1} & 0 & 5/2 & 0 & 8/3 & 2 \\ 0 & \boxed{1} & -1/2 & 0 & -5/3 & -1 \\ 0 & 0 & -1 & 0 & 0 & 0 \\ 0 & 0 & 0 & \boxed{1} & 4/3 & 1 \\ 0 & 0 & 0 & 0 & -1 & 0 \end{array}\right] = [\tilde{B}|\tilde{c}].$$

The nonpivotal columns in $\tilde{B}$ are the third and the fifth, say, v_3 and v_5. The solution set is given by $\text{Sol}(A, b) = \{\tilde{c} + \alpha_1 v_1 + \alpha_2 v_5 : \alpha_1, \alpha_2 \in \mathbb{F}\}$. That is,

$$\text{Sol}(A, b) = \left\{ \begin{bmatrix} 2 \\ -1 \\ 0 \\ 1 \\ 0 \end{bmatrix} + \alpha_1 \begin{bmatrix} 5/2 \\ -1/2 \\ -1 \\ 0 \\ 0 \end{bmatrix} + \alpha_2 \begin{bmatrix} 8/3 \\ -5/3 \\ 0 \\ 4/3 \\ -1 \end{bmatrix} : \alpha_1, \alpha_2 \in \mathbb{F} \right\}. \qquad \square$$

Example 3.41 Consider the system of linear equations:

$$x_1 + x_2 - 4x_3 = 1,\ 2x_1 - x_2 - x_4 = 2,\ x_2 + 4x_3 - x_4 = -4.$$

Using Gaussian elimination on the augmented matrix, we obtain

$$\left[\begin{array}{cccc|c} \boxed{1} & 1 & -4 & 0 & 1 \\ 2 & -1 & 0 & -1 & 2 \\ 0 & 1 & 4 & -1 & -4 \end{array}\right] \xrightarrow{E_{-2}[2,1]} \left[\begin{array}{cccc|c} \boxed{1} & 1 & -4 & 0 & 1 \\ 0 & \boxed{-3} & 8 & -1 & 0 \\ 0 & 1 & 4 & -1 & -4 \end{array}\right]$$

$$\xrightarrow{E_{1/3}[3,2]} \left[\begin{array}{cccc|c} \boxed{1} & 1 & -4 & 0 & 1 \\ 0 & \boxed{-3} & 8 & -1 & 0 \\ 0 & 0 & \boxed{20/3} & -4/3 & -4 \end{array}\right] \xrightarrow{R1} \left[\begin{array}{cccc|c} \boxed{1} & 1 & -4 & 0 & 1 \\ 0 & \boxed{1} & -8/3 & 1/3 & 0 \\ 0 & 0 & \boxed{1} & -1/5 & -3/5 \end{array}\right]$$

$$\xrightarrow{R2} \left[\begin{array}{cccc|c} \boxed{1} & 1 & 0 & -4/5 & -7/5 \\ 0 & \boxed{1} & 0 & -1/5 & -8/5 \\ 0 & 0 & \boxed{1} & -1/5 & -3/5 \end{array}\right] \xrightarrow{E_{-1}[1,2]} \left[\begin{array}{cccc|c} \boxed{1} & 0 & 0 & -3/5 & 1/5 \\ 0 & \boxed{1} & 0 & -1/5 & -8/5 \\ 0 & 0 & \boxed{1} & -1/5 & -3/5 \end{array}\right]$$

$$\xrightarrow{ins-0} \left[\begin{array}{cccc|c} \boxed{1} & 0 & 0 & -3/5 & 1/5 \\ 0 & \boxed{1} & 0 & -1/5 & -8/5 \\ 0 & 0 & \boxed{1} & -1/5 & -3/5 \\ 0 & 0 & 0 & 0 & 0 \end{array}\right] \xrightarrow{ch(-1)} \left[\begin{array}{cccc|c} \boxed{1} & 0 & 0 & -3/5 & 1/5 \\ 0 & \boxed{1} & 0 & -1/5 & -8/5 \\ 0 & 0 & \boxed{1} & -1/5 & -3/5 \\ 0 & 0 & 0 & -1 & 0 \end{array}\right].$$

Here, $R1$ is $E_{-1/3}[2]$, $E_{3/20}[3]$; $R2$ is $E_{8/3}[2,3]$, $E_4[1,3]$. The solution set is given by

$$\mathrm{Sol}(A,b) = \left\{ \begin{bmatrix} 1/5 \\ -8/5 \\ -3/5 \\ 0 \end{bmatrix} + \alpha_1 \begin{bmatrix} -3/5 \\ -1/5 \\ -1/5 \\ -1 \end{bmatrix} : \alpha_1 \in \mathbb{F} \right\} = \left\{ \frac{1}{5} \begin{bmatrix} 1 \\ -8 \\ -3 \\ 0 \end{bmatrix} + \alpha \begin{bmatrix} 3 \\ 1 \\ 1 \\ 5 \end{bmatrix} : \alpha \in \mathbb{F} \right\}.$$

That is,

$$x_1 = \tfrac{1}{5} + 3\alpha,\ x_2 = -\tfrac{8}{5} + \alpha,\ x_3 = -\tfrac{3}{5} + \alpha,\ x_4 = 5\alpha,$$

where α is any number in $\mathbb{F}$. Verify that it is a solution. The null space of the system matrix has the basis $\{(3,1,1,5)^T\}$. □

In Steps (5)–(6) of the Gaussian elimination, we make the pivots as 1 and then zero-out the nonpivotal entries above the pivot in all pivotal columns. This signals at using the row reduced echelon form instead of using the row echelon form. The result is the *Gauss–Jordan elimination* method.

In this method, we convert the augmented matrix $[A\,|\,b]$ to its row reduced echelon form, using elementary row operations. Thus back substitution of Gaussian elimination is done within the pivoting steps. All other steps taken in Gaussian elimination are followed as they are.

We describe Gauss–Jordan method as an algorithm with input $Ax = b$ and output $\mathrm{Sol}(A,b)$, where $A \in \mathbb{F}^{m\times n}$.

Gauss–Jordan Elimination

1. Construct the augmented matrix $[A\,|\,b]$.
2. Use the algorithm *Reduction to Row Reduced Echelon Form* on $[A\,|\,b]$; call the row reduced echelon form as $[A'\,|\,b']$.
3. If a pivot occurs in b', then the system does not have a solution. Stop.
4. Else, delete the zero rows in $[A'\,|\,b']$, all of which occur at the bottom.
5. If every column of A' is a pivotal column, then $\mathrm{Sol}(A,b) = \{b'\}$. Stop.
6. Else, insert required number of zero rows to $[A'\,|\,b']$ so that in the resulting matrix $[B\,|\,c]$, each pivot is a diagonal entry.
7. Adjoin required number of zero rows at the bottom of $[B\,|\,c]$ to make the B portion a square matrix.
8. Change all nonpivotal diagonal entries to -1 to obtain $[\tilde{B}\,|\,\tilde{c}]$.
9. Suppose the nonpivotal columns in $[\tilde{B}\,|\,\tilde{c}]$ are $v_1, \ldots, v_k$. Then

$$\mathrm{Sol}(A,b) = \{\tilde{c} + \alpha_1 v_1 + \cdots + \alpha_k v_k : \alpha_1, \ldots, \alpha_k \in \mathbb{F}\}.$$

Example 3.42 Consider the system of linear equations in Example 3.39:

$$\begin{aligned}
x_1 + x_2 + 2x_3 + \quad\quad x_5 &= 1\\
3x_1 + 5x_2 + 5x_3 + x_4 + x_5 &= 2\\
4x_1 + 6x_2 + 7x_3 + x_4 + 2x_5 &= 3\\
x_1 + 5x_2 + \quad\quad 5x_4 + x_5 &= 2\\
2x_1 + 8x_2 + x_3 + 6x_4 + 0x_5 &= 2
\end{aligned}$$

The row echelon form of the corresponding augmented matrix is computed as follows:

$$\left[\begin{array}{ccccc|c} 1&1&2&0&1&1\\ 3&5&5&1&1&2\\ 4&6&7&1&2&3\\ 1&5&0&5&1&2\\ 2&8&1&6&0&2 \end{array}\right] \xrightarrow{R1} \left[\begin{array}{ccccc|c} \boxed{1}&1&2&0&1&1\\ 0&2&-1&1&-2&-1\\ 0&2&-1&1&-2&-1\\ 0&4&-2&5&0&1\\ 0&6&-3&6&-2&0 \end{array}\right]$$

$$\xrightarrow{R2} \left[\begin{array}{ccccc|c} \boxed{1}&0&5/2&-1/2&2&3/2\\ 0&\boxed{1}&-1/2&1/2&-1&-1/2\\ 0&0&0&0&0&0\\ 0&0&0&3&4&3\\ 0&0&0&3&4&3 \end{array}\right] \xrightarrow{R3} \left[\begin{array}{ccccc|c} \boxed{1}&0&5/2&-1/2&2&3/2\\ 0&\boxed{1}&-1/2&1/2&-1&-1/2\\ 0&0&0&\boxed{1}&4/3&1\\ 0&0&0&0&0&0\\ 0&0&0&3&4&3 \end{array}\right]$$

$$\xrightarrow{R4} \left[\begin{array}{ccccc|c} \boxed{1}&0&5/2&0&8/3&2\\ 0&\boxed{1}&-1/2&0&-5/3&-1\\ 0&0&0&\boxed{1}&4/3&1\\ 0&0&0&0&0&0\\ 0&0&0&0&0&0 \end{array}\right].$$

Here,

$R1 = E_{-3}[2, 1],\ E_{-4}[3, 1],\ E_{-1}[4, 1],\ E_{-2}[5, 1]$;
$R2 = E_{1/2}[2],\ E_{-1}[1, 2],\ E_{-2}[3, 2],\ E_{-4}[4, 2],\ E_{-6}[5, 2]$;
$R3 = E[3, 4],\ E_{1/3}[3]$; and
$R4 = E_{1/2}[1, 3],\ E_{-1/2}[2, 3],\ E_{-3}[5, 3]$.

The basic variables are x_1, x_2 and x_4; the free variables are x_3 and x_5. The rewritten form of the equations are

$$\begin{aligned}
x_1 &= 2 - \tfrac{5}{2}x_3 - \tfrac{8}{3}x_5\\
x_2 &= -1 + \tfrac{1}{2}x_3 + \tfrac{5}{3}x_5\\
x_4 &= 1 \quad\quad\quad\ - \tfrac{4}{3}x_5
\end{aligned}$$

These equations provide the set of all solutions as the free variables x_4 and x_5 can take any arbitrary values. Notice that rank $(A) = 3$ which is equal to the number of basic variables, and the number of linearly independent solutions is the null $(A) = 2$, the number of free variables.

Following Gauss-Jordan method, we delete the last two zero rows, and insert zero rows so that pivots appear on the diagonal. Next, we adjoin suitable zero rows to make the portion before "|" a square matrix. And then, we change the diagonal entries on the added zero rows to -1 to obtain the following:

$$\xrightarrow{del-0;\, ins-0} \left[\begin{array}{ccccc|c} \boxed{1} & 0 & 5/2 & 0 & 8/3 & 2 \\ 0 & \boxed{1} & -1/2 & 0 & -5/3 & -1 \\ 0 & 0 & 0 & 0 & 0 & 0 \\ 0 & 0 & 0 & \boxed{1} & 4/3 & 1 \\ 0 & 0 & 0 & 0 & 0 & 0 \end{array}\right] \xrightarrow{ch(-1)} \left[\begin{array}{ccccc|c} \boxed{1} & 0 & 5/2 & 0 & 8/3 & 2 \\ 0 & \boxed{1} & -1/2 & 0 & -5/3 & -1 \\ 0 & 0 & -1 & 0 & 0 & 0 \\ 0 & 0 & 0 & \boxed{1} & 4/3 & 1 \\ 0 & 0 & 0 & 0 & -1 & 0 \end{array}\right].$$

Then the nonpivotal columns in the A portion form a basis for $N(A)$, and the solution set is given by

$$\mathrm{Sol}(A, b) = \left\{ \begin{bmatrix} 2 \\ -1 \\ 0 \\ 1 \\ 0 \end{bmatrix} + \alpha \begin{bmatrix} 5/2 \\ -1/2 \\ -1 \\ 0 \\ 0 \end{bmatrix} + \beta \begin{bmatrix} 8/3 \\ -5/3 \\ 0 \\ 4/3 \\ -1 \end{bmatrix} : \alpha, \beta \in \mathbb{F} \right\}. \qquad \square$$

Exercises for Sect. 3.7

1. Apply Gaussian elimination to the following linear systems. Decide whether each is solvable; in that case, determine the solution set.

   (a) $x_1 + 2x_2 - x_3 = 2,\ 2x_1 - x_2 - 2x_3 = 4,\ x_1 + 12x_2 - x_3 = 2.$
   (b) $x_1 + x_2 + x_3 = 1,\ x_1 - x_2 + 2x_3 = 1,\ 2x_1 + 3x_3 = 2,\ 2x_1 + 6x_2 = 2.$
   (c) $x_1 - x_2 + 2x_3 - 3x_4 = 7,\ 4x_1 + 3x_3 + x_4 = 9,\ 2x_1 - 5x_2 + x_3 = -2,$
   $3x_1 - x_2 - x_3 + 2x_4 = -2.$
   (d) $x_1 - x_2 - x_3 = 1,\ 2x_1 - x_2 + 2x_3 = 7,\ x_1 + x + 2 + x + 3 = 5,$
   $x_1 - 2x_2 - x_3 = 0.$
   (e) $4x_1 + 3x_3 + x_4 = 9,\ x_1 - x_2 + 2x_3 - 3x_4 = 7,\ 2x_1 - 5x_2 + x_3 = -2,$
   $3x_1 - x_2 - x_3 + 2x_4 = -2.$

   Also solve all those which are solvable by Gauss–Jordan elimination.

2. Let $A = \begin{bmatrix} 1 & 2 & 0 & 3 & 0 \\ 1 & 2 & -1 & -1 & 0 \\ 0 & 0 & 1 & 4 & 0 \\ 2 & 4 & 1 & 10 & 1 \\ 0 & 0 & 0 & 0 & 1 \end{bmatrix}$.

   (a) Find an invertible matrix P such that PA is in row reduced echelon form.
   (b) Find a basis B for the space U spanned by the rows of A.
   (c) Describe any typical vector in U. Determine whether $(-5, -10, 1, -11, 20)$ and $(1, 2, 3, 4, 5)$ are linear combinations of rows of A or not.
   (d) Find $[v]_B$ for any $v \in U$.

(e) Describe the solution space W of $Ax = 0$.
(f) Find a basis for W.
(g) For which $y \in \mathbb{R}^{5\times 1}$, $Ax = y$ has a solution?

3. Solve the system $\begin{bmatrix} 3 & -2 & -2 \\ 0 & 6 & -3 \\ 0 & 0 & 1 \end{bmatrix} \begin{bmatrix} 1 & 0 & 0 \\ 4 & 1 & 0 \\ -1 & 6 & 2 \end{bmatrix} \begin{bmatrix} x_1 \\ x_2 \\ x_3 \end{bmatrix} = \begin{bmatrix} 2 \\ 0 \\ 2 \end{bmatrix}$.
Notice that it is in the form $ULx = b$.
4. Using Gaussian elimination, determine the values of k for which the system
$$x_1 + x_2 + kx_3 = 1,\ x_1 - x_2 - x_3 = 2,\ 2x_1 + x_2 - 2x_3 = 3.$$
has (a) no solution, (b) at least one solution, or (c) a unique solution.
5. Using Gaussian elimination, find all possible values of $k \in \mathbb{R}$ such that the following system of linear equations has more than one solution:
$$x + y + 2z - 5w = 3,\ 2x + 5y - z - 9w = -3,$$
$$x - 2y + 6z - 7w = 7, 2x + 2y + 2z + kw = -4.$$
6. Find α, β, γ so that $2\sin\alpha - \cos\beta + 3\tan\gamma = 0$, $4\sin\alpha + 2\cos\beta - 2\tan\gamma = 0$, and $6\sin\alpha - 3\cos\beta + \tan\gamma = 9$.
7. Find the coefficients $a, b, c, d \in \mathbb{R}$ so that the graph of $y = a + bx + cx^2 + dx^3$ passes through the points $(-1, 6)$, $(0, 1)$, $(1, 2)$, and $(2, 3)$.
[Hint: Set up a system of linear equations in a, b, c, and d.]
8. A function $f : S \to S$ is said to have a *fixed point* $s \in S$ if and only if $f(s) = s$. Find a polynomial $p(t) \in \mathcal{P}_3(\mathbb{R})$ having fixed points as 1, 2, and 3.
9. Let $u = (-1, 0, 1, 2)$, $v = (3, 4, -2, 5)$, and $w = (1, 4, 0, 9)$. Construct a linear system whose solution space is $\text{span}\{u, v, w\}$.

3.8 Problems

1. Let B and D be two bases for a subspace of $\mathbb{F}^{1\times n}$. Prove that D can be obtained from B by a finite sequence of elementary row operations on the vectors of B.
2. Let A be a square matrix. Prove that elementary matrices $E_1, \ldots, E_k$ exist such that $E_1 \cdots E_k A$ is either I or has the last row a zero row.
3. Show that if $A \in \mathbb{F}^{2\times 2}$ is invertible, then A can be written as a product of at most four elementary matrices.

4. Let $i, j, r, s \in \{1, \ldots, m\}$ satisfy $r < s$ and $r < i < j$. Let $\alpha \in \mathbb{F}$, $\alpha \neq 0$. Prove the following:

   (a) If $s \neq i$ and $s \neq j$, then $E[i, j]\, E_\alpha[s, r] = E_\alpha[s, r]\, E[i, j]$.
   (b) If $s = i$, then $E[i, j]\, E_\alpha[s, r] = E_\alpha[j, r]\, E[i, j]$.
   (c) If $s = j$, then $E[i, j]\, E_\alpha[s, r] = E_\alpha[i, r]\, E[i, j]$.

5. Use Problem 4 to prove (formally) that for any $m \times n$ matrix A, there exists a permutation matrix P such that in reducing PA to row echelon form, no row exchanges are required.
6. Let U be subspace of $\mathbb{F}^n$ of dimension $m \leq n$. Prove that there exists a unique row reduced echelon matrix $A \in \mathbb{F}^{m\times n}$ whose rows span U.
7. Let $A, B \in \mathbb{F}^{m\times n}$. Prove that the following are equivalent:

   (a) B can be obtained from A by a sequence of elementary row operations.
   (b) The space spanned by the rows of A is the same as that spanned by the rows of B.
   (c) $B = QA$ for some invertible matrix $Q \in \mathbb{F}^{m\times m}$.

8. Let $P \in \mathbb{F}^{n\times n}$. Show that P is a permutation matrix if and only if each row of P contains exactly one nonzero entry, and it is equal to 1.
9. Give a proof of Theorem 3.14 analogous to that of Theorem 3.7.
10. If $A \in \mathbb{F}^{m\times n}$ is reduced to its echelon form $B \in \mathbb{F}^{m\times n}$, then show that

$$B = L_k\, P_k\, L_{k-1}\, P_{k-1} \cdots L_1\, P_1\, A = L_k\, (P_k L_{k-1} P_k^{-1})\, (P_k P_{k-1} L_{k-1} P_{k-1}^{-1} P_k^{-1}) \cdots (P_k \cdots P_2 L_1 P_2^{-1} \cdots P_k^{-1})(P_k \cdots P_1)\, A$$

    for some lower triangular matrices L_r and some permutation matrices P_s. Use this to derive the LU-factorization of A.
11. Prove that an elementary matrix of Type 1 can be expressed as a (finite) product of elementary matrices of other two types.
12. Show that column equivalence is an equivalence relation on the set of matrices of the same size.
13. Let $A, B \in \mathbb{F}^{m\times n}$ be in row reduced echelon form. Show that if A and B are row equivalent, then $A = B$.
14. Show that each matrix is row equivalent to a unique matrix in row reduced echelon form.
15. Let $A \in \mathbb{F}^{m\times n}$ be of rank r. Let $\{u_1, \ldots, u_r\}$ be a basis for $R(A)$. Then the columns of A are linear combinations of $u_1, \ldots, u_r$. Use this to give an alternative proof of the rank decomposition theorem.
16. Let $A \in \mathbb{F}^{m\times n}$. Let B be its row reduced echelon matrix. Mark the pivotal columns. Construct $C \in \mathbb{F}^{m\times r}$ by removing all nonpivotal columns from A. Also, construct F by removing all zero rows from B. Prove that $A = CF$ is a rank decomposition of A.
17. Deduce $\operatorname{rank}(A^T) = \operatorname{rank}(A)$ from the rank decomposition of A.
18. Let $A \in \mathbb{F}^{m\times n}$. Prove that $\operatorname{rank}(A^*) = \operatorname{rank}(A) = \operatorname{rank}(A^T)$.

19. Let $A, B \in \mathbb{F}^{m\times n}$. Show that $\text{rank}\,(A+B) \le \text{rank}\,(A) + \text{rank}\,(B)$. Construct an example where equality is achieved.
20. Construct three pairs of row equivalent matrices. Also, construct three pairs of matrices which are not row equivalent.
21. If a matrix A is equivalent to a matrix B, is it necessary that either A is row equivalent to B or A is column equivalent to B?
22. Prove that a square matrix A is not invertible if and only if there exists a nonzero square matrix B such that $AB = 0$.
23. Let $A, B \in \mathbb{F}^{n\times n}$. Show that AB is invertible if and only if both A and B are invertible.
24. Let $A, B \in \mathbb{R}^{n\times n}$. Suppose that there exists an invertible matrix $P \in \mathbb{C}^{n\times n}$ such that $PA = BP$. Show that there exists an invertible matrix $Q \in \mathbb{R}^{n\times n}$ such that $QA = BQ$.
25. Let $A \in \mathbb{F}^{n\times n}$. Prove that the following are equivalent:
    (a) There exists $B \in \mathbb{F}^{n\times n}$ such that $AB = I$.
    (b) A is row equivalent to I.
    (c) A is a product of elementary matrices.
    (d) A is invertible.
    (e) A^T is invertible.
    (f) There exists $B \in \mathbb{F}^{n\times n}$ such that $BA = I$.
    (g) $Ax = 0$ has no nontrivial solution.
    (h) $Ax = b$ has a unique solution for some $b \in \mathbb{F}^{n\times 1}$.
    (i) $Ax = b$ is consistent for each $b \in \mathbb{F}^{n\times 1}$.
    (j) $Ax = e_i$ has a unique solution for each standard basis vector e_i of $\mathbb{F}^{n\times 1}$.
    (k) $\text{rank}\,(A) = n$.
    (l) The row reduced echelon form of A is I.
    (m) $Ax = b$ has a unique solution for each $b \in \mathbb{F}^{n\times 1}$.
26. Let A, B, C, and D be $n \times n$ matrices, where D is invertible and $CD = DC$. Show that $\det\left(\begin{bmatrix} A & B \\ C & D \end{bmatrix}\right) = \det(AD - BC)$. What happens if $CD \ne DC$? What happens if D is not invertible?
27. Suppose $X = \begin{bmatrix} A & B \\ C & D \end{bmatrix}$, where $A \in \mathbb{F}^{m\times m}$, $B \in \mathbb{F}^{m\times n}$, $C \in \mathbb{F}^{n\times m}$, and $D \in \mathbb{F}^{n\times n}$. Show the following:
    (a) If $B = 0$ or $C = 0$, then $\det(X) = \det(A)\det(D)$.
    (b) If A is invertible, then $\det(X) = \det(A)\det(D - CA^{-1}B)$.
    (c) $X^{-1} = \begin{bmatrix} (A - BD^{-1}C)^{-1} & A^{-1}B(CA^{-1}B - D)^{-1} \\ (CA^{-1}B - D)^{-1}CA^{-1} & (D - CA^{-1}B)^{-1} \end{bmatrix}$ provided that all the inverses exist.
28. Prove that if A is an $n \times k$ matrix and B is a $k \times n$ matrix, where $k < n$, then AB is not invertible.
29. Let $A \in \mathbb{F}^{m\times n}$ and $B \in \mathbb{F}^{n\times m}$. Prove that $I_m - AB$ is invertible if and only if $I_n - BA$ is invertible.

30. Suppose that $A \in \mathbb{F}^{m\times n}$ and $y \in \mathbb{F}^{m\times 1}$. Assume that there exists a matrix B such that $BA = I$. Now, $Ax = y \Rightarrow BAx = By \Rightarrow Ix = By \Rightarrow x = By$. When verifying our solution, we see that $x = By \Rightarrow Ax = ABy \Rightarrow x = y$. But the last $\Rightarrow$ is questionable since we have no information whether $AB = I$. Explain the situation.
31. Let $A \in \mathbb{F}^{n\times n}$. If $\operatorname{tr}(AB) = 0$ for each $B \in \mathbb{F}^{n\times n}$, then prove that $A = 0$.
32. Since traces of similar matrices coincide, trace is well defined for a linear operator on any finite dimensional vector space. Let $P \in \mathbb{F}^{2\times 2}$. Define a linear operator T on $\mathbb{F}^{2\times 2}$ by $T(X) = PX$. Show that $\operatorname{tr}(T) = 2\operatorname{tr}(P)$.
33. Let f be a linear functional on $\mathbb{F}^{n\times n}$ that satisfies $f(AB) = f(BA)$ for all $A, B \in \mathbb{F}^{n\times n}$ and $f(I) = n$. Prove that $f(X) = \operatorname{tr}(X)$ for all $X \in \mathbb{F}^{n\times n}$.
34. Let W be the subspace of $\mathbb{F}^{n\times n}$ which is spanned by matrices of the form $AB - BA$. Prove that W is the subspace of matrices with trace 0.
35. Let $A \in \mathbb{F}^{n\times n}$ have more than $n^2 - n$ number of entries as 0. Does it follow that $\det(A) = 0$?
36. Does the set of all $n \times n$ matrices with complex entries whose determinant is 0 form a vector space with the usual operations of addition and scalar multiplication?
37. Determine $\mathrm{Sol}(A, 0)$, where $A \in \mathbb{F}^{n\times n}$ has all diagonal entries as $1 - n$ and all other entries as 1.
38. Let A be an $n \times n$ matrix with integer entries. Show that $\det(A) = \pm 1$ if and only if A is invertible and all entries of A^{-1} are integers.
39. Let $A = [a_{ij}] \in \mathbb{R}^{n\times n}$, where $a_{ij} = 1$ for $j = n + 1 - i$, and $a_{ij} = 0$ otherwise. Compute $\det(A)$.
40. Compute $\det(A)$ and $\det(B)$, where

$$A = \begin{bmatrix} & & & 1 \\ & & 1 & \\ & \cdot^{\cdot^{\cdot}} & & \\ & 1 & & \\ 1 & & & \end{bmatrix} \quad \text{and} \quad B = \begin{bmatrix} 1 & 2 & \cdots & n \\ & & \vdots & \\ 1 & 2 & \cdots & n \end{bmatrix}.$$

41. Let $A \in \mathbb{C}^{n\times n}$ be of rank r. A $k \times k$ submatrix of A is obtained from A by deleting some $m - k$ rows and $n - k$ columns. Prove that some $r \times r$ submatrix of A is invertible and that no $(r + 1) \times (r + 1)$ submatrix of A is invertible.
42. Define a **box** in $\mathbb{R}^n$ determined by the vectors $v_1, \ldots, v_n$ as the set

$$\{\alpha_1 v_1 + \cdots + \alpha_n v_n : \alpha_i \in \mathbb{R} \text{ and } 0 \le \alpha_i \le 1 \text{ for each } i\}.$$

The **volume** of such a box is defined as $|\det([v_1 \ \cdots \ v_n])|$. Let $A \in \mathbb{R}^{n\times n}$ and let B be a box in $\mathbb{R}^n$. Let $A(B)$ be the image of the box B under the map A. Show that $A(B)$ is a box and that the volume of the box $A(B)$ is equal to $|\det(A)|$ times the volume of the box B. Deduce that an orthogonal matrix preserves volumes of boxes.

43. If ω is a primitive cube root of unity, then show that

$$\begin{vmatrix} a & b & c \\ c & a & b \\ b & c & a \end{vmatrix} = (a+b+c)(a+b\omega+c\omega^2)(a+b\omega^2+c\omega).$$

44. Obtain a result for $\begin{vmatrix} a & b & c & d \\ d & a & b & c \\ c & d & a & b \\ b & c & d & a \end{vmatrix}$ similar to the matrix in Exercise 43 using primitive fourth roots of unity. Such matrices are called *circulant matrices.* Can you generalize the result for an $n \times n$ circulant matrix?

45. The *classical Vandermonde matrix* with distinct numbers $x_1, \ldots, x_{n-1}$ is the $n \times n$ matrix $A = [a_{ij}]$, where $a_{ij} = x_i^{j-1}$. Show the following:

    (a) $\det(A) = \Pi_{1 \le i < j \le n}(x_i - x_j)$.
    (b) Let $A^{-1} = [b_{ij}]$. Let $p_j(t)$ be a polynomial of degree $n-1$ such that $p_j(x_i) = \delta_{ij}$ for $1 \le i \le n$, with δ_{ij} as Kronecker's delta. Then

$$p_j(t) = \sum_{k=1}^{n} b_{kj} t^{k-1} = \Pi_{1 \le k \le n, k \ne j} \frac{t - x_k}{x_j - x_k}.$$

    Can you give an exact expression for b_{ij}? And what is the sum $\sum_i \sum_j b_{ij}$?

46. The *Vandermonde matrix* with given numbers $x_1, \ldots, x_n$ is the $n \times n$ matrix $A = [a_{ij}]$ where $a_{ij} = x_j^i$. Write $p_i(t) = \Pi_{1 \le k \le n, k \ne i}(t - x_k)$, a polynomial of degree $n-1$. Show the following:

    (a) $\det(A)$ is the product $\Pi_{1 \le j \le n} \Pi_{1 \le i, j \le n}(x_j - x_i)$.
    (b) If $\det(A) \ne 0$, then $A^{-1} = [b_{ij}]$ where

$$b_{ij} = (-1)^{j+1} \frac{\text{Coefficient of } t^{i-1} \text{ in } p_i(t)}{x_i \, p_i(x_k)}.$$

    (c) If A is invertible, then what is the sum of all n^2 entries in A^{-1}?

47. (*Hilbert matrix*) Let $A = [a_{ij}] \in \mathbb{R}^{n \times n}$ with $a_{ij} = 1/(i+j-1)$. Show:

    (a) A is invertible and the inverse of A has only integer entries.
    (b) The sum of all n^2 entries of A^{-1} is n^2. (See [4].)

48. The *combinatorial matrix* with given numbers x and y is the $n \times n$ matrix $A = [a_{ij}]$, where $a_{ij} = y + \delta_{ij} x$. Show the following:

    (a) $\det(A) = x^{n-1}(x + ny)$.
    (b) If $\det(A) \ne 0$, then $A^{-1} = [b_{ij}]$, where $b_{ij} = \dfrac{-y + \delta_{ij}(x + ny)}{x(x + ny)}$.

    What is the sum of all n^2 entries in A^{-1}?

49. The *Cauchy matrix* with given numbers $x_1, \ldots, x_n, y_1, \ldots, y_n$ with $x_i + y_j \neq 0$ for any i, j, is the $n \times n$ matrix $A = [a_{ij}]$, where $a_{ij} = (x_i + y_j)^{-1}$. Show the following:

   (a) $\det(A) = \dfrac{\prod_{1 \le i < j \le n}(x_j - x_i)(y_j - y_i)}{\prod_{1 \le i, j \le n}(x_i + y_j)}$.

   (b) $A^{-1} = [b_{ij}]$, where
$$b_{ij} = \frac{\prod_{1 \le k \le n}(x_j + y_k)(x_k + y_i)}{(x_j + y_i)\big(\prod_{1 \le k \le n, k \neq j}(x_j - x_k)\big)\big(\prod_{1 \le k \le n, k \neq i}(y_i - y_k)\big)}.$$

   (c) What is the sum of all n^2 entries of A^{-1}?

50. (*Hadamard Matrix*) A Hadamard matrix of order n is a matrix $H \in \mathbb{R}^{n \times n}$ with each entry as ± 1 that satisfies $HH^T = nI$. Show the following:

   (a) If the entries of $A = [a_{ij}] \in \mathbb{F}^{n \times n}$ satisfy $|a_{ij}| \le 1$, then $|\det(A)| \le n^{n/2}$. Further, equality holds when A is a Hadamard matrix.

   (b) If a Hadamard matrix of order n exists, then either $n = 1$ or $n = 2$, or n is a multiple of 4.

   (c) If B is a Hadamard matrix of order n, then $\begin{bmatrix} B & B \\ B & -B \end{bmatrix}$ is a Hadamard matrix of order $2n$. (This way Sylvester shows how to construct Hadamard matrices of order 2^n.)

   It is not yet known how to construct Hadamard matrices of arbitrary order.

Chapter 4
Inner Product Spaces

4.1 Inner Products

In Chap. 1 we defined a vector space as an abstraction of the familiar Euclidean space. In doing so, we took into account only two aspects of the set of vectors in a plane, namely the vector addition and scalar multiplication. Now, we consider the third aspect, namely the *angle* between vectors.

Recall that if $\vec{x}$ and $\vec{y}$ are two nonzero vectors in the plane $\mathbb{R}^2$, then the angle $\theta(\vec{x}, \vec{y})$ between $\vec{x}$ and $\vec{y}$ is given by

$$\cos\theta(\vec{x}, \vec{y}) = \frac{\vec{x} \cdot \vec{y}}{|\vec{x}|\,|\vec{y}|},$$

where for a vector $\vec{u} = (a, b) \in \mathbb{R}^2$, $|\vec{u}|$ denotes the *absolute value* of the vector $\vec{u}$, that is,

$$|\vec{u}| = \sqrt{a^2 + b^2} = \sqrt{\vec{u} \cdot \vec{u}}.$$

This is the distance of the point $(a, b) \in \mathbb{R}^2$ from the origin.

Observe that the angle between two vectors is completely determined by the dot product. The dot product satisfies the following properties for all $\vec{x}, \vec{y}, \vec{u} \in \mathbb{R}^2$ and for all $\alpha \in \mathbb{R}$:

$$\begin{gathered}
(\vec{x} + \vec{y}) \cdot \vec{u} = \vec{x} \cdot \vec{u} + \vec{y} \cdot \vec{u}, \\
(\alpha\vec{x}) \cdot \vec{y} = \alpha(\vec{x} \cdot \vec{y}), \\
\vec{x} \cdot \vec{y} = \vec{y} \cdot \vec{x}, \\
\vec{x} \cdot \vec{x} \geq 0, \\
\vec{x} \cdot \vec{x} = 0 \text{ if and only if } \vec{x} = \vec{0}.
\end{gathered}$$

M. T. Nair and A. Singh, *Linear Algebra*,
https://doi.org/10.1007/978-981-13-0926-7_4

When a vector has complex coordinates, some of the above properties may require modifications. Since dot product must give rise to the absolute value of a vector by taking $|\vec{x}|^2 = \vec{x} \cdot \vec{x}$, the modified definition must keep the property

$$\vec{x} \cdot \vec{x} \geq 0$$

intact. Taking into account the above requirements, we generalize the notion of a dot product to any vector space V with the notation $\langle u, v \rangle$ for $u, v \in V$, in place of the dot product in $\mathbb{R}^2$.

Definition 4.1 Let V be a vector space over $\mathbb{F}$. An **inner product** on V is a function from $V \times V$ to $\mathbb{F}$, denoted by $\langle \cdot, \cdot \rangle$, that is $(x, y) \mapsto \langle x, y \rangle$, satisfying the following properties for all $x, y, z \in V$ and for all $\alpha \in \mathbb{F}$:

(1) $\langle x, x \rangle \geq 0$.
(2) $\langle x, x \rangle = 0$ if and only if $x = 0$.
(3) $\langle x + y, z \rangle = \langle x, z \rangle + \langle y, z \rangle$.
(4) $\langle \alpha x, y \rangle = \alpha \langle x, y \rangle$.
(5) $\langle x, y \rangle = \overline{\langle y, x \rangle}$.

A vector space together with an inner product is called an **inner product space**. If $\mathbb{F} = \mathbb{R}$, the inner product space V is called a **real inner product space**, and if $\mathbb{F} = \mathbb{C}$, then V is called a **complex inner product space**.

A real inner product space is also called a *Euclidean space*, and a complex inner product space is also called a *unitary space*.

A simple example of an inner product is the dot product in $\mathbb{R}^2$. In the following examples, it can be easily verified that the map $\langle \cdot, \cdot \rangle$ is indeed an inner product.

Example 4.2 (1) For $x = (\alpha_1, \ldots, \alpha_n)$ and $y = (\beta_1, \ldots, \beta_n)$ in $\mathbb{F}^n$, define

$$\langle x, y \rangle := \sum_{j=1}^{n} \alpha_j \overline{\beta}_j = \overline{y} x^T.$$

Then $\langle \cdot, \cdot \rangle$ is an inner product on $\mathbb{F}^n$, called the **standard inner product** on $\mathbb{F}^n$.

(2) For $x = (\alpha_1, \ldots, \alpha_n)^T$ and $y = (\beta_1, \ldots, \beta_n)^T$ in $\mathbb{F}^{n \times 1}$, define

$$\langle x, y \rangle := \sum_{j=1}^{n} \alpha_j \overline{\beta}_j = y^* x.$$

Recall that y^* is the conjugate transpose of y. Then $\langle \cdot, \cdot \rangle$ is an inner product on $\mathbb{F}^{n \times 1}$. It is called the **standard inner product** on $\mathbb{F}^{n \times 1}$.

(3) Recall that a matrix $A \in \mathbb{F}^{n \times n}$ is called *hermitian*, when $A^* = A$. A hermitian matrix is called *positive definite* if for each nonzero $x \in \mathbb{F}^{n \times 1}$, $x^* A x > 0$.

Let A be a positive definite matrix. Then

$$\langle x, y \rangle = y^* A x$$

defines an inner product on $\mathbb{F}^{n\times 1}$.

(4) For $p(t) = a_0 + a_1 t + \cdots + a_n t^n$ and $q(t) = b_0 + b_1 t + \cdots + b_n t^n$ in $\mathcal{P}_n(\mathbb{F})$, define

$$\langle p, q \rangle = a_0 \overline{b}_0 + a_1 \overline{b}_1 + \cdots + a_n \overline{b}_n.$$

Then $\langle \cdot, \cdot \rangle$ is an inner product on $\mathcal{P}_n(\mathbb{F})$.

(5) Let V be a finite dimensional vector space. Let $E = \{v_1, \ldots, v_n\}$ be an ordered basis of V. For $x = \sum_{i=1}^n \alpha_i v_i$ and $y = \sum_{i=1}^n \beta_i v_i$ in V, define

$$\langle x, y \rangle_E = \sum_{i=1}^n \alpha_i \overline{\beta}_i.$$

Then $\langle \cdot, \cdot \rangle_E$ is an inner product on V.

(6) For $f, g \in C([a, b], \mathbb{R})$, the vector space of all continuous real-valued functions with domain as the interval $[a, b]$, define

$$\langle f, g \rangle = \int_a^b f(t)\, g(t)\, dt.$$

Clearly,

$$\langle f, f \rangle = \int_a^b |f(t)|^2\, dt \geq 0 \quad \text{for all } f \in C([a, b], \mathbb{R}).$$

By the continuity of the function f,

$$\langle f, f \rangle = \int_a^b |f(t)|^2\, dt = 0 \quad \text{if and only if} \quad f(t) = 0 \text{ for all } t \in [a, b].$$

Other conditions can be verified to show that $\langle \cdot, \cdot \rangle$ is an inner product.

Similarly, on $C([a, b], \mathbb{C})$, the vector space of all continuous complex-valued functions with domain as the interval $[a, b]$,

$$\langle f, g \rangle = \int_a^b f(t) \overline{g(t)}\, dt \quad \text{for } f, g \in C([a, b], \mathbb{C})$$

defines an inner product.

(7) For polynomials $p, q \in \mathcal{P}(\mathbb{R})$, define

$$\langle p, q \rangle = \int_0^1 p(t)\, q(t)\, dt.$$

It can be verified that it is an inner product on $\mathcal{P}(R)$.

Similarly, on $\mathcal{P}(\mathbb{C})$,

$$\langle p, q\rangle = \int_0^1 p(t)\overline{q(t)}\,dt \quad \text{for } p, q \in \mathcal{P}(\mathbb{C})$$

defines an inner product. It may be checked that this inner product restricted to $\mathcal{P}_n(\mathbb{F})$ is different from the one discussed in (4) above. □

The inner product in Example 4.2(5) is related to the standard Inner product on $\mathbb{F}^n$ by the canonical isomorphism, which gives the coordinate vectors. That is,

$$\langle u, v\rangle = \langle [u]_E, [v]_E\rangle.$$

In fact, any finite dimensional vector space can be made into an inner product space by fixing an ordered basis.

If V is a vector space over $\mathbb{F}$, W is an inner product space over $\mathbb{F}$ with an inner product $\langle\cdot,\cdot\rangle_W$, and $T : V \to W$ is an injective linear transformation, then the map $\langle\cdot,\cdot\rangle_V : V \times V \to \mathbb{F}$ defined by

$$\langle x, y\rangle_V = \langle Tx, Ty\rangle_W \quad \text{for } x, y \in V$$

is an inner product on V.

It is easy to verify that the restriction of an inner product on a vector space to a subspace is an inner product on the subspace.

Theorem 4.3 *Let V be an inner product space. Then, for all $x, y, u, v \in V$ and $\alpha \in \mathbb{F}$,*

$$\langle x, u+v\rangle = \langle x, u\rangle + \langle x, v\rangle \quad \textit{and} \quad \langle x, \alpha y\rangle = \overline{\alpha}\langle x, y\rangle.$$

Proof Let x, y, u, v in V and $\alpha \in \mathbb{F}$. Due to Properties (3)–(5) in Definition 4.1,

$$\langle x, u+v\rangle = \overline{\langle u+v, x\rangle} = \overline{\langle u, x\rangle + \langle v, x\rangle} = \overline{\langle u, x\rangle} + \overline{\langle v, x\rangle} = \langle x, u\rangle + \langle x, v\rangle,$$

$$\langle x, \alpha y\rangle = \overline{\langle \alpha y, x\rangle} = \overline{\alpha\langle y, x\rangle} = \overline{\alpha}\langle x, y\rangle.$$ ■

Property (2) of the inner product provides a standard technique in showing uniqueness results in inner product spaces.

Theorem 4.4 *Let V be an inner product space and let $x, y \in V$. If $\langle x, u\rangle = \langle y, u\rangle$ for all $u \in V$, then $x = y$.*

Proof Suppose $\langle x, u\rangle = \langle y, u\rangle$ for all $u \in V$. Then $\langle x - y, u\rangle = 0$ for all $u \in V$. In particular, with $u = x - y$, we have $\langle x - y, x - y\rangle = 0$. This implies, by Property (2) of the inner product, that $x - y = 0$. ■

Exercises for Sect. 4.1

1. In each of the following check whether $\langle \cdot, \cdot \rangle$ is an inner product on the given vector space.

   (a) $\langle x, y \rangle = ac$ for $x = (a, b),\ y = (c, d)$ in $\mathbb{R}^2$.
   (b) $\langle x, y \rangle = ac - bd$ for $x = (a, b),\ y = (c, d)$ in $\mathbb{R}^2$.
   (c) $\langle x, y \rangle = y^T Ax$ for $x, y \in \mathbb{R}^{2\times 1}$, where A is a given 2×2 real symmetric matrix.
   (d) $\langle f, g \rangle = \int_0^1 f'(t)g(t)\,dt$ on $\mathcal{P}(\mathbb{R})$.
   (e) $\langle f, g \rangle = \int_0^{1/2} f(t)g(t)$ on $\mathcal{C}[0, 1]$.
   (f) $\langle x, y \rangle = \int_0^1 x(t)y(t)\,dt$ for x, y in $\mathcal{R}([0, 1], \mathbb{R})$.
   (g) $\langle x, y \rangle = \int_0^1 x'(t)y'(t)\,dt$ for x, y in $\mathcal{C}^1([0, 1], \mathbb{R})$.
   (h) $\langle x, y \rangle = x(0)y(0) + \int_0^1 x'(t)y'(t)\,dt$ for x, y in $\mathcal{C}^1([0, 1], \mathbb{R})$.
   (i) $\langle x, y \rangle = \int_0^1 x(t)y(t)\,dt + \int_0^1 x'(t)y'(t)\,dt$ for x, y in $\mathcal{C}^1([0, 1], \mathbb{R})$.
   (j) $\langle A, B \rangle = \mathrm{tr}(A + B)$ for A, B in $\mathbb{R}^{2\times 2}$.
   (k) $\langle A, B \rangle = \mathrm{tr}(A^T B)$ for A, B in $\mathbb{R}^{3\times 3}$.

2. Let $t_1, \ldots, t_{n+1}$ be distinct real numbers. Let $\langle p, q \rangle := \sum_{i=1}^{n+1} p(t_i)\overline{q(t_i)}$ for $p, q \in \mathcal{P}_n(\mathbb{C})$. Show that $\langle \cdot, \cdot \rangle$ is an inner product on $\mathcal{P}_n(\mathbb{C})$.
3. Let $\gamma_1, \ldots, \gamma_n > 0$. Let $\langle (a_1, \ldots, a_n), (b_1, \ldots, b_n) \rangle = \gamma_1 a_1 b_1 + \cdots + \gamma_n a_n b_n$. Show that $\langle \cdot, \cdot \rangle$ is an inner product on $\mathbb{R}^n$.
4. Let $A = [a_{ij}] \in \mathbb{R}^{2\times 2}$. For $x, y \in \mathbb{R}^{2\times 1}$, let $f_A(x, y) = y^T Ax$. Show that f_A is an inner product on $\mathbb{R}^{2\times 1}$ if and only if $a_{12} = a_{21},\ a_{11} > 0,\ a_{22} > 0$ and $\det(A) > 0$.
5. For $x = (\alpha_1, \alpha_2, \alpha_3)$ and $y = (\beta_1, \beta_2, \beta_3)$ in $\mathbb{R}^3$, define $\langle x, y \rangle = \sum_{i,j=1}^{3} a_{ij}\alpha_i\beta_j$, where $A = [a_{ij}]$ is given by $A = \begin{bmatrix} 2 & 1 & 0 \\ 1 & 2 & 0 \\ 0 & 1 & 4 \end{bmatrix}$. Show that this defines an inner product on $\mathbb{R}^3$. Compute cosines of the angles between the standard basis vectors of $\mathbb{R}^3$ using this inner product.
6. Let $\langle \cdot, \cdot \rangle$ be the standard inner product on $\mathbb{F}^n$, as defined in Example 4.2(1). Let $T : V \to \mathbb{F}^n$ be an isomorphism. Define $\langle x, y \rangle_T := \langle Tx, Ty \rangle$ for all $x, y \in V$. Show that $\langle \cdot, \cdot \rangle_T$ is also an inner product on $\mathbb{F}^n$.
7. Suppose V is a vector space over $\mathbb{F}$, W is an inner product space over $\mathbb{F}$ with an inner product $\langle \cdot, \cdot \rangle_W$, and $T : V \to W$ is an injective linear transformation. Define $\langle x, y \rangle_V = \langle Tx, Ty \rangle_W$ for all $x, y \in V$. Show that $\langle \cdot, \cdot \rangle$ is an inner product on V.
8. Let V_1 and V_2 be inner product spaces with inner products $\langle \cdot, \cdot \rangle_1$ and $\langle \cdot, \cdot \rangle_2$, respectively. Show that $\langle \cdot, \cdot \rangle$ as given below is an inner product on $V = V_1 \times V_2$:

$$\langle (x_1, x_2), (y_1, y_2) \rangle := \langle x_1, y_1 \rangle_1 + \langle x_2, y_2 \rangle_2 \quad \text{for all } (x_1, x_2), (y_1, y_2) \in V.$$

9. Let $a, b, c, d \in \mathbb{C}$. For $u = (\alpha, \beta)$ and $v = (\gamma, \delta)$ in $\mathbb{C}^2$, define

$$\langle u, v\rangle = a\alpha\overline{\gamma} + b\beta\overline{\gamma} + c\alpha\overline{\delta} + d\beta\overline{\delta}.$$

Under what conditions on a, b, c, d, $\langle\cdot,\cdot\rangle$ is an inner product?

10. Let E be a basis for a finite dimensional inner product space V. Let $x, y \in V$. Prove that if $\langle x, u\rangle = \langle y, u\rangle$ for all $u \in E$, then $x = y$.
11. Suppose V is a complex inner product space. Prove that $\mathrm{Re}\langle ix, y\rangle = -\mathrm{Im}\langle x, y\rangle$ for all $x, y \in V$.
12. Let $u \in \mathbb{R}^2$. For any $\alpha \in \mathbb{R}$, let $V_\alpha = \{x \in \mathbb{R}^2 : \langle x, u\rangle = \alpha\}$. Show that V_α is a subspace of $\mathbb{R}^2$ if and only if $\alpha = 0$.

4.2 Norm and Angle

Recall that the length or absolute value of a vector $\vec{x} = (a, b)$ in $\mathbb{R}^2$ is given by

$$|\vec{x}| = \sqrt{a^2 + b^2} = \sqrt{\vec{x}\cdot\vec{x}} = \sqrt{\langle\vec{x}, \vec{x}\rangle},$$

where $\langle\vec{x}, \vec{x}\rangle$ is the Euclidean inner product of $\vec{x}$ with itself. For defining the notion of length of a vector in a general inner product space, now called the *norm* of the vector, we use the analogous definition.

Definition 4.5 Let V be an inner product space. Let $x \in V$.

(a) The **norm of** x, denoted by $\|x\|$, is the nonnegative square root of $\langle x, x\rangle$. That is, $\|x\| := \sqrt{\langle x, x\rangle}$.
(b) The map $x \mapsto \|x\|$, also denoted by $\|\cdot\|$, is called a **norm** on V.
(c) A **unit vector** in V is any vector whose norm is 1.

It is easy to see that if x is any nonzero vector, then $u := x/\|x\|$ is a unit vector; we say that this u is a *unit vector in the direction of* x.

An inequality involving both inner product and the norm is given in the following theorem.

Theorem 4.6 (Cauchy–Schwarz inequality) *Let V be an inner product space and let $x, y \in V$. Then*

(1) $|\langle x, y\rangle| \le \|x\|\,\|y\|$, *and*
(2) $|\langle x, y\rangle| = \|x\|\,\|y\|$ *if and only if x and y are linearly dependent.*

Proof If $y = 0$, then both (1) and (2) are obvious.

Assume that $y \ne 0$. For any $\alpha \in \mathbb{F}$,

$$\begin{aligned}\langle x - \alpha y, x - \alpha y\rangle &= \langle x, x\rangle - \langle x, \alpha y\rangle - \langle\alpha y, x\rangle + \langle\alpha y, \alpha y\rangle\\ &= \|x\|^2 - \overline{\alpha}\langle x, y\rangle - \alpha\langle y, x\rangle + |\alpha|^2\,\|y\|^2.\end{aligned}$$

In particular, taking $\alpha = \dfrac{\langle x, y\rangle}{\|y\|^2}$, we obtain $\overline{\alpha} = \dfrac{\langle y, x\rangle}{\|y\|^2}$ and

$$\overline{\alpha}\langle x, y\rangle = \alpha\langle y, x\rangle = |\alpha|^2\|y\|^2 = \frac{|\langle x, y\rangle|^2}{\|y\|^2}.$$

Consequently,

$$\langle x - \alpha y, x - \alpha y\rangle = \|x\|^2 - \frac{|\langle x, y\rangle|^2}{\|y\|^2}.$$

(1) Since $0 \le \langle x - \alpha y, x - \alpha y\rangle$, we have $|\langle x, y\rangle| \le \|x\|\,\|y\|$.
(2) If $x = \alpha y$ for some $\alpha \in \mathbb{F}$, then $|\langle x, y\rangle| = |\langle \alpha y, y\rangle| = |\alpha|\,\|y\|^2 = \|x\|\,\|y\|$.
Conversely, assume that $x, y \in V$ are such that $|\langle x, y\rangle| = \|x\|\,\|y\|$. With $y \ne 0$ and $\alpha = \dfrac{\langle x, y\rangle}{\|y\|^2}$, we see that

$$\langle x - \alpha y, x - \alpha y\rangle = \|x\|^2 - \frac{|\langle x, y\rangle|^2}{\|y\|^2} = \|x\|^2 - \|x\|^2 = 0.$$

Therefore, $x = \alpha y$. ■

Some useful properties of the norm are listed in the following theorem.

Theorem 4.7 *Let V be an inner product space and let $x, y \in V$. Then the following are true:*

(1) $\|x\| \ge 0$.
(2) $\|x\| = 0$ *if and only if* $x = 0$.
(3) $\|\alpha x\| = |\alpha|\,\|x\|$ *for all* $\alpha \in \mathbb{F}$.
(4) (Triangle inequality) $\|x + y\| \le \|x\| + \|y\|$.
(5) (Parallelogram law) $\|x + y\|^2 + \|x - y\|^2 = 2(\|x\|^2 + \|y\|^2)$.

Proof The statements in (1)–(3) and (5) follow directly from the properties of the inner product. For (4), observe that

$$\|x + y\|^2 = \|x\|^2 + \|y\|^2 + 2\mathrm{Re}\langle x, y\rangle.$$

$$(\|x\| + \|y\|)^2 = \|x\|^2 + \|y\|^2 + 2\|x\|\|y\|.$$

Due to Cauchy–Schwarz inequality,

$$\mathrm{Re}\,\langle x, y\rangle \le |\langle x, y\rangle| \le \|x\|\,\|y\|.$$

Therefore, $\|x + y\|^2 \le (\|x\| + \|y\|)^2$. ■

Similar to Cauchy–Schwarz inequality, the triangle inequality becomes an equality when one vector is a nonnegative multiple of the other; prove it!

The triangle inequality is so named because if the two sides of a triangle in the plane are vectors x and y, then the third side is $x + y$, and the inequality says that the sum of the lengths of any two sides of a triangle is at least that of the third. Similarly, the parallelogram law in the plane says that the sum of the areas of the squares on the sides of a parallelogram is equal to the sum of the areas of the squares on its diagonals.

We define the *distance between* vectors u and v in an inner product space as

$$d(x, y) := \|u - v\|.$$

Using Theorem 4.7(1)–(4), the following may be easily seen:

1. For all $x, y \in V,\ d(x, y) \geq 0$.
2. For each $x \in V,\ d(x, x) = 0$ if and only if $x = 0$.
3. For all $x, y, z \in V,\ d(x, z) \leq d(x, y) + d(y, z)$.

Since $\|x\|$ is the distance of x from the origin, it is the length of x, as it is supposed to be.

In fact, a norm can be defined on a vector space without recourse to an inner product. In that case, a norm is taken as any function from V to $\mathbb{R}$ mapping x to $\|x\|$ satisfying the properties listed in Theorem 4.7(1)–(4). The parallelogram law is satisfied by only those norms which come from inner products. For example, on $\mathbb{R}^2$, $\|(a, b)\| = |a| + |b|$ defines a norm but this norm does not come from an inner product; show it!

In an inner product space, if a norm comes from an inner product there should be a way of computing the inner product from the norm. In this regard, we have the following theorem; its proof is left as an exercise.

Theorem 4.8 (Polarization identities) *Let V be an inner product space over $\mathbb{F}$. Let $x, y \in V$.*

(1) *If $\mathbb{F} = \mathbb{R}$, then* $4\langle x, y\rangle = \|x + y\|^2 - \|x - y\|^2$.
(2) *If $\mathbb{F} = \mathbb{C}$, then* $4\langle x, y\rangle = \|x + y\|^2 - \|x - y\|^2 + i\|x + iy\|^2 - i\|x - iy\|^2$.

Recall that the acute (nonobtuse) angle $\theta(\vec{x}, \vec{y})$ between two vectors $\vec{x}, \vec{y} \in \mathbb{R}^2$ is given by

$$\cos\theta(\vec{x}, \vec{y}) = \frac{|\vec{x} \cdot \vec{y}|}{|\vec{x}|\,|\vec{y}|}.$$

Replacing the dot product by the inner product yields a definition of angle between two vectors in an inner product space. Observe that due to Cauchy–Schwarz inequality,

$$0 \leq \frac{|\langle x, y\rangle|}{\|x\|\,\|y\|} \leq 1 \quad \text{for } x \neq 0,\ y \neq 0.$$

Definition 4.9 Let x and y be nonzero vectors in an inner product space V. The **angle $\theta(x, y)$ between x and y** is defined by

$$\cos\theta(x, y) = \frac{|\langle x, y\rangle|}{\|x\|\,\|y\|}.$$

Let x and y be nonzero vectors. If one of them is a scalar multiple of the other, say, $x = \alpha y$ for some nonzero scalar α, then $\cos\theta(x, y) = 1$; consequently, $\theta(x, y) = 0$. Conversely, if $\theta(x, y) = 0$, then $\cos\theta(x, y) = 1$; and then $|\langle x, y\rangle| = \|x\|\,\|y\|$. Due to Theorem 4.6(2), one of x, y is a scalar multiple of the other. In the other extreme, if $\langle x, y\rangle = 0$, then $\theta(x, y) = \pi/2$. Moreover, $\langle x, y\rangle$ can be 0 when $x = 0$ or $y = 0$. We single out this important case.

Definition 4.10 Let x and y be vectors in an inner product space V. We say that x is **orthogonal to** y, written as $x \perp y$, if $\langle x, y\rangle = 0$.

Thus, the zero vector is orthogonal to every vector. For a nontrivial example, consider $\{e_1, \ldots, e_n\}$, the standard basis of $\mathbb{F}^n$. Suppose $i, j \in \{1, \ldots, n\}$ with $i \neq j$. Then $e_i \perp e_j$. Moreover, $e_i + e_j \perp e_i - e_j$ since

$$\langle e_i + e_j, e_i - e_j\rangle = \langle e_i, e_i\rangle - \langle e_i, e_j\rangle + \langle e_j, e_i\rangle - \langle e_j, e_j\rangle = 0.$$

We also read $x \perp y$ as "x is perpendicular to y" or "x perp y" for short. Notice that if $x \perp y$, then $y \perp x$. Once orthogonality is present, many geometrical facts can be proved effortlessly.

Theorem 4.11 (Pythagoras theorem) *Let V be an inner product space over $\mathbb{F}$ and let $x, y \in V$.*

(1) *If $x \perp y$, then $\|x + y\|^2 = \|x\|^2 + \|y\|^2$.*
(2) *If $\mathbb{F} = \mathbb{R}$, then $\|x + y\|^2 = \|x\|^2 + \|y\|^2$ implies that $x \perp y$.*

Proof Since $\|x + y\|^2 = \|x\|^2 + \|y\|^2 + 2\mathrm{Re}\langle x, y\rangle$, we see that

$$\|x + y\|^2 = \|x\|^2 + \|y\|^2 \quad \text{if and only if} \quad \mathrm{Re}\langle x, y\rangle = 0.$$

(1) If $x \perp y$, then $\langle x, y\rangle = 0$. Thus $\mathrm{Re}\langle x, y\rangle = 0$.
(2) If $\mathbb{F} = \mathbb{R}$, then $\langle x, y\rangle = \mathrm{Re}\langle x, y\rangle$. Thus $\mathrm{Re}\langle x, y\rangle = 0$ implies that $x \perp y$. ■

If the underlying field is $\mathbb{C}$, then a statement similar to Theorem 4.11(2) need not hold. For example, take $V = \mathbb{C}$ with the standard inner product $\langle x, y\rangle = x\overline{y}$. With $x = 1$ and $y = i$, we see that

$$\|x + y\|^2 = \|1 + i\|^2 = 2, \quad \|x\|^2 + \|y\|^2 = |1|^2 + |i|^2 = 2.$$

Thus $\|x + y\|^2 = \|x\|^2 + \|y\|^2$; but $\langle x, y\rangle = \langle 1, i\rangle = \overline{i} = -i \neq 0$.

Observe that for any $x, y \in V$ with $y \neq 0$ and $\alpha = \frac{\langle x, y\rangle}{\|y\|^2}$, we have $x - \alpha y \perp \alpha y$. Now, writing $x = (x - \alpha y) + \alpha y$, you can use Pythagoras theorem for an alternative proof of Cauchy–Schwarz inequality. See Exercise 9 below.

Exercises for Sect. 4.2

1. Let u and v be vectors in an inner product space V. If $\|u\| = 1,\ \|v\| = 2$ and $\|u - v\| = 3$, then what is $\|u + v\|$?
2. With the inner product $\langle p, q\rangle = \int_0^1 p(t)\overline{q(t)}dt$ on $\mathcal{P}_{n+1}(\mathbb{C})$, find a polynomial of degree n orthogonal to all of $1, t, t^2, \ldots, t^{n-1}$.
3. Let V be an inner product space of dimension 2. Let $x \in V$ be a nonzero vector. Find a nonzero vector orthogonal to x.
4. For vectors $x, y \in V$, a real inner product space, prove that $x + y \perp x - y$ if and only if $\|x\| = \|y\|$. What happens in a complex inner product space?
5. Show that $\left(\sum_{k=1}^{n} a_k b_k\right)^2 \le \left(\sum_{k=1}^{n} k a_k^2\right)\left(\sum_{k=1}^{n} (b_k/k)^2\right)$ for $a_k, b_k \in \mathbb{R}$.
6. Show that $\langle\cdot,\cdot\rangle : \mathbb{R}^3 \times \mathbb{R}^3 \to \mathbb{R}$ defined by $\langle x, y\rangle = y \begin{bmatrix} 2 & 1 & 0 \\ 1 & 2 & 1 \\ 0 & 1 & 4 \end{bmatrix} x^T$ is an inner product on $\mathbb{R}^3$. Using this inner product, compute the cosines of angles between the standard vectors of $\mathbb{R}^3$.
7. Show that $\langle A, B\rangle = \mathrm{tr}(B^T A)$ defines an inner product on $\mathbb{R}^{3\times 3}$. In this inner product space, determine which of the following matrices are orthogonal to which one. Also find the norm of the matrices.
$$\begin{bmatrix} 1 & 1 & 0 \\ -1 & 1 & 1 \\ 0 & 1 & 0 \end{bmatrix}, \quad \begin{bmatrix} 0 & 1 & 0 \\ 1 & -1 & 0 \\ 0 & 0 & 1 \end{bmatrix}, \quad \begin{bmatrix} 2 & -1 & 0 \\ -1 & 0 & 1 \\ 0 & 1 & 0 \end{bmatrix}.$$
8. *(Appolonius Identity)*: Let V be an inner product space. Let $x, y, z \in V$. Show that
$$\|z - x\|^2 + \|z - y\|^2 = \tfrac{1}{2}\|x - y\|^2 + 2\|z - \tfrac{1}{2}(x + y)\|^2.$$
9. Derive Cauchy–Schwarz inequality from Pythagoras theorem. (Hint: Write $u = y/\|y\|$. Next, for any vector $x \in V$, write $x = v + w$ with $v = \langle x, u\rangle u$ and $w = x - \langle x, u\rangle u$. Observe that $v \perp w$.)
10. Define $\|\cdot\| : C[a, b] \to \mathbb{R}$ by $\|f\| = \max\{f(t) : a \le t \le b\}$. Show that $\|\cdot\|$ is a norm on $C[a, b]$ but it does not satisfy the parallelogram law.

4.3 Orthogonal and Orthonormal Sets

Subsets in which each vector is orthogonal to the other can be interesting. For example, the vectors in the standard basis for $\mathbb{F}^n$ are orthogonal to each other. Moreover, each standard basis vector is a unit vector.

Definition 4.12 Let S be a nonempty subset of an inner product space V. We say that

(a) S is an **orthogonal set** in V if each vector in S is orthogonal to every other vector in S.

(b) S is a **proper orthogonal set** in V if S is an orthogonal set in V and $0 \notin S$.
(c) S is an **orthonormal set** in V if S is a proper orthogonal set in V and each vector in S is a unit vector.

Example 4.13 (1) The standard basis $E = \{e_1, \ldots, e_n\}$ is an orthonormal set in $\mathbb{F}^n$ since $\langle e_i, e_j \rangle = e_j^* e_i = \delta_{ij}$ for any $i, j \in \{1, \ldots, n\}$.

(2) Consider $\mathcal{P}_n(\mathbb{R})$ as an inner product space with the inner product as given in Example 4.2(4). $\{u_1, \ldots, u_{n+1}\}$ with $u_j(t) := t^{j-1}$, $j = 1, \ldots, n+1$, is an orthonormal subset of $\mathcal{P}_n(\mathbb{R})$.

Next, consider $\mathcal{P}_n(\mathbb{R})$ as an inner product space with the inner product

$$\langle p, q \rangle = \int_0^1 p(t)q(t)\,dt.$$

We find that the set $\{u_1, \ldots, u_{n+1}\}$ is not an orthogonal set. For example,

$$\langle u_1, u_2 \rangle = \int_0^1 t\,dt = \frac{1}{2} \neq 0.$$

(3) Consider the vector space $C([0, 2\pi], \mathbb{C})$ with the inner product defined by

$$\langle f, g \rangle := \int_0^{2\pi} f(t)\overline{g(t)}\,dt \quad \text{for } f, g \in C([0, 2\pi], \mathbb{C}).$$

For $n \in \mathbb{N}$, let $u_n, v_n \in C([0, 2\pi], \mathbb{C})$ be given by

$$u_n(t) := \sin(nt), \ v_n(t) := \cos(nt), \quad 0 \le t \le 2\pi.$$

For each $k \in \mathbb{N}$, $\int_0^{2\pi} \cos(kt)\,dt = 0 = \int_0^{2\pi} \sin(kt)\,dt$. Thus, for $n \neq m$, (work it out)

$$\langle u_n, u_m \rangle = \langle v_n, v_m \rangle = \langle u_n, v_n \rangle = \langle u_n, v_m \rangle = 0, \quad \langle u_n, u_n \rangle = \langle v_n, v_n \rangle = \pi.$$

So, $\left\{ \frac{u_n}{\sqrt{\pi}} : n \in \mathbb{N} \right\} \cup \left\{ \frac{v_n}{\sqrt{\pi}} : n \in \mathbb{N} \right\}$ is an orthonormal set in $C([0, 2\pi], \mathbb{C})$. □

Vacuously, a singleton set is an orthogonal set, and a singleton having a nonzero vector is a proper orthogonal set. If $v \neq 0$, then $\{v/\|v\|\}$ is an orthonormal set. Moreover, if S is a proper orthogonal set, then for each $x \in S$, $\langle x, x \rangle > 0$.

Theorem 4.14 *Each proper orthogonal set in an inner product space is linearly independent.*

Proof Let S be a proper orthogonal set in an inner product space V. For $u_1, \ldots, u_n \in S$ and $\alpha_1, \ldots, \alpha_n \in \mathbb{F}$, suppose $\alpha_1 u_1 + \cdots + \alpha_n u_n = 0$. Then, for each $j \in \{1, \ldots, n\}$,

$$0 = \Big\langle \sum_{i=1}^{n} \alpha_i u_i, u_j \Big\rangle = \sum_{i=1}^{n} \langle \alpha_i u_i, u_j \rangle = \sum_{i=1}^{n} \alpha_i \langle u_i, u_j \rangle = \alpha_j \langle u_j, u_j \rangle.$$

Since $u_j \neq 0$, $\alpha_j = 0$. Therefore, S is linearly independent. ■

As a corollary to Theorem 4.14, we obtain the following result.

Theorem 4.15 *Let V be an inner product space of dimension n. Any orthonormal set in V having n vectors is a basis of V.*

Theorem 4.16 *Let $S = \{u_1, \ldots, u_n\}$ be an orthonormal set in an inner product space V. Then the following are true:*

(1) (Fourier expansion) *For each* $x \in \text{span}(S)$, $x = \sum_{j=1}^{n} \langle x, u_j \rangle u_j$.

(2) (Parseval identity) *For each* $x \in \text{span}(S)$, $\|x\|^2 = \sum_{j=1}^{n} |\langle x, u_j \rangle|^2$.

(3) (Bessel inequality) *For each* $y \in V$, $\sum_{j=1}^{n} |\langle y, u_j \rangle|^2 \leq \|y\|^2$.

In particular, if $\dim(V) = n$, *then* (1) *and* (2) *hold for every* $x \in V$.

Proof (1) Since $\{u_1, \ldots, u_n\}$ is an orthonormal set, $\langle u_i, u_j \rangle = \delta_{ij}$ for any $i, j \in \{1, \ldots, n\}$. As $x \in \text{span}(S)$, $x = \alpha_1 u_1 + \cdots + \alpha_n u_n$ for some scalars $\alpha_1, \alpha_2, \ldots, \alpha_n$. Then, for each $i \in \{1, \ldots, n\}$,

$$\langle x, u_i \rangle = \alpha_1 \langle u_1, u_i \rangle + \cdots + \alpha_n \langle u_n, u_i \rangle = \alpha_i$$

Therefore, $x = \sum_{j=1}^{n} \langle x, u_j \rangle u_j$.
(2) Using (1), we obtain

$$\|x\|^2 = \langle x, x \rangle = \Big\langle \sum_{i=1}^{n} \langle x, u_i \rangle u_i, \sum_{j=1}^{n} \langle x, u_j \rangle u_j \Big\rangle = \sum_{i=1}^{n} \sum_{j=1}^{n} \langle x, u_i \rangle \overline{\langle x, u_j \rangle} \langle u_i, u_j \rangle$$
$$= \sum_{i=1}^{n} \sum_{j=1}^{n} |\langle x, u_i \rangle|^2 \delta_{ij} = \sum_{i=1}^{n} |\langle x, u_i \rangle|^2.$$

(3) Write $z := \sum_{i=1}^{n} \langle y, u_i \rangle u_i$. We observe that, for each $j \in \{1, \ldots, n\}$,

$$\langle z, u_j \rangle = \sum_{i=1}^{n} \langle y, u_i \rangle \langle u_i, u_j \rangle = \langle y, u_j \rangle$$

so that $\langle y - z, u_j \rangle = 0$. Therefore, $\langle y - z, z \rangle = 0$. By Pythagoras theorem and (2), it follows that

$$\|y\|^2 = \|z\|^2 + \|y - z\|^2 \geq \|z\|^2 = \sum_{i=1}^{n} |\langle z, u_i \rangle|^2 = \sum_{i=1}^{n} |\langle y, u_i \rangle|^2.$$

■

In the Fourier expansion of a vector $x \in \text{span}\{u_1, \ldots, u_n\}$, the scalars $\langle x, u_j \rangle$ are called the **Fourier coefficients** of x.

Example 4.17 Consider $\mathcal{C}([0, 2\pi], \mathbb{C})$ with the inner product

$$\langle x, y \rangle := \int_0^{2\pi} x(t)\overline{y(t)}dt \quad \text{for } x, y \in \mathcal{C}([0, 2\pi], \mathbb{C}).$$

For $n \in \mathbb{Z}$, define u_n by

$$u_n(t) = \frac{e^{i\,nt}}{\sqrt{2\pi}} \quad \text{for } t \in [0, 2\pi].$$

We see that

$$\langle u_n, u_m \rangle = \frac{1}{2\pi} \int_0^{2\pi} e^{i\,(n-m)t} dt = \begin{cases} 1 & \text{if } n = m, \\ 0 & \text{if } n \neq m. \end{cases}$$

Hence, $\{u_n : n \in \mathbb{Z}\}$ is an orthonormal set in $\mathcal{C}([0, 2\pi], \mathbb{C})$. Let $m \in \mathbb{N}$. By Theorem 4.16, if $x \in \text{span}\{u_j : j = -m, \ldots, -1, 0, 1, \ldots, m\}$, then

$$x = \sum_{k=-m}^{m} a_k\, e^{i\,kt} \quad \text{with} \quad a_k = \frac{1}{2\pi} \int_0^{2\pi} x(t) e^{-i\,kt} dt.$$

□

Notice that if $\{u_1, \ldots, u_n\}$ is an orthogonal set and $\alpha_1, \ldots, \alpha_n$ are scalars, then a slightly generalized form of Pythagoras theorem would look like:

$$\|\alpha_1 u_1 + \cdots + \alpha_n u_n\|^2 = |\alpha_1|^2 \|u_1\|^2 + \cdots + |\alpha_n|^2 \|u_n\|^2.$$

And if $\{u_1, \ldots, u_n\}$ is an orthonormal set, then

$$\|\alpha_1 u_1 + \cdots + \alpha_n u_n\|^2 = |\alpha_1|^2 + \cdots + |\alpha_n|^2.$$

The last equality is Parseval identity in disguise. These equalities provide alternative proofs of the facts that proper orthogonal sets and orthonormal sets are linearly independent.

To see what is going on in the proof of Bessel inequality, consider the standard basis vectors e_1 and e_2, and the vector $(1, 2, 3)$ in $\mathbb{R}^3$. Projections of $(1, 2, 3)$ on the x-axis and y-axis are, respectively, e_1 and $2e_2$. Thus the vector $e_1 + 2e_2 = (1, 2, 0)$ is the projection of $(1, 2, 3)$ on the xy-plane. Moreover, the vector $(1, 2, 3) - (1, 2, 0) = (0, 0, 3)$ is orthogonal to the xy-plane, and $\|(1, 2, 0)\| \leq \|(1, 2, 3)\|$.

Definition 4.18 Let $S = \{u_1, \ldots, u_n\}$ be an orthonormal set in an inner product space V, and let $U = \text{span}(S)$. For each $x \in V$, the vector

$$\text{proj}_U(x) := \sum_{i=1}^{n} \langle x, u_i \rangle u_j$$

is called the **projection of x on the subspace U**.

The proof of Bessel inequality yields the following result.

Theorem 4.19 *Let V be a finite dimensional inner product space, $S = \{u_1, \ldots, u_n\}$ be an orthonormal set in V and let $U = \text{span}(S)$. Then for each $x \in V$, $\text{proj}_U(x) \in U$, and $x - \text{proj}_U(x)$ is orthogonal to every vector in U.*

Intuitively, $x - \text{proj}_U(x)$ minimizes the length $\|x - y\|$ where y varies over U. We will prove this fact in Sect. 4.7.

Notice that the Fourier expansion of a vector $u \in U := \text{span}\{u_1, \ldots, u_n\}$ for the orthonormal set $\{u_1, \ldots, u_n\}$ shows that the $\text{proj}_U(u) = u$, as expected.

In fact, if we define a function $P : V \to V$ by $P(x) = \text{proj}_U(x)$ for each $x \in V$, then P is a linear operator with $R(P) = U$ and $P^2 = P$. We will discuss projection operators later.

Exercises for Sect. 4.3

1. In an inner product space, what is the distance between two orthogonal unit vectors?
2. Let $\langle \cdot, \cdot \rangle$ be an inner product on $\mathbb{R}^{n\times 1}$. Let $A \in \mathbb{R}^{n\times n}$. Let $b \in \mathbb{R}^{n\times 1}$ be a nonzero vector. If each column of A is orthogonal to b, then show that the system $Ax = b$ has no solution.
3. Let $n \in \mathbb{N}$. Show that $\left\{\frac{1}{\sqrt{\pi}}, \frac{\sin t}{\sqrt{\pi}}, \ldots, \frac{\sin nt}{\sqrt{\pi}}, \frac{\cos t}{\sqrt{\pi}}, \ldots, \frac{\cos nt}{\sqrt{\pi}}\right\}$ is an orthonormal set in $C[-\pi, \pi]$ where $\langle f, g\rangle = \int_{-\pi}^{\pi} f(t)g(t)dt$.
4. Show that $U = \{x \in \mathbb{R}^4 : x \perp (1, 0, -1, 1),\ x \perp (2, 3, -1, 2)\}$ is a subspace of $\mathbb{R}^4$, where $\mathbb{R}^4$ has the standard inner product. Find a basis for U.
5. Consider $\mathcal{P}_4(\mathbb{R})$ as an inner product space with $\langle p, q\rangle = \int_{-1}^{1} p(t)q(t)\,dt$. Let $U = \text{span}\{1, t, t^2 - \frac{1}{3}\}$. Determine $\text{proj}_U(t^3)$.

4.4 Gram–Schmidt Orthogonalization

Consider two nonzero linearly independent vectors u, v in $\mathbb{R}^2$ or $\mathbb{R}^3$. Can we choose a vector $w \in \mathbb{R}^3$ such that $w \perp u$ and $\text{span}\{u, w\} = \text{span}\{u, v\}$?

Let $\hat{u}$ denote the unit vector in the direction of u. The vector $(v \cdot \hat{u})\hat{u}$ is the projection of v on the subspace $\text{span}(\{u\})$. Then the vector

$$w = v - (v \cdot \hat{u})\hat{u}$$

does the job. This method can be generalized to any finite number of linearly independent vectors, in any inner product space.

Let u_1, u_2, u_3 be linearly independent vectors in an inner product space V. Suppose that we have already obtained nonzero vectors v_1, v_2 orthogonal to each other so that $\text{span}\{u_1, u_2\} = \text{span}\{v_1, v_2\}$. Let $\hat{v}_1$ and $\hat{v}_2$ be the unit vectors in the directions of

v_1 and v_2, respectively; that is, $\hat{v}_1 := v_1/\|v_1\|$ and $\hat{v}_2 := v_2/\|v_2\|$. We consider the projection of u_3 onto the subspace spanned by $\hat{v}_1$ and $\hat{v}_2$, namely $\langle u_3, \hat{v}_1\rangle\hat{v}_1 + \langle u_3, \hat{v}_2\rangle\hat{v}_2$, and define the required vector as

$$v_3 := u_3 - (\langle u_3, \hat{v}_1\rangle\hat{v}_1 + \langle u_3, \hat{v}_2\rangle\hat{v}_2).$$

Notice that v_3 is orthogonal to both v_1 and v_2; and $\text{span}\{u_1, u_2, u_3\} = \text{span}\{v_1, v_2, v_3\}$. This process can be continued if more than three vectors are initially given.

We consolidate this procedure in the following theorem.

Theorem 4.20 (Gram–Schmidt orthogonalization) *Let $\{u_1, \ldots, u_n\}$ be a linearly independent ordered subset of an inner product space V. Define $v_1, \ldots, v_n$ inductively by*

$$\begin{aligned} v_1 &:= u_1 \\ v_{k+1} &:= u_{k+1} - \sum_{j=1}^{k} \frac{\langle u_{k+1}, v_j\rangle}{\langle v_j, v_j\rangle} v_j, \quad k = 1, \ldots, n-1. \end{aligned}$$

Then $\{v_1, \ldots, v_n\}$ is a proper orthogonal ordered set in V satisfying

$$\text{span}\{u_1, \ldots, u_k\} = \text{span}\{v_1, \ldots, v_k\} \quad \textit{for each } k \in \{1, \ldots, n\}.$$

Proof Since $u_1 = v_1$, the case $n = 1$ is obvious. So, let $1 \le m < n$. For the induction step, assume that $\{v_1, \ldots, v_m\}$ is an orthogonal set of nonzero vectors with

$$\text{span}\{u_1, \ldots, u_m\} = \text{span}\{v_1, \ldots, v_m\}.$$

Now,

$$v_{m+1} := u_{m+1} - \sum_{j=1}^{m} \frac{\langle u_{k+1}, v_j\rangle}{\langle v_j, v_j\rangle} v_j.$$

Notice that $u_{m+1} \in \text{span}\{v_1, \ldots, v_m, v_{m+1}\}$ and $v_{m+1} \in \text{span}\{u_1, \ldots, u_m, u_{m+1}\}$. Thus

$$\text{span}\{u_1, \ldots, u_m, u_{m+1}\} = \text{span}\{v_1, \ldots, v_m, v_{m+1}\}.$$

Since $\{u_1, \ldots, u_m, u_{m+1}\}$ is linearly independent, from the above equality, it also follows that $v_{m+1} \neq 0$. For the orthogonality of $\{v_1, \ldots, v_{m+1}\}$, let $i \in \{1, \ldots, m\}$. Since $\{v_1, \ldots, v_m\}$ is an orthogonal set, $\langle v_j, v_i\rangle = 0$ for $j \neq i$. Hence

$$\langle v_{m+1}, v_i\rangle = \langle u_{m+1}, v_i\rangle - \sum_{j=1}^{m} \frac{\langle u_{m+1}, v_j\rangle}{\langle v_j, v_j\rangle}\langle v_j, v_i\rangle = \langle u_{m+1}, v_i\rangle - \frac{\langle u_{m+1}, v_i\rangle}{\langle v_i, v_i\rangle}\langle v_i, v_i\rangle = 0.$$

Thus $v_{m+1} \perp v_i$ for each $i \in \{1, \ldots, m\}$. Therefore, $\{v_1, \ldots, v_{m+1}\}$ is an orthogonal set of nonzero vectors. ■

Observe that the span condition in Theorem 4.20 asserts that the set $\{v_1, \ldots, v_k\}$ is an orthogonal basis for $\operatorname{span}\{u_1, \ldots, u_k\}$. The proof of Gram–Schmidt orthogonalization process reveals that if $u_1, \ldots, u_k$ are linearly independent vectors and $u_{k+1} \in \operatorname{span}\{u_1, \ldots, u_k\}$, then v_{k+1} becomes 0. Conversely, if v_{k+1} becomes 0, then $u_{k+1} \in \operatorname{span}\{u_1, \ldots, u_k\}$. Therefore, Gram–Schmidt orthogonalization can also be used to determine whether a given list of vectors is linearly independent or not.

By ignoring all those v_js which become 0 in the process, we arrive at a basis for the span of $u_1, \ldots, u_n$. Therefore, Gram–Schmidt procedure is our second tool to extract a basis from a given list of vectors keeping the span unchanged. Recall that the first tool of elementary operations was discussed in Sect. 3.7.

Example 4.21 Consider $\mathbb{F}^3$ with its standard inner product. Take the vectors $u_1 = (1, 0, 0)$, $u_2 = (1, 1, 0)$ and $u_3 = (1, 1, 1)$. Clearly, u_1, u_2, u_3 are linearly independent in $\mathbb{F}^3$. Gram–Schmidt orthogonalization yields the following:

$$v_1 = u_1, \quad v_2 = u_2 - \frac{\langle u_2, v_1\rangle}{\langle v_1, v_1\rangle} v_1.$$

Notice that $\langle v_1, v_1\rangle = 1$ and $\langle u_2, v_1\rangle = 1$. Hence, $v_2 = u_2 - v_1 = (0, 1, 0)$. Next,

$$v_3 = u_3 - \frac{\langle u_3, v_1\rangle}{\langle v_1, v_1\rangle} v_1 - \frac{\langle u_3, v_2\rangle}{\langle v_2, v_2\rangle} v_2.$$

Since $\langle v_2, v_2\rangle = 1$, $\langle u_3, v_1\rangle = 1$, $\langle u_3, v_2\rangle = 1$, we have $v_3 = u_3 - v_1 - v_2 = (0, 0, 1)$. Thus, Gram–Schmidt orthogonalization of $\{u_1, u_2, u_3\}$ results in the ordered set $\{(1, 0, 0), (0, 1, 0), (0, 0, 1)\}$. □

Example 4.22 Consider the vectors $u_1 = (1, 1, 0)$, $u_2 = (0, 1, 1)$ and $u_3 = (1, 0, 1)$ in $\mathbb{F}^3$ with the standard inner product. Clearly, u_1, u_2, u_3 are linearly independent in $\mathbb{F}^3$. Gram–Schmidt orthogonalizaion gives:

$$v_1 = u_1, \quad v_2 = u_2 - \frac{\langle u_2, v_1\rangle}{\langle v_1, v_1\rangle} v_1.$$

Notice that $\langle v_1, v_1\rangle = 2$ and $\langle u_2, v_1\rangle = 1$. So, $v_2 = (0, 1, 1) - \frac{1}{2}(1, 1, 0) = (-\frac{1}{2}, \frac{1}{2}, 1)$. Next,

$$v_3 = u_3 - \frac{\langle u_3, v_1\rangle}{\langle v_1, v_1\rangle} v_1 - \frac{\langle u_3, v_2\rangle}{\langle v_2, v_2\rangle} v_2.$$

We find that $\langle v_2, v_2\rangle = \frac{3}{2}$, $\langle u_3, v_1\rangle = 1$ and $\langle u_3, v_2\rangle = \frac{1}{2}$. Hence,

$$v_3 = (1, 0, 1) - \tfrac{1}{2}(1, 1, 0) - \tfrac{1}{3}(-\tfrac{1}{2}, \tfrac{1}{2}, 1) = (\tfrac{2}{3}, -\tfrac{2}{3}, \tfrac{2}{3}).$$

Therefore, $\{(1, 1, 0), (-\frac{1}{2}, \frac{1}{2}, 1), (\frac{2}{3}, -\frac{2}{3}, \frac{2}{3})\}$ is the Gram–Schmidt orthogonalization of $\{u_1, u_2, u_3\}$. □

Example 4.23 Consider the inner product space $\mathcal{P}(\mathbb{R})$, where the inner product is given by

$$\langle p, q\rangle = \int_{-1}^{1} p(t)\, q(t)\, dt \quad \text{for } p, q \in \mathcal{P}(\mathbb{R}).$$

Let $u_j(t) = t^{j-1}$ for $j = 1, 2, 3$. Consider the linearly independent set $\{u_1, u_2, u_3\}$. Then $v_1(t) = u_1(t) = 1$ for all $t \in [-1, 1]$; and

$$v_2 = u_2 - \frac{\langle u_2, v_1\rangle}{\langle v_1, v_1\rangle} v_1.$$

We see that

$$\langle v_1, v_1\rangle = \int_{-1}^{1} v_1(t)\, v_1(t)\, dt = \int_{-1}^{1} dt = 2,$$
$$\langle u_2, v_1\rangle = \int_{-1}^{1} u_2(t)\, v_1(t)\, dt = \int_{-1}^{1} t\, dt = 0.$$

Hence, $v_2(t) = u_2(t) = t$ for all $t \in [-1, 1]$. Next,

$$v_3 = u_3 - \frac{\langle u_3, v_1\rangle}{\langle v_1, v_1\rangle} v_1 - \frac{\langle u_3, v_2\rangle}{\langle v_2, v_2\rangle} v_2.$$

Here,

$$\langle u_3, v_1\rangle = \int_{-1}^{1} u_3(t)\, v_1(t)\, dt = \int_{-1}^{1} t^2\, dt = \frac{2}{3},$$
$$\langle u_3, v_2\rangle = \int_{-1}^{1} u_3(t)\, v_2(t)\, dt = \int_{-1}^{1} t^3\, dt = 0.$$

Hence, $v_3(t) = t^2 - \frac{1}{3}$ for all $t \in [-1, 1]$. Thus, Gram–Schmidt orthogonalization yields $\{1,\ t,\ t^2 - \frac{1}{3}\}$ as the orthogonal set of polynomials whose span equals $\operatorname{span}\{1, t, t^2\}$. □

The polynomials $p_0(t),\ p_1(t),\ p_2(t), \ldots$ obtained by orthogonalizing the polynomials $1,\ t,\ t^2, \ldots$ using the inner product

$$\langle p, q\rangle = \int_{-1}^{1} p(t)\, q(t)\, dt \quad \text{for } p, q \in \mathcal{P}(\mathbb{R}),$$

as in Example 4.23 are called **Legendre polynomials**.

Gram–Schmidt orthogonalization can easily be adopted for orthonormalization. For this, we normalize the newly obtained vectors at each step. Thus, instead of $v_1 := u_1$, we would take $v_1 := u_1/\|u_1\|$.

Next, we take $w_2 := u_2 - \langle u_2, v_1\rangle v_1$ and $v_2 := w_2/\|w_2\|$. In general, with $U_k = \text{span}\{v_1, \ldots, v_k\}$, we construct

$$w_{k+1} = u_{k+1} - \text{proj}_{U_k}(u_{k+1}) \quad \text{and} \quad v_{k+1} := w_{k+1}/\|w_{k+1}\| \quad \text{for } k > 1.$$

This procedure of constructing an orthonormal set $\{v_1, \ldots, v_n\}$ from a linearly independent set $\{u_1, \ldots, u_n\}$ such that

$$\text{span}\{v_1, \ldots, v_k\} = \text{span}\{u_1, \ldots, u_k\} \quad \text{for } k = 1, \ldots, n,$$

is called **Gram–Schmidt orthonormalization**. As an application of orthonormalization, we obtain the following result.

Theorem 4.24 (QR-factorization) *If the columns of a matrix $A \in \mathbb{F}^{m\times n}$ are linearly independent, then there exist a matrix $Q \in \mathbb{F}^{m\times n}$ with orthonormal columns and an invertible upper triangular matrix $R \in \mathbb{F}^{n\times n}$ such that $A = QR$.*

Proof Let $u_1, \ldots, u_n \in \mathbb{F}^{m\times 1}$ be the columns of $A \in \mathbb{F}^{m\times n}$. Assume that $u_1, \ldots, u_n$ are linearly independent. Then $m \geq n$. It is understood that the inner product in $\mathbb{F}^{m\times 1}$ is the standard inner product $\langle x, y\rangle = y^*x$. Applying Gram–Schmidt orthonormalization on the ordered set $\{u_1, \ldots, u_n\}$ we obtain an orthonormal ordered set, say, $\{v_1, \ldots, v_n\}$ so that

$$\text{span}\{u_1, \ldots, u_k\} = \text{span}\{v_1, \ldots, v_k\} \quad \text{for } k \in \{1, \ldots, n\}.$$

Thus, there exist scalars r_{ij} such that

$$\begin{aligned} u_1 &= r_{11}v_1 \\ u_2 &= r_{12}v_1 + r_{22}v_2 \\ &\vdots \\ u_n &= r_{1n}v_1 + r_{2n}v_2 + \cdots + r_{nn}v_n. \end{aligned}$$

For $i, j = 1, \ldots, n$ and $i > j$, take $r_{ij} = 0$; and construct the matrices $R = [r_{ij}] \in \mathbb{F}^{n\times n}$ and $Q = [v_1, \cdots, v_n] \in \mathbb{F}^{m\times n}$. The above equalities show that

$$A = [u_1, \cdots, u_n] = QR.$$

Here, the columns of Q are orthonormal and R is upper triangular. Moreover, the scalars r_{ii} that appear on the diagonal of R are nonzero. Thus, R is invertible. ■

Notice that since the columns of Q are orthonormal, $Q^*Q = I$. Then $A = QR$ implies that $R = Q^*A$. Further, Q need not be an orthogonal matrix since $Q^*Q = I$ does not imply that $QQ^* = I$ unless $m = n$.

Example 4.25 Let $A = \begin{bmatrix} 1 & 1 \\ 0 & 1 \\ 1 & 1 \end{bmatrix}$. Its columns $u_1 := \begin{bmatrix} 1 \\ 0 \\ 1 \end{bmatrix}$ and $u_2 := \begin{bmatrix} 1 \\ 1 \\ 1 \end{bmatrix}$ are linearly independent. Using Gram–Schmidt orthogonalization on the ordered set $\{u_1, u_2\}$ we obtain the vectors

$$v_1 = \begin{bmatrix} 1/\sqrt{2} \\ 0 \\ 1/\sqrt{2} \end{bmatrix} \quad \text{and} \quad v_2 = \begin{bmatrix} 0 \\ 1 \\ 0 \end{bmatrix}.$$

Thus

$$Q = \begin{bmatrix} 1/\sqrt{2} & 0 \\ 0 & 1 \\ 1/\sqrt{2} & 0 \end{bmatrix}, \quad R = Q^*A = Q^TA = \begin{bmatrix} \sqrt{2} & \sqrt{2} \\ 0 & 1 \end{bmatrix}.$$

We see that $A = QR$.

Notice that $Q^*Q = Q^TQ = I$ but $QQ^* = QQ^T = \begin{bmatrix} 1 & 0 & 1 \\ 0 & 1 & 0 \\ 1 & 0 & 1 \end{bmatrix} \neq I$. □

In the context of Theorem 4.24, it is assumed that the inner product is the standard inner product in $\mathbb{F}^{m\times 1}$. If you choose a different inner product, say, $\langle\cdot,\cdot\rangle$, then you will end up with $A = QR$, where Q has orthonormal columns with respect to $\langle\cdot,\cdot\rangle$. In this case, $v_i^*v_j$ need not be equal to δ_{ij}; consequently, Q^*Q need not be equal to I. However, R is still invertible. If $m = n$, then Q is invertible. In this case, the formula $R = Q^{-1}A$ holds but $R = Q^*A$ need not hold. In fact, the (i, j)th entry in R is equal to $\langle v_i, u_j\rangle$, where v_i is the ith column of Q and u_j is the jth column of A.

Exercises for Sect. 4.4

1. Suppose that the vectors $v_1, \ldots, v_n$ have been obtained by Gram–Schmidt orthogonalization of the vectors $u_1, \ldots, u_n$. Show that $u_1, \ldots, u_n$ are linearly dependent if and only if $v_n = 0$.
2. Consider $\mathbb{R}^3$ with the standard inner product. In each of the following, find orthogonal vectors obtained from the given vectors using Gram–Schmidt orthogonalization:
   (a) $(1, 2, 0),\ (2, 1, 0),\ (1, 1, 1)$ (b) $(1, 1, 1),\ (1, -1, 1),\ (1, 1, -1)$
   (c) $(1, 0, 1),\ (0, 1, 1),\ (1, 1, 0)$ (d) $(0, 1, 1),\ (0, 1, -1), (-1, 1, -1)$.
3. Consider $\mathbb{R}^3$ with the standard inner product. In each of the following, find a unit vector orthogonal to the given two vectors:
   (a) $(2, 1, 0),\ (1, 2, 1)$ (b) $(1, 2, 3),\ (2, 1, -2)$
   (c) $(1, 0, 1),\ (1, 0, -1)$ (d) $(0, 2, -1),\ (-1, 2, -1)$.
4. Using Gram–Schmidt orthogonalization find an orthogonal basis for the subspace $\text{span}\{(4, -2, 0, 6), (3, 3, -3, -3), (5, 5, -7, -7)\}$ of $\mathbb{R}^4$.
5. In $\mathbb{R}^3$, define the inner product by $\langle(a, b, c), (\alpha, \beta, \gamma)\rangle = a\alpha + 2b\beta + 3c\gamma$. Use Gram–Schmidt procedure on the list of vectors $(1, 1, 1), (1, 0, 1), (0, 1, 2)$.

6. Consider the polynomials $u_1(t) = 1,\ u_2(t) = t,\ u_3(t) = t^2$ in the vector space of all polynomials with real coefficients. Find orthogonal polynomials by Gram–Schmidt orthogonalization of $\{u_1, u_2, u_3\}$ with respect to the following inner products:
   (a) $\langle p, q\rangle = \int_0^1 p(t)q(t)\,dt$ (b) $\langle p, q\rangle = \int_{-1}^1 p(t)q(t)\,dt$.
7. Consider $\mathbb{C}^3$ with the standard inner product. Find an orthonormal basis for the subspace spanned by the vectors $(1, 0, i)$ and $(2, 1, 1+i)$.
8. Consider $\mathbb{R}^{3\times 3}$ with the inner product $\langle A, B\rangle = \mathrm{tr}(B^T A)$. Using Gram–Schmidt orthogonalization procedure, find a nonzero matrix which is orthogonal to both the matrices $\begin{bmatrix} 1 & 1 & 1\\ 1 & -1 & 1\\ 1 & 1 & -1\end{bmatrix}$ and $\begin{bmatrix} 1 & 0 & 1\\ 1 & 1 & 0\\ 0 & 1 & 1\end{bmatrix}$.
9. Find a QR-factorization of each of the following matrices:
   (a) $\begin{bmatrix} 0 & 1\\ 1 & 1\\ 0 & 1\end{bmatrix}$ (b) $\begin{bmatrix} 1 & 0 & 2\\ 0 & 1 & 1\\ 1 & 2 & 0\end{bmatrix}$ (c) $\begin{bmatrix} 1 & 1 & 2\\ 0 & 1 & -1\\ 1 & 1 & 0\\ 0 & 0 & 1\end{bmatrix}$.
10. Let $Q \in \mathbb{F}^{m\times n}$, where $m > n$. Show that $QQ^* \neq I$.
11. Let $Q \in \mathbb{F}^{m\times n}$, where $m > n$ and $Q^*Q = I$. What is rank (QQ^*)?

4.5 Orthogonal and Orthonormal Bases

Recall that a linearly independent set which cannot be further extended to a bigger linearly independent set is a basis. Since an orthogonal set is linearly independent, we look for extending it to a larger orthogonal set.

Definition 4.26 Let V be an inner product space. Let S be a proper orthogonal set in V.

(a) The set S is called an **orthogonal basis** of V if no proper superset of S is a proper orthogonal set in V.
(b) The set S is called an **orthonormal basis** of V if S is an orthonormal set and no proper superset of S is an orthonormal set in V.

Informally, an orthogonal basis is a maximal orthogonal set and an orthonormal basis is a maximal orthonormal set. For example, the standard basis of $\mathbb{F}^n$ is an orthogonal basis; it is also an orthonormal basis. An immediate consequence of the definition is the following.

Theorem 4.27 *A proper orthogonal (orthonormal) set S in an inner product space V is an orthogonal (orthonormal) basis for V if and only if 0 is the only vector orthogonal to all vectors in S.*

Proof Let S be an orthogonal basis. Clearly, S is a proper orthogonal set and 0 is orthogonal to all vectors in S. If x is a nonzero vector orthogonal to all vectors in S, then $x \notin S$. Then, $S \cup \{x\}$ would be a proper orthogonal set in V, which is not possible. Therefore, 0 is the only vector orthogonal to all vectors in S.

Conversely, suppose that S is a proper orthogonal set and 0 is the only vector orthogonal to all vectors in S. Then the only orthogonal proper superset of S is $S \cup \{0\}$. But this is not a proper orthogonal set in V. So, no proper superset of S is a proper orthogonal set in V.

The case of orthonormal basis is proved similarly. ■

By Theorem 4.27, it is easy to see that if an orthogonal set S in an inner product space V is a basis of V, then S is an orthogonal basis. However, an orthogonal basis of an inner product space need not be a (Hamel) basis. The same comments apply to orthonormal sets. Here is an example to this effect.

Example 4.28 Let $\ell^2(\mathbb{N}, \mathbb{R})$ be the set of square-summable real sequences, that is,

$$\ell^2(\mathbb{N}, \mathbb{R}) := \Big\{(a_1, a_2, \ldots) : a_k \in \mathbb{R}, \sum_k a_k^2 \text{ converges} \Big\}.$$

Define both addition and scalar multiplication on V component-wise. It can be verified easily that V is a real vector space. Further, $\ell^2(\mathbb{N}, \mathbb{R})$ is an inner product space with the inner product given by

$$\langle (a_1, a_2, \ldots),\ (b_1, b_2, \ldots) \rangle = \sum_{k \in \mathbb{N}} a_k b_k.$$

Let $B := \{e_1, e_2, \ldots\}$, where e_k is the sequence whose kth term is 1 and all other terms are 0. Clearly, B is an orthonormal set. Let $v = (b_1, b_2, \ldots) \in \ell^2(\mathbb{N}, \mathbb{R})$. Since $\langle v,\ e_k \rangle = b_k$, if $v \perp e_k$ for each $k \in \mathbb{N}$, then each term b_k of v becomes 0; that is, the zero vector is the only vector that is orthogonal to B. Therefore, by Theorem 4.27, B is an orthonormal basis of $\ell^2(\mathbb{N}, \mathbb{R})$.

Notice that $\text{span}(B) = c_{00}(\mathbb{R})$, which is a proper subspace of $\ell^2(\mathbb{N}, \mathbb{R})$. Thus, B is not a basis of $\ell^2(\mathbb{N}, \mathbb{R})$. □

The situation is different if the inner product space has a countable basis.

Theorem 4.29 *For an inner product space V having a countable basis, the following are true:*

(1) *V has an orthogonal (orthonormal) basis.*
(2) *Each proper orthogonal (orthonormal) set in V can be extended to an orthogonal (orthonormal) basis of V.*
(3) *Each orthogonal (orthonormal) basis of V is a basis of V.*

Proof (1) Suppose $\{u_1, u_2, \ldots\}$ is a basis of V. Then using Gram–Schmidt orthogonalization we obtain a countable orthogonal set $S = \{v_1, v_2, \ldots, \}$ such that

$$\text{span}\{u_1, \ldots, u_k\} = \text{span}\{v_1, \ldots, v_k\} \quad \text{for } k \geq 1.$$

It follows that S is a spanning set so that it is a basis of V; hence it is an orthogonal basis of V.
(2) Due to Theorem 4.14, the given orthogonal set is linearly independent. We first extend it to a basis of V and then orthogonalize it to obtain the required orthogonal basis.
(3) Let B be an orthogonal basis of V. Then B is linearly independent. Since V has a countable basis, B is a countable set. If B is not a (Hamel) basis of V, then there exists $u \in V$ such that $u \notin \text{span}(B)$. Since B is orthogonal, Gram–Schmidt orthogonalization on $B \cup \{u\}$ yields an orthogonal set $B \cup \{v\}$. But this is an orthogonal proper superset of B. It contradicts the assumption that B is a maximal orthogonal set.

Using Gram–Schmidt orthonormalization, instead of orthogonalization, we obtain (1)–(3) for orthonormal sets in place of orthogonal sets. ■

Theorem 4.29 implies that in a finite dimensional inner product space, a proper orthogonal set is an orthogonal basis if and only if it spans the given space.

Further, $\{e_1, e_2, \ldots\}$ is an orthonormal basis as well as a basis of $c_{00}(\mathbb{F})$. The Legendre polynomials form an orthogonal basis and also a basis for the infinite dimensional inner product space $\mathcal{P}(\mathbb{F})$. Similar statements hold for orthonormal bases also. Theorem 4.29 and Example 4.28 show that $\ell^2(\mathbb{N}, \mathbb{R})$ does not have a countable basis.

Example 4.30 Continuing with the orthogonalization process in Example 4.23, we end up with the nth Legendre polynomial $p_n(t)$, which is of degree n for each $n \in \mathbb{N}$. The set

$$S = \{p_n(t) : p_n \text{ is the Legendre polynomial of degree } n \text{ for } n \geq 0\}$$

is a proper orthogonal set in $\mathcal{P}(\mathbb{R})$. The set S is also a proper orthogonal set in the inner product space $C([-1, 1], \mathbb{R})$, where the inner product is

$$\langle f, g \rangle = \int_{-1}^{1} f(t)g(t)\, dt.$$

Further, there is no proper superset of S, which is a proper orthogonal set in $C([-1, 1], \mathbb{R})$. The proof of this fact relies on a result in Analysis, namely the *Weierstrass approximation theorem*, which states that every function in $C([-1, 1], \mathbb{R})$ is a uniform limit of a sequence of polynomials.

Thus, S is an orthogonal basis of $C([-1, 1], \mathbb{R})$. But the function f defined by $f(t) = \sin t$ for $t \in [-1, 1]$ is in $C([-1, 1], \mathbb{R})$ and $f \notin \mathcal{P}(\mathbb{R}) = \text{span}(S)$. Therefore, S is not a basis of $C([-1, 1], \mathbb{R})$. □

We tolerate the linguistic anomaly that an orthogonal (orthonormal) basis may fail to be a basis. To pre-empt such a situation, sometimes a basis is called a *Hamel basis*.

In an inner product space with a countable basis, this linguistic anomaly does not exist due to Theorem 4.29. When the set $\{u_1, \ldots, u_n\}$ is an orthonormal basis of V, the Fourier coefficients $\langle x, u_j\rangle$ comprise the coordinate vector of x with respect to this basis, and Bessel inequality becomes an equality, which is nothing but the Parseval identity.

Using an orthonormal basis in a finite dimensional inner product space amounts to working in $\mathbb{F}^{n\times 1}$. An orthonormal basis converts the inner product to the dot product. For, suppose that $B = \{v_1, \ldots, v_n\}$ is an orthonormal basis for V. Let $u, v \in V$. Then $u = \sum_{i=1}^{n}\langle u, v_i\rangle v_i$ and $v = \sum_{j=1}^{n}\langle v, v_j\rangle v_j$. We find that

$$\langle u, v\rangle = \sum_{i=1}^{n}\sum_{j=1}^{n}\langle u, v_i\rangle\overline{\langle v, v_j\rangle}\langle v_i, v_j\rangle = \sum_{i=1}^{n}\langle u, v_i\rangle\overline{\langle v, v_i\rangle} = [u]_B \cdot [v]_B.$$

Moreover, orthonormal bases allow writing the entries of the matrix representation of a linear transformation by using the inner products; see the following theorem.

Theorem 4.31 *Let $B = \{u_1, \ldots, u_n\}$ and $E = \{v_1, \ldots, v_m\}$ be ordered bases of inner product spaces U and V, respectively. Let $T : U \to V$ be a linear transformation. If E is an orthonormal basis of V, then the (ij)th entry of $[T]_{E,B}$ is equal to $\langle Tu_j, v_i\rangle$.*

Proof Let $i \in \{1, \ldots, m\}$ and $j \in \{1, \ldots, n\}$. Let a_{ij} denote the (ij)th entry in the matrix $[T]_{E,B}$. Then $Tu_j = a_{1j}v_1 + \cdots + a_{ij}v_i + \cdots + a_{mj}v_m$. Therefore,

$$\langle Tu_j, v_i\rangle = \langle a_{1j}v_1 + \cdots + a_{ij}v_i + \cdots + a_{mj}v_m,\ v_i\rangle = a_{ij}\langle v_i, v_i\rangle = a_{ij},$$

due to the orthonormality of E. ■

If $B = \{u_1, \ldots, u_n\}$ and $E = \{v_1, \ldots, v_m\}$ are bases but not orthonormal bases for the inner product spaces U and V, respectively, then the matrix $[\langle Tu_j, v_i\rangle]$ need not be equal to $[T]_{E,B}$. However, the matrix $[\langle Tu_j, v_i\rangle]$ shares a nice property with $[T]_{E,B}$; see the following theorem.

Theorem 4.32 *Let $\{u_1, \ldots, u_n\}$ and $\{v_1, \ldots, v_m\}$ be bases for the inner product spaces U and V, respectively. Let $S, T : U \to V$ be linear transformations. Then*

$$S = T \text{ if and only if } \langle Su_j, v_i\rangle = \langle Tu_j, v_i\rangle \text{ for all } i = 1, \ldots, m,\ j = 1, \ldots, n.$$

Proof If $S = T$, then clearly, $\langle Su_j, v_i\rangle = \langle Tu_j, v_i\rangle$ for all $i = 1, \ldots, m$, $j = 1, \ldots, n$. Conversely, suppose that this condition holds. Suppose $u \in U$ and $v \in V$. Then we have scalars α_j and β_i such that

$$u = \alpha_1 u_1 + \cdots + \alpha_n u_n \quad\text{and}\quad v = \beta_1 v_1 + \cdots + \beta_m v_m.$$

Due to the linearity of the map T the condition $\langle Su_j, v_i\rangle = \langle Tu_j, v_i\rangle$ for all i, j implies that $\langle Su, v\rangle = \langle Tu, v\rangle$ for all $u \in U$ and all $v \in V$. Hence, by Theorem 4.4, $Su = Tu$ for all $u \in U$ so that $S = T$. ■

Exercises for Sect. 4.5

1. Let $\{u_1, u_2, u_3\}$ be an orthonormal basis of $\mathbb{R}^3$. Let $x = 3u_1 + 4u_2$. Let $y \in \mathbb{R}^3$ satisfy $\langle y, u_3\rangle = 0$ and $\|y\| = 5$. What is the cosine of the angle between $x + y$ and $x - y$? What can go wrong if $\langle y, u_3\rangle \neq 0$?
2. Let $u = \left({}^1/\sqrt{3},\ {}^1/\sqrt{3},\ {}^1/\sqrt{3}\right)^T$ and $v = \left({}^1/\sqrt{2},\ 0,\ -{}^1/\sqrt{2}\right)^T$. Find $w \in \mathbb{R}^{3\times 1}$ so that the matrix $A = [u\ v\ w]$ is orthogonal, that is, $A^*A = I$. Verify that the rows of A are orthonormal.
3. With the notation in Example 4.28, show that $\ell^2(\mathbb{N}, \mathbb{R})$ is an inner product space. Further, show that $({}^1/2,\ {}^1/4,\ \ldots) \notin \text{span}\{u_1, u_2, \ldots\}$, where u_k is the sequence whose kth term is 1 and all other terms 0.
4. Let $E = \{u_1, \ldots, u_n\}$ be an orthonormal set in an inner product space V. Prove that the following statements are equivalent:

   (a) $\text{span}(E) = V$.
   (b) E is an orthonormal basis of V.
   (c) For any $v \in V$, if $\langle v, u_j\rangle = 0$ for each j, then $v = 0$.
   (d) If $v \in V$, then $v = \sum_{i=1}^n \langle v, u_i\rangle u_i$.
   (e) If $x, y \in V$, then $\langle x, y\rangle = \sum_{i=1}^n \langle x, u_i\rangle\langle u_i, y\rangle$.
   (f) If $v \in V$, then $\|v\|^2 = \sum_{i=1}^n |\langle v, u_i\rangle|^2$.

4.6 Orthogonal Complement

We generalize the notion of orthogonality a bit.

Definition 4.33 Let S be a nonempty subset of an inner product space V.

(a) A vector $x \in V$ is said to be **orthogonal to** S if $\langle x, y\rangle = 0$ for all $y \in S$; and in that case, we write $x \perp S$.
(b) The set of all vectors in V that are orthogonal to S is called the **orthogonal complement of** S; it is written as $S^\perp$. That is,

$$S^\perp := \{x \in V : \langle x, y\rangle = 0 \text{ for all } y \in S\}.$$

Also, $\varnothing^\perp = V$. And, $(S^\perp)^\perp$ is written as $S^{\perp\perp}$.

Example 4.34 Let $V = \mathbb{R}^2$ and let $S = \{(1, 2)\}$. Then

$$S^\perp = \{(a, b) \in \mathbb{R}^2 : a + 2b = 0\} = \{(2\alpha, -\alpha) : \alpha \in \mathbb{R}\}.$$
$$S^{\perp\perp} = \{(c, d) \in \mathbb{R}^2 : 2\alpha c - \alpha d = 0,\ \text{for all } \alpha \in \mathbb{R}\} = \{(\beta, 2\beta) : \beta \in \mathbb{R}\}. \quad \square$$

Notice that in Example 4.34, $S^\perp$ and $S^{\perp\perp}$ are subspaces of V, and S is a proper subset of $S^{\perp\perp}$.

You should be able to read the symbolism correctly. For example, the statement

If x is a vector such that $\langle x, y\rangle = 0$ for each $y \in V$, then $x = 0$.

is now written as $V^\perp \subseteq \{0\}$. Similarly, $S^\perp = \{0\}$ asserts that (See Theorem 4.27)

the only vector orthogonal to all the vectors in S is the zero vector.

We now list some useful properties of orthogonal complements.

Theorem 4.35 *Let V be an inner product space. Let S, S_1 and S_2 be nonempty subsets of V.*

(1) $V^\perp = \{0\}$ *and* $\{0\}^\perp = V = \varnothing^\perp$.
(2) $S^\perp$ *is a subspace of* V.
(3) *If* $S_1 \subseteq S_2$, *then* $S_2^\perp$ *is a subspace of* $S_1^\perp$.
(4) $(\text{span}(S))^\perp = S^\perp$.
(5) *If* S *is a basis of* V, *then* $S^\perp = \{0\}$.
(6) $S \subseteq S^{\perp\perp}$.
(7) $(S_1 \cup S_2)^\perp = (\text{span}(S_1) + \text{span}(S_2))^\perp = S_1^\perp \cap S_2^\perp$.
(8) $S_1^\perp + S_2^\perp$ *is a subspace of* $(S_1 \cap S_2)^\perp$.

Proof (1) If $x \in V^\perp$, then $\langle x, y\rangle = 0$ for each $y \in V$. By Theorem 4.4, $x = 0$. Conversely, $\langle 0, y\rangle = 0$ for each $y \in V$. Thus $V^\perp = \{0\}$. Again, since $\langle 0, v\rangle = 0$ for each $v \in V$, $\{0\}^\perp = V$. Also, by definition, $\varnothing^\perp = V$.
(2) Let $x, y \in S^\perp$ and let $\alpha \in \mathbb{F}$. Then $\langle x, u\rangle = 0 = \langle y, u\rangle$ for each $u \in S$. Hence $\langle x + \alpha y, u\rangle = \langle x, u\rangle + \alpha\langle y, u\rangle = 0$. That is, $x + \alpha y \in S^\perp$.
(3) Suppose $S_1 \subseteq S_2$. Let $x \in S_2^\perp$. Then $\langle x, y\rangle = 0$ for each $y \in S_2$. In particular, $\langle x, y\rangle = 0$ for each $y \in S_1$. That is, $x \in S_1^\perp$. So, $S_2^\perp \subseteq S_1^\perp$. By (2), $S_1^\perp$ and $S_2^\perp$ are subspaces. Hence, $S_2^\perp$ is a subspace of $S_1^\perp$.
(4) Since $S \subseteq \text{span}(S)$, by (3), $(\text{span}(S))^\perp \subseteq S^\perp$. Conversely, let $x \in S^\perp$. Suppose $y \in \text{span}(S)$. Then $y = \alpha_1 u_1 + \cdots + \alpha_n u_n$ for some scalars $\alpha_1, \ldots, \alpha_n$ and vectors $u_1, \ldots, u_n \in S$. Since $\langle x, u_i\rangle = 0$ for each $i \in \{1, \ldots, n\}$, we have

$$\langle x, y\rangle = \overline{\alpha}_1\langle x, u_1\rangle + \cdots + \overline{\alpha}_n\langle x, u_n\rangle = 0.$$

Therefore, $x \in (\text{span}(S))^\perp$.
(5) If S is a basis of V, then $V = \text{span}(S)$. The result follows from (4) and (1).
(6) Let $x \in S$. Then $\langle x, y\rangle = 0$ for each $y \in S^\perp$. That is, $x \in S^{\perp\perp}$.
(7) From (4), we have $(S_1 \cup S_2)^\perp = (\text{span}(S_1 \cup S_2))^\perp = (\text{span}(S_1) + \text{span}(S_2))^\perp$. For the second equality, observe that both $\text{span}(S_1)$ and $\text{span}(S_2)$ are subsets of $\text{span}(S_1) + \text{span}(S_2)$. Using (3) twice, we get

$$(\text{span}(S_1) + \text{span}(S_2))^\perp \subseteq (\text{span}(S_1))^\perp \cap (\text{span}(S_2))^\perp.$$

Conversely, let $x \in (\text{span}(S_1))^\perp \cap (\text{span}(S_2))^\perp$. Then

$$\langle x, y\rangle = 0 \text{ for each } y \in \text{span}(S_1) \text{ and } \langle x, y\rangle = 0 \text{ for each } y \in \text{span}(S_2).$$

Each vector in $\text{span}(S_1) + \text{span}(S_2)$ can be written as sum of a vector form $\text{span}(S_1)$ and a vector form $\text{span}(S_2)$. Therefore, for each $y \in \text{span}(S_1) + \text{span}(S_2)$, $\langle x, y\rangle = 0$. That is, $x \in (\text{span}(S_1) + \text{span}(S_2))^{\perp}$.

(8) Since $S_1 \cap S_2$ is a subset of both S_1 and S_2, by (3), $S_1^{\perp} \subseteq (S_1 \cap S_2)^{\perp}$ and also $S_2^{\perp} \subseteq (S_1 \cap S_2)^{\perp}$. Hence $S_1^{\perp} + S_2^{\perp} \subseteq (S_1 \cap S_2)^{\perp}$. By (2), $S_1^{\perp} + S_2^{\perp}$ is a subspace of $(S_1 \cap S_2)^{\perp}$. ■

To see that $(S_1 \cap S_2)^{\perp} \subseteq S_1^{\perp} + S_2^{\perp}$ is not true in general, take $S_1 = \{(1, 0)\}$ and $S_2 = \{(2, 0)\}$, then $(S_1 \cap S_2)^{\perp} = \varnothing^{\perp} = \mathbb{R}^2$ whereas $S_1^{\perp} + S_2^{\perp} = \{(0, a) : a \in \mathbb{R}\}$. The equality is achieved for finite dimensional subspaces S_1 and S_2, as Theorem 4.37 below shows.

Theorem 4.35(4) says that in an inner product space V, if a vector v is orthogonal to a subset S, then v must be orthogonal to $U := \text{span}(S)$. Can we construct such a vector which is orthogonal to U? Yes, provided $\{u_1, \ldots, u_n\}$ is an orthonormal basis of U (see Theorem 4.19).

The following theorem uses this fact to give an orthogonal decomposition of V. It justifies why orthogonal complements are so named.

Theorem 4.36 (Projection theorem) *Let U be a finite dimensional subspace of an inner product space V. Then*

$$V = U \oplus U^{\perp} \text{ and } U^{\perp\perp} = U.$$

In fact, for each $v \in V$, the unique vectors $u \in U$ and $w \in U^{\perp}$ that satisfy $v = u + w$ are given by $u = \text{proj}_U(v)$ and $w = v - u$.

Proof Since $U, U^{\perp} \subseteq V$, $U + U^{\perp} \subseteq V$. For the other inclusion $V \subseteq U + U^{\perp}$, let $B = \{u_1, \ldots, u_n\}$ be an orthonormal basis of U. Let $v \in V$. Write

$$u := \text{proj}_U(v) = \sum_{i=1}^{n} \langle v, u_i\rangle u_i, \quad w := v - u.$$

Now, for any $j \in \{1, \ldots, n\}$,

$$\langle w, u_j\rangle = \langle v, u_j\rangle - \langle u, u_j\rangle = \langle v, u_j\rangle - \sum_{i=1}^{n} \langle v, u_i\rangle\langle u_i, u_j\rangle = \langle v, u_j\rangle - \langle v, u_j\rangle = 0.$$

Hence, $w \in U^{\perp}$. Then $v = u + w$ shows that $V \subseteq U + U^{\perp}$. Hence $V = U + U^{\perp}$. Further, $U \cap U^{\perp} = \{0\}$. Therefore, $V = U \oplus U^{\perp}$.

Next, due to Theorem 4.35(6), $U \subseteq U^{\perp\perp}$. For the other inclusion, let $x \in U^{\perp\perp}$. Since $V = U + U^{\perp}$, there exists $u \in U$ and $y \in U^{\perp}$ such that $x = u + y$. Then

$$0 = \langle x, y\rangle = \langle u + y, y\rangle = \langle u, y\rangle + \langle y, y\rangle = \langle y, y\rangle.$$

That is, $y = 0$. So, $x = u \in U$. We conclude that $U^{\perp\perp} = U$. ■

When V is finite dimensional, an alternative proof can be given for the projection theorem using basis extension. It is as follows:

If $U = \{0\}$, then $U^{\perp} = V$ and $U^{\perp\perp} = U$; all the requirements are satisfied. Otherwise, let $E = \{u_1, \ldots, u_k\}$ be an orthonormal basis of U. Extend E to an orthonormal basis $E \cup \{v_1, \ldots, v_m\}$ of V. Then $U^{\perp} = \text{span}\{v_1, \ldots, v_m\}$; $V = U \oplus U^{\perp}$ and $U^{\perp\perp} = U$.

In general, if S is a subset of V, neither $S^{\perp\perp} = S$ nor $V = S + S^{\perp}$ need hold; see Example 4.34. To see why finite dimension of the subspace U is important in the projection theorem, consider $V = \ell^2(\mathbb{N}, \mathbb{R})$ and $U = c_{00}(\mathbb{R})$. We see that $U^{\perp} = \{0\}$ so that $U^{\perp\perp} = V \neq U$ and $V \neq U + U^{\perp}$. However, if U is a *complete* subspace of V, then both $U^{\perp\perp} = U$ and $V = U + U^{\perp}$ hold; see for example, [15]. Further, the projection theorem implies that the orthogonal complement of each finite dimensional proper subspace of an inner product space is a nontrivial subspace.

As a corollary to the projection theorem, we see that equality in Theorem 4.35(8) can be achieved for finite dimensional subspaces.

Theorem 4.37 *Let U and W be finite dimensional subspaces of an inner product space V. Then $U^{\perp} + W^{\perp} = (U \cap W)^{\perp}$.*

Proof By Theorem 4.35(8), $U^{\perp} + W^{\perp} \subseteq (U \cap W)^{\perp}$. To prove the other inclusion $(U \cap W)^{\perp} \subseteq U^{\perp} + W^{\perp}$, we consider the orthogonal complements and show that $(U^{\perp} + W^{\perp})^{\perp} \subseteq U \cap W$.

Towards this, let $x \in (U^{\perp} + W^{\perp})^{\perp}$. Then $\langle x, y\rangle = 0$ for each $y \in U^{\perp} + W^{\perp}$. As $U^{\perp} \subseteq U^{\perp} + W^{\perp}$, we see that $\langle x, y\rangle = 0$ for each $y \in U^{\perp}$. So, $x \in U^{\perp\perp} = U$, by the projection theorem. Similarly, it follows that $x \in W$. Hence $x \in U \cap W$. Therefore, $(U^{\perp} + W^{\perp})^{\perp} \subseteq U \cap W$.

Then, by Theorem 4.35(3) and the projection theorem, we conclude that

$$(U \cap W)^{\perp} \subseteq (U^{\perp} + W^{\perp})^{\perp\perp} = U^{\perp} + W^{\perp}.$$

■

Exercises for Sect. 4.6

1. In $\mathbb{R}^4$, find $U^{\perp}$ where $U = \text{span}\{u_1, u_2\}$:
   (a) $u_1 = (1, 0, 1, 0)$, $u_2 = (0, 1, 0, 1)$.
   (b) $u_1 = (1, 2, 0, 1)$, $u_2 = (2, 1, 0, -1)$.
   (c) $u_1 = (1, 1, 1, 0)$, $u_2 = (1, -1, 1, 1)$.
   (d) $u_1 = (0, 1, 1, -1)$, $u_2 = (0, 1, -1, 1)$.
2. Using the inner product $\langle p, q\rangle = \int_0^1 p(t)q(t)dt$ on $\mathcal{P}_3(\mathbb{R})$, find the orthogonal complement of the subspace of constant polynomials.
3. Let V be the the set of all bounded sequences $x := (x_n)$ of real numbers.
   (a) Show that V is an inner product space with the inner product $\langle x, y\rangle = \sum_{n=1}^{\infty} x_n y_n / n^2$.
   (b) Find a proper subspace U of V such that $U^{\perp} = \{0\}$.

(c) Show that the inner product in (a) is induced by an appropriate injective linear transformation from V into $\ell^2(\mathbb{N}, \mathbb{R})$.

4. Let $V = \mathbb{C}^{n\times n}$ with the inner product $\langle A, B\rangle = \text{tr}(A^*B)$. Find the orthogonal complement of the subspace of diagonal matrices.
5. Let S be a nonempty subset of an inner product space V. Prove or disprove: (a) $S^{\perp\perp\perp} = S^{\perp}$ (b) $S^{\perp\perp\perp\perp} = S^{\perp\perp}$.

4.7 Best Approximation and Least Squares

Orthogonality can be used to answer a geometrical problem: suppose U is a subspace of an inner product space V. Given a vector $v \in V$, how to determine a vector from u that is closest to v? Since it is closest to v, it may be considered as the best possible approximation of v from U. Since $\|x - y\|$ is the distance between the vectors x and y, such a vector u would minimize $\|v - w\|$ as w varies over U.

Definition 4.38 Let U be a subspace of an inner product space V. Let $v \in V$. A vector $u \in U$ is called a **best approximation** of v from U if

$$\|v - u\| \le \|v - x\| \quad \text{for all } x \in U.$$

Example 4.39 Consider $\mathbb{R}^2$ as an inner product space with the standard inner product. Let $v := (1, 0) \in \mathbb{R}^2$. Let $U = \{(x, x) : x \in \mathbb{R}\}$. To find the best approximation of v from U, we seek a real number b such that

$$\|(1, 0) - (b, b)\| \le \|(1, 0) - (x, x)\| \quad \text{for all } x \in \mathbb{R}.$$

That is, we require an $x \in \mathbb{R}$ that minimizes $\|(1, 0) - (x, x)\|$. Equivalently, we minimize the function $f : \mathbb{R} \to \mathbb{R}$, where

$$f(x) = \|(1, 0) - (x, x)\|^2 = (1 - x)^2 + x^2 \quad \text{for } x \in \mathbb{R}.$$

Using the methods of calculus, it may be seen that f attains its minimum at $x = \frac{1}{2}$. Then the best approximation of $(1, 0)$ from U is $(1/2, 1/2)$. □

Theorem 4.40 *Let V be an inner product space. Let $v \in V$ and let U be a subspace of V.*

(1) *A vector $u \in U$ is a best approximation of v from U if and only if $v - u \perp U$.*
(2) *If a best approximation of v from U exists, then it is unique.*
(3) *If $\dim(U) < \infty$, then $\text{proj}_U(v)$ is the best approximation of v from U.*

Proof (1) Let $u \in U$ satisfy $v - u \perp U$. Let $x \in U$. Then $v - u \perp u - x$. By Pythagoras theorem,

$$\|v - x\|^2 = \|(v - u) + (u - x)\|^2 = \|v - u\|^2 + \|u - x\|^2 \geq \|v - u\|^2.$$

Therefore, u is a best approximation of v from U.

Conversely, suppose that $u \in U$ is a best approximation of v. Let $x \in U$. If $x = 0$, then $v - u \perp x$. Assume that $x \neq 0$. For any scalar α, we have

$$\begin{aligned}\|v - u\|^2 \leq \|v - u - \alpha x\|^2 &= \langle v - u - \alpha x, v - u - \alpha x\rangle \\ &= \|v - u\|^2 - \langle v - u, \alpha x\rangle - \langle \alpha x, v - u\rangle + |\alpha|^2\|x\|^2 \end{aligned} \tag{4.1}$$

In particular, for $\alpha = \dfrac{\langle v - u, x\rangle}{\langle x, x\rangle}$, we obtain

$$\begin{aligned}\langle v - u, \alpha x\rangle &= \overline{\alpha}\langle v - u, x\rangle = \overline{\alpha}\alpha\langle x, x\rangle = |\alpha|^2\|x\|^2 ; \\ \langle \alpha x, v - u\rangle &= \overline{\langle v - u, \alpha x\rangle} = |\alpha|^2\|x\|^2.\end{aligned}$$

From (4.1) it follows that

$$\|v - u\|^2 \leq \|v - u\|^2 - |\alpha|^2\|x\|^2 \leq \|v - u\|^2.$$

It implies that $|\alpha|^2\|x\|^2 = 0$. As $x \neq 0$, $\alpha = 0$. That is, $\langle v - u, x\rangle = 0$. Therefore, $v - u \perp U$.

(2) Suppose $u, w \in U$ are best approximations of v from U. Then

$$\|v - u\|^2 \leq \|v - w\|^2 \quad \text{and} \quad \|v - w\|^2 \leq \|v - u\|^2.$$

So, $\|v - w\|^2 = \|v - u\|^2$. Due to (1), $v - u \perp U$. But $u - w \in U$. By Pythagoras theorem, we have

$$\|v - w\|^2 = \|v - u + u - w\|^2 = \|v - u\|^2 + \|u - w\|^2 = \|v - w\|^2 + \|u - w\|^2.$$

Thus $\|u - w\|^2 = 0$. That is, $u = w$.

(3) Let U be a finite dimensional subspace of V. Then $u = \text{proj}_U(v) \in U$. Due to the projection theorem, $v - u \perp U$. By (1)-(2), u is the best approximation of v from U. ■

Example 4.41 (1) Consider Example 4.39 once more, in the light of Theorem 4.40. Suppose $u = (\alpha, \alpha)$ is the best approximation of $v = (1, 0) \in \mathbb{R}^2$ from $\{(x, x) : x \in \mathbb{R}\}$. Then

$$(1, 0) - (\alpha, \alpha) \perp (\beta, \beta) \quad \text{for all } \beta \in \mathbb{R}.$$

In particular, $(1, 0) - (\alpha, \alpha) \perp (1, 1)$. It leads to $\alpha = \frac{1}{2}$. Therefore the best approximation of $(1, 0)$ from $\{(x, x) : x \in \mathbb{R}\}$ is $(1/2, 1/2)$.

(2) Consider $V = C([0, 1], \mathbb{R})$ as a real inner product space, with the inner product

$$\langle f, g\rangle = \int_0^1 f(t)g(t)\,dt \quad \text{for } f, g \in V.$$

To find the best approximation of $u(t) := t^2$ from $U = \mathcal{P}_1(\mathbb{R})$ we determine $\alpha, \beta \in \mathbb{R}$ so that $v(t) := \alpha + \beta t$ satisfies $u - v \perp u_1$ and $u - v \perp u_2$, where $u_1(t) = 1$ and $u_2(t) = t$. That is,

$$\int_0^1 (t^2 - \alpha - \beta t)\,dt = 0 = \int_0^1 (t^3 - \alpha t - \beta t^2)\,dt.$$

This gives

$$\frac{1}{3} - \alpha - \frac{\beta}{2} = 0 = \frac{1}{4} - \frac{\alpha}{2} - \frac{\beta}{3}.$$

Therefore, the best approximation of $u(t) := t^2$ from $\mathcal{P}_1(\mathbb{R})$ is $v(t) = -\frac{1}{6} + t$.

(3) Let $V = \ell^2(\mathbb{N}, \mathbb{R})$ as in Example 4.28 and let $U = \text{span}\{e_{2n} : n \in \mathbb{N}\}$. Let $v = (\alpha_1, \alpha_2, \alpha_3, \ldots) \in V$. Consider $u = (0, \alpha_2, 0, \alpha_4, 0, \ldots) \in U$.

If $x = (0, \beta_2, 0, \beta_4, \ldots)$ is any vector in U, then we find that

$$\langle v - u, x\rangle = \big\langle (\alpha_1, 0, \alpha_3, 0, \alpha_4, \ldots), (0, \beta_2, 0, \beta_4, \ldots)\big\rangle = 0.$$

That is, $v - u \perp U$. Therefore, u is the best approximation of v from U. □

Observe that Theorem 4.40 guarantees the existence of the best approximation of a vector in V from the subspace U when U is finite dimensional. If U is infinite dimensional, then a best approximation to any given vector in V from U may not exist. See the following example.

Example 4.42 Let $V = \ell^2(\mathbb{N}, \mathbb{R})$ as in Example 4.28, $U = c_{00}(R)$, and let $v := \big(1, 1/2, 1/3, \ldots\big)$. Clearly $v \in V \setminus U$. Suppose $u \in U$ is the best approximation of v.

With $v_n = \big(1, 1/2, \ldots, 1/n, 0, 0 \ldots\big) \in U$, we have

$$\|v - u\|^2 \le \|v - v_n\|^2 = \frac{1}{(n+1)^2} + \frac{1}{(n+2)^2} + \cdots.$$

This is true for all $n \in \mathbb{N}$. As $n \to \infty$, we obtain $\|v - u\| = 0$. So that $v = u$, which is not possible. Therefore, v does not have a best approximation from U.

In fact, this happens for any $v \in V \setminus U$. To see this, let $v = \big(\alpha_1, \alpha_2, \ldots, \big) \in V \setminus U$. Let $e_j \in U$ be the sequence whose jth term is 1, and all other terms equal to 0. The set $\{e_j : j \in \mathbb{N}\}$ is an orthonormal basis of V. Thus for any $u \in U$,

$$\langle v - u, w\rangle = 0 \text{ for all } w \in U \text{ iff } \langle v - u, e_j\rangle = 0 \text{ for all } j \in \mathbb{N} \text{ iff } v - u = 0.$$

This is impossible since $u \in U$ but $v \in V \setminus U$. Hence, by Theorem 4.40, u cannot be a best approximation of v from U. Therefore, vectors in $V \setminus U$ do not have best approximations from U. □

If the subspace U of V has an orthonormal basis $\{u_1, \ldots, u_n\}$, then by Theorem 4.40, the best approximation $u \in U$ of v is given by

$$u = \text{proj}_U(v) = \sum_{j=1}^{k} \langle v, u_j \rangle u_j.$$

It is not necessary to start with an orthonormal basis of U for computing u. If we have a basis of U, then the orthogonality condition given in Theorem 4.40(1) can be used to determine u. We have followed this method in Example 4.41. Let us review the general situation.

Let $v \in V$. Let $\{u_1, \ldots, u_n\}$ be any basis of U, which is a subspace of V. Let u be the best approximation of v from U. Suppose

$$u = \alpha_1 u_1 + \cdots + \alpha_n u_n$$

for some scalars $\alpha_1, \ldots, \alpha_n$. Using the orthogonality condition that $v - u \perp u_j$ for each $j \in \{1, \ldots, n\}$, we obtain the system of linear equations

$$\alpha_1 \langle u_1, u_j \rangle + \cdots + \alpha_n \langle u_n, u_j \rangle = \langle v, u_j \rangle \quad \text{for } j = 1, \ldots, n.$$

We may write this linear system as $Ax = y$, where

$$A = [\langle u_i, u_j \rangle], \quad x = [\alpha_1 \ \cdots \ \alpha_n]^T, \quad y = [\langle v, u_1 \rangle \ \cdots \ \langle v, u_n \rangle]^T.$$

This matrix A is called the **Gram matrix** of the basis $\{u_1, \ldots, u_n\}$. The Gram matrix is invertible since there exists a unique best approximation. (This fact can also be established using the linear independence of $\{u_1, \ldots, u_n\}$.) Then the solution of this linear system gives the coefficients $\alpha_1, \ldots, \alpha_n$; and consequently, the best approximation u is determined.

Best approximation can be used for computing approximate solutions of linear systems. If u is a solution of the linear system $Ax = b$, then the *residual* $\|Au - b\|$ must be zero. Thus, if $Ax = b$ does not have a solution, then an approximate solution is obtained by choosing a vector u so that the residual $\|Au - b\|$ is minimized. Equivalently, we seek a minimizer of $\|Ax - b\|^2$. Such a minimizer is named as a *least squares* solution of the linear system. More generally, we have the following definition.

Definition 4.43 Let $T : U \to V$ be a linear transformation, where U is a vector space and V is an inner product space. Let $y \in V$. A vector $u \in U$ is called a **least squares solution** of the equation $Tx = y$ if $\|Tu - y\| \le \|Tz - y\|$ for all $z \in U$.

Observe that if v is a solution of $Tx = y$, then $\|Tv - y\| = 0$. Thus v minimizes the norm $\|Tx - y\|$. That is, each solution of $Tx = y$ is a least squares solution. But, a least squares solution need not be a solution; see the following example.

Example 4.44 Let $A = \begin{bmatrix} 1 & 1 \\ 0 & 0 \end{bmatrix}$ and $b = \begin{bmatrix} 0 \\ 1 \end{bmatrix}$. If $u = \begin{bmatrix} \alpha \\ \beta \end{bmatrix} \in \mathbb{R}^{2\times 1}$, then

$$\|Au - b\|^2 = (\alpha + \beta)^2 + 1 \geq 1.$$

Hence, minimum value of $\|Au - b\|^2$ is attained for all $\alpha, \beta \in \mathbb{R}$ with $(\alpha + \beta)^2 = 0$. Therefore, $u = (\alpha, -\alpha)^T$ is a least squares solution for any $\alpha \in \mathbb{R}$.

However, the linear system $Ax = b$ does not have a solution.

It also shows that a least squares solution is not necessarily unique. □

It follows from Definition 4.43 that

u is a least squares solution of $Tx = y$ if and only if $v = Tu$ is a best approximation of y from $R(T)$, the range space of T.

Thus a least squares solution is also called a *best approximate solution.* Notice that a least squares solution of $Ax = y$ is in the domain space of T whereas the best approximation of y is in the range space of T. Therefore, uniqueness of a least squares solution will depend upon the injectivity of T. Theorem 4.40 entails the following.

Theorem 4.45 *Let $T : U \to V$ be a linear transformation, where U is a vector space and V is an inner product space.*

(1) *A vector $u \in U$ is a least squares solution of $Tx = y$ if and only if $Tu - y$ is orthogonal to $R(T)$.*
(2) *A least squares solution of $Tx = y$ is unique if and only if T is injective.*
(3) *If $R(T)$ is finite dimensional, then $Tx = y$ has a least squares solution.*
(4) *Let T be injective. Let $u_1, \ldots, u_k$ be distinct vectors in U. If $\{Tu_1, \ldots, Tu_k\}$ is an orthonormal basis of $R(T)$, then $u := \sum_{i=1}^{n} \langle y, Tu_i \rangle u_i$ is the least squares solution of $Tx = y$.*

Least squares solutions of linear systems can be computed in a simpler way by using the standard inner product on $\mathbb{F}^{m\times 1}$. Further, QR-factorization can be employed for this purpose.

Theorem 4.46 *Let $A \in \mathbb{F}^{m\times n}$ and let $b \in \mathbb{F}^{m\times 1}$. Then the following are true:*

(1) *A vector $u \in \mathbb{F}^{n\times 1}$ is a least squares solution of $Ax = b$ if and only if $A^*Au = A^*b$.*
(2) *If the columns of A are linearly independent, then $u = R^{-1}Q^*b$ is the least squares solution of $Ax = b$, where $A = QR$ is the QR-factorization of A.*

Proof (1) The columns $v_1, \ldots, v_n$ of A span $R(A)$. Thus
u is a least squares solution of $Ax = b$
if and only if $\langle Au - b, v_i\rangle = 0$, for each $i \in \{1, \ldots, n\}$
if and only if $v_i^*(Au - b) = 0$ for each $i \in \{1, \ldots, n\}$
if and only if $A^*(Au - b) = 0$
if and only if $A^*Au = A^*b$.
(2) Suppose the columns of A are linearly independent. Let $A = QR$ be the QR-factorization of A. Write $u := R^{-1}Q^*b$. Since $Q^*Q = I$,

$$A^*Au = R^*Q^*QRR^{-1}Q^*b = R^*Q^*b = A^*b.$$

Due to (1), u is a least squares solution of $Ax = b$.

Since the columns of $A \in \mathbb{F}^{m\times n}$ are linearly independent, rank $(A) = n$, which equals the dimension of the domain space $\mathbb{F}^{n\times 1}$ of the linear transformation A. Due to Theorem 2.26(1), A is injective. By Theorem 4.45(2), $u = R^{-1}Q^*b$ is the (only) least squares solution of $Ax = b$. ■

In Example 4.44, $A = \begin{bmatrix} 1 & 1 \\ 0 & 0 \end{bmatrix}$ and $b = \begin{bmatrix} 0 \\ 1 \end{bmatrix}$. Thus $u = \begin{bmatrix} \alpha \\ \beta \end{bmatrix}$ is a least squares solution of $Ax = b$ if and only if $A^*Au = A^*b$, or,

$$\begin{bmatrix} 1 & 1 \\ 1 & 1 \end{bmatrix}\begin{bmatrix} \alpha \\ \beta \end{bmatrix} = \begin{bmatrix} 0 \\ 0 \end{bmatrix}.$$

Clearly, its solution set is $\{(\alpha, -\alpha)^T : \alpha \in \mathbb{R}\}$ as found earlier.

For the linear system $Ax = b$ with columns of A being linearly independent, we have a formula $u = R^{-1}Q^*b$ for the least squares solution. In actual computation we rather solve the associated linear system $Ru = Q^*b$. It is easier to do so since R is upper triangular.

Exercises for Sect. 4.7

1. Find the best approximation of $v \in V$ from U, where
   (a) $V = \mathbb{R}^2$, $v = (1, 0)$, $U = \{(a, a) : a \in \mathbb{R}\}$.
   (b) $V = \mathbb{R}^3$, $v = (1, 2, 1)$, $U = \text{span}\{(3, 1, 2), (1, 0, 1)\}$.
   (c) $V = \mathbb{R}^3$, $v = (1, 2, 1)$, $U = \{(a, b, c) \in \mathbb{R}^3 : a + b + c = 0\}$.
   (d) $V = \mathbb{R}^4$, $v = (1, 0, -1, 1)$, $U = \text{span}\{(1, 0, -1, 1), (0, 0, 1, 1)\}$.
   (e) $V = \mathbb{R}^4$, $v = (1, 2, 3, 4)$, $U = \text{span}\{(1, 1, 0, 0), (0, 0, 1, 2)\}$.
2. In the real inner product space $C([0, 1], \mathbb{R})$, with $\langle f, g\rangle = \int_0^1 f(t)g(t)\,dt$, find the best approximation of
   (a) $u(t) := t^2$ from $\mathcal{P}_1(\mathbb{R})$ (b) $u(t) := \exp(t)$ from $\mathcal{P}_4(\mathbb{R})$.
3. Determine a polynomial $p(t)$ of degree at most 3 such that $p(0) = 0 = p'(0)$ and $\int_0^1 |p(t) - 2 - 3t|^2 dt$ is as small as possible.
4. Determine the polynomial $p(t) \in \mathcal{P}_5(\mathbb{R})$ minimizing $\int_{-\pi}^{\pi} |\sin t - p(t)|^2 dt$.

5. Find the least squares solution for the system $Ax = b$, where

$$\text{(a) } A = \begin{bmatrix} 3 & 1 \\ 1 & 2 \\ 2 & -1 \end{bmatrix},\ b = \begin{bmatrix} 1 \\ 0 \\ -2 \end{bmatrix} \quad \text{(b) } A = \begin{bmatrix} 1 & 1 & 1 \\ -1 & 0 & 1 \\ 1 & -1 & 0 \\ 0 & 1 & -1 \end{bmatrix},\ b = \begin{bmatrix} 0 \\ 1 \\ -1 \\ -2 \end{bmatrix}.$$

6. Let $A \in \mathbb{R}^{m\times n}$ and $b \in \mathbb{R}^m$. If columns of A are linearly independent, then show that there exists a unique $x \in \mathbb{R}^n$ such that $A^T Ax = A^T b$.

4.8 Riesz Representation and Adjoint

Let V be an inner product space and let $y \in V$. The inner product defines a function $f : V \to \mathbb{F}$, given by

$$f(x) = \langle x, y\rangle \quad \text{for } x \in V.$$

For all $u, v \in V$ and each $\alpha \in \mathbb{F}$, we see that

$$f(u+v) = \langle u+v, y\rangle = \langle u, y\rangle + \langle v, y\rangle = f(u) + f(v),$$
$$f(\alpha u) = \langle \alpha u, y\rangle = \alpha\langle u, y\rangle = \alpha f(u).$$

That is, the map $x \mapsto \langle x, y\rangle$ is a linear functional on V.

Similarly, if we define a function $g : V \to \mathbb{F}$ by fixing the first vector, that is, by taking $g(y) = \langle x, y\rangle$, then $g(u+v) = g(u) + g(v)$ but $g(\alpha y) = \overline{\alpha} g(y)$. Such a function is called a *conjugate linear functional* on V. In a real inner product space, a conjugate linear functional is also a linear functional.

We show that every linear functional can be written via the inner product if the inner product space is finite dimensional. It yields a representation of functionals by vectors. Again, orthonormal bases come in handy.

Theorem 4.47 (Riesz representation) *Let V be a finite dimensional inner product space. For each linear functional $f : V \to \mathbb{F}$, there exists a unique $y \in V$ such that*

$$f(x) = \langle x, y\rangle \quad \textit{for all } x \in V,$$

which is given by $y = \sum_{j=1}^{n} \overline{f(u_j)}\, u_j$ for any orthonormal basis $\{u_1, \ldots, u_n\}$ of V.

Proof Let $f : V \to \mathbb{F}$ be a linear functional. Let $\{u_1, \ldots, u_n\}$ be an orthonormal basis of V and let $x \in V$. By Fourier expansion, $x = \sum_{j=1}^{n} \langle x, u_j\rangle u_j$. Consequently,

$$f(x) = \sum_{j=1}^{n} \langle x, u_j\rangle f(u_j) = \sum_{j=1}^{n} \langle x, \overline{f(u_j)}\, u_j\rangle = \Big\langle x, \sum_{j=1}^{n} \overline{f(u_j)}\, u_j\Big\rangle.$$

Thus, $f(x) = \langle x, y\rangle$ for all $x \in V$, where $y = \sum_{j=1}^{n} \overline{f(u_j)}\, u_j$.

For uniqueness of such a vector y, let $y_1, y_2 \in V$ be such that

$$f(x) = \langle x, y_1 \rangle, \quad f(x) = \langle x, y_2 \rangle \quad \text{for all } x \in V.$$

Then $\langle x,\ y_1 - y_2 \rangle = 0$ for all $x \in V$. Therefore, $y_1 = y_2$. ■

The vector y in the equation $f(x) = \langle x, y \rangle$ is called the **Riesz representer** of the functional f, and we denote it by v_f. That is, if $\{u_1, \ldots, u_n\}$ is an orthonormal basis of V, then

$$v_f = \sum_{j=1}^{n} \overline{f(u_j)}\, u_j.$$

The map $f \mapsto v_f$ is a function from V' to V. Write this map as χ. That is, let

$$\chi(f) = v_f \quad \text{for } f \in V'.$$

Suppose that $f, g \in V'$ and $\alpha \in \mathbb{F}$. Then $\chi(f)$, $\chi(g)$, $\chi(f + g)$ and $\chi(\alpha f)$ are vectors in V that satisfy

$$\langle x, \chi(f + g) \rangle = (f + g)(x) = f(x) + g(x) = \langle x, \chi(f) \rangle + \langle x, \chi(g) \rangle = \langle x, \chi(f) + \chi(g) \rangle,$$

$$\langle x, \chi(\alpha f) \rangle = (\alpha f)(x) = \alpha f(x) = \alpha \langle x, \chi(f) \rangle = \langle x, \overline{\alpha}\, \chi(f) \rangle,$$

for every $x \in V$. Thus,

$$\chi(f + g) = \chi(f) + \chi(g) \quad \text{and} \quad \chi(\alpha f) = \overline{\alpha}\, \chi(f),$$

for every $f, g \in V'$ and $\alpha \in \mathbb{F}$. That is, the map $f \mapsto v_f$ is conjugate linear. Also, this map is injective.

Now, going a bit further to the dual of the space V', we see that each functional $\phi \in V''$ has a Riesz representer f_ϕ in V'. Again, the map $\phi \mapsto f_\phi$ is conjugate linear. Then the composition map

$$\phi \mapsto f_\phi \mapsto \chi(f_\phi) = v_{f_\phi}$$

from V'' to V is a linear transformation. Since both $\phi \mapsto f_\phi$ and χ are injective and conjugate linear, the composition map is injective and linear, and with the finite dimensionality of V, this map becomes an isomorphism. This is the natural isomorphism between V and V'' that we mentioned in Sect. 2.6.

Here is an important application of Riesz representation theorem. To keep the notation simple, we use the same notation $\langle \cdot, \cdot \rangle$ for inner products on both the inner product spaces involved.

Theorem 4.48 *Let $T : V \to W$ be a linear transformation, where V and W are finite dimensional inner product spaces. Then there exists a unique linear transformation $S : W \to V$ such that $\langle Tx, y \rangle = \langle x, Sy \rangle$ for all $x \in V,\ y \in W$.*

Proof Let $w \in W$. Consider the map $g : V \to \mathbb{F}$ defined by

$$g(x) = \langle Tx, w\rangle \quad \text{for each } x \in V.$$

Since $T : V \to W$ is a linear transformation, g is a linear functional on V. Let v be the Riesz representer of the functional g. Then

$$g(x) = \langle x, v\rangle \quad \text{for each } x \in V.$$

That is, for each $w \in W$, we have a corresponding vector $v \in V$ such that

$$\langle Tx, w\rangle = \langle x, v\rangle \quad \text{for each } x \in V.$$

Due to the uniqueness of the Riesz representer, the correspondence $w \mapsto v$ defines a function from W to V. Call this function as S. That is,

$$S : W \to V \quad \text{with} \quad S(w) = v.$$

To see that S is a linear transformation, let $x \in V,\ y, z \in W$, and let $\alpha \in \mathbb{F}$. Then

$$\langle x, S(y+z)\rangle = \langle Tx, y+z\rangle = \langle Tx, y\rangle + \langle Tx, z\rangle = \langle x, Sy\rangle + \langle x, Sz\rangle = \langle x, Sy + Sz\rangle,$$
$$\langle x, S(\alpha y)\rangle = \langle Tx, \alpha y\rangle = \overline{\alpha}\langle Tx, y\rangle = \overline{\alpha}\langle x, Sy\rangle = \langle x, \alpha Sy\rangle.$$

Hence for all $x \in V$, $S(y+z) = Sy + Sz$ and $S(\alpha y) = \alpha Sy$. That is, S is a linear transformation.

In fact, $S(w)$ is the Riesz representer of the linear functional $x \mapsto \langle Tx, w\rangle$. Thus $S : W \to V$ is a linear transformation satisfying the property that

$$\langle Tx, w\rangle = \langle x, Sw\rangle \quad \text{for each } x \in V.$$

For the uniqueness, suppose $S_1, S_2 : W \to V$ are linear transformations satisfying

$$\langle Tx, y\rangle = \langle x, S_1 y\rangle = \langle x, S_2 y\rangle \quad \text{for all } x \in V,\ y \in W.$$

Then for all $x \in V,\ y \in W,\ \langle x, (S_1 - S_2)y\rangle = 0$. In particular, for $x = (S_1 - S_2)y$, we have

$$\langle (S_1 - S_2)y,\ (S_1 - S_2)y\rangle = 0 \quad \text{for all } y \in W.$$

It implies that $S_1 - S_2$ is the zero linear transformation. ■

The above proof shows that if $\{v_1, \ldots, v_n\}$ is an orthonormal basis of V, and $y \in W$, then $S(y)$ is the Riesz representor of the functional $f(\cdot) := \langle T(\cdot), y\rangle$. That is,

$$S(y) = \sum_{j=1}^{n} \overline{f(v_j)}\, v_j = \sum_{j=1}^{n} \overline{\langle T v_j, y\rangle}\, v_j = \sum_{j=1}^{n} \langle y, T v_j, y\rangle\, v_j \quad \text{for } y \in W.$$

In fact, this gives rise to the following direct proof of Theorem 4.48 without using the Riesz representer, where finite dimensionality of W is not required.

A direct proof of Theorem 4.48:

Let $\{v_1, \ldots, v_n\}$ be an orthonormal basis of V. Let $x \in V$. Due to Theorem 4.16, $x = \sum_{j=1}^{n} \langle x, v_j\rangle\, v_j$. For any $y \in W$, we have

$$\begin{aligned}\langle T x, y\rangle &= \Big\langle \sum_{j=1}^{n} \langle x, v_j\rangle\, T v_j,\ y\Big\rangle = \sum_{j=1}^{n} \langle x, v_j\rangle\, \langle T v_j, y\rangle \\ &= \sum_{j=1}^{n} \big\langle x,\ \overline{\langle T v_j, y\rangle}\, v_j\big\rangle = \Big\langle x,\ \sum_{j=1}^{n} \langle y, T v_j\rangle\, v_j\Big\rangle.\end{aligned}$$

The assignment $y \mapsto \sum_{j=1}^{n} \langle y, T v_j\rangle\, v_j$ is uniquely determined from the orthonormal basis, the given linear transformation T, and the inner product. Therefore, $S : W \to V$ defined by

$$S(y) = \sum_{j=1}^{n} \langle y, T v_j\rangle\, v_j \quad \text{for } y \in W \tag{4.2}$$

is a well-defined function. Clearly, S is a linear transformation from W to V satisfying

$$\langle T x, y\rangle = \langle x, S y\rangle \quad \text{for } x \in V,\ y \in W.$$

The uniqueness of such an S follows as in the first proof given above. ■

Example 4.49 Let $A := [a_{ij}] \in \mathbb{F}^{m\times n}$. Then A can be thought of as a linear transformation from $\mathbb{F}^n$ to $\mathbb{F}^m$, defined by

$$A(\alpha_1, \ldots, \alpha_n) = (\beta_1, \ldots, \beta_m), \quad \beta_i = \sum_{j=1}^{n} a_{ij}\alpha_j.$$

With the standard inner products on $\mathbb{F}^n$ and $\mathbb{F}^m$, we have

$$\begin{aligned}\langle A(\alpha_1, \ldots, \alpha_n), (\gamma_1, \ldots, \gamma_m)\rangle &= \sum_{i=1}^{m} \Big(\sum_{j=1}^{n} a_{ij}\alpha_j \Big) \overline{\gamma}_i = \sum_{j=1}^{n} \alpha_j \Big(\sum_{i=1}^{m} a_{ij}\overline{\gamma}_i \Big) \\ &= \sum_{j=1}^{n} \alpha_j \overline{\Big(\sum_{i=1}^{m} \overline{a}_{ij}\gamma_i \Big)} = \sum_{i=1}^{n} \alpha_i \overline{\Big(\sum_{j=1}^{m} \overline{a}_{ji}\gamma_j \Big)}.\end{aligned}$$

Thus, $\langle A(\alpha_1, \ldots, \alpha_n), (\gamma_1, \ldots, \gamma_m)\rangle = \langle(\alpha_1, \ldots, \alpha_n), B(\gamma_1, \ldots, \gamma_m)\rangle$, where

$$B(\gamma_1, \ldots, \gamma_m) = \sum_{j=1}^{m} \overline{a}_{ji}\gamma_j.$$

Notice that the matrix $B := [\overline{a}_{ji}] \in \mathbb{F}^{m\times n}$ is the conjugate transpose, that is, the *adjoint* of the matrix A. □

In view of Theorem 4.48 and Example 4.49, we give a notation and a name to the linear transformation $S : W \to V$ corresponding to the linear transformation $T : V \to W$, in Theorem 4.48.

Definition 4.50 Let V and W be finite dimensional inner product spaces. Let $T : V \to W$ be a linear transformation. The unique linear transformation $T^* : W \to V$ that satisfies

$$\langle Tx, y\rangle = \langle x, T^*y\rangle \quad \text{for all } x \in V,\ y \in W$$

is called the **adjoint of** T.

Due to Theorem 4.48, each linear transformation from a finite dimensional inner product space to another has a unique adjoint.

We have seen in Example 4.49 that the adjoint of a matrix in $\mathbb{F}^{m\times n}$ is its conjugate transpose, and we know that a matrix is the matrix representation of a linear transformation with respect tot the standard bases. In general, if orthonormal bases are chosen for the spaces, then the matrix representation of T^* happens to be the conjugate transpose (adjoint) of the matrix representation of T. We prove this fact in the following theorem.

Theorem 4.51 *Let $B = \{v_1, \ldots, v_n\}$ and $E = \{w_1, \ldots, w_m\}$ be orthonormal bases for the inner product spaces V and W, respectively. Let $T : V \to W$ be a linear transformation. Then $[T^*]_{B,E} = ([T]_{E,B})^*$.*

Proof Let $i \in \{1, \ldots, m\}$ and let $j \in \{1, \ldots, n\}$. Denote the (i, j)th entry of $[T]_{E,B}$ by a_{ij} and that of $[T^*]_{B,E}$ by b_{ij}. By Theorem 4.31,

$$b_{ij} = \langle T^*w_j, v_i\rangle = \overline{\langle v_i, T^*w_j\rangle} = \overline{\langle Tv_i, w_j\rangle} = \overline{a_{ji}}.$$

Therefore, $[T^*]_{B,E} = ([T]_{E,B})^*$. ■

A commutative diagram may be helpful. Let $T : V \to W$ be a linear transformation, where $\dim(V) = n$ and $\dim(W) = m$. Suppose $[T]_{E,B}$ is the matrix representation of T with respect to the orthonormal bases B for V and E for W. Let ϕ_B be the canonical basis isomorphism from V to $\mathbb{F}^{n\times 1}$ and let ψ_E be the canonical basis isomorphism from W to $\mathbb{F}^{m\times 1}$. Then the map T^* is that linear transformation from W to V such that the following happens for the commutative diagrams:

$$\text{If}\quad \begin{array}{ccc} V_B & \xrightarrow{\;T\;} & W_E \\ \phi_B \big\downarrow \simeq & & \simeq \big\downarrow \psi_E \\ \mathbb{F}^{n\times 1} & \xrightarrow[{[T]_{E,B}}]{} & \mathbb{F}^{m\times 1} \end{array} \quad \text{then} \quad \begin{array}{ccc} V_B & \xleftarrow{\;T^*\;} & W_E \\ \phi_B \big\downarrow \simeq & & \simeq \big\downarrow \psi_E \\ \mathbb{F}^{n\times 1} & \xleftarrow[{[T]^*_{E,B}}]{} & \mathbb{F}^{m\times 1} \end{array}.$$

In fact, the adjoint of a linear transformation can be defined alternatively through the above property of the commutative diagrams.

We notice that if the bases are not orthonormal, then the conjugate transpose may not represent the adjoint; see the following example.

Example 4.52 Consider $E = \{u_1, u_2, u_3\}$ as a basis for $\mathbb{R}^3$, where $u_1 = (1, 1, 0)$, $u_2 = (1, 0, 1)$ and $u_3 = (0, 1, 1)$; and $B = \{e_1, e_2, e_3, e_4\}$ as the standard basis for $\mathbb{R}^4$. Use the standard inner products (the dot products) on these spaces. Consider the linear transformation $T : \mathbb{R}^4 \to \mathbb{R}^3$ given by

$$T(a, b, c, d) = (a + c, b - 2c + d, a - b + c - d).$$

For obtaining $T^* : \mathbb{R}^3 \to \mathbb{R}^4$, let $y = (\alpha, \beta, \gamma) \in \mathbb{R}^3$. We compute the inner product $\langle Tx, y\rangle$ for any $x = (a, b, c, d) \in \mathbb{R}^4$ as in the following:

$$\begin{aligned}\langle T(a, b, c, d), (\alpha, \beta, \gamma)\rangle &= \langle (a + c, b - 2c + d, a - b + c - d), (\alpha, \beta, \gamma)\rangle \\ &= (a + c)\alpha + (b - 2c + d)\beta + (a - b + c - d)\gamma \\ &= a(\alpha + \gamma) + b(\beta - \gamma) + c(\alpha - 2\beta + \gamma) + d(\beta - \gamma) \\ &= \langle (a, b, c, d), (\alpha + \gamma, \beta - \gamma, \alpha - 2\beta + \gamma, \beta - \gamma)\rangle \\ &= \langle (a, b, c, d), T^*(\alpha, \beta, \gamma)\rangle.\end{aligned}$$

Therefore, $T^* : \mathbb{R}^3 \to \mathbb{R}^4$ is given by

$$T^*(\alpha, \beta, \gamma) = (\alpha + \gamma, \beta - \gamma, \alpha - 2\beta + \gamma, \beta - \gamma).$$

To determine the matrix representations of T and T^*, we proceed as follows:

$$\begin{aligned} Te_1 &= T(1, 0, 0, 0) = (1, 0, 1) &&= 0\,u_1 + 1\,u_2 + 0\,u_3 \\ Te_2 &= T(0, 1, 0, 0) = (0, 1, -1) &&= 1\,u_1 - 1\,u_2 + 0\,u_3 \\ Te_3 &= T(0, 0, 1, 0) = (1, -2, 1) &&= -1\,u_1 + 2\,u_2 - 1\,u_3 \\ Te_4 &= T(0, 0, 0, 1) = (0, 1, -1) &&= 1\,u_1 - 1\,u_2 + 0\,u_3 \end{aligned}$$

$$\begin{aligned} T^*u_1 &= T^*(1, 1, 0) = (1, 1, -1, 1) &&= 1\,e_1 + 1\,e_2 - 1\,e_3 + 1\,e_4 \\ T^*u_2 &= T^*(1, 0, 1) = (2, -1, 2, -1) &&= 2\,e_1 - 1\,e_2 + 2\,e_3 - 1\,e_4 \\ T^*u_3 &= T^*(0, 1, 1) = (1, 0, -1, 0) &&= 1\,e_1 + 0\,e_2 - 1\,e_3 + 0\,e_4 \end{aligned}$$

Therefore, the matrices are

$$[T]_{E,B} = \begin{bmatrix} 0 & 1 & -1 & 1 \\ 1 & -1 & 2 & -1 \\ 0 & 0 & -1 & 0 \end{bmatrix}, \qquad [T^*]_{B,E} = \begin{bmatrix} 1 & 2 & 1 \\ 1 & -1 & 0 \\ -1 & 2 & -1 \\ 1 & -1 & 0 \end{bmatrix}.$$

Notice that $[T^*]_{B,E} \neq ([T]_{E,B})^*$. □

The following theorems state some facts about the map that associates a linear transformation to its adjoint. We also mention some useful facts concerning the range spaces, the null spaces, and their orthogonal complements, of a linear transformation and its adjoint.

Theorem 4.53 *Let U, V and W be finite dimensional inner product spaces. Let $S : U \to V$ and $T, T_1, T_2 : V \to W$ be linear transformations. Let $I : V \to V$ be the identity operator and let $\alpha \in \mathbb{F}$. Then*

$$(T_1 + T_2)^* = T_1^* + T_2^*, \quad (\alpha T)^* = \overline{\alpha}\, T^*, \quad (T^*)^* = T, \quad I^* = I, \quad (TS)^* = S^* T^*.$$

Proof $\langle x, (\alpha T)^* y\rangle = \langle \alpha T x, y\rangle = \alpha\langle Tx, y\rangle = \alpha\langle x, T^* y\rangle = \langle x, \overline{\alpha}\, T^* y\rangle$. Therefore, $(\alpha T)^* = \overline{\alpha}\, T^*$. Other equalities are proved similarly. ■

Theorem 4.54 *Let $T : V \to W$ be a linear transformation, where V and W are finite dimensional inner product spaces. Then*

(1) $N(T^*) = R(T)^\perp$, $R(T^*)^\perp = N(T)$,
(2) $R(T^*) = N(T)^\perp$, $N(T^*)^\perp = R(T)$,
(3) $N(T^*T) = N(T)$, $N(TT^*) = N(T^*)$,
(4) $R(T^*T) = R(T^*)$, $R(TT^*) = R(T)$, *and*
(5) $\text{rank}\,(T^*) = \text{rank}\,(T)$, $\text{null}\,(T^*) = \text{null}\,(T) + \dim(W) - \dim(V)$.

Proof (1) $w \in N(T^*)$

if and only if $T^* w = 0$

if and only if $\langle v, T^* w\rangle = 0$ for all $v \in V$

if and only if $\langle Tv, w\rangle = 0$ for all $v \in V$

if and only if $w \in R(T)^\perp$.

Therefore, $N(T^*) = R(T)^\perp$. Replacing T by T^* in this equality, and using the fact that $(T^*)^* = T$, we obtain $N(T) = R(T^*)^\perp$.
(2) By (1), $N(T^*) = R(T)^\perp$ and $R(T^*)^\perp = N(T)$. Taking orthogonal complements and using the projection theorem, we get $R(T^*) = N(T)^\perp$ and $N(T^*)^\perp = R(T)$.

(3) If $x \in N(T)$, then $Tx = 0$. Then $T^*Tx = 0$, which implies that $x \in N(T^*T)$. Conversely, let $y \in N(T^*T)$. Then $\langle Ty, Ty\rangle = \langle y, T^*Ty\rangle = 0$. Hence $Ty = 0$. That is, $y \in N(T)$. Therefore, $N(T^*T) = N(T)$. The other equality is proved similarly.
(4) Using (1)–(3), and $(T^*T)^* = T^*T^{**} = T^*T$, we obtain:

$$R(T^*T) = N((T^*T)^*)^\perp = N(T^*T)^\perp = N(T)^\perp = R(T^*).$$

The other equality is proved similarly.
(5) Using (4) and the rank-nullity theorem, we obtain

$$\text{rank}\,(T^*) = \text{rank}\,(T^*T) = \dim(V) - \text{null}\,(T^*T) = \dim(V) - \text{null}\,(T) = \text{rank}\,(T).$$

Then

$$\begin{aligned}\text{null}\,(T^*) &= \dim(W) - \text{rank}\,(T^*) = \dim(W) - \text{rank}\,(T)\\ &= \dim(W) - (\dim(V) - \text{null}\,(T)) = \text{null}\,(T) + \dim(W) - \dim(V).\end{aligned}$$ ■

Notice that $\text{rank}\,(T^*) = \text{rank}\,(T)$ provides another proof that the row rank and the column rank of a matrix are equal! It also follows that T^* is invertible if and only if T is invertible. In this case, $(T^*)^{-1} = (T^{-1})^*$. Similarly, $\text{rank}\,(T^*T) = \text{rank}\,(T)$ implies that T is invertible iff T^*T is invertible.

Inverse of a canonical basis isomorphism coincides with its adjoint as the following theorem shows.

Theorem 4.55 *Let $\{v_1, \ldots, v_n\}$ be an orthonormal basis of an inner product space V. Let ϕ be the canonical basis isomorphism from V onto $\mathbb{F}^{n\times 1}$. Then $\phi^{-1} = \phi^*$.*

Proof Let $j \in \{1, \ldots, n\}$. We have $\phi(v_j) = e_j$. Let $v \in V$. There exist scalars $\alpha_1, \ldots, \alpha_n$ such that $v = \sum_{i=1}^n \alpha_i v_i$; so that $\phi(v) = \sum_{i=1}^n \alpha_i e_i$. Moreover, $\langle e_i, e_j\rangle = \delta_{ij} = \langle v_i, v_j\rangle$ for each $i \in \{1, \ldots, n\}$. Now,

$$\langle v, \phi^*(e_j)\rangle = \langle \phi(v), e_j\rangle = \sum_{i=1}^n \alpha_i \langle e_i, e_j\rangle = \sum_{i=1}^n \alpha_i \langle v_i, v_j\rangle = \Big\langle \sum_{i=1}^n \alpha_i v_i, v_j\Big\rangle = \langle v, \phi^{-1}e_j\rangle.$$

Hence $\phi^* e_j = \phi^{-1} e_j$ for each j. Therefore, $\phi^* = \phi^{-1}$. ■

In general, if a square matrix $A \in \mathbb{F}^{n\times n}$ maps an orthonormal basis of $\mathbb{F}^{n\times 1}$ to another orthonormal basis, then it is necessarily unitary, that is, $A^{-1} = A^*$. Along with this, the following theorem generalizes the result to any $m \times n$ matrix.

Theorem 4.56 *Let $\{u_1, \ldots, u_n\}$ and $\{v_1, \ldots, v_m\}$ be orthonormal bases for $\mathbb{F}^{n\times 1}$ and $\mathbb{F}^{m\times 1}$, respectively. Let $A \in \mathbb{F}^{m\times n}$ satisfy $Au_i = v_i$ for $1 \le i \le \min\{m, n\}$.*

(1) *If $n \le m$, then A has orthonormal columns; thus, $A^*A = I$.*
(2) *If $m \le n$, then A has orthonormal rows; thus, $AA^* = I$.*
(3) *If $m = n$, then A is unitary.*

Proof (1) Assume that $n \le m$. That is, the number of columns in A is less than or equal to the number of rows in A. The standard basis vectors of $\mathbb{F}^{n\times 1}$ can be expressed as linear combinations of u_js. That is, for any $i, j \in \{1, \ldots, n\}$, there exist scalars $\alpha_1, \ldots, \alpha_n, \beta_1, \ldots, \beta_n$ such that $e_i = \sum_{k=1}^{n} \alpha_k u_k$ and $e_j = \sum_{r=1}^{n} \beta_r u_r$. Due to orthonormality of u_is and of v_js, we have

$$\begin{aligned}\langle Ae_i, Ae_j\rangle &= \sum_{k=1}^{n}\sum_{r=1}^{n} \alpha_k\overline{\beta}_r\langle Au_k, Au_r\rangle = \sum_{k=1}^{n}\sum_{r=1}^{n} \alpha_k\overline{\beta}_r\langle v_r, v_k\rangle \\ &= \sum_{k=1}^{n}\sum_{r=1}^{n} \alpha_k\overline{\beta}_r\langle u_r, u_k\rangle = \langle e_i, e_j\rangle = \delta_{ij}.\end{aligned}$$

Therefore, the columns of A are orthonormal. If b_{ij} is the (i, j)th entry in A^*A, then it is equal to the dot product of the ith row of A^* with the jth column of A. That is, $b_{ij} = (Ae_i)^*(Ae_j) = \langle Ae_j, Ae_i\rangle = \delta_{ji}$. Therefore, $A^*A = I$.
(2) Suppose that $m \le n$. Then the number of columns in A^* is less than or equal to the number of rows in A^*. Using (1) for A^* in place of A, we conclude that the columns of A^* are orthonormal. Therefore, the rows of A are orthonormal. Also, (1) implies that $(A^*)^*A^* = I$. That is, $AA^* = I$.
(3) It follows from (1) and (2). ■

For instance, take $m = 3$ and $n = 2$. Consider the standard basis vectors $u_1 = e_1$, $u_2 = e_2$ for $\mathbb{F}^{2\times 1}$ and $v_1 = e_1'$, $v_2 = e_2'$, $v_3 = e_3'$ for $\mathbb{F}^{3\times 1}$. Then the condition $Au_i = v_i$ for $1 \le i \le \min\{m, n\}$ says that the columns of A are e_1' and e_2', which are clearly orthonormal.

Exercises for Sect. 4.8

1. On $\mathbb{C}^3$, consider the linear functional f defined by $f(a, b, c) = (a + b + c)/3$. Find a vector $y \in \mathbb{C}^3$ such that $f(x) = \langle x, y\rangle$ for each $x \in \mathbb{C}^3$.
2. Let T be a linear operator on a complex inner product space V. Let $v \in V$. Define the linear functional f on V by $f(x) = \langle v, Tx\rangle$ for $x \in V$. Determine a vector $y \in V$ such that $f(x) = \langle x, y\rangle$ for each $x \in V$.
3. Determine a polynomial $q(t) \in \mathcal{P}_2(\mathbb{R})$ so that for every $p(t) \in \mathcal{P}_2(\mathbb{R})$,
   (a) $\int_0^1 p(t)q(t)dt = p(1/2)$ (b) $\int_0^1 \cos(\pi t)p(t)dt = \int_0^1 q(t)p(t)dt$.
4. Fix a vector u in an inner product space. Define a linear functional T on V by $Tv = \langle v, u\rangle$. What is $T^*(\alpha)$ for a scalar α?
5. Define a linear operator T on $\mathbb{F}^n$ by $T(a_1, \ldots, a_n) = (0, a_1, \ldots, a_{n-1})$. What is $T^*(a_1, \ldots, a_n)$?
6. Let U be a subspace of a finite dimensional inner product space V. Let $v \in V$. Show the following:

   (a) The function $f : U \to \mathbb{F}$ given by $f(x) = \langle x, v\rangle$ for $x \in U$, is a linear functional on U.
   (b) The Riesz representor of f in (a) is given by $\mathrm{proj}_U(v)$.
   (c) If $u \in U$ is the Riesz representor of v, then $v - u \in U^\perp$.

7. Let T be a linear operator on a finite dimensional inner product space. Prove that T^* is invertible if and only if T is invertible. In that case, show that $(T^*)^{-1} = (T^{-1})^*$.

4.9 Problems

1. Let V be an inner product space, and let $x, y \in V$. Show the following:
   (a) If $\mathrm{Re}\langle x, y\rangle = \|x\|\,\|y\|$, then $\mathrm{Re}\langle x, y\rangle = |\langle x, y\rangle|$.
   (b) For each $\alpha \in \mathbb{F}$, $\|x + \alpha y\| = \|x - \alpha y\|$ if and only if $x \perp y$.
   (c) For each $\alpha \in \mathbb{F}$, $\|x\| \le \|x + \alpha y\|$ if and only if $x \perp y$.
2. Is it true that two vectors u and v in a complex inner product space are orthogonal if and only if for all scalars α, β, $\|\alpha u + \beta v\|^2 = \|\alpha u\|^2 + \|\beta v\|^2$?
3. Show that there does not exist an inner product on $\mathbb{F}^2$ such that the corresponding norm is $\|(a, b)\| = |a| + |b|$.
4. For $x = (\alpha_1, \ldots, \alpha_n) \in \mathbb{R}^n$, define $\|x\| = \max_i |\alpha_i|$. Show that there exists no inner product on $\mathbb{R}^n$ such that $\langle x, x\rangle = \|x\|^2$ for all $x \in \mathbb{R}^n$.
5. Recall that any function $\|\cdot\|$ from an inner product space to $\mathbb{R}$ satisfying the properties proved in Theorem 4.7, except the parallelogram law, is called a norm. A norm need not come from an inner product. Using this generalized sense of a norm, answer the following:
   (a) On $\mathbb{R}^2$, define $\|(a, b)\| = (|a|^\alpha + |b|^\alpha)^{1/\alpha}$ for $\alpha > 0$. For which α, this defines a norm?
   (b) Let $\|\cdot\|$ be a norm on $\mathbb{R}^2$. Is it true that $\|x\|^2 = \langle x, x\rangle$ for some inner product $\langle\cdot,\cdot\rangle$ on $\mathbb{R}^2$ if and only if $\{x \in \mathbb{R}^2 : \|x\| = 1\}$ is an ellipse or a circle?
6. Let $\|\cdot\|$ be a norm on a vector space V. Prove that there exists an inner product $\langle\cdot,\cdot\rangle$ on V with $\langle u, u\rangle = \|u\|^2$ for all $u \in V$ if and only if $\|\cdot\|$ satisfies the parallelogram law. (Hint: Use polarization identities).
7. Let V be an inner product space. Let $x, y \in V$. Prove that $\|x + y\| = \|x\| + \|y\|$ if and only if one of x, y is a nonnegative multiple of the other.
8. Construct an orthonormal basis for $\mathcal{P}_2(\mathbb{R})$ in which the matrix of the differentiation operator is upper triangular.
9. Let T be linear operator on an inner product space V. Prove that $\langle Tx, Ty\rangle = \langle x, y\rangle$ for all $x, y \in V$ if and only if $\|Tx\| = \|x\|$ for all $x \in V$.
10. Derive Cauchy–Schwarz inequality from Bessel's inequality.
11. *(Minimum property of Fourier coefficients)* Let $\{u_1, \ldots, u_n\}$ be an orthonormal set in an inner product space V. Let $y = \alpha_1 u_1 + \cdots + \alpha_n u_n$ for some scalars $\alpha_1, \ldots, \alpha_n$. Let $x \in V$. Then $\|x - y\|$ depends on the scalars $\alpha_1, \ldots, \alpha_n$. Show that $\|x - y\|$ is minimum if and only if $\alpha_i = \langle x, u_i\rangle$ for $i \in \{1, \ldots, n\}$.
12. Let $(v_1, \ldots, v_n)$ be an n-tuple of linearly independent vectors in a real inner product space V. Let $V_m = \mathrm{span}\{v_1, \ldots, v_m\}$ for $1 \le m \le n$. How many orthonormal

n-tuples of vectors $(u_1, \ldots, u_n)$ exist so that $\text{span}\{u_1, \ldots, u_m\} = V_m$ for each m?

13. Let V be a real inner product space of dimension 2. Let $T : V \to V$ be a linear transformation satisfying $\langle Tx, Ty \rangle = \langle x, y \rangle$ for all $x, y \in V$. Show that with respect to any orthonormal basis of V, the matrix of T is in the form

$$\begin{bmatrix} \cos\theta & \sin\theta \\ \sin\theta & -\cos\theta \end{bmatrix}.$$

14. Recall that $A \in \mathbb{R}^{n\times n}$ is called an orthogonal matrix if it respects the inner product on $\mathbb{R}^n$, i.e. when $\langle Ax, Ay \rangle = \langle x, y \rangle$ holds for all $x, y \in \mathbb{R}^n$. Prove that the following are equivalent:

   (a) A is orthogonal.
   (b) A preserves length, i.e. $\|Ax\| = \|x\|$ for each $x \in \mathbb{R}^n$.
   (c) A is invertible and $A^{-1} = A^T$.
   (d) The rows of A form an orthonormal basis for $\mathbb{R}^n$.
   (e) The columns of A form an orthonormal basis for $\mathbb{R}^{n\times 1}$.

15. Show that orthogonal matrices preserve angles. That is, if A is an orthogonal matrix, then the angle between Ax and Ay is same as that between x and y.
16. Let $A \in \mathbb{C}^{n\times n}$. Show that the following are equivalent:

   (a) A is unitary, i.e. $A^*A = AA^* = I$.
   (b) $\|Av\| = \|v\|$ for each $v \in \mathbb{R}^{n\times 1}$.
   (c) The rows of A form an orthonormal basis for $\mathbb{C}^n$.
   (d) The columns of A form an orthonormal basis for $\mathbb{C}^{n\times 1}$.

17. Let $\{u_1, \ldots, u_n\}$ and $\{v_1, \ldots, v_n\}$ be orthonormal bases of an inner product space V. Let T be a linear operator on V. Prove that $\sum_{j=1}^{n} \|Tu_j\|^2 = \sum_{j=1}^{n} \|Tv_j\|^2$. Can you identify this quantity?
18. Let u, v be vectors in an inner product space V. Define the linear operator T on V by $Tx = \langle x, u \rangle v$. Compute the trace of T. [Note: $\text{tr}(T) = \text{tr}([T])$, where $[T]$ is a matrix representation of T with respect to any ordered basis.]
19. Construct an example to show that a linear system $Ax = b$ has a unique least squares solution, which is not a solution. Explain the situation in terms of the QR-factorization of A observing that $QQ^* \neq I$.
20. Let V be an inner product space with an ordered basis $B = \{u_1, \ldots, u_n\}$. Show the following:

   (a) The Gram matrix $[a_{ij}]$ of B, where $a_{ij} = \langle u_i, u_j \rangle$, is invertible.
   (b) If $\alpha_1, \ldots \alpha_n \in \mathbb{F}$, then there is exactly one vector $x \in V$ such that $\langle x, u_j \rangle = \alpha_j$, for $j = 1, 2, \ldots, n$.

21. Let T be a linear operator of rank r on a finite dimensional vector space V. Prove that $T = T_1 + \cdots + T_r$, where each of $T_1, \ldots, T_r$ is a linear operator of rank 1, on V.

22. Let V be a finite dimensional real inner product space. A function $h : \mathbb{R} \to V$ is said to be differentiable at $a \in \mathbb{R}$ if the limit $\lim_{t \to a} \dfrac{h(t) - h(a)}{t - a}$ exists. In that case, we write the value of this limit as $(Dh)(a)$.

    (a) Let $f, g : \mathbb{R} \to V$ be differentiable at $a \in \mathbb{R}$. Show that the real-valued function $t \mapsto \langle f(t), g(t) \rangle$ is also differentiable at a.
    (b) Show that $(\frac{d}{dt}\langle f(t), g(t)\rangle)(a) = \langle (Df)(a), g(a)\rangle + \langle f(a), (Dg)(a)\rangle$.
    (c) If $f : \mathbb{R} \to V$ is differentiable at all $t \in \mathbb{R}$ and $\|f(t)\|$ is constant for all $t \in \mathbb{R}$, then show that $f(t) \perp (Df)(t)$ for all $t \in \mathbb{R}$.

23. Let $\langle \cdot, \cdot \rangle$ be any inner product on $\mathbb{F}^{n \times 1}$. Show that there exists a hermitian matrix A such that $\langle x, y \rangle = x^* A y$.
24. Let $T : V \to W$ be a linear transformation, where V and W are finite dimensional inner product spaces. Prove the following:

    (a) T^* is injective if and only if T is surjective.
    (b) T^* is surjective if and only if T is injective.

25. *(Fredholm Alternative)* Given $A \in \mathbb{F}^{n \times n}$ and $b \in \mathbb{F}^{n \times 1}$, prove that either the linear system $Ax = b$ has a solution or the system $A^* x = 0$ has a solution x with $x^* b \neq 0$.
26. Let $f : \mathbb{R}^{n \times 1} \to \mathbb{R}^{n \times 1}$ be a function. Prove that the following are equivalent:

    (a) For all $x, y \in \mathbb{R}^{n \times 1}$, $\|f(x) - f(y)\| = \|x - y\|$; and $f(0) = 0$.
    (b) For all $x, y \in \mathbb{R}^{n \times 1}$, $\langle f(x), f(y) \rangle = \langle x, y \rangle$.
    (c) There exists an orthogonal matrix $A \in \mathbb{R}^{n \times n}$ such that $f(x) = Ax$.

    It says that a distance preserving map that fixes the origin is precisely one that preserves the dot product; also, such maps can be described as orthogonal matrices.
27. *Rigid motions* are distance preserving maps. Prove that every rigid motion is the composition of an orthogonal operator and a translation. That is, if $f : \mathbb{R}^{n \times 1} \to \mathbb{R}^{n \times 1}$ satisfies $\|f(x) - f(y)\| = \|x - y\|$ for all $x, y \in \mathbb{R}^{n \times 1}$, then prove that there exists an orthogonal matrix A and a vector $b \in \mathbb{R}^{n \times 1}$ such that $f(x) = Ax + b$ for all $x \in \mathbb{R}^{n \times 1}$. (See Problem 26.)
28. Let T be a linear operator on a finite dimensional inner product space V. Prove the following:

    (a) There exists $k \in \mathbb{R}$ such that $\|Tv\|/\|v\| \le k$ for every nonzero $v \in V$.
    (b) Define $\|T\| = \sup\{\|Tv\|/\|v\| : v \neq 0,\ v \in V\}$. Then $\|\cdot\| : \mathcal{L}(V, V) \to \mathbb{R}$ is a norm on $\mathcal{L}(V, V)$.
    (c) $\|T\| = \sup\{\|Tv\| : v \in V,\ \|v\| = 1\}$.
    (d) $\|I\| = 1$.
    (e) Let $v \in V$, and let $T \in \mathcal{L}(V, V)$. Then $\|Tv\| \le \|T\|\,\|v\|$.
    (f) Let $S, T \in \mathcal{L}(V, V)$. Then $\|ST\| \le \|S\|\,\|T\|$.

29. Let T be a linear operator on a finite dimensional inner product space V. Let $\{\alpha_n\}$ be sequence of scalars. We say that the power series $\sum_{n=0}^{\infty} \alpha_n T^n$ *converges*

if and only if the power series of scalars $\sum_{n=0}^{\infty} \alpha_n \|T^n\|$ converges. Show that $\sum_{n=0}^{\infty}(1/n!)T^n$ converges. Note: using this, we define the *exponential* of T by $\exp(T) = \sum_{n=0}^{\infty}(1/n!)T^n$.

30. Let V be a finite dimensional inner product space. Let V' be the dual of V. For any $f \in V'$, let $v_f \in V$ be its Riesz representer. Show that the function $\langle\, , \,\rangle : V' \times V' \to \mathbb{F}$ by $\langle f, g\rangle = \langle v_g, v_f\rangle$ defines an inner product on V'.

Chapter 5
Eigenvalues and Eigenvectors

5.1 Existence of Eigenvalues

Consider the linear operator T on $\mathbb{R}^3$ defined by

$$T(a, b, c) = (a + b, b + c, c + a) \quad \text{for } (a, b, c) \in \mathbb{R}^3.$$

We notice that $T(\alpha, \alpha, \alpha) = 2(\alpha, \alpha, \alpha)$. Therefore, T maps any point on the line containing the points $(0, 0, 0)$ and $(1, 1, 1)$ to another point on the same line. That is, the line $L := \{(\alpha, \alpha, \alpha) : \alpha \in \mathbb{R}\}$ remains invariant (fixed) under this linear operator T. Given any linear operator on a vector space, which lines remain invariant?

Definition 5.1 Let T be a linear operator on a vector space V over $\mathbb{F}$. For a scalar $\lambda \in \mathbb{F}$ and a nonzero vector $v \in V$, the pair (λ, v) is called an **eigenpair** of T if the equation $Tv = \lambda v$ is satisfied.

In such a case, λ is called an **eigenvalue** of T and v is called an **eigenvector** of T corresponding to the eigenvalue λ.

To break them apart, $\lambda \in \mathbb{F}$ is an eigenvalue of T if and only if there exists a nonzero vector $v \in V$ such that $Tv = \lambda v$. Similarly, $v \in V$ is an eigenvector of T if and only if $v \neq 0$, and there exists a scalar $\lambda \in \mathbb{F}$ such that $Tv = \lambda v$.

Example 5.2 Consider the linear operator $T : \mathbb{C}^3 \to \mathbb{C}^3$ defined by

$$T(a, b, c) = (a + b, b + c, c + a).$$

Since $T(1, 1, 1) = (1 + 1, 1 + 1, 1 + 1) = 2(1, 1, 1)$, the scalar 2 is an eigenvalue of T with a corresponding eigenvector $(1, 1, 1)$. We also see that

$$\begin{aligned} T(-2,\ 1 - i\sqrt{3},\ 1 + i\sqrt{3}) &= (-1 - i\sqrt{3},\ 2,\ -1 + i\sqrt{3}) \\ &= \tfrac{1+i\sqrt{3}}{2}(-2,\ 1 - i\sqrt{3},\ 1 + i\sqrt{3}). \end{aligned}$$

M. T. Nair and A. Singh, *Linear Algebra*,
https://doi.org/10.1007/978-981-13-0926-7_5

Hence, $(-2,\ 1-i\sqrt{3},\ 1+i\sqrt{3})$ is an eigenvector corresponding to the eigenvalue $(1+i\sqrt{3})/2$. Is $(1-i\sqrt{3})/2$ also an eigenvalue of T? □

If λ is an eigenvalue of T, then $(T-\lambda I)v=0$ for a nonzero vector (an eigenvector) v. That is, the linear operator $T-\lambda I$ is not injective. Conversely, if $T-\lambda I$ is not injective, then there exists a nonzero vector v such that $(T-\lambda I)v=0$. That is, $Tv=\lambda v$. Therefore, we have the following characterization of an eigenvalue:

A scalar λ is an eigenvalue of T if and only if $T-\lambda I$ is not injective.

Moreover, if the vector space is finite dimensional, then $T-\lambda I$ is not injective if and only if it is not surjective, due to Theorem 2.26. Hence we also obtain the following:

If T is a linear operator on a finite dimensional vector space over $\mathbb{F}$, then a scalar $\lambda \in \mathbb{F}$ is an eigenvalue of T if and only if $T-\lambda I$ is not invertible.

We will often use this condition for eigenvalues since it gives a way for ascertaining a scalar to be an eigenvalue without recourse to an eigenvector. This characterization immediately connects eigenvalues of a linear operator with those of its adjoint, whenever the vector space is an inner product space. Observe that $T-\lambda I$ is not invertible if and only if $(T-\lambda I)^*$ is not invertible due to Theorem 4.54. Further, Theorem 4.53 implies that $(T-\lambda I)^* = T^* - \overline{\lambda} I$. Therefore,

if T is a linear operator on a finite dimensional inner product space, then a scalar λ is an eigenvalue of T if and only if $\overline{\lambda}$ is an eigenvalue of T^*.

Given a linear operator T if there exists a scalar that is an eigenvalue of T, then we say that T **has an eigenvalue**. In that case, we also say that the *eigenvalue problem* for T, that is, $Tv=\lambda v$, is *solvable*. If T has an eigenvalue with a corresponding eigenvector v, then the line $\{\alpha v : \alpha \in \mathbb{F}\}$ is invariant under T. In general, a linear operator T need not have an eigenvalue; then no line is invariant under such a linear operator T.

Example 5.3 Let $T : \mathbb{R}^2 \to \mathbb{R}^2$ be defined by $T(a,b)=(-b,a)$. If λ is an eigenvalue of T with corresponding eigenvector (α,β), then

$$T(\alpha,\beta)=(-\beta,\alpha)=\lambda(\alpha,\beta).$$

This implies that $-\beta=\lambda\alpha$ and $\alpha=\lambda\beta$. That is, $(\lambda^2+1)\beta=0$. As $(\alpha,\beta)=(\lambda\beta,\beta)$ is required to be nonzero, $\beta \neq 0$. Then we must have $\lambda^2+1=0$. There exists no $\lambda \in \mathbb{R}$ with $\lambda^2+1=0$. Therefore, T does not have an eigenvalue.

However, for the same map $(a,b) \mapsto (-b,a)$ with $\mathbb{C}^2$ in place of $\mathbb{R}^2$, $\lambda=i$ and $\lambda=-i$ satisfy the equation $\lambda^2+1=0$. It can be verified that $(1,-i)$ and $(1,i)$ are eigenvectors corresponding to the eigenvalues i and $-i$, respectively. □

Example 5.3 shows that we may need to solve for zeros of certain polynomials for obtaining eigenvalues of a linear operator. To investigate the issue of existence of eigenvalues, we will use the following fact, called the *Fundamental Theorem of Algebra*:

Any nonconstant polynomial of degree n with complex coefficients has exactly n number of complex zeros, counting multiplicities.

If α is a zero of a polynomial $p(t)$ then $p(t) = (t - \alpha)^m q(t)$, for some $m \in \mathbb{N}$, and some polynomial $q(t)$. In addition, if α is not a zero of $q(t)$, then we say that α as a zero of $p(t)$ has **multiplicity** m. In such a case, we also say that α is a root of multiplicity m, of the equation $p(t) = 0$.

The proof of the fundamental theorem of algebra may be found in any text on complex analysis. The fundamental theorem is used to factor polynomials with complex coefficients in a natural way. A polynomial with real coefficients may have complex zeros. In that case, the complex zeros occur in conjugate pairs. That is, if $\alpha + i\beta$ is a zero of a polynomial with real coefficients, then $\alpha - i\beta$ is also a zero of the same polynomial. Hence, whenever $t - (\alpha + i\beta)$ is a factor of such a polynomial, $t - (\alpha - i\beta)$ is also a factor. Then their product $(t - \alpha)^2 + \beta^2$ is a factor of such a polynomial for real numbers α and β. These facts are summarized in the following statement.

Proposition 5.1 *Let* $p(t) = a_0 + a_1 t + \cdots + a_k t^k$ *be a nonconstant polynomial with* $a_0, \ldots, a_k \in \mathbb{C}$, *and* $a_k \neq 0$. *Then the following are true:*

(1) *There exist* $\lambda_1, \ldots, \lambda_k \in \mathbb{C}$ *such that* $p(t) = a_k(t - \lambda_1) \cdots (t - \lambda_k)$.
(2) *If* $a_0, \ldots, a_k \in \mathbb{R}$, *then there exist* $\lambda_1, \ldots, \lambda_j \in \mathbb{R}$ *and nonzero* $\alpha_1, \ldots, \alpha_m$, $\beta_1, \ldots, \beta_m \in \mathbb{R}$ *with* $k = j + 2m$, $j \geq 0$, $m \geq 0$ *such that*

$$p(t) = a_k(t - \lambda_1) \cdots (t - \lambda_j)\big((t - \alpha_1)^2 + \beta_1^2\big) \cdots \big((t - \alpha_m)^2 + \beta_m^2\big).$$

(3) *The factorizations in* (1)–(2) *are unique up to ordering of the factors in the product. Further, in* (2), *if* k *is odd, then* $j \geq 1$; *that is,* $p(t)$ *has a real zero.*

In Proposition 5.1(2), if $p(t)$ has no real zeros, then the linear terms are absent; and if all zeros of $p(t)$ are real, then the quadratic terms are absent in the product.

Theorem 5.4 *Every linear operator on a finite dimensional complex vector space has an eigenvalue.*

Proof Let V be a vector space of dimension n over $\mathbb{C}$. Let $T : V \to V$ be a linear operator. Let x be a nonzero vector in V. The list $x, Tx, T^2x, \ldots, T^n x$ has $n + 1$ vectors, and hence, it is linearly dependent. So, there exist $a_0, a_1 \ldots, a_n \in \mathbb{C}$, not all zero, such that

$$a_0 x + a_1 T x + \cdots + a_n T^n x = 0.$$

Let k be the maximum index such that $a_k \neq 0$. That is, if $a_j \neq 0$, then $j \leq k$. Write $p(t) = a_0 + a_1 t + \cdots + a_k t^k$. Then

$$p(T) = a_0 I + a_1 T + \cdots + a_k T^k, \quad a_k \neq 0 \quad \text{and} \quad p(T)(x) = 0.$$

By Proposition 5.1, there exist complex numbers $\lambda_1, \ldots, \lambda_k$ such that

$$p(t) = a_k(t - \lambda_1)(t - \lambda_2) \ldots (t - \lambda_k).$$

Then

$$a_k(T - \lambda_1 I)(T - \lambda_2 I) \ldots (T - \lambda_k I)(x) = p(T)(x) = 0.$$

Since $x \neq 0$, $p(T)$ is not injective. Hence, at least one of $T - \lambda_1 I, \ldots, T - \lambda_k I$ is not injective. Therefore, at least one of $\lambda_1, \ldots, \lambda_k$ is an eigenvalue of A. ■

Linear operators on infinite dimensional spaces need not have eigenvalues even if the underlying field is $\mathbb{C}$; see the following example.

Example 5.5 (1) Consider the vector space $\mathcal{P}(\mathbb{C})$. Define $T : \mathcal{P}(\mathbb{C}) \to \mathcal{P}(\mathbb{C})$ by

$$Tp(t) = tp(t), \quad \text{for } p \in \mathcal{P}(\mathbb{C}).$$

As $tp(t) = \lambda p(t)$ is impossible for any scalar λ and a nonzero polynomial $p(t)$, T has no eigenvalue.

(2) Let $V := c_{00}(\mathbb{N}, \mathbb{C})$, the space of all complex sequences having only a finite number of nonzero entries. Let T be the right shift linear operator on V, that is,

$$T(\alpha_1, \alpha_2, \ldots) = (0, \alpha_1, \alpha_2, \ldots).$$

If $\lambda \in \mathbb{C}$ and $v := (\alpha_1, \alpha_2, \ldots) \in V$ are such that $Tv = \lambda v$, then $(0, \alpha_1, \alpha_2, \ldots) = \lambda(\alpha_1, \alpha_2, \ldots)$. It implies $\lambda\alpha_1 = 0$ and $\lambda\alpha_{n+1} = \alpha_n$ for each $n \in \mathbb{N}$. This is possible only for the zero vector v. Therefore, T does not have an eigenvalue. □

Exercises for Sect. 5.1

1. Find eigenvalues and eigenvectors of the following linear operators:
   (a) $T : \mathbb{F}^2 \to \mathbb{F}^2$ defined by $T(a, b) = (b, a)$.
   (b) $T : \mathbb{F}^3 \to \mathbb{F}^3$ defined by $T(a, b, c) = (0, 5a, 2c)$.
   (c) $T : \mathbb{R}^4 \to \mathbb{R}^4$ defined by $T(a, b, c, d) = (b, -a, d, -c)$.
   (d) $T : \mathbb{F}^n \to \mathbb{F}^n$ defined by $T(a_1, \ldots, a_n) = (\sum_{i=1}^n a_i, \ \ldots, \ \sum_{i=1}^n a_i)$.
   (e) $T : \mathbb{F}^\infty \to \mathbb{F}^\infty$ defined by $T(a_1, a_2, \ldots) = (a_2, a_3, \ldots)$.
   (f) $T : \mathbb{F}^\infty \to \mathbb{F}^\infty$ defined by $T(a_1, a_2, \ldots) = (0, a_1, a_2, \ldots)$.
2. Let T be a linear operator on a finite dimensional complex vector space V. Show that each nonzero $v \in V$ is an eigenvector of T if and only if $T = \alpha I$ for some $\alpha \in \mathbb{C}$.
3. Suppose S and T are linear operators on V, λ is an eigenvalue of S, and μ is an eigenvalue of T. Is it necessary that $\mu\lambda$ is an eigenvalue of ST?
4. Let A be an $n \times n$ matrix and α be a scalar such that each row (or each column) sums to α. Show that α is an eigenvalue of A.
5. Give an example of a linear operator T on a real inner product space, where $(T - \alpha I)^2 + \beta^2 I$ is not invertible for some real numbers α and β.

6. If $(1, 2)^T$ and $(2, 1)^T$ are eigenvectors corresponding to eigenvalues 1 and 2 of a 2×2 matrix A, then what is A?
7. Let $B = \{v_1, \ldots, v_n\}$ be a basis of a vector space V. Let T be a linear operator on V. Let $A = [T]_{B,B}$. Prove that v_j is an eigenvector of T with eigenvalue λ if and only if the jth column of A has the form λv_j.
8. If a square matrix is of rank 1, then show that its trace is one of its eigenvalues. What are its other eigenvalues?
9. Let T be a linear operator on a real vector space. Suppose $T^2 + \alpha T + \beta = 0$ for some $\alpha, \beta \in \mathbb{R}$. Prove that T has an eigenvalue if and only if $\alpha^2 \geq 4\beta$.
10. If x and y are eigenvectors corresponding to distinct eigenvalues of a real symmetric matrix of order 3, then show that the cross product of x and y is a third eigenvector linearly independent with x and y.

5.2 Characteristic Polynomial

By eigenvalues and eigenvectors of a matrix $A \in \mathbb{C}^{n\times n}$, we mean eigenvalues and eigenvectors of the linear operator $A : \mathbb{C}^{n\times 1} \to \mathbb{C}^{n\times 1}$. It means that $\lambda \in \mathbb{C}$ is an eigenvalue of $A \in \mathbb{C}^{n\times n}$ with a corresponding eigenvector $v \in \mathbb{C}^{n\times 1}$ if and only if $v \neq 0$ and $Av = \lambda v$.

Let $A \in \mathbb{R}^{n\times n}$. As a linear operator on $\mathbb{R}^{n\times 1}$, it may or may not have an eigenvalue. Non-existence of an eigenvalue for $A \in \mathbb{R}^{n\times n}$ means that there does not exist a real number λ such that $Av = \lambda v$ for some nonzero vector $v \in \mathbb{R}^{n\times 1}$. On the other hand, A is also in $\mathbb{C}^{n\times n}$. If we consider it as a linear operator on $\mathbb{C}^{n\times 1}$, then Theorem 5.4 guarantees that it has at least one eigenvalue $\lambda \in \mathbb{C}$ with a corresponding eigenvector $v \in \mathbb{C}^{n\times 1}$, $v \neq 0$. In case $\lambda \in \mathbb{R}$, write $v = x + iy$ with $x, y \in \mathbb{R}^{n\times 1}$. Then $Ax = \lambda x$ and $Ay = \lambda y$. As $v \neq 0$, at least one of x or y is nonzero. Therefore, λ is also an eigenvalue of the linear operator $A : \mathbb{R}^{n\times 1} \to \mathbb{R}^{n\times 1}$. Therefore,

> we view a matrix $A \in \mathbb{R}^{n\times n}$ as a linear operator on $\mathbb{C}^{n\times 1}$ so that it has an eigenvalue $\lambda \in \mathbb{C}$ with a corresponding eigenvector $v \in \mathbb{C}^{n\times 1}$.

In fact, a real vector space can be embedded in a complex vector space; then a linear operator is identified with its incarnation on the complex vector space. Consequently, the eigenvalues of the old linear operator are precisely the real eigenvalues of the new one. Further, the new linear operator may have more nonreal eigenvalues. For details on this technique, called *complexification*, see Problem 50.

Observe that similar matrices have the same eigenvalues. Indeed, if $B = P^{-1}AP$ for some invertible matrix $P \in \mathbb{F}^{n\times n}$, then for $\lambda \in \mathbb{F}$ and $v \in \mathbb{F}^{n\times n}$,

$$Av = \lambda v \quad \text{iff} \quad APP^{-1}v = \lambda PP^{-1}v \quad \text{iff} \quad (P^{-1}AP)P^{-1}v = \lambda P^{-1}v.$$

Since similar matrices represent the same linear transformation, we ask whether there is a connection between the eigenvalues of a linear operator and those of its matrix representation.

Theorem 5.6 *Let T be a linear operator on a finite dimensional vector space V over $\mathbb{F}$. Let B be an ordered basis of V. Let $\lambda \in \mathbb{F}$ be a scalar and let $v \in V$ be a nonzero vector. Then (λ, v) is an eigenpair of T if and only if $(\lambda, [v]_B)$ is an eigenpair of $[T]_{B,B}$.*

Proof Since $[Tv]_B = [T]_{B,B}[v]_B$ and $[\lambda v]_B = \lambda[v]_B$, we have $Tv = \lambda v$ if and only if $[T]_{B,B}[v]_B = \lambda[v]_B$. ■

We emphasize that in Theorem 5.6, if $\mathbb{F} = \mathbb{R}$, then the eigenvalues of T are precisely the real eigenvalues of $[T]_{B,B}$.

Suppose λ is an eigenvalue of an $n \times n$ matrix A. Then $A - \lambda I$, as a linear operator on $\mathbb{C}^{n\times 1}$, is not injective. Then, it is not surjective; consequently, $\det(A - \lambda I) = 0$. Expansion of this determinant shows that it is a polynomial in λ and it has degree n, where the coefficient of λ^n in this polynomial is $(-1)^n$. By multiplying such a polynomial with $(-1)^n$, it can be made to be a monic polynomial. Recall that the coefficient of the highest degree term in a *monic* polynomial is 1; the degree of the zero polynomial is zero. Thus we give the following definition.

Definition 5.7 Let T be a linear operator on a finite dimensional vector space.

(a) The monic polynomial $\chi_T(t) := (-1)^n \det(T - t\,I)$ is called the **characteristic polynomial** of T.
(b) The equation $\chi_T(t) = 0$ is called the **characteristic equation** of T.
(c) Any complex number which is a root of the characteristic equation of T is called a **characteristic value** of T.

The definition makes sense since the determinant of a linear operator is the same as the determinant of any matrix representation of the linear operator. Further, the characteristic equation of T may also be given by $\det(T - t\,I) = 0$. Notice that the characteristic polynomial of T can also be written as $\det(t\,I - T)$. It plays the role of the polynomial of degree k that we had met in the proof of Theorem 5.4.

A proof of the fact that similar matrices have the same eigenvalues can be given using the characteristic polynomial as well; it is as follows:

$$\begin{aligned}\det(P^{-1}AP - t\,I) &= \det(P^{-1}(A - t\,I)P)\\ &= \det(P^{-1})\det(A - t\,I)\det(P) = \det(A - t\,I).\end{aligned}$$

Two matrices of the same size can have same eigenvalues but different characteristic polynomials. For instance, take $A = \text{diag}(1, 1, 2)$, the diagonal matrix of order 3 with the diagonal entries as 1, 1, and 2, and take $B = \text{diag}(1, 2, 2)$. Then $\chi_A(t) = (t-1)^2(t-2)$ and $\chi_B(t) = (t-1)(t-2)^2$. Both have the same eigenvalues 1 and 2. But the number of times they are the zeros of the respective characteristic polynomials are different.

Definition 5.8 Let T be a linear operator on a finite dimensional vector space.

(a) For an eigenvalue λ of T, the largest positive integer m such that $(t-\lambda)^m$ divides the characteristic polynomial of T is called the **algebraic multiplicity** of the eigenvalue λ; it is denoted by $\mu_T(\lambda)$, or as $\mu(\lambda)$.
(b) If $\lambda_1, \ldots, \lambda_k$ are the distinct eigenvalues of T with algebraic multiplicities $m_1, \ldots, m_k$, respectively, then we say that T **has** $m := m_1 + \cdots + m_k$ **number of eigenvalues counting multiplicities.**

If the characteristic polynomial of T can be written as $(t-\lambda)^m q(t)$, where $t-\lambda$ does not divide $q(t)$, then $\mu(\lambda) = m$, and vice versa. For example, the matrix $\mathrm{diag}(I_m, 2I_n)$ has eigenvalues 1 and 2 with $\mu(1) = m$ and $\mu(2) = n$.

Existence of an eigenvalue depends on the linear operator and also the underlying field. Unless the field is $\mathbb{C}$, it cannot be guaranteed that the characteristic polynomial is a product of linear factors using only numbers from the field; see Proposition 5.1. We wish to isolate this convenient case from others.

Definition 5.9 A polynomial $p(t) = a_0 + a_1 t + \cdots + a_k t^k$ of degree k with coefficients from $\mathbb{F}$ is said to **split over** $\mathbb{F}$ if there exist scalars $\lambda_1, \ldots, \lambda_k \in \mathbb{F}$ such that

$$p(t) = a_k(t-\lambda_1)\cdots(t-\lambda_k).$$

By Proposition 5.1, each polynomial with complex coefficients splits over $\mathbb{C}$. If a polynomial with real coefficients has only real zeros, then it splits over $\mathbb{R}$. There can exist polynomials with real coefficients which do not split over $\mathbb{R}$; for example, $p(t) = 1 + t^2$.

The characteristic polynomial $\chi_T(t)$ of T is a polynomial of degree n. When it splits over $\mathbb{F}$, all its zeros are in $\mathbb{F}$. In such a case, if $\chi_T(t) = (t-\lambda_1)\cdots(t-\lambda_n)$, then $\lambda_1, \ldots, \lambda_n$ are the n eigenvalues of T, counting multiplicities. Notice that the multiplicity of such an eigenvalue λ as a root of the characteristic equation is same as its algebraic multiplicity. We thus obtain the following result.

Theorem 5.10 *Let T be a linear operator on a vector space V of dimension n over $\mathbb{F}$. If the characteristic polynomial of T splits over $\mathbb{F}$, then each characteristic value of T is an eigenvalue of T. In this case, T has exactly n eigenvalues, counting multiplicities.*

Observe that if $\mathbb{F} = \mathbb{C}$, then the characteristic polynomial of T splits; then by Theorem 5.10, each characteristic value of T is an eigenvalue of T. If $\mathbb{F} = \mathbb{R}$ and $\lambda \in \mathbb{R}$ is a characteristic value of T, then by Theorem 3.34(2), $\det(T - \lambda I) = 0$ implies that $T - \lambda I$ is not injective. Thus, λ is an eigenvalue of T. Therefore, every characteristic value of T that is in $\mathbb{F}$, is an eigenvalue of T. In particular, we obtain the following result.

Theorem 5.11 *Let T be a linear operator on a finite dimensional vector space V. Then, each real characteristic value of T is an eigenvalue of T.*

As we see, splitting of the characteristic polynomial is equivalent to each of the following statements:

1. All characteristic values of T are eigenvalues of T.
2. Either $\mathbb{F} = \mathbb{C}$ or, $\mathbb{F} = \mathbb{R}$ and T has only real characteristic values.

Moreover, Theorem 5.10 suggests the following method for computing the eigenvalues and eigenvectors of a linear operator T on a vector space V of dimension n:

1. Choose any basis B of V;
2. Determine the matrix representation $[T]_{B,B}$ of T with respect to B;
3. Compute the characteristic polynomial $\chi_T(t) := (-1)^n \det([T]_{B,B} - tI)$;
4. Find the roots of the characteristic equation $\det([T]_{B,B} - tI) = 0$;
5. Keep only those roots which are in the underlying field of V. These are the eigenvalues of T along with their multiplicities.

For computing eigenvalues of a matrix $A \in \mathbb{F}^{n\times n}$, we take B as the standard basis of $\mathbb{F}^{n\times 1}$ so that $[A]_{B,B} = A$. Be aware that this procedure becomes unmanageable when $n = \dim(V) > 4$. We rework Example 5.2.

Example 5.12 Let $T : \mathbb{C}^3 \to \mathbb{C}^3$ be defined by $T(a, b, c) = (a + b, b + c, c + a)$. Choosing B as the standard basis for $\mathbb{C}^3$, we have

$$[T]_{B,B} = \begin{bmatrix} 1 & 1 & 0 \\ 0 & 1 & 1 \\ 1 & 0 & 1 \end{bmatrix}.$$

The characteristic polynomial of T is

$$\chi_T(t) = (-1)^3 \det(T - tI) = -\begin{vmatrix} 1-t & 1 & 0 \\ 0 & 1-t & 1 \\ 1 & 0 & 1-t \end{vmatrix} = -[(1-t)^3 - 1(-1)].$$

The roots of the characteristic equation

$$(t-1)^3 - 1 = (t-2)(t^2 - t + 1) = 0$$

are 2 and $(1 \pm i\sqrt{3})/2$. These are the characteristic values and eigenvalues of T and of $[T]_{B,B}$. If T were a linear operator on $\mathbb{R}^3$, its only eigenvalue would be 2. □

Each matrix in $\mathbb{F}^{n\times n}$ has a characteristic polynomial, which is a monic polynomial. Conversely each monic polynomial is the characteristic polynomial of some matrix. To see this, let $p(t)$ be a monic polynomial with coefficients from $\mathbb{F}$. By Proposition 5.1, $p(t)$ can be written as

$$p(t) = (t - \lambda_1) \cdots (t - \lambda_j)\big((t - \alpha_1)^2 + \beta_1^2\big) \cdots \big((t - \alpha_m)^2 + \beta_m^2\big),$$

where $\lambda_1, \ldots, \lambda_j, \alpha_1, \ldots, \alpha_m, \beta_1, \ldots, \beta_m \in \mathbb{F}$. We then take

$$M = \text{diag}\left(\lambda_1, \ldots, \lambda_k, \begin{bmatrix} \alpha_1 & \beta_1 \\ -\beta_1 & \alpha_1 \end{bmatrix}, \ldots, \begin{bmatrix} \alpha_m & \beta_m \\ -\beta_m & \alpha_m \end{bmatrix}\right).$$

It is easy to verify that $p(t) = \chi_M(t)$.

However, a simpler matrix using the coefficients in $p(t)$ can be constructed so that $p(t)$ is the characteristic polynomial of the matrix. It is as follows.

If the monic polynomial is $p(t) = a_0 + t$, then we take $C_p = [-a_0]$.

If $p(t) = a_0 + a_1 t + \cdots + a_{n-1}t^{n-1} + t^n$ for $n \geq 2$, then we take $C_p \in \mathbb{F}^{n \times n}$ as

$$C_p := \begin{bmatrix} 0 & & & & -a_0 \\ 1 & 0 & & & -a_1 \\ & 1 & \ddots & & -a_2 \\ & & \ddots & & \vdots \\ & & & 0 & -a_{n-2} \\ & & & 1 & -a_{n-1} \end{bmatrix}.$$

It may be verified that $\det(C_p - t\,I) = (-1)^n(a_0 + a_1 t + \cdots + a_{n-1}t^{n-1} + t^n)$. That is, the characteristic polynomial of C_p is $p(t)$. Accordingly, the matrix C_p is called the **companion matrix** of the polynomial $p(t) = a_0 + a_1 t + \cdots + a_{n-1}t^{n-1} + t^n$.

Exercises for Sect. 5.2

1. Determine the eigenvalues and corresponding eigenvectors for the following matrices:

(a) $\begin{bmatrix} 3 & 2 \\ -1 & 0 \end{bmatrix}$ (b) $\begin{bmatrix} -2 & -1 \\ 5 & 2 \end{bmatrix}$ (c) $\begin{bmatrix} 1 & 1 \\ i & i \end{bmatrix}$ (d) $\begin{bmatrix} 0 & 1 & 0 \\ 1 & 0 & 0 \\ 0 & 0 & 1 \end{bmatrix}$

(e) $\begin{bmatrix} 2 & 6 & -1 \\ 0 & 1 & 3 \\ 0 & 3 & 1 \end{bmatrix}$ (f) $\begin{bmatrix} 1 & 1 & 0 \\ i & i & 0 \\ 0 & 0 & 1+i \end{bmatrix}$ (g) $\begin{bmatrix} 1 & 0 & 1 & 0 \\ 0 & 0 & 1 & 1 \\ 1 & 0 & 1 & 0 \\ 0 & 0 & 1 & 1 \end{bmatrix}$.

2. Let $A = [a_{ij}] \in \mathbb{F}^{2 \times 2}$. Show the following:
   (a) If λ is an eigenvalue of A, then $(a_{12}, \lambda - a_{11})^T$ is a corresponding eigenvector.
   (b) A has a real eigenvalue if and only if $(a_{11} - a_{22})^2 + 4a_{12}a_{21} \geq 0$.
3. Give a linear operator whose characteristic polynomial is $(t-5)^2(t-6)^2$.
4. Let $T : \mathbb{R}^2 \to \mathbb{R}^2$ be defined by $T(a, b) = (0, b)$. Show that $t^2 - 2t$ is not the characteristic polynomial of T, and $T^2 - 2T$ is not invertible.
5. Let $T : \mathbb{R}^2 \to \mathbb{R}^2$ be the linear operator given by $T(a, b) = (-b, a)$. Let B be any basis for $\mathbb{R}^2$. If $[T]_{B,B} = [a_{ij}]$, then show that $a_{12}a_{21} \neq 0$.

6. Find the eigenvalues and corresponding eigenvectors of the differentiation operator $d/dt : \mathcal{P}_3(\mathbb{R}) \to \mathcal{P}_3(\mathbb{R})$.
7. Let V be the space of all twice differentiable functions from $\mathbb{R}$ to $\mathbb{R}$. Determine all eigenvalues of the second derivative $d^2/dt^2 : V \to V$.
8. Let T be a linear operator on a vector space of dimension n. Let $\lambda_1, \ldots, \lambda_n$ be all characteristic roots of T. Prove that $\det(T) = \Pi_{i=1}^n \lambda_i$ and $\operatorname{tr}(T) = \sum_{i=1}^n \lambda_i$.

5.3 Eigenspace

If λ_1 and λ_2 are distinct eigenvalues of a linear operator $T : V \to V$, then their corresponding eigenvectors must be distinct. For, if a vector v is an eigenvector corresponding to both the eigenvalues λ_1 and λ_2, then $\lambda_1 v = Tv = \lambda_2 v$ implies that $(\lambda_1 - \lambda_2)v = 0$. As $v \neq 0$, $\lambda_1 = \lambda_2$. A stronger conclusion holds.

Theorem 5.13 *Eigenvectors corresponding to distinct eigenvalues of any linear operator are linearly independent.*

Proof Suppose, on the contrary, that there exists a linear operator T on a vector space V, with distinct eigenvalues $\lambda_1, \ldots, \lambda_k$, $k \geq 2$, (among others) and corresponding eigenvectors $v_1, \ldots, v_k$, which are linearly dependent. For each $i \in \{1, \ldots, k\}$, $v_i \neq 0$ and $Tv_i = \lambda_i v_i$. By Theorem 1.29, there exists $j \in \{2, \ldots, k\}$ such that $v_1, \ldots, v_{j-1}$ are linearly independent and

$$v_j = \alpha_1 v_1 + \cdots + \alpha_{j-1} v_{j-1}$$

for some scalars $\alpha_1, \ldots, \alpha_{j-1}$. Then

$$\lambda_j v_j = \alpha_1 \lambda_j v_1 + \cdots + \alpha_{j-1} \lambda_j v_{j-1}$$
$$Tv_j = \alpha_1 T v_1 + \cdots + \alpha_{j-1} T v_{j-1} = \alpha_1 \lambda_1 v_1 + \cdots + \alpha_{j-1} \lambda_{j-1} v_{j-1}.$$

Now, $\lambda_j v_j - Tv_j = 0$ implies

$$\alpha_1(\lambda_j - \lambda_1)v_1 + \cdots + \alpha_{j-1}(\lambda_j - \lambda_{j-1})v_{j-1} = 0.$$

Since $v_1, \ldots, v_{j-1}$ are linear independent,

$$\alpha_1(\lambda_j - \lambda_1) = \cdots = \alpha_{j-1}(\lambda_j - \lambda_{j-1}) = 0.$$

As the eigenvalues λ_i are distinct, $\alpha_1 = \cdots = \alpha_{j-1} = 0$. But then $v_j = 0$. This is impossible since v_j is an eigenvector. ∎

More than one linearly independent eigenvector may exist corresponding to the same eigenvalue. For example, the identity operator $I : \mathbb{R}^3 \to \mathbb{R}^3$ has eigenvectors e_1, e_2, e_3 corresponding to the same eigenvalue 1.

Recall that if λ is an eigenvalue of a linear operator T, then the set of all eigenvectors corresponding to the eigenvalue λ along with the zero vector is $N(T - \lambda I)$, the null space of $T - \lambda I$. The number of linearly independent eigenvectors corresponding to the eigenvalue λ of T is the nullity of the linear operator $T - \lambda I$. Moreover, Theorem 5.13 implies that if λ_1 and λ_2 are distinct eigenvalues of a linear operator, then $N(T - \lambda_1 I) \cap N(T - \lambda_2 I) = \{0\}$.

Definition 5.14 Let T be a linear operator on a vector space V.

(a) The set of all eigenvalues of T is called the **eigenspectrum** of T and is denoted by $\mathbf{eig}(T)$.
(b) The subspace $N(T - \lambda I)$ of V is called the **eigenspace** of T with respect to the eigenvalue λ and is denoted by $E(\lambda)$.
(c) For $\lambda \in \text{eig}(T)$, $\gamma(\lambda) := \dim(E(\lambda))$ is called the **geometric multiplicity** of the eigenvalue λ.

We read $\lambda \in \text{eig}(T)$ as "λ is an eigenvalue of T". If more than one linear operator are involved in a certain context, we may denote $E(\lambda)$ and $\gamma(\lambda)$ as $E_T(\lambda)$ and $\gamma_T(\lambda)$, respectively.

Example 5.15 The conclusions drawn in each of the following cases may be easily verified:

(1) Let $T : \mathbb{F}^2 \to \mathbb{F}^2$ be defined by $T(a, b) = (a + b, b)$. Then $\lambda = 1$ is the only eigenvalue; so, $\text{eig}(T) = \{1\}$. $E(1) = N(T - I) = \text{span}\{(1, 0)\}$, and $\gamma(1)$, the geometric multiplicity of the eigenvalue 1, is 1.

(2) Define $T : \mathbb{R}^2 \to \mathbb{R}^2$ by $T(a, b) = (b, -a)$. Then $\text{eig}(A) = \varnothing$.

(3) Let $T : \mathbb{C}^2 \to \mathbb{C}^2$ be defined by $T(a, b) = (b, -a)$. Then $\text{eig}(T) = \{i, -i\}$. $E(i) = N(T - iI) = \text{span}\{(1, i)\}$, $E(-i) = N(T + iI) = \text{span}\{(1, -i)\}$, and $\gamma(i) = \gamma(-i) = 1$.

(4) Define $T : \mathbb{F}^3 \to \mathbb{F}^3$ by $T(a, b, c) = (a, a + b, a + b + c)$. Then $\text{eig}(T) = \{1\}$, $E(1) = N(T - I) = \text{span}\{(0, 0, 1)\}$, and $\gamma(1) = 1$.

(5) Let $T : \mathcal{P}(\mathbb{F}) \to \mathcal{P}(\mathbb{F})$ be defined by $(Tp)(t) = tp(t)$ for $p \in \mathcal{P}(\mathbb{F})$. Then $\text{eig}(T) = \varnothing$.

(6) Let $T : \mathcal{P}([a, b], \mathbb{R}) \to \mathcal{P}([a, b], \mathbb{R})$, where $(Tp)(t) = \frac{d}{dt}p(t)$ for $p \in \mathcal{P}([a, b], \mathbb{R})$. Then $\text{eig}(T) = \{0\}$ and $E(0) = N(T) = \text{span}\{p_0\}$, where $p_0(t) = 1$ for all $t \in [a, b]$. Consequently, $\gamma(0) = 1$. □

The following theorem generalizes Theorem 5.13.

Theorem 5.16 *Let $\lambda_1, \ldots, \lambda_k$, for $k \geq 2$, be distinct eigenvalues of a linear operator T on a vector space V. Then the following are true:*

(1) $E(\lambda_j) \cap \sum_{\substack{i=1 \\ i \neq j}}^{k} E(\lambda_i) = \{0\}$ *for each* $j \in \{1, \ldots, k\}$.
(2) *For each $i \in \{1, \ldots, k\}$, let B_i be a linearly independent subset of $E(\lambda_i)$. Then $\cup_{i=1}^{k} B_i$ is linearly independent. Further, if $i, j \in \{1, \ldots, k\}$ and $i \neq j$, then $B_i \cap B_j = \varnothing$.*

Proof (1) Suppose $E(\lambda_j) \cap \sum_{\substack{i=1 \\ i \neq j}}^k E(\lambda_i) \neq \{0\}$ for some $j \in \{1, \ldots, k\}$. Let v_j be a nonzero vector in $E(\lambda_j) \cap \sum_{\substack{i=1 \\ i \neq j}}^k E(\lambda_i)$. Then

$$v_j = v_1 + \cdots + v_{j-1} + v_{j+1} + \cdots + v_k$$

for some vectors $v_1 \in E(\lambda_1), \ldots, v_{j-1} \in E(\lambda_{j-1}), v_{j+1} \in E(\lambda_{j+1}), \ldots, v_k \in E(\lambda_k)$. Since $v_j \neq 0$, at least one of the vectors on the right hand side is nonzero. Thus

$$v_j = v_{m_1} + \cdots + v_{m_n}$$

for nonzero vectors $v_{m_1}, \ldots, v_{m_n}$ with $m_1, \ldots, m_n \in \{1, \ldots, k\} \setminus \{j\}$. This contradicts Theorem 5.13, since $v_j, v_{m_1}, \ldots, v_{m_n}$ are eigenvectors corresponding to the eigenvalues $\lambda_j, \lambda_{m_1}, \ldots, \lambda_{m_n}$, respectively.

(2) To show that $\cup_{i=1}^k B_i$ is linearly independent, we take any linear combination of vectors from $\cup_{i=1}^k B_i$ and equate it to 0. Such an equation can be written in the form

$$\sum_{i=1}^{k} (\alpha_{i1} v_{i1} + \cdots + \alpha_{im_i} v_{im_i}) = 0,$$

where $v_{i1}, \ldots, v_{im_i} \in B_i$ and $\alpha_{i1}, \ldots, \alpha_{im_i} \in \mathbb{F}$ for $i \in \{1, \ldots, k\}$.

Since $\alpha_{i1} v_{i1} + \cdots + \alpha_{im_i} v_{im_i} \in E(\lambda_i)$, due to (1), we conclude that

$$\alpha_{i1} v_{i1} + \cdots + \alpha_{im_i} v_{im_i} = 0 \quad \text{for each } i \in \{1, \ldots, k\}.$$

Now that B_i is linearly independent, it follows that $\alpha_{i1} = \cdots = \alpha_{im_i} = 0$ for each $i \in \{1, \ldots, k\}$. Hence $\cup_{i=1}^k B_i$ is linearly independent.

For the second part of the conclusion, suppose that $B_i \cap B_j \neq \emptyset$ for some $i \neq j$. Let $v \in B_i \cap B_j$. Then $v \neq 0$ and $\lambda_i v = Tv = \lambda_j v$. Hence, $(\lambda_i - \lambda_j)v = 0$. This is not possible since $\lambda_i \neq \lambda_j$. ■

If λ is an eigenvalue of T with corresponding eigenvector v, then $T^2 v = \lambda T v = \lambda^2 v$. That is, λ^2 is an eigenvalue of T^2. Will all eigenvalues of T^2 come only this way? In general, if $p(t)$ is any polynomial, then is there any relation between eigenvalues of T and those of $p(T)$?

Theorem 5.17 (Spectral mapping theorem) *Let T be a linear operator on a vector space V over $\mathbb{F}$. Let $p(t)$ be a polynomial with coefficients from $\mathbb{F}$. Then the following are true:*

(1) $\{p(\lambda) : \lambda \in \text{eig}(T)\} \subseteq \text{eig}(p(T))$.
(2) *If* $\mathbb{F} = \mathbb{C}$, *then* $\text{eig}(p(T)) = \{p(\lambda) : \lambda \in \text{eig}(T)\}$.
(3) *Let* $\delta \in \mathbb{F}$. *If* $T - \delta I$ *is invertible then* $\delta \notin \text{eig}(T)$ *and*

$$\text{eig}\big((T - \delta I)^{-1}\big) = \{(\lambda - \delta)^{-1} : \lambda \in \text{eig}(T)\}.$$

(4) *If T is invertible, then* $0 \notin \text{eig}(T)$ *and* $\text{eig}(T^{-1}) = \{\lambda^{-1} : \lambda \in \text{eig}(T)\}$.

Proof (1) Let $\lambda \in \mathbb{F}$. Then $p(t) - p(\lambda)$ is a polynomial which vanishes at $t = \lambda$. Thus $p(t) - p(\lambda) = (t - \lambda)q(t) = q(t)(t - \lambda)$ for some polynomial $q(t)$. Hence

$$p(T) - p(\lambda)I = (T - \lambda I)q(T) = q(T)(T - \lambda I).$$

If $(T - \lambda I)v = 0$ for some $v \neq 0$, then $(p(T) - p(\lambda))v = q(T)(T - \lambda I)v = 0$. That is, if λ is an eigenvalue of T, then $p(\lambda)$ is an eigenvalue of $p(T)$.

(2) Suppose $\mathbb{F} = \mathbb{C}$. Let $\beta \in \mathbb{C}$ be an eigenvalue of $p(T)$. Then $p(T) - \beta I$ is not injective. If $p(t)$ is a polynomial of degree n, then so is $p(t) - \beta$. Let a be the coefficient of t^n in $p(t)$. Due to Proposition 5.1,

$$p(t) - \beta = a(t - \lambda_1) \cdots (t - \lambda_n) \quad \text{for some } \lambda_1, \ldots, \lambda_n \in \mathbb{C}. \tag{5.1}$$

Therefore,

$$p(T) - \beta I = a(T - \lambda_1 I) \cdots (T - \lambda_n I).$$

Since $p(T) - \beta I$ is not injective, $T - \lambda_j I$ is not injective for some $j \in \{1, \ldots, n\}$. That is, λ_j is an eigenvalue of T. Substituting $t = \lambda_j$ in (5.1), we have $\beta = p(\lambda_j)$. That is, each eigenvalue of $p(T)$ is equal to $p(\lambda)$ for some eigenvalue λ of T. Conversely, due to (1), if λ is an eigenvalue of T, then $p(\lambda)$ is an eigenvalue of $p(T)$. This completes the proof.

(3) Let $\lambda \in \mathbb{F}$, $\lambda \neq \delta$. Suppose that $T - \delta I$ is invertible. Then $T - \delta I$ is injective; hence δ is not an eigenvalue of T. Now,

$$(\lambda - \delta)(T - \delta I)\big((T - \delta I)^{-1} - (\lambda - \delta)^{-1} I\big) = -(T - \lambda I).$$

That is,

$$(T - \delta I)^{-1} - (\lambda - \delta)^{-1} I = -(\lambda - \delta)^{-1}(T - \delta I)^{-1}(T - \lambda I).$$

Therefore, $T - \lambda I$ is not injective if and only if $(T - \delta I)^{-1} - (\lambda - \delta)^{-1} I$ is not injective. That is, $\lambda \in \text{eig}(T)$ if and only if $(\lambda - \delta)^{-1} \in \text{eig}\big((T - \delta I)^{-1}\big)$. It completes the proof.

(4) It is a particular case of (3). ■

If $\mathbb{F} = \mathbb{R}$, then all eigenvalues of $p(T)$ may not be in the form $p(\lambda)$ for eigenvalues λ of T. See the following example.

Example 5.18 (1) Let $T : \mathbb{R}^2 \to \mathbb{R}^2$ be given by $T(a, b) = (-b, a)$. We have seen in Example 5.3 that T has no eigenvalue. That is, $\text{eig}(T) = \varnothing$. But $T^2(a, b) = T(-b, a) = -(a, b)$ for each $(a, b) \in \mathbb{R}^2$. Hence $\text{eig}(T^2) = \{-1\}$. Here, $\text{eig}(T^2) \supsetneqq \{\lambda^2 : \lambda \in \text{eig}(T)\}$.

(2) Let $T : \mathbb{R}^3 \to \mathbb{R}^3$ be given by $T(a, b, c) = (-b, a, 0)$. Then $\text{eig}(T) = \{0\}$ but $\text{eig}(T^2) = \{-1, 0\}$. That is, $\text{eig}(T^2) \supsetneqq \{\lambda^2 : \lambda \in \text{eig}(T)\}$. □

Theorem 5.17(4) can be seen more directly. Suppose T is invertible. If $0 \in \text{eig}(T)$, then for a nonzero vector v, we have $Tv = 0$. Then T would not be injective. Thus $0 \notin \text{eig}(T)$. Now, (λ, v) is an eigenpair of T if and only if $v \neq 0$ and $Tv = \lambda v$ if and only if $v \neq 0$ and $T^{-1}v = \lambda^{-1}v$ if and only if (λ^{-1}, v) is an eigenpair of T^{-1}.

Moreover, if V is finite dimensional, then $0 \notin \text{eig}(T)$ implies that T is invertible due to Theorem 2.26(4). Therefore, for linear operators T on finite dimensional vector spaces, $\lambda \notin \text{eig}(T)$ is equivalent to $T - \lambda I$ is invertible.

Recall that if v is an eigenvector of a linear operator T, then the line $\{\alpha v : \alpha \in \mathbb{F}\}$ is fixed by T. In other words, the image of every vector in the $\text{span}\{v\}$ is inside $\text{span}\{v\}$. We generalize this notion a bit to include arbitrary subspaces in our ambit, which will help us in understanding the eigenspaces.

Definition 5.19 Let T be a linear operator on a vector space V. A subspace U of V is said to be T-**invariant** (or, invariant under T) if $T(U) \subseteq U$.

For instance, if T is a linear operator on a vector space V, then its range space and null space are T-invariant. For, if $y \in R(T)$, then $Ty \in R(T)$; and if $x \in N(T)$, then $Tx = 0 \in N(T)$.

Observe that when a subspace U is T-invariant, for each $x \in U$, we have $Tx \in U$. Therefore, $T|_U$, the restriction of T to the subspace U is well-defined as a linear operator on U.

Invariant subspaces are related to the factorization of the characteristic polynomial of a linear operator.

Theorem 5.20 *Let T be a linear operator on a finite dimensional vector space V.*

(1) *If U is a nontrivial T-invariant subspace of V, then the characteristic polynomial of $T|_U$ divides the characteristic polynomial of T.*
(2) *If U and W are nontrivial proper T-invariant subspaces of V such that $V = U \oplus W$, then the characteristic polynomial of T is the product of the characteristic polynomials of $T|_U$ and that of $T|_W$.*

Proof (1) If $U = V$, then there is nothing to prove. Assume that U is a proper subspace of V. Let $E_0 := \{u_1, \ldots, u_k\}$ be a basis for U. Extend E_0 to a basis $E := E_0 \cup \{v_1, \ldots, v_m\}$ for V. Notice that $m \geq 1$. Let $v \in V$. Since U is T-invariant, $Tu_j \in U$ for each $j \in \{1, \ldots, k\}$. So, there exist scalars a_{ij}, b_{ij} and c_{ij} such that

$$
\begin{aligned}
Tu_1 &= a_{11}u_1 + \cdots + a_{k1}u_k \\
&\vdots \\
Tu_k &= a_{1k}u_1 + \cdots + a_{kk}u_k \\
Tv_1 &= b_{11}u_1 + \cdots + b_{k1}u_k + c_{11}v_1 + \cdots + c_{m1}v_m \\
&\vdots \\
Tv_m &= b_{1m}u_1 + \cdots + b_{km}u_k + c_{1m}v_1 + \cdots + c_{mm}v_m
\end{aligned}
$$

We see that the matrix $[T]_{E,E}$ has the block form:

$$[T]_{E,E} = \begin{bmatrix} A & B \\ 0 & C \end{bmatrix}$$

where $A = [T|_U]_{E_0,E_0} \in \mathbb{F}^{k\times k}$, $B \in \mathbb{F}^{k\times m}$, $0 \in \mathbb{F}^{m\times k}$ and $C \in \mathbb{F}^{m\times m}$. Therefore, $\det([T]_{E,E} - t\,I) = \det([T|_U]_{E_0,E_0} - t\,I_k)\ \det(C - t\,I_m)$.

(2) Let $V = U \oplus W$, where both U and W are T-invariant nontrivial proper subspaces of V. In the proof of (1), take the same E_0 as a basis for U and choose $E_1 = \{v_1, \ldots, v_m\}$ as a basis for W. Then $E = E_0 \cup E_1$ is a basis for V. Now, the T-invariance of W implies that the scalars b_{ij} in the proof of (1) are all 0. Thus the matrix $[T]_{E,E}$ has the block form:

$$[T]_{E,E} = \begin{bmatrix} A & 0 \\ 0 & C \end{bmatrix},$$

where $A = [T|_U]_{E_0,E_0} \in \mathbb{F}^{k\times k}$, $C = [T|_W]_{E_1,E_1} \in \mathbb{F}^{m\times m}$, and the zero matrices are of suitable size. Then

$$\det([T]_{E,E} - t\,I) = \det([T|_U]_{E_0,E_0} - t\,I_k)\ \det([T|_W]_{E_1,E_1} - t\,I_m).$$

This completes the proof. ■

Exercises for Sect. 5.3

1. Determine the eigenvalues and the associated eigenspaces for the following matrices, considering first, $\mathbb{F} = \mathbb{C}$, and then $\mathbb{F} = \mathbb{R}$.
(a) $\begin{bmatrix} 0 & 0 \\ 0 & 0 \end{bmatrix}$ (b) $\begin{bmatrix} 0 & 0 \\ 1 & 0 \end{bmatrix}$ (c) $\begin{bmatrix} 0 & 1 \\ 0 & 0 \end{bmatrix}$ (d) $\begin{bmatrix} 0 & 1 \\ 1 & 0 \end{bmatrix}$ (e) $\begin{bmatrix} 0 & -1 \\ 1 & 0 \end{bmatrix}$.
2. Suppose A is a block upper triangular matrix with the blocks on the diagonal as the square matrices $A_1, \ldots, A_m$. Show that $\operatorname{eig}(A) = \cup_{i=1}^{m} \operatorname{eig}(A_i)$.
3. Let $V := \ell^1(\mathbb{N}, \mathbb{C})$, the vector space of all absolutely summable complex sequences. Let S_L be the left shift operator and let S_R be the right shift operator on V. That is,
$S_L(\alpha_1, \alpha_2, \alpha_3, \ldots) = (\alpha_2, \alpha_3, \alpha_4, \ldots)$, $S_R(\alpha_1, \alpha_2, \alpha_3, \ldots) = (0, \alpha_1, \alpha_2, \ldots)$.
Determine the eigenvalues and eigenspaces of S_L and S_R.
4. Let T be a linear operator on a vector space V of dimension 2. If T has only one eigenvalue λ, then show that $Av - \lambda v \in E(\lambda)$ for all $v \in V$.
5. Let $A, B, P \in \mathbb{R}^{n\times n}$ be such that P is invertible and $B = P^{-1}AP$. Show that a vector $v \in \mathbb{R}^{n\times 1}$ is an eigenvector of B corresponding to the eigenvalue λ if and only if Pv is an eigenvector of A corresponding to the same eigenvalue λ. Show also that the geometric multiplicity of the eigenvalue λ of A is same as the geometric multiplicity of the eigenvalue λ of B.
6. Show that $\operatorname{eig}(A^T) = \operatorname{eig}(A)$ for each $n \times n$ matrix A. What about $\operatorname{eig}(A^*)$ and $\operatorname{eig}(A)$?

7. Let $A \in \mathbb{F}^{5\times 5}$ satisfy $A^3 = A$. How many distinct eigenvalues can A have?
8. Determine all T-invariant subspaces when the linear operator T is given by $T(a, b) = (a - b, 2a + 2b)$ for
   (a) $(a, b) \in \mathbb{R}^2$ (b) $(a, b) \in \mathbb{C}^2$.
9. Let T be the linear operator given by $T(a, b, c) = (a + b, 2b, 3c)$ for $(a, b, c) \in \mathbb{R}^3$. Determine all T-invariant subspaces of $\mathbb{R}^3$.
10. Let $T : V \to V$ be a linear operator. Let $n \in \mathbb{N}$. Show that the following are T-invariant:
    (a) $N(T^n)$ (b) $R(T^n)$ (c) Any eigenspace of T^n.
11. Give an example of a linear operator T on a finite dimensional vector space V so that the T-invariant subspaces are $\{0\}$ and V, and nothing else.
12. Prove that the only subspaces invariant under every linear operator on a finite dimensional vector space V are $\{0\}$ or V.
13. Let T be a linear operator on a vector space V. Prove that if $U_1, \ldots, U_n$ are T-invariant subspaces of V, then $U_1 + \cdots + U_n$ is also T-invariant.

5.4 Generalized Eigenvectors

Let T be a linear operator on a vector space V. Observe that $R(T)$ and $N(T)$ are T-invariant subspaces of V. Something more can be told along this line.

Theorem 5.21 *Let T be a linear operator on a vector space V. Then for every $\alpha \in \mathbb{F}$ and $j \in \mathbb{N}$,*

(1) *$R(T - \alpha I)^j \supseteq R(T - \alpha I)^{j+1}$, and $R(T - \alpha I)^j$ is T-invariant*
(2) *$N(T - \alpha I)^j \subseteq N(T - \alpha I)^{j+1}$, and $N(T - \alpha I)^j$ is T-invariant. In particular, for an eigenvalue λ of T, the eigenspace $E(\lambda)$ is T-invariant.*

Proof (1) If $y \in R(T - \alpha I)^{j+1}$, then there exists a vector $x \in V$ such that

$$y = (T - \alpha I)^{j+1} x = (T - \alpha I)^j ((T - \alpha I)x).$$

This shows that $R(T - \alpha I)^j \supseteq R(T - \alpha I)^{j+1}$.

For T-invariance of $R(T - \alpha I)^j$, let $z \in R(T - \alpha I)^j$. Then there exists $w \in V$ such that $z = (T - \alpha I)^j w$. Now,

$$Tz = T(T - \alpha I)^j w = (T - \alpha I)^j T w \in R(T - \alpha I)^j.$$

Therefore, $R(T - \alpha I)^i$ is T-invariant.

(2) Let $x \in N(T - \alpha I)^j$. Then $(T - \alpha I)^j x = 0$. So, $(T - \alpha I)^{j+1} x = 0$. That is, $N(T - \alpha I)^j \subseteq N(T - \alpha I)^{j+1}$. Further, $(T - \alpha I)^j T x = T(T - \alpha I)^j x = 0$. That is, $Tx \in N(T - \alpha I)^j$. Therefore, $N(T - \alpha I)^j$ is T-invariant.

In particular, $E(\lambda) = N(T - \lambda I)$ is T-invariant for an eigenvalue λ of T. ■

Example 5.22 Let $T : \mathbb{R}^3 \to \mathbb{R}^3$ be the linear operator defined by $T(a, b, c) = (a + b, b + c, 2c)$. With respect to the standard basis, its matrix representation is

$$[T] = \begin{bmatrix} 1 & 1 & 0 \\ 0 & 1 & 1 \\ 0 & 0 & 2 \end{bmatrix}$$

Here, $[T]$ is a real matrix. Its characteristic equation is $(t - 1)^2(t - 2) = 0$. To get the eigenvectors corresponding to the eigenvalue 1, we solve the linear system

$$(T - I)(a, b, c) = (b, c, c) = 0.$$

This yields $b = 0 = c$, whereas a is arbitrary. Thus $\gamma(1) = 1$ and the eigenspace is $E(1) = \text{span}\{(1, 0, 0)\}$, which is T-invariant. The basis $\{(1, 0, 0)\}$ of the eigenspace can be extended to a basis of $\mathbb{R}^3$, the domain space of T. One way is to extend it by using the basis vectors of $N(T - I)^2$, if possible. In this case,

$$(T - I)^2(a, b, c) = (c, c, c).$$

Then $(T - I)^2(a, b, c) = 0$ implies that $c = 0$, whereas a and b can be any real numbers. That is, $N(T - I)^2 = \text{span}\{(1, 0, 0), (0, 1, 0)\}$. This basis of $N(T - I)^2$ is an extension of the basis of $N(T - I)$. We see that $N(T - I)^2 \neq \mathbb{R}^3$.

We try extending it to a basis for $\mathbb{R}^3$ by choosing a vector from $N(T - I)^3$. Now,

$$(T - I)^3(a, b, c) = (c, c, c) = (T - I)^2(a, b, c).$$

We get no more vector from $N(T - I)^3$ for extending the earlier basis. We see that $N(T - I) \subsetneqq N(T - I)^2 = N(T - I)^3 \subsetneqq \mathbb{R}^3$.

So, we try a basis of the eigenspace $E(2)$. For this, we solve the linear system

$$(T - 2I)(a, b, c) = (b - a, c - b, 0) = 0.$$

It gives $a = b,\ b = c$ and c is arbitrary. For example, $(1, 1, 1)$ is such a vector.

It results in the basis $\{(1, 0, 0), (0, 1, 0), (1, 1, 1)\}$ of $\mathbb{R}^3$, which come from the bases of the T-invariant subspaces $N(T - I)^2$ and $N(T - 2I)$. □

We look at the T-invariant subspaces $N(T - \lambda I)^j$ more closely.

Theorem 5.23 *Let T be a linear operator on a vector space V and let λ be an eigenvalue of T.*

(1) *If there exists $k \in \mathbb{N}$ such that $N(T - \lambda I)^k = N(T - \lambda I)^{k+1}$, then*

$$N(T - \lambda I)^k = N(T - \lambda I)^{k+j} \quad \textit{for all } j \in \mathbb{N}.$$

(2) *If* $\dim(V) = n$, *then there exists* $k \in \{1, \ldots, n\}$ *such that*

$$0 \subsetneqq N(T - \lambda I) \subsetneqq \cdots \subsetneqq N(T - \lambda I)^k = N(T - \lambda I)^{k+1} \subseteq V.$$

Proof (1) Suppose there exists $k \in \mathbb{N}$ such that $N(T - \lambda I)^k = N(T - \lambda I)^{k+1}$. By Theorem 5.21, we know that $N(T - \lambda I)^{k+1} \subseteq N(T - \lambda I)^{k+2}$. To show the other way inclusion, let $x \in N(T - \lambda I)^{k+2}$. Then,

$$(T - \lambda I)x \in N(T - \lambda I)^{k+1} = N(T - \lambda I)^k.$$

Thus, $(T - \lambda I)^{k+1}x = N(T - \lambda I)^k(T - \lambda I)x = 0$ so that $x \in N(T - \lambda I)^{k+1}$. Hence $N(T - \lambda I)^{k+2} \subseteq N(T - \lambda I)^{k+1}$.

Therefore, $N(T - \lambda I)^{k+2} = N(T - \lambda I)^{k+1} = N(T - \lambda I)^k$. By induction, it follows that $N(T - \lambda I)^{k+j} = N(T - \lambda I)^k$ for all $j \in \mathbb{N}$.

(2) Let V be of dimension n. On the contrary, assume that for each $j \in \{1, \ldots, n\}$, $N(T - \lambda I)^j \subsetneqq N(T - \lambda I)^{j+1} \subseteq V$. Due to Theorem 5.21, we obtain

$$\{0\} \subsetneqq N(T - \lambda I) \subsetneqq \cdots \subsetneqq N(T - \lambda I)^n \subsetneqq N(T - \lambda I)^{n+1} \subsetneqq \cdots \subseteq V.$$

Then, writing $d_j = \dim(N(T - \lambda I)^j)$, we see that $d_{j+1} \geq d_j + 1$; consequently, $n = \dim(V) \geq d_{n+1} \geq n + 1$. This is impossible. ■

In view of Theorem 5.23, we introduce the following terminology.

Definition 5.24 Let V be a vector space of dimension n. Let T be a linear operator on V and let λ be an eigenvalue of T.

(a) The least $k \in \{1, \ldots, n\}$ such that $N(T - \lambda I)^k = N(T - \lambda I)^{k+1}$ is called the **ascent** of λ; and it is denoted by $\ell(\lambda)$.
(b) The subspace $N(T - \lambda I)^{\ell(\lambda)}$ of V is called the **generalized eigenspace** of T corresponding to the eigenvalue λ and is denoted by $G(\lambda)$.
(c) Any nonzero vector in $G(\lambda)$ is called a **generalized eigenvector** of T corresponding to the eigenvalue λ.

Our notation does not show the operator T explicitly. If need arises, we may write $\ell(\lambda)$ as $\ell_T(\lambda)$, and $G(\lambda)$ as $G_T(\lambda)$. To emphasize, if $\ell := \ell(\lambda)$ is the ascent of the eigenvalue λ of a linear operator T on a vector space V of dimension n, then

$$\{0\} \subsetneqq N(T - \lambda I) \subsetneqq \cdots \subsetneqq N(T - \lambda I)^\ell = G(\lambda) = N(T - \lambda I)^{\ell+1} = \cdots \subseteq V.$$

Further, by Theorem 5.21, $G(\lambda)$ is T-invariant.

Example 5.25 Revisit Examples 5.15 and 5.18.
(1) For the linear operator $T : \mathbb{F}^2 \to \mathbb{F}^2$ defined by $T(a, b) = (a + b, b)$ we have

$$(T - I)(a, b) = (b, 0), \quad (T - I)^2(a, b) = (0, 0), \quad N(T - I)^2 = \mathbb{R}^2 = N(T - I)^3.$$

The characteristic polynomial of T is $(-1)^2 \det(T - t\,I) = (t-1)^2$. Then

$$\text{eig}(T) = \{1\}, \quad \gamma(1) = 1, \quad \mu(1) = 2, \quad \ell(1) = 2, \quad G(1) = \mathbb{R}^2.$$

(2) Let $T : \mathbb{C}^2 \to \mathbb{C}^2$ be defined by $T(a, b) = (b, -a)$. The characteristic polynomial of T is $(-1)^2 \det(T - t\,I) = (t - i)(t + i)$. Then $\text{eig}(T) = \{i, -i\}$, and $\gamma(i) = \gamma(-i) = 1$. Now,

$$(T - iI)^2(a, b) = -2i(-ia + b, -a - ib) = -2i(T - iI)(a, b).$$

So, $N(T - iI)^2 = N(T - iI)$. Then

$$G(i) = E(i) = \text{span}\{(1, i)\}, \quad \ell(i) = 1 = \mu(i).$$

It may be verified that $G(-i) = E(-i) = \text{span}\{(1, -i)\}$, and $\ell(-i) = 1 = \mu(-i)$.
(3) For $T : \mathbb{F}^3 \to \mathbb{F}^3$ defined by $T(a, b, c) = (a, a + b, a + b + c)$, the characteristic polynomial is $(t - 1)^3$. So, $\text{eig}(T) = \{1\}$. Now,

$$\begin{aligned}(T - I)(a, b, c) = (0, a, a + b),\ & (T - I)^2(a, b, c) \\ &= (0, 0, a),\ (T - I)^3(a, b, c) = (0, 0, 0).\end{aligned}$$

Therefore, $N(T - I) \subsetneqq N(T - I)^2 \subsetneqq N(T - I)^3 = \mathbb{R}^3$. We see that

$$\gamma(1) = 1, \quad \mu(1) = 3, \quad \ell(1) = 3, \quad G(1) = \mathbb{R}^3.$$

(4) For the linear operator $T : \mathbb{R}^3 \to \mathbb{R}^3$ defined by $T(a, b, c) = (a + b, b, c)$, the characteristic polynomial is $(t - 1)^3$. Thus $\mu(1) = 3$. Here,

$$(T - I)(a, b, c) = (b, 0, 0), \ (T - I)^2(a, b, c) = (0, 0, 0).$$

Consequently, $E(1) = N(T - I) = \text{span}\{(1, 0, 0)\}$ and $\gamma(1) = 1$. Hence

$$N(T - I) \subsetneqq N(T - I)^2 = \mathbb{R}^3, \quad \ell(1) = 2, \quad G(1) = \mathbb{R}^3. \qquad \Box$$

If V is any vector space, T is a linear operator on V, and λ is an eigenvalue of T, then the ascent $\ell(\lambda)$ of λ and the generalized eigenspace $G(\lambda)$ of T corresponding to the eigenvalue λ are defined as follows:

$$\ell(\lambda) := \inf\{k \in \mathbb{N} : N(T - \lambda\,I)^k = N(T - \lambda\,I)^{k+1}\}; \quad G(\lambda) := \bigcup_{j \in \mathbb{N}} N(T - \lambda I)^j.$$

Notice that if V is finite dimensional, then $\ell(\lambda)$ and $G(\lambda)$ as defined above coincide with the those in Definition 5.24. If V is infinite dimensional, then it can happen that $\ell(\lambda) = \infty$ as the following example shows.

Example 5.26 Let T be the left shift operator on the vector space $\ell^1(\mathbb{N}, \mathbb{F})$ of all absolutely summable sequences of scalars, that is,

$$T(\alpha_1, \alpha_2, \ldots) = (\alpha_2, \alpha_3, \ldots), \quad (\alpha_1, \alpha_2, \ldots) \in \ell^1(\mathbb{N}, \mathbb{F}).$$

For $n \in \mathbb{N}$, let $e_n = (\delta_{1n}, \delta_{2n}, \ldots)$, the sequence with its nth entry 1 and all other entries 0. Then we see that $T^n e_n = 0$ and $T^n e_{n+1} = e_n$ for all $n \in \mathbb{N}$. Thus, $\lambda = 0$ is an eigenvalue of T and

$$\{0\} \subsetneqq N(T - \lambda I) \subsetneqq \cdots \subsetneqq N(T - \lambda I)^n \subsetneqq N(T - \lambda I)^{n+1} \subsetneqq \cdots \subseteq \ell^1(\mathbb{N}, \mathbb{F}).$$

Therefore, $\ell(\lambda) = \infty$. □

In Example 5.25, we observe that the algebraic multiplicity of an eigenvalue is equal to the dimension of the generalized eigenspace. We would like to confirm that it is indeed the case for any linear operator on a finite dimensional vector space. It shows that the generalized eigenspace is not an artificial construction.

Theorem 5.27 *Let λ be an eigenvalue of a linear operator T on a vector space V of dimension n. Let γ, μ and ℓ be the geometric multiplicity of λ, the algebraic multiplicity of λ, and the ascent of λ, respectively. Let $T_G := T|_{G(\lambda)}$ be the restriction of T to the generalized eigenspace $G(\lambda) := N(T - \lambda I)^\ell$. Then the following are true:*

(1) $V = G(\lambda) \oplus R(T - \lambda I)^\ell$.
(2) *The characteristic polynomial of T is the product of the characteristic polynomials of $T|_{G(\lambda)}$ and of $T|_{R(T-\lambda I)^\ell}$.*
(3) $\text{eig}(T_G) = \{\lambda\}$, *and* $\lambda \notin \text{eig}(T|_{R(T-\lambda I)^\ell})$.
(4) $\dim(G(\lambda)) = \mu = \text{null}\,(T - \lambda I)^n$.
(5) *If $\alpha \in \mathbb{F}$ and $\alpha \neq \lambda$, then for each $j \in \mathbb{N}$, $N(T - \alpha I)^j \cap G(\lambda) = \{0\}$.*
(6) *Suppose $\ell \geq 2$, $j \in \{1, \ldots, \ell - 1\}$, and $\{v_1, \ldots, v_r\}$ is a linearly independent subset of $N(T - \lambda I)^{j+1}$ such that* $\text{span}\{v_1, \ldots, v_r\} \cap N(T - \lambda I)^j = \{0\}$. *Then $\{(T - \lambda I)v_1, \ldots, (T - \lambda I)v_r\}$ is a linearly independent subset of $N(T - \lambda I)^j$ and* $\text{span}\{(T - \lambda I)v_1, \ldots, (T - \lambda I)v_r\} \cap N(T - \lambda I)^{j-1} = \{0\}$.
(7) *Let $\ell \geq 2$. Then for each $j \in \{1, \ldots, \ell - 1\}$,*

   (a) $\text{null}\,(T - \lambda I)^j + 1 \leq \text{null}\,(T - \lambda I)^{j+1} \leq \text{null}\,(T - \lambda I)^j + \gamma$.
   (b) $\text{null}\,(T - \lambda I)^{j+1} - \text{null}\,(T - \lambda I)^j \leq \text{null}\,(T - \lambda I)^j - \text{null}\,(T - \lambda I)^{j-1}$.

(8) $\ell + \gamma - 1 \leq \mu \leq \ell\gamma$.

Proof (1) Due to Theorem 2.25, $\dim(N(T - \lambda I)^\ell) + \dim(R(T - \lambda I)^\ell) = \dim(V)$. To show $N(T - \lambda I)^\ell \cap R(T - \lambda I)^\ell = \{0\}$, let $x \in N(T - \lambda I)^\ell \cap R(T - \lambda I)^\ell$. Then, $(T - \lambda I)^\ell x = 0$ and there exists $u \in V$ such that $x = (T - \lambda I)^\ell u$. Now, $(T - \lambda I)^{2\ell} u = (T - \lambda I)^\ell x = 0$; so that $u \in N(T - \lambda I)^{2\ell} = N(T - \lambda I)^\ell$, since ℓ is the ascent of λ. Then $x = (T - \lambda I)^\ell u = 0$.

(2) This follows from Theorem 5.20 and (1).

(3) If v is an eigenvector of T corresponding to λ, then $v \in E(\lambda) \subseteq G(\lambda)$. For such a vector v, $T_G(v) = Tv = \lambda v$. Therefore, $\lambda \in \text{eig}(T_G)$.

Conversely, suppose that the scalar $\alpha \in \text{eig}(T_G)$. Let $v \in G$, $v \neq 0$ be an eigenvector corresponding to α. Then $T_G(v) = \alpha v$. Also, $T_G(v) = Tv = \lambda v$. This implies $(\alpha - \lambda)v = 0$ for nonzero v. Therefore, $\lambda = \alpha$.

For the second statement, if $\lambda \in \text{eig}(T|_{R(T-\lambda I)^\ell})$, then there exists a nonzero $v \in R(T - \lambda I)^\ell$ such that $T|_{R(T-\lambda I)^\ell}(v) = Tv = \lambda v$. Then $v \in N(T - \lambda I) \subseteq G(\lambda)$. That is, $v \in N(T - \lambda I)^\ell \cap R(T - \lambda I)^\ell$. By (1), $v = 0$, a contradiction.

(4) Let $p(t)$, $q(t)$ and $r(t)$ be the characteristic polynomials of the linear operators T, $T|_{G(\lambda)}$ and of $T|_{R(T-\lambda I)^\ell}$, respectively. By (2), $p(t) = q(t)r(t)$. Since λ is the only eigenvalue of $T|_{G(\lambda)}$ and λ is not an eigenvalue of $T|_{R(T-\lambda I)^\ell}$, $(t - \lambda)$ is a factor of $q(t)$ and $(t - \lambda)$ is not a factor of $r(t)$. Moreover $(t - \lambda)^\mu$ divides $p(t)$, and $(t - \lambda)^{\mu+1}$ is not a factor of $p(t)$. As $q(t)$ is a monic polynomial, $q(t) = (t - \lambda)^\mu$. That is, the characteristic polynomial of $T|_G$ is $(t - \lambda)^\mu$. It shows that $\mu = \dim(G(\lambda))$.

Since $G(\lambda) = N(T - \lambda I)^\ell = N(T - \lambda I)^n$, it follows that $\mu = \text{null}\,(T - \lambda I)^n$.

(5) Let $\alpha \in \mathbb{F}$ and $\alpha \neq \lambda$. Let $v \in N(T - \alpha I)^j \cap G(\lambda)$. We prove by induction on j that $v = 0$ for each $j \in \mathbb{N}$.

For $j = 1$, let $v \in N(T - \alpha I) \cap G(\lambda)$. That is, $Tv = \alpha v$ where $v \in G(\lambda)$. If $v \neq 0$, then α is an eigenvalue of T_G with eigenvector v. This contradicts (4) since $\alpha \neq \lambda$.

Assume that $N(T - \alpha I)^k \cap G(\lambda) = \{0\}$ for some $k \in \mathbb{N}$. We need to show that $N(T - \alpha I)^{k+1} \cap G(\lambda) = \{0\}$. For this, let $v \in N(T - \alpha I)^{k+1} \cap G(\lambda)$. Then

$$0 = (T - \alpha I)^{k+1} v = (T - \alpha I)^k (T - \alpha I) v.$$

That is $u := (T - \alpha I)v \in N(T - \alpha I)^k$. Moreover, $v \in G(\lambda) = N(T - \lambda I)^\ell$ implies $(T - \lambda I)^\ell v = 0$. Then

$$(T - \lambda I)^\ell (T - \lambda I) v = (T - \lambda I)(T - \lambda I)^\ell v = 0.$$

So, $u = (T - \lambda I)v \in G(\lambda)$. Thus, $u = (T - \lambda I)v \in N(T - \alpha I)^k \cap G(\lambda) = \{0\}$. Hence, $(T - \alpha I)v = u = 0$. Then $v \in N(T - \lambda I) \cap G(\lambda) \subseteq N(T - \lambda I)^k \cap G(\lambda) = \{0\}$; and $v = 0$.

(6) Clearly, $(T - \lambda I)v_1, \ldots, (T - \lambda I)v_r$ are in $N(T - \lambda I)^j$. Suppose that for some scalars $\alpha_1, \ldots, \alpha_r$, $\sum_{i=1}^r \alpha_i (T - \lambda I)v_i = 0$. Then

$$\sum_{i=1}^r \alpha_i v_i \in N(T - \lambda I) \subseteq N(T - \lambda I)^j.$$

As $\operatorname{span}\{v_1, \ldots, v_r\} \cap N(T - \lambda I)^j = \{0\}$, we have $\sum_{i=1}^r \alpha_i v_i = 0$. Since $v_1, \ldots, v_r$ are linearly independent, $\alpha_i = 0$ for each i. Thus $(T - \lambda I)v_1, \ldots, (T - \lambda I)v_r$ are linearly independent vectors in $N(T - \lambda I)^j$.

Next, let $x \in \operatorname{span}\{(T - \lambda I)v_1, \ldots, (T - \lambda I)v_r\} \cap N(T - \lambda I)^{j-1}$. Then there exist scalars $\beta_1, \ldots, \beta_r$ such that

$$x = \sum_{i=1}^r \beta_i (T - \lambda I)v_i, \quad (T - \lambda I)^{j-1} x = 0.$$

It follows that

$$(T - \lambda I)^j \Big(\sum_{i=1}^r \beta_i v_i \Big) = \sum_{i=1}^r \beta_i (T - \lambda I)^j v_i = (T - \lambda I)^{j-1} x = 0.$$

That is, $\sum_{i=1}^r \beta_i v_i \in \operatorname{span}\{v_1, \ldots, v_r\} \cap N(T - \lambda I)^j$. Consequently, $\sum_{i=1}^r \beta_i v_i = 0$. Due to linear independence of $v_1, \ldots, v_r$, each β_i is zero. It then follows that $x = 0$. Therefore, $\operatorname{span}\{(T - \lambda I)v_1, \ldots, (T - \lambda I)v_r\} \cap N(T - \lambda I)^{j-1} = \{0\}$.

(7) Let $j \in \{1, \ldots, \ell - 1\}$ for $\ell \geq 2$. Then $\dim(N(T - \lambda I)^j) < \dim(N(T - \lambda I)^{j+1}$. This proves the first inequality.

For the second inequality, let E be a basis of $N(T - \lambda I)^j$. Extend this to a basis $E \cup \{u_1, \ldots, u_k\}$ for $N(T - \lambda I)^{j+1}$. Then $\{u_1, \ldots, u_k\} \cap N(T - \lambda I)^j = \{0\}$. Therefore, by (6) above, $(T - \lambda I)^j u_1, \ldots, (T - \lambda I)^j u_k$ are linearly independent in $N(T - \lambda I)$. Consequently, $k \leq \dim(E(\lambda)) = \gamma$ and

$$\dim(N(T - \lambda I)^{j+1}) = \dim(N(T - \lambda I)^j + k \leq \dim(N(T - \lambda I)^j + \gamma.$$

This proves (a). For (b), notice that $|E| = \operatorname{null} N(T - \lambda I)^j$ and

$$k = \operatorname{null} N(T - \lambda I)^{j+1} - \operatorname{null} N(T - \lambda I)^j.$$

By (6), $\operatorname{span}\{(T - \lambda I)u_1, \ldots, (T - \lambda I)u_k\} \cap N(T - \lambda I)^{j-1} = \{0\}$. Hence if B is any basis of $N(T - \lambda I)^{j-1}$, then the basis extension theorem implies that there exists a basis for $N(T - \lambda I)^j$ that contains $B \cup \{(T - \lambda I)u_1, \ldots, (T - \lambda I)u_k\}$. It follows that $|B| + k \leq \operatorname{null}(T - \lambda I)^j$. Therefore,

$$\operatorname{null}(T - \lambda I)^{j-1} + (\operatorname{null}(T - \lambda I)^{j+1} - \operatorname{null}(T - \lambda I)^j) \leq \operatorname{null}(T - \lambda I)^j.$$

(8) If $\ell = 1$, then $G(\lambda) = E(\lambda)$. By (4), $\mu = \gamma$. Therefore, $\ell + \gamma - 1 \le \mu \le \ell\gamma$. So, assume that $\ell \ge 2$. Look at the dimensions of the subspaces in

$$E(\lambda) = N(T - \lambda I) \subsetneqq N(T - \lambda I)^2 \subsetneqq \cdots \subsetneqq N(T - \lambda I)^\ell = G(\lambda).$$

For each i, $\dim\big(N(T - \lambda I)^{i+1}\big) \ge \dim\big(N(T - \lambda I)^i\big) + 1$; there are $\ell - 1$ such containments. Hence $\gamma + \ell - 1 \le \mu$. Due to (6),

$$\begin{aligned}\mu = \dim(G(\lambda)) &= \dim(N(T - \lambda I)^\ell) \le \dim(N(T - \lambda I)^{\ell-1} + \gamma \\ &\le \dim(N(T - \lambda I)^{\ell-2} + 2\gamma \le \ \cdots \ \le \dim(N(T - \lambda I) + (\ell - 1)\gamma \le \ell\gamma. \blacksquare\end{aligned}$$

Notice that the ascent of an eigenvalue λ is not known beforehand. Applying Theorem 5.27(6) one constructs bases for $N(T - \lambda I)$, $N(T - \lambda I)^2, \ldots,$ in succession, where a basis for $N(T - \lambda I)^i$ is extended to a basis for $N(T - \lambda I)^{i+1}$. By Theorems 5.23(1) and 5.27(4), if $\text{null}\,(T - \lambda I)^i < \mu$, then we have $\text{null}\,(T - \lambda I)^i < \text{null}\,(T - \lambda I)^{i+1}$. The process of extending a basis for $N(T - \lambda I)^i$ to that of $N(T - \lambda I)^{i+1}$ stops when for some $j \le \mu$, the basis for $N(T - \lambda I)^j$ contains μ number of vectors. In that case, j is the ascent of λ. We interpret this fact for $n \times n$ matrices.

Theorem 5.28 *Let $A \in \mathbb{C}^{n\times n}$. Let λ be an eigenvalue of A of algebraic multiplicity μ. Let $j, k \in \{1, \ldots, \mu - 1\}$. If the linear system $(A - \lambda I)^j x = 0$ has k linearly independent solutions, then the linear system $(A - \lambda I)^{j+1} x = 0$ has at least $k + 1$ linearly independent solutions. Further, the linear system $(A - \lambda I)^\mu x = 0$ has μ linearly independent solutions.*

Theorem 5.28 helps in constructing the exponential of a matrix, which is used in solving a system of linear ordinary differential equations; see Problem 29.

Due to Theorem 5.27, the restricted linear operator $T|_{G(\lambda)}$ maps $G(\lambda)$ into $G(\lambda)$. This has bearing on the matrix representation of $T|_{G(\lambda)}$. In any chosen basis of $T|_{G(\lambda)}$, the matrix representation $[T|_{G(\lambda)}]$ of $T|_{G(\lambda)}$ is an $\mu \times \mu$ matrix.

On the other hand, $T|_{G(\lambda)}$ is also a map from $G(\lambda)$ to V. If $\dim(V) = n$ and $\dim(G(\lambda)) = \mu$, then in any chosen bases for the spaces, the matrix representation of $T|_{G(\lambda)}$ is an $n \times \mu$ matrix. Also, $R(T|_{G(\lambda)}) \subseteq G(\lambda)$ says that we may as well start with a basis of $G(\lambda)$ and extend it to a basis of V; in these bases, the $n \times \mu$ matrix will have zero entries on $(\mu + 1)$th row onwards. In fact, in this view, the linear transformation $T|_{G(\lambda)} : G(\lambda) \to V$ is nothing but the linear transformation $T : G(\lambda) \to V$.

Thus, we stick to the former view of $T|_{G(\lambda)}$ as a linear operator on $G(\lambda)$ and put forth the following convention.

Convention 5.1 Let λ be an eigenvalue of a linear operator T on a finite dimensional vector space V. The restriction map $T|_{G(\lambda)}$ is regarded as a linear operator on $G(\lambda)$. Consequently, if λ has algebraic multiplicity μ, then with respect to any basis of $G(\lambda)$, $[T|_{G(\lambda)}] \in \mathbb{F}^{\mu\times\mu}$.

Splitting of the characteristic polynomial of a linear operator allows writing the vector space as a direct sum of the generalized eigenspaces; see the following theorem.

Theorem 5.29 *Let $\lambda_1, \ldots, \lambda_k$ be the distinct eigenvalues of a linear operator T on a vector space V of dimension n over $\mathbb{F}$. If the characteristic polynomial of T splits over $\mathbb{F}$, then*

$$V = G(\lambda_1) \oplus \cdots \oplus G(\lambda_k) \quad and \quad \mu(\lambda_1) + \cdots + \mu(\lambda_k) = n.$$

Proof Assume that the characteristic polynomial of T splits over $\mathbb{F}$. We prove the theorem by induction on k, the number of distinct eigenvalues of T. For $k = 1$, λ_1 is the only eigenvalue of T. Since $\chi_T(t)$ splits, $\mu(\lambda_1) = n$. But $G(\lambda_1) \subseteq V$, $\dim(G(\lambda_1)) = \mu(\lambda_1)$, and $\dim(V) = n$. Therefore, $G(\lambda_1) = V$.

Assume the induction hypothesis that for $k = m$, the theorem is true. Let T be a linear operator on a vector space V of dimension n having distinct eigenvalues $\lambda_1, \ldots, \lambda_m, \lambda$, with algebraic multiplicities $\mu_1, \ldots, \mu_m, \mu$, respectively. Let ℓ be the ascent of λ. By Theorem 5.27, $V = G(\lambda) \oplus H(\lambda)$, where $H(\lambda) = R(T - \lambda I)^\ell$; and both $G(\lambda)$ and $H(\lambda)$ are T-invariant subspaces of V. Consider the restriction linear operators $T_G := T|_{G(\lambda)}$ and $T_H := T|_{H(\lambda)}$. Again, by Theorem 5.27, the characteristic polynomial of T is the product of those of T_G and T_H; λ is not an eigenvalue of T_H, and no λ_j is an eigenvalue of T_G. Hence T_G has single eigenvalue λ with algebraic multiplicity μ, and T_H has m distinct eigenvalues $\lambda_1, \ldots, \lambda_m$ with algebraic multiplicities $\mu_1, \ldots, \mu_m$, respectively. By the induction hypothesis,

$$H(\lambda) = G_{T_H}(\lambda_1) \oplus \cdots \oplus G_{T_H}(\lambda_m),$$

where $G_{T_H}(\lambda_j)$ is the generalized eigenspace of T_H corresponding to the eigenvalue λ_j. Since T_H is a restriction of T, for each $j \in \{1, \ldots, m\}$, we have

$$G_{T_H}(\lambda_j) = N(T_H - \lambda_j I)^{\mu_j} \subseteq N(T - \lambda_j I)^{\mu_j} = G_T(\lambda_j) = G(\lambda_j).$$

Further, $\dim(G_{T_H}(\lambda_j)) = \mu_j = \dim(G(\lambda_j))$. Hence $G_{T_H}(\lambda_j) = G(\lambda_j)$. Consequently, $H(\lambda) = G(\lambda_1) \oplus \cdots \oplus G(\lambda_m)$. Therefore,

$$V = G(\lambda) \oplus H(\lambda) = G(\lambda_1) \oplus \cdots \oplus G(\lambda_m) \oplus G(\lambda).$$

This completes the induction proof of $V = G(\lambda_1) \oplus \cdots \oplus G(\lambda_k)$. Their dimensions yield $\mu(\lambda_1) + \cdots + \mu(\lambda_k) = n$. ■

Some of the important equalities and inequalities concerning various parameters can be summarized as follows. Let $T : V \to V$ be a linear operator, where $\dim(V) = n$. If $\chi_T(t) = (t - \lambda_1)^{m_1} \cdots (t - \lambda_k)^{m_k}$, then for $i \in \{1, \ldots, k\}$, we have

$$
\begin{aligned}
m_1 + \cdots + m_k &= n, \\
\gamma(\lambda_i) &= \text{null}\,(T - \lambda_i I), \\
1 \le \ell\,(\lambda_i),\ \gamma(\lambda_i) &\le \mu(\lambda_i) = m_i \le n, \\
\ell(\lambda_i) + \gamma(\lambda_i) - 1 &\le m_i \le \ell(\lambda_i)\gamma(\lambda_i), \\
\text{null}\,(T - \lambda_i I)^{j+1} - \text{null}\,(T - \lambda_i I)^j &\le \text{null}\,(T - \lambda_i I)^j - \text{null}\,(T - \lambda_i I)^{j-1}, \\
\mu(\lambda_i) = \dim\big(N(T - \lambda_i I)^{\ell(\lambda_i)}\big) &= \dim\big(N(T - \lambda_i I)^{m_i}\big) = \dim\big(N(T - \lambda_i I)^n\big).
\end{aligned}
$$

Since the finite dimensional vector space V is a direct sum of the generalized eigenspaces and the generalized eigenspaces are T-invariant subspaces, we see that the linear operator T can be written as a direct sum of its restrictions to the generalized eigenspaces. We write it as

$$T = T|_{G(\lambda_1)} \oplus \cdots \oplus T|_{G(\lambda_k)},$$

where $\lambda_1, \ldots, \lambda_k$ are the distinct eigenvalues of T. It says that any vector $v \in V$ and its image under T can be written uniquely as

$$v = v_1 + \cdots + v_k, \quad Tv = T|_{G(\lambda_1)}(v_1) + \cdots + T|_{G(\lambda_k)}(v_k),$$

where $v_i \in G(\lambda_i)$ and $T|_{G(\lambda_i)}(v_i) \in G(\lambda_i)$. This gives rise to the block-diagonal form of the matrix representation of T, which we will discuss later so that more structure may be given to the blocks. Note that this direct sum decomposition of T is guaranteed when the characteristic polynomial of T splits; it may break down when T is a linear operator on a real vector space having a nonreal characteristic value.

Exercises for Sect. 5.4

1. Determine the generalized eigenvectors of the linear operator T on $\mathbb{C}^2$, where (a) $T(\alpha, \beta) = (\beta, 0)$ (b) $T(\alpha, \beta) = (-\beta, \alpha)$.
2. Let T be a linear operator on a 2-dimensional vector space V. Suppose that the matrix of T with respect to some basis is equal to $\begin{bmatrix} 0 & 0 \\ 0 & 1 \end{bmatrix}$. What are the T-invariant subspaces of V?
3. Let D be the differentiation operator on $\mathcal{P}_n(\mathbb{R})$. Let $m \le n$. Is $\mathcal{P}_m(\mathbb{R})$ invariant under D? Is D invertible on $\mathcal{P}_m(\mathbb{R})$?
4. Let U and W be T-invariant subspaces of a vector space V for a linear operator T on V. Is $U + W$ invariant under T?
5. Let V be a finite dimensional vector space. Let $S, T : V \to V$ be linear operators such that $ST = TS$. Show that for every $\alpha \in \mathbb{F}$, $N(T - \alpha I)$ is S-invariant.
6. Let T be a linear operator on a vector space V of dimension n. Show that $V = R(T^0) \supseteq R(T^1) \supseteq R(T^2) \supseteq \cdots \supseteq R(T^n) = R(T^{n+i})$ for each $i \in \mathbb{N}$.

5.5 Two Annihilating Polynomials

Let T be a linear operator on a finite dimensional vector space V. In the proof of Theorem 5.4, we have shown that corresponding to a nonzero vector v, a polynomial $p(t)$ exists such that $p(T)v = 0$. This technique can be extended to show that there exists a polynomial $p(t)$ such that $p(T) = 0$, the zero operator. To see this, let T be a nonzero linear operator on a vector space V of dimension n. Since $\mathcal{L}(V, V)$ is of dimension $m = n^2$, the linear operators $I, T, \ldots, T^m$ are linearly dependent. Then we have scalars $\alpha_0, \alpha_1, \ldots, \alpha_m$, not all zero, such that

$$\alpha_0 I + \alpha_1 T + \cdots + \alpha_m T^m = 0.$$

With $p(t) = \alpha_0 + \alpha_1 t + \cdots + \alpha_m t^m$, we see that $p(T) = 0$. Informally, we say that the polynomial $p(t)$ *annihilates* the linear operator T. Notice that starting with a nonzero vector v, when we obtain $p(t)v = 0$, the degree of such a polynomial $p(t)$ does not exceed n. Whereas the polynomial $p(t)$ that annihilates T can be of degree anywhere between 1 to n^2. Can we get a polynomial of degree at most n that annihilates T?

Theorem 5.30 (Cayley–Hamilton) *If T is a linear operator on a finite dimensional vector space, then $\chi_T(T) = 0$.*

Proof Let V be a vector space of dimension n. Fix a basis E for V. Let $A = [T]_{E,E}$. Write the characteristic polynomial $\chi_T(t)$ as $p(t)$. Then

$$p(t) = (-1)^n \det(A - t\,I) \quad \text{and} \quad [p(T)]_{E,E} = p(A).$$

We show that $p(A) = 0$, the zero matrix.

Theorem 3.30(14) with the matrix $(A - t\,I)$, in place of A there, asserts that

$$p(t)\,I = (-1)^n \det(A - tI)\,I = (-1)^n (A - tI)\,\mathrm{adj}\,(A - tI).$$

The entries in adj $(A - tI)$ are polynomials in t of degree at most $n - 1$. Therefore, adj $(A - tI)$ can be written as

$$\mathrm{adj}\,(A - tI) := B_0 + tB_1 + \cdots + t^{n-1} B_{n-1},$$

where $B_0, \ldots, B_{n-1} \in \mathbb{F}^{n \times n}$. Then

$$p(t)I = (-1)^n (A - t\,I)(B_0 + tB_1 + \cdots + t^{n-1} B_{n-1}).$$

This is an identity in polynomials, where the coefficients of t^j are matrices. Substituting t by any matrix of the same order will satisfy the equation. In particular, substitute A for t to obtain $p(A) = 0$. It then follows that $p(T) = 0$. ■

Using Cayley–Hamilton theorem, inverse of a linear operator T on a vector space of dimension n can be expressed as a polynomial in T of degree $n-1$. Note that the characteristic polynomial of T, which has degree n, annihilates T. It is quite possible that there exists an annihilating polynomial of yet smaller degree. While searching for such annihilating polynomials we may restrict our attention to monic polynomials.

Definition 5.31 Let T be a linear operator on a finite dimensional vector space. A monic polynomial $q(t)$ with coefficients from $\mathbb{F}$ is called a **minimal polynomial** of T if $q(T)=0$, and whenever $p(t)$ is any polynomial with $p(T)=0$, the degree of $q(t)$ is less than or equal to the degree of $p(t)$.

A minimal polynomial is a monic polynomial of least degree that annihilates T. Here is a basic result on minimal polynomials.

Theorem 5.32 *Let T be a linear operator on a finite dimensional vector space. Let $q(t)$ be a minimal polynomial of T. If $p(t)$ is any polynomial with $p(T)=0$, then $q(t)$ divides $p(t)$. Further, a minimal polynomial of T is unique.*

Proof Let $p(t)$ be a polynomial with $p(T)=0$. Since $q(t)$ is a minimal polynomial of T, degree of $q(t)$ is less than or equal to degree of $p(t)$. Assume the contrary that $q(t)$ does not divide $p(t)$. We can write $p(t)$ in the following form:

$$p(t)=d(t)q(t)+\alpha r(t),$$

for some monic polynomials $d(t)$, $r(t)$, and a nonzero scalar α. Also, the degree of $r(t)$ is less than the degree of $q(t)$. Notice that since $p(T)=q(T)=0$, the polynomial $r(t)$ is not a constant polynomial and $r(T)=0$. This contradicts the fact that $q(t)$ is the minimal polynomial of T.

Further, if $q(t)$ and $s(t)$ are two minimal polynomials, then both of them have the same degree, each divides the other, and each is a monic polynomial. Therefore, $s(t)=q(t)$. ■

It now follows from Theorem 5.30 that the minimal polynomial divides the characteristic polynomial. Sometimes, this is referred to as the Cayley–Hamilton theorem. There are linear operators for which the minimal polynomial coincides with the characteristic polynomial. For instance, the companion matrix of any polynomial $p(t)$ has both the characteristic polynomial and the minimal polynomial as $p(t)$; prove it!

The minimal polynomial is a theoretical tool since eigenvalues of a linear operator are also the zeros of the minimal polynomial; see the following theorem.

Theorem 5.33 *Let $q(t)$ be the minimal polynomial of a linear operator T on a finite dimensional vector space. A scalar λ is an eigenvalue of T if and only if $q(\lambda)=0$.*

Proof Let λ be an eigenvalue of T. By Theorem 5.17(1), $q(\lambda)$ is an eigenvalue of $q(T)$. Since $q(T)=0$, $q(\lambda)=0$.

Conversely, suppose α is a scalar such that $q(\alpha) = 0$. Then $q(t) = (t - \alpha)r(t)$, for some polynomial $r(t)$. If $r(T) = 0$, then $q(t)$ must divide $r(t)$. But this is impossible since degree of $r(t)$ is less than that of $q(t)$. Thus, $r(T)$ is not the zero linear operator. Therefore, there exists a vector $v \in V$ such that $r(T)v \neq 0$. Since $q(T) = 0$, we have $q(T)v = 0$. That is, $(T - \alpha I)r(T)v = 0$. So, α is an eigenvalue of T with a corresponding eigenvector $r(T)v$. ■

Theorem 5.33 does not say that each zero of the minimal polynomial is an eigenvalue. If the characteristic polynomial splits, then each zero of the minimal polynomial is an eigenvalue. In the interesting case of the splitting of the characteristic polynomial of T, the minimal polynomial can be written down explicitly.

Theorem 5.34 *Let T be a linear operator on a finite dimensional vector space V over $\mathbb{F}$. Let $\lambda_1, \ldots, \lambda_k$ be the distinct eigenvalues of T with ascents $\ell_1, \ldots, \ell_k$, respectively. If the characteristic polynomial of T splits over $\mathbb{F}$, then the minimal polynomial of T is $(t - \lambda_1)^{\ell_1} \cdots (t - \lambda_k)^{\ell_k}$.*

Proof Let $i \in \{1, \ldots, k\}$. Write $h(t) = (t - \lambda_1)^{\ell_1} \cdots (t - \lambda_k)^{\ell_k}$. For each generalized eigenvector $x \in G(\lambda_i)$, $(T - \lambda_i I)^{\ell_i}(x) = 0$. Since for each $j \in \{1, \ldots, k\}$,

$$(T - \lambda_i I)^{\ell_i}(T - \lambda_j I)^{\ell_j} = (T - \lambda_j I)^{\ell_j}(T - \lambda_i I)^{\ell_i},$$

we see that $h(T)x = 0$. Since the characteristic polynomial of T splits over $\mathbb{F}$, by Theorem 5.29, $V = G(\lambda_i) \oplus \cdots \oplus G(\lambda_k)$. Therefore, $h(T)v = 0$ for each $v \in V$. That is, $h(T) = 0$. Theorem 5.33 implies that the minimal polynomial $q(t)$ of T is in the form

$$q(t) = (t - \lambda_1)^{n_1} \cdots (t - \lambda_k)^{n_k}, \quad \text{where } 1 \leq n_1 \leq \ell_1, \ldots, 1 \leq n_k \leq \ell_k.$$

We need to show that $n_1 = \ell_1, \ldots, n_k = \ell_k$. On the contrary, suppose that for some i, $1 \leq n_i < \ell_i$. Then

$$N(T - \lambda_i I)^{n_i} \subseteq N(T - \lambda_i I)^{\ell_i - 1} \subsetneqq N(T - \lambda_i I)^{\ell_i} = G(\lambda_i).$$

Consider a vector $u \in G(\lambda_i) \setminus N(T - \lambda_i I)^{n_i}$. Then $v := (T - \lambda_i I)^{n_i} u \neq 0$. Notice that $v \in G(\lambda_i)$. Theorem 5.29 implies that for any $j \neq i$,

$$G(\lambda_i) \cap N(T - \lambda_j I)^{n_j} \subseteq G(\lambda_i) \cap G(\lambda_j) = \{0\}.$$

Thus, if $x \in G(\lambda_i)$, then $(T - \lambda_j I)^{n_j} x = 0$ implies $x = 0$. That is, the map $(T - \lambda_j I)^{n_j}$ is injective on $G(\lambda_i)$. Then the composition map

$$S := (T - \lambda_1 I)^{n_1} \cdots (T - \lambda_{i-1} I)^{n_{i-1}} (T - \lambda_{i+1} I)^{n_{i+1}} \cdots (T - \lambda_k I)^{n_k}$$

is injective on $G(\lambda_i)$. So, $v \neq 0$ implies $Sv \neq 0$. But $Sv = S(T - \lambda_i I)^{n_i} u = q(T)u$. Thus, $q(T)u \neq 0$. This contradicts the fact that $q(T) = 0$. ■

The minimal polynomial can be constructed by determining the least k such that the ordered set of linear operators $\{I, T, T^2, \ldots, T^k\}$ is linearly dependent. In that case, we have the scalars α_i such that

$$T^k + \alpha_{k-1}T^{k-1} + \cdots + \alpha_1 T + \alpha_0 I = 0.$$

Then the polynomial $t^k + \alpha_{k-1}t^{k-1} + \cdots + \alpha_1 t + \alpha_0$ is the minimal polynomial. Due to Cayley–Hamilton Theorem, this k cannot exceed n, the dimension of the domain space V. Computation may become easier if we use a matrix representation $[T]$ of T instead.

Also, due to Cayley–Hamilton Theorem, the minimal polynomial can be determined by writing down all the factors of the characteristic polynomial systematically. For example, if the characteristic polynomial of T is $t^2(t-1)^3$, then the minimal polynomial can be one of

$$t(t-1),\ t^2(t-1),\ t(t-1)^2,\ t^2(t-1)^2.$$

Then, one checks starting from degree one polynomials from among these, which one annihilates the linear operator.

Another alternative is to compute the ascents of each eigenvalue of the matrix $[T]$ and then form the minimal polynomial using the expression in Theorem 5.34. Unfortunately, no better method for determining the minimal polynomial is available.

Exercises for Sect. 5.5

1. Verify Cayley–Hamilton theorem for the matrix $\begin{bmatrix} 7 & -1 \\ 2 & 4 \end{bmatrix}$.
2. Let A be a 2×2 matrix with $A^2 = I$ but $A \neq \pm I$. Show that $\operatorname{tr}(A) = 0$ and $\det(A) = -1$.
3. Show that if a matrix is invertible, then its inverse can be expressed as a polynomial in the matrix.
4. Compute A^{30}, where $A = \begin{bmatrix} 2 & 1 \\ 1 & 2 \end{bmatrix}$.
5. Let T be a linear operator on a vector space of dimension n. Suppose that 1 and 2 are the only eigenvalues of T. Show that $(T - I)^{n-1}(T - 2I)^{n-1} = 0$.
6. What are the minimal polynomials of the zero operator and the identity operator on an n-dimensional vector space?
7. Show that $t(t+2)(t-2)$ is the minimal polynomial of $\begin{bmatrix} 0 & 1 & 0 & 1 \\ 1 & 0 & 1 & 0 \\ 0 & 1 & 0 & 1 \\ 1 & 0 & 1 & 0 \end{bmatrix}$.
8. Find the minimal polynomials of the following matrices:

$$\text{(a)} \begin{bmatrix} 0 & 0 & 3 \\ 0 & 0 & 0 \\ 0 & 0 & 4 \end{bmatrix} \quad \text{(b)} \begin{bmatrix} 1 & 0 & 1 & 0 \\ 0 & 1 & 0 & 1 \\ 1 & 0 & 1 & 0 \\ 0 & 1 & 0 & 1 \end{bmatrix} \quad \text{(c)} \begin{bmatrix} 0 & 0 & 0 & 0 \\ 1 & 1 & 1 & 1 \\ 1 & 1 & 1 & 1 \\ 1 & 1 & 1 & 1 \end{bmatrix} \quad \text{(d)} \begin{bmatrix} 0 & 2 & 0 & 0 \\ 0 & 0 & 2 & 0 \\ 0 & 0 & 0 & 5 \\ 2 & 1 & 0 & 0 \end{bmatrix}.$$

9. Find the minimal polynomial of the matrix $\begin{bmatrix} 0 & 0 & c \\ 1 & 0 & b \\ 0 & 1 & a \end{bmatrix}$, where $a, b, c \in \mathbb{R}$.
10. Give a linear operator on $\mathbb{R}^3$ whose minimal polynomial is t^2.
11. Give a linear operator on $\mathbb{C}^4$ whose minimal polynomial is $t(t-1)^2$.
12. Give a linear operator on $\mathbb{R}^4$ whose characteristic polynomial and the minimal polynomial are equal to $t(t-1)^2(t-2)$.
13. Construct a matrix in $\mathbb{C}^{4\times 4}$ whose minimal polynomial is t^3.
14. Give a linear operator on $\mathbb{R}^4$ whose minimal polynomial is $t(t-1)(t-2)$ and characteristic polynomial is $t(t-1)^2(t-2)$.
15. What could be the minimal polynomial of a diagonal matrix?
16. Let $a_1, \ldots, a_n \in \mathbb{F}$. Determine the characteristic and minimal polynomials of the linear operator $T : \mathbb{F}^n \to \mathbb{F}^n$ given by

$$T(b_1, \ldots, b_n) = (-a_1 b_n, b_1 - a_2 b_n, b_2 - a_3 b_n, \ldots, b_{n-1} - a_n b_n).$$

17. What is the minimal polynomial of the differentiation operator on $\mathcal{P}_n(\mathbb{R})$?
18. Find the minimal polynomial of the operataor $T : \mathcal{P}_n(\mathbb{R}) \to \mathcal{P}_n(\mathbb{R})$ given by $Tp(t) = p(t+1)$.
19. Is it true that a linear operator is invertible if and only if the constant term in its minimal polynomial is nonzero?
20. Let $A, B \in \mathbb{C}^{n\times n}$ be two similar matrices. Show that their characteristic polynomials coincide, and that their minimal polynomials coincide.

5.6 Problems

1. Given the scalars $a_0, a_1, \ldots, a_{n-1}$, let

$$C = \begin{bmatrix} -a_{n-1} & -a_{n-2} & \cdots & -a_1 & -a_0 \\ 1 & 0 & \cdots & 0 & 0 \\ 0 & 1 & \cdots & 0 & 0 \\ \vdots & \vdots & & \vdots & \vdots \\ 0 & 0 & \cdots & 1 & 0 \end{bmatrix}.$$

Show that $\det(C - tI) = p(t) := (-1)^n(a_0 + a_1 t + \cdots + a_{n-1}t^{n-1} + t^n)$. This matrix C is yet another form of the companion matrix of $p(t)$.

2. For a matrix $A \in \mathbb{C}^{n\times n}$, are the following statements true?
(a) If A is hermitian, then $\det(A) \in \mathbb{R}$. (b) If A is unitary, then $|\det(A)| = 1$.
3. Without using the characteristic polynomial, prove that every linear operator on a vector space of dimension n can have at most n distinct eigenvalues. Deduce that every polynomial of degree n with complex coefficients can have at most n complex zeros.

4. Show that all zeros of the polynomial $p(t) = a_0 + a_1 t + \cdots + a_n t^n$, where $a_0, \ldots, a_n$ are complex numbers, lie in a disc (in the complex plane) with centre 0 and radius $\max\{|a_0|, 1 + |a_1|, \ldots, 1 + |a_n|\}$.
5. Let $t^3 - \alpha t^2 + \beta t - \gamma = 0$ be the characteristic equation of a 3×3 matrix $[a_{ij}]$. Show that
$$\beta = \begin{vmatrix} a_{11} & a_{12} \\ a_{21} & a_{22} \end{vmatrix} + \begin{vmatrix} a_{11} & a_{13} \\ a_{31} & a_{33} \end{vmatrix} + \begin{vmatrix} a_{22} & a_{23} \\ a_{32} & a_{33} \end{vmatrix}.$$
Generalize the result to $n \times n$ matrices.
6. Let $T : V \to V$ be a linear operator. Prove the following without using the spectral mapping theorem:
   (a) If λ is an eigenvalue of T then λ^k is an eigenvalue of T^k.
   (b) If λ is an eigenvalue of T and $\alpha \in \mathbb{F}$, then $\lambda + \alpha$ is an eigenvalue of $T + \alpha I$.
   (c) If $p(t) = a_0 + a_1 t + \cdots + a_k t^k$ for some $a_0, a_1, \ldots, a_k$ in $\mathbb{F}$, and if λ is an eigenvalue of T then $p(\lambda)$ is an eigenvalue of $p(T)$.
7. An *idempotent* matrix is a square matrix whose square is itself. Show that an eigenvalue of an idempotent matrix can be 0 or 1, nothing else.
8. A linear operator T is *nilpotent* if and only if $T^k = 0$ for some $k \in \mathbb{N}$. Show that 0 is the only eigenvalue of a nilpotent linear operator whereas a noninvertible linear operator has 0 as an eigenvalue. Construct noninvertible linear operators which are not nilpotent.
9. Let $A \in \mathbb{F}^{n \times n}$ satisfy $A^n = I$. Prove that the eigenvalues of A are the nth roots of unity.
10. Let $A \in \mathbb{F}^{n \times n}$ be such that $A^k = I$ for some $k \in \mathbb{N}$. Prove that if A has only one eigenvalue λ, then $A = \lambda I$.
11. Let A be an orthogonal 3×3 matrix with $\det(A) = -1$. Show that -1 is an eigenvalue of A.
12. Find the eigenvalues of the $n \times n$ *permutation matrix* whose ith column is $e_{\pi(i)}$, where $\pi : \{1, \ldots, n\} \to \{1, \ldots, n\}$ is a bijection.
13. Let $A \in \mathbb{R}^{n \times n}$ satisfy $A^T = A^2$. What are possible eigenvalues of A?
14. Let $T : V \to V$ be a linear operator of rank r. Show the following:
   (a) If T has k number of distinct nonzero eigenvalues, then $k \le r$.
   (b) If $\dim(V) = n < \infty$ and T has m number of distinct eigenvalues, then $m \le \min\{r + 1, n\}$.
15. Let T be a linear operator on a finite dimensional vector space V. Show that for any polynomial $p(t) \in \mathcal{P}(\mathbb{F})$, $N(p(T))$ is a T-invariant subspace of V.
16. Let W be a nontrivial proper subspace of a finite dimensional complex vector space V. Show that there exists a nonzero linear operator T on V such that W is T-invariant. Is such a T unique?
17. Prove that each linear operator on a real finite dimensional nonzero vector space has an invariant subspace of dimension at most 2.
18. Prove that the linear operator T on $C(\mathbb{R}, \mathbb{R})$ given by $(Tf)(t) = \int_0^t f(s)ds$ does not have an eigenvalue.

19. Let U be a finite dimensional subspace of an inner product space V. Let T be a linear operator on V. Prove that U is T-invariant if and only if $U^\perp$ is T^*-invariant.
20. Let T be a linear operator on a vector space V of dimension n. Prove that $V = R(T^n) \oplus N(T^n)$.
21. Let T be a linear operator on a finite dimensional vector space. Prove that there exists an $n \in \mathbb{N}$ such that $N(T^n) \cap R(T^n) = \{0\}$.
22. Let T be a linear operator on a complex vector space V of dimension n. Prove that if $N(T^{n-1}) \neq N(T^{n-2})$, then T has at most two distinct eigenvalues.
23. Let T be a linear operator on a finite dimensional real vector space V. If $T^2 = I$, then show that $V = \{v \in V : Tv = v\} \oplus \{v \in V : Tv = -v\}$.
24. Let S and T be linear operators on a finite dimensional vector space. Prove that ST and TS have the same eigenvalues.
25. Let $A, B \in \mathbb{F}^{n\times n}$ be such that $I - AB$ is invertible. Prove that $I - BA$ is also invertible. Then show that AB and BA have the same eigenvalues.
26. Let $A, B \in \mathbb{F}^{n\times n}$. Show that the characteristic polynomial of AB is same as that of BA.
27. Let S and T be linear operators on a vector space V of dimension n. Suppose S has n distinct eigenvalues and each eigenvector of S is also an eigenvector of T. Show that $ST = TS$.
28. Let $A \in \mathbb{R}^{n\times n}$ have all entries positive. Prove that A has an eigenvector all of whose entries are positive.
29. Let A be a 2×2 matrix with positive entries. Prove that A has two distinct real eigenvalues; and that the larger eigenvalue has an eigenvector in the first quadrant and the smaller eigenvalue has an eigenvector in the second quadrant.
30. Let A be a 2×2 matrix with positive entries. Let u and v be the first and second columns of A, respectively. Let S denote the first quadrant in the plane $\mathbb{R}^2$.

    (a) Show that if $v \in S$, then Av lies in the sector between the lines span$\{u\}$ and span$\{v\}$.
    (b) Show that $S \supseteq A(S) \supseteq A^2(S) \supseteq \cdots$.
    (c) Let $U = \cap_j A^j(S)$. Show that $A(U) = U$.
    (d) Draw a diagram showing above facts by taking a particular matrix A. See that U is a half-line.
    (e) Conclude that any nonzero vector in U is an eigenvector of A.

31. Let λ be an eigenvalue of a linear operator T on a vector space V. For $k \in \mathbb{N}$, Prove that $N((T - \lambda I)^k)$ and $R((T - \lambda I)^k)$ are T-invariant subspaces of V.
32. Let $\mathcal{C}$ be a collection of T-invariant subspaces of a vector space V, where $T \in \mathcal{L}(V, V)$. Show that $\cap_{X\in\mathcal{C}} X$ is T-invariant.
33. Let λ be an eigenvalue of a matrix $A \in \mathbb{F}^{n\times n}$. Show that any column of adj$(A - \lambda I)$ is an eigenvector of A corresponding to λ.
34. Let T be a linear operator on a finite dimensional vector space. Prove that either there exists a linear operator S on V such that $ST = I$ or there exists a linear operator S on V such that $ST = 0$.

35. Let T be a linear operator on a real vector space of dimension 2. Let $p(t)$ be a polynomial of degree 2. Prove that if T does not have an eigenvalue and if $p(t) \neq \alpha \chi_T(t)$ for any scalar α, then $p(T)$ is invertible.
36. Let T be a linear operator on a vector space V. Prove that if each nonzero vector in V is an eigenvector of T, then $T = \alpha I$ for some $\alpha \in \mathbb{F}$.
37. Let T be a linear operator on a finite dimensional vector space V. Show that if T has the same matrix with respect to every basis of V, then $T = \alpha I$ for some scalar α.
38. Let T be a linear operator on a real vector space V of dimension 1 or 2. Let $\alpha, \beta \in \mathbb{R}$. Show that $N((T - \alpha I)^2 + \beta^2)$ is either $\{0\}$ or V.
39. Show that the minimal polynomial of $A := \begin{bmatrix} 1 & 1 & 0 & 0 \\ -1 & -1 & 0 & 0 \\ -2 & -2 & 2 & 1 \\ 1 & 1 & -1 & 0 \end{bmatrix}$ is $\chi_A(t)$.
40. Let T be a linear operator on a finite dimensional vector space. Let v be a nonzero vector and let $p(t)$ be a nonzero polynomial of least degree such that $p(T)v = 0$. Show that $p(t)$ divides the minimal polynomial of T.
41. Show that the minimal polynomial of the companion matrix C_p of a monic polynomial $p(t)$ is $p(t)$ itself.
42. Let λ be an eigenvalue of a linear operator T on a finite dimensional vector space V. What is the linear operator $T|_{E(\lambda)}$?
43. Let T be a linear operator on a finite dimensional vector space V. Let U be a T-invariant subspace of V. Prove that the minimal polynomial of $T|_U$ divides the minimal polynomial of T.
44. Let $\lambda_1, \ldots, \lambda_k$ be the distinct eigenvalues of a diagonal matrix $D \in \mathbb{C}^{n \times n}$. Let V be the subspace of $\mathbb{C}^{n \times n}$ consisting of all matrices X such that $DX = XD$. Prove that $\dim(V) = (\mu(\lambda_1))^2 + \cdots + (\mu(\lambda_k))^2$.
45. Let T be a linear operator on a finite dimensional complex vector space V. Prove that for each $k \in \{1, \ldots, \dim(V)\}$, there exists a T-invariant subspace of V with dimension k.
46. Let $A \in \mathbb{C}^{n \times n}$. A *left eigenvalue* of A is a complex number λ for which there exists a nonzero vector $u \in \mathbb{C}^{1 \times n}$ such that $uA = \lambda u$. Such a row vector u is called a *left eigenvector* of A. Show that if x is a left eigenvector of A, then x^* is an eigenvector of A^*. Show also that the set of left eigenvalues is same as the eigenspectrum of A.
47. Let $A \in \mathbb{C}^{m \times m}$ and let $B \in \mathbb{C}^{n \times n}$. Define a linear operator T on $\mathbb{C}^{m \times n}$ by $T(X) = AXB$.

    (a) Let x and y be eigenvectors of A and B^T, respectively. Show how to construct an eigenvector of T.
    (b) Show how to determine eigenvalues of T from those of A and B.

48. Let $A \in \mathbb{C}^{n \times n}$. Define a linear operator T on $\mathbb{C}^{n \times n}$ by $T(X) = AX - XA$. Prove that rank $(T) \leq n^2 - n$. Also, determine the eigenvalues of T in terms of the eigenvalues of A.

49. Let $A \in \mathbb{C}^{n \times n}$. Prove that there exists a matrix $B \in \mathbb{C}^{n \times n}$ such that B has distinct eigenvalues and that A is arbitrarily close to B; that is, $|a_{ij} - b_{ij}|$ can be made as small as possible for all i, j.
50. *(Complexification)* Let T be a linear operator on a finite dimensional real vector space V. Define

$$\tilde{V} := \{u + iv : u, v \in V\}.$$

Here, i is the imaginary square root of -1, and the sum $u + iv$ is a formal sum. In $\tilde{V}$, define addition and scalar multiplication by

$$(u + iv) + (x + iy) = (u + x) + i(v + y) \quad \text{for } u, v, x, y \in V;$$
$$\alpha(u + iv) = (\alpha u) + i(\alpha v) \quad \text{for } \alpha \in \mathbb{C}.$$

Further, define $\tilde{T} : \tilde{V} \to \tilde{V}$ by $\tilde{T}(u + iv) = T(u) + iT(v)$ for $u, v \in V$. Prove the following:

(a) $\tilde{V}$ is a complex vector space.
(b) Any basis of V is also a basis of $\tilde{V}$. Thus, $\dim(\tilde{V}) = \dim(V)$.
(c) $\tilde{T}$ is a linear operator on $\tilde{V}$.
(d) If B is a basis of V, then $[\tilde{T}]_{B,B} = [T]_{B,B}$.
(e) $\chi_{\tilde{T}}(t) = \chi_T(t)$.
(f) The real eigenvalues of $\tilde{T}$ are precisely the eigenvalues of T.

Chapter 6
Block-Diagonal Representation

6.1 Diagonalizability

In this chapter, we are concerned with special types of matrix representations of linear operators on finite dimensional spaces. The simplest matrix representation of a linear operator that one would like to have is a diagonal matrix. In such a case, the eigenvalues can be read off the diagonal.

Definition 6.1 A linear operator T on a finite dimensional vector space V is called **diagonalizable** if and only if there exists a basis B for V such that $[T]_{B,B}$ is a diagonal matrix.

It may be seen that a matrix in $\mathbb{F}^{n \times n}$ is diagonalizable if and only if it is similar to a diagonal matrix.

Eigenvectors come of help in characterizing diagonalizability.

Theorem 6.2 *A linear operator T on a finite dimensional vector space V is diagonalizable if and only if there exists a basis of V consisting of eigenvectors of T.*

Proof Let $B = \{v_1, \ldots, v_n\}$ be a basis of V with $[T]_{B,B} = \operatorname{diag}(\lambda_1, \ldots, \lambda_n)$. By the definition of matrix representation, $Tv_1 = \lambda_1 v_1, \ \ldots, \ Tv_n = \lambda_n v_n$. That is, each λ_i is an eigenvalue of T with corresponding eigenvector v_i.

Conversely, let $B = \{v_1, \ldots, v_n\}$ be a basis of V, where v_i is an eigenvector of T corresponding to the eigenvalue λ_i for $i \in \{1, \ldots, n\}$. Then

$$Tv_1 = \lambda_1 v_1, \quad \cdots, \quad Tv_n = \lambda_n v_n.$$

Consequently, $[T]_{B,B} = \operatorname{diag}(\lambda_1, \ldots, \lambda_n)$. ■

Since eigenvectors corresponding to distinct eigenvalues are linearly independent, we obtain the following useful corollary to Theorem 6.2.

M. T. Nair and A. Singh, *Linear Algebra*,
https://doi.org/10.1007/978-981-13-0926-7_6

Theorem 6.3 *If a linear operator on an n-dimensional vector space has n distinct eigenvalues, then it is diagonalizable.*

Proof of Theorem 6.2 says that if the ordered set $B = \{v_1, \ldots, v_n\}$ is a basis for V, and $v_1, \ldots, v_n$ are eigenvectors corresponding to the distinct eigenvalues $\lambda_1, \ldots, \lambda_n$ of $T : V \to V$, then $[T]_{B,B}$ is the diagonal matrix $\operatorname{diag}(\lambda_1, \ldots, \lambda_n)$.

Example 6.4 Consider $T : \mathbb{R}^3 \to \mathbb{R}^3$ defined by $T(a, b, c) = (a + b + c, 2b + c, 3c)$. We see that

$$T(1, 0, 0) = 1\,(1, 0, 0),\ T(1, 1, 0) = 2\,(1, 1, 0),\ T(1, 1, 1) = 3\,(1, 1, 1).$$

Hence T has three distinct eigenvalues 1, 2, and 3. A basis consisting of eigenvectors corresponding to these eigenvalues is $B = \{(1, 0, 0), (1, 1, 0), (1, 1, 1)\}$. For $[T]_{B,B}$, we express images of the basis vectors in the same basis:

$$\begin{aligned} T(1, 0, 0) &= 1\,(1, 0, 0) + 0\,(1, 1, 0) + 0\,(1, 1, 1) \\ T(1, 1, 0) &= 0\,(1, 0, 0) + 2\,(1, 1, 0) + 0\,(1, 1, 1) \\ T(1, 1, 1) &= 0\,(1, 0, 0) + 0\,(1, 1, 0) + 3\,(1, 1, 1) \end{aligned}$$

Therefore, $[T]_{B,B} = \operatorname{diag}(1, 2, 3)$. □

Generalized eigenvectors help in characterizing diagonalizability. In addition to Theorems 6.2 and 6.3, we have the following characterization result.

Theorem 6.5 (Diagonalizabilty) *Let T be a linear operator on a vector space V of dimension n over $\mathbb{F}$. Let $\lambda_1, \ldots, \lambda_k$ be the distinct eigenvalues of T. Then the following are equivalent:*

(1) *T is diagonalizable.*
(2) $V = E(\lambda_1) \oplus \cdots \oplus E(\lambda_k)$.
(3) $\gamma(\lambda_1) + \cdots + \gamma(\lambda_k) = n$.
(4) $V = U_1 \oplus \cdots \oplus U_n$, *where* $\dim(U_j) = 1$ *and* $T(U_j) = U_j$ *for each* $j \in \{1, \ldots, n\}$.
(5) *The minimal polynomial of T is $(t - \lambda_1) \cdots (t - \lambda_k)$.*
(6) *The characteristic polynomial of T splits over $\mathbb{F}$, and for each eigenvalue λ of T, $\mu(\lambda) = \gamma(\lambda)$.*
(7) *The characteristic polynomial of T splits over $\mathbb{F}$, and each generalized eigenvector of T is an eigenvector of T.*

Proof Let $\lambda_1, \ldots, \lambda_k$ be the distinct eigenvalues of T with geometric multiplicities $\gamma_1, \ldots, \gamma_k$, and algebraic multiplicities $\mu_1, \ldots, \mu_k$ and ascents $\ell_1, \ldots, \ell_k$, respectively. Let $i \in \{1, \ldots, k\}$.

(1) ⇒ (2) : Due to Theorem 6.2, let B be a basis of eigenvectors of T for V. Each eigenvector is in $\cup_i E(\lambda_i)$. Thus $B \subseteq \cup_i E(\lambda_i)$. Consequently,

$$V = \operatorname{span}(B) \subseteq \operatorname{span}\big(\cup_i E(\lambda_i)\big) = E(\lambda_1) + \cdots + E(\lambda_k) \subseteq V.$$

Due to Theorem 5.16, this sum is a direct sum.

$(2) \Rightarrow (3)$: Suppose that $V = E(\lambda_1) \oplus \cdots \oplus E(\lambda_k)$. Dimensions of the subspaces $E(\lambda_j)$ add up, yielding (3).

$(3) \Rightarrow (4)$: Suppose $\gamma(\lambda_1) + \cdots + \gamma(\lambda_k) = n$. Let $B_j := \{u_{j1}, \ldots u_{j\gamma_j}\}$ be a basis of $E(\lambda_j)$ for $j \in \{1, \ldots, k\}$. Due to Theorem 5.16, $B := B_1 \cup \cdots \cup B_k$ is linearly independent. Thus $|B| = \gamma(\lambda_1) + \cdots + \gamma(\lambda_k) = n = \dim(V)$. Write

$$\begin{aligned} V_{11} &:= \operatorname{span}(u_{11}), \quad \ldots, \quad V_{1\gamma_1} := \operatorname{span}(u_{1\gamma_1}), \\ &\vdots \\ V_{k1} &:= \operatorname{span}(u_{k1}), \quad \ldots, \quad V_{k\gamma_k} := \operatorname{span}(u_{k\gamma_k}). \end{aligned}$$

Then, relabelling the subspaces $V_{11}, \ldots, V_{k\gamma_k}$ as $U_1, \ldots, U_n$, respectively, all the requirements in (4) are fulfilled.

$(4) \Rightarrow (5)$: Assume (4). Let $j \in \{1, \ldots, n\}$. Since $\dim(U_j) = 1$, $U_j = \operatorname{span}\{v_j\}$ for some nonzero vector $v_j \in U_j$. Since $Tv_j \in U_j = \operatorname{span}\{v_j\}$, there exists a scalar α_j such that $Tv_j = \alpha_j v_j$. Thus the eigenvalues are these α_js. That is, in the list $\alpha_1, \ldots, \alpha_n$, each eigenvalue λ_i occurs some n_i times so that $n_1 + \cdots + n_k = n$. Without loss of generality, suppose that the first n_1 scalars $\alpha_1, \ldots, \alpha_{n_1}$ are all equal to λ_1, next n_2 scalars $\alpha_{n_1+1}, \ldots, \alpha_{n_1+n_2}$ are all λ_2, and etc. Then

$$(T - \lambda_1 I)v_1 = \cdots = (T - \lambda_1 I)v_{n_1} = 0, \ldots, (T - \lambda_k I)v_{n-n_k+1} = \cdots = (T - \lambda_k I)v_n = 0.$$

Since the factors in the product $(T - \lambda_1 I) \cdots (T - \lambda_k I)$ commute, we have

$$(T - \lambda_1 I) \cdots (T - \lambda_k I)v_j = 0 \quad \text{for each } j \in \{1, \ldots, n\}.$$

Now, each $v \in V$ can be written as $v = \beta_1 v_1 + \cdots + \beta_n v_n$ for scalars β_j. Hence

$$(T - \lambda_1 I) \cdots (T - \lambda_k I)v = 0 \quad \text{for each } v \in V.$$

That is, $(T - \lambda_1 I) \cdots (T - \lambda_k I) = 0$. Hence the minimal polynomial of T divides $(t - \lambda_1) \cdots (t - \lambda_k)$. However, each eigenvalue of T is a zero of the minimal polynomial of T. Since $\lambda_1, \ldots, \lambda_k$ are all the eigenvalues of T, it follows that $(t - \lambda_1) \cdots (t - \lambda_k)$ is the minimal polynomial.

$(5) \Rightarrow (6)$: Suppose the minimal polynomial of T is $(t - \lambda_1) \cdots (t - \lambda_k)$. Since the minimal polynomial divides the characteristic polynomial, the characteristic polynomial is $(t - \lambda_1)^{\mu_i} \cdots (t - \lambda_k)^{\mu_k}$; obviously it splits. Also, $\ell_i = 1$ for each i. Consequently, $\gamma_i = \mu_i$ for each i.

$(6) \Rightarrow (7)$: Suppose that the characteristic polynomial of T splits and for each eigenvalue λ of T, $\mu(\lambda) = \gamma(\lambda)$. By Theorem 5.27(4), $\mu(\lambda) = \dim(N(T - \lambda I)^n)$. Therefore, $\dim(N(T - \lambda I)^n) = \dim(N(T - \lambda I))$. But $N(T - \lambda I)$ is a subspace of $N(T - \lambda I)^n$. Hence $N(T - \lambda I)^n = N(T - \lambda I)$. That is, each generalized eigenvector is an eigenvector.

(7) $\Rightarrow$ (1) : Suppose that the characteristic polynomial of T splits and each generalized eigenvector of T is an eigenvector of T. However, V always has a basis consisting of generalized eigenvectors of T. In this case, the same basis consists of eigenvectors of T. Therefore, T is diagonalizable. ■

Example 6.6 (1) **(Rotation)** Let $\theta \in \mathbb{R}$. Define the linear operator $T_\theta : \mathbb{R}^2 \to \mathbb{R}^2$ by

$$T_\theta(a, b) = (a\cos\theta - b\sin\theta,\ a\sin\theta + b\cos\theta).$$

It describes rotation about the origin at an angle θ. With respect to the standard basis E on both the copies of $\mathbb{R}^2$, T_θ has the matrix representation

$$[T_\theta]_{E,E} = \begin{bmatrix} \cos\theta & -\sin\theta \\ \sin\theta & \cos\theta \end{bmatrix}.$$

The characteristic values of T_θ are $\cos\theta \pm i\sin\theta$. Hence T_θ has no eigenvalues if $\sin\theta \neq 0$. Therefore, rotation is not diagonalizable when θ is a not an integer multiple of π.

(2) **(Reflection)** Let $\theta \in \mathbb{R}$. Define the linear operator $R_\theta : \mathbb{R}^2 \to \mathbb{R}^2$ by

$$R_\theta(a, b) = (a\cos\theta + b\sin\theta,\ a\sin\theta - b\cos\theta).$$

It describes reflection in the straight line (axis of reflection) that makes an angle $\theta/2$ with the x-axis. With respect to the standard basis E on both the copies of $\mathbb{R}^2$, R_θ has the matrix representation

$$[R_\theta]_{E,E} = \begin{bmatrix} \cos\theta & \sin\theta \\ \sin\theta & -\cos\theta \end{bmatrix}.$$

The characteristic values are ± 1. R_θ has two distinct eigenvalues and $\dim(R^2) = 2$. Thus it is diagonalizable. Indeed, the matrix $[R_\theta]_{E,E}$ is similar to the diagonal matrix $\mathrm{diag}(1, -1)$. The basis of eigenvectors is

$$\big\{\big(\cos\tfrac{\theta}{2}, \sin\tfrac{\theta}{2}\big),\ \big(-\sin\tfrac{\theta}{2}, \cos\tfrac{\theta}{2}\big)\big\}.$$

(3) **(Shear)** Let the linear operator $S : \mathbb{R}^2 \to \mathbb{R}^2$ be defined by

$$S(a, b) = (a + b, b).$$

With respect to the standard basis E on $\mathbb{R}^2$, it has the matrix representation

$$[S]_{E,E} = \begin{bmatrix} 1 & 1 \\ 0 & 1 \end{bmatrix}$$

The characteristic values of S are 1, 1. These are the eigenvalues of S. To find the eigenvectors, we set $S(a,b) = 1\,(a,b)$. Solving this, we see that all eigenvectors are in the form $(a,-a)$ for $a \in \mathbb{R} \setminus \{0\}$. Hence Shear has eigenvalue 1 occurring twice, but it has only one linearly independent eigenvector; $\gamma(1) = 1$. The eigenvectors do not form a basis for $\mathbb{R}^2$; thus Shear is not diagonalizable. □

If $[T]_{E,E} = \text{diag}(d_1, \ldots, d_n)$ with respect to some basis E, then the characteristic polynomial of T is equal to $(t-d_1)\cdots(t-d_n)$. The characteristic polynomial of a diagonalizable operator splits. However, splitting of the characteristic polynomial does not imply diagonalizability. For instance, the characteristic polynomial of the Shear in Example 6.6 splits but the Shear is not diagonalizable.

Exercises for Sect. 6.1

1. Diagonalize the following matrices, if possible:

(a) $\begin{bmatrix} 2 & 3 \\ 6 & -1 \end{bmatrix}$ (b) $\begin{bmatrix} 1 & -10 & 0 \\ -1 & 3 & 1 \\ -1 & 0 & 4 \end{bmatrix}$ (c) $\begin{bmatrix} 2 & -1 & 0 \\ -1 & 2 & 0 \\ 2 & 2 & 3 \end{bmatrix}$

(d) $\begin{bmatrix} 7 & -2 & 0 \\ -2 & 6 & -2 \\ 0 & -2 & 5 \end{bmatrix}$ (e) $\begin{bmatrix} 7 & -5 & 15 \\ 6 & -4 & 15 \\ 0 & 0 & 1 \end{bmatrix}$ (f) $\begin{bmatrix} 2 & 1 & 0 & 0 \\ 0 & 2 & 0 & 0 \\ 0 & 0 & 2 & 0 \\ 0 & 0 & 0 & 5 \end{bmatrix}$.

2. Show that the matrix $\begin{bmatrix} 3 & 2 & 1 & 0 \\ 0 & 1 & 3 & 0 \\ 0 & 0 & 2 & 1 \\ 0 & 0 & 1 & 2 \end{bmatrix}$ is not diagonalizable.
3. Diagonalize the rotation matrix $\begin{bmatrix} \cos\theta & -\sin\theta \\ \sin\theta & \cos\theta \end{bmatrix}$.
4. Let $A = \begin{bmatrix} 6 & -3 & -2 \\ 4 & -1 & -2 \\ 10 & -5 & -3 \end{bmatrix}$. Does there exist a matrix $P \in \mathbb{R}^{3\times 3}$ such that $P^{-1}AP$ is diagonal? What about $P \in \mathbb{C}^{3\times 3}$?
5. Let λ and μ be distinct eigenvalues of a 2×2 matrix $A = [a_{ij}]$. Determine a matrix P such that $P^{-1}AP$ is diagonal. (Hint: See Exercise 2 of Sect. 5.2.)
6. In the following, a linear operator T on $\mathbb{R}^3$ is specified. Determine whether T is diagonalizable. If it is so, then find a basis B of $\mathbb{R}^3$ so that $[T]_{B,B}$ is a diagonal matrix.

   (a) $Te_1 = 0, \quad Te_2 = e_1, \quad Te_3 = e_2$.
   (b) $Te_1 = e_2, \quad Te_2 = e_3, \quad Te_3 = 0$.
   (c) $Te_1 = e_3, \quad Te_2 = e_2, \quad Te_3 = e_1$.
   (d) $T(a,b,c) = (a+b+c, a+b-c, a-b+c)$.
   (e) $T(a,b,c) = (-9a+4b+4c, -8a+3b+4c, -16a+8b+7c)$.

7. Let $T : \mathcal{P}_3(\mathbb{C}) \to \mathcal{P}_3(\mathbb{C})$ be given by $T(a_0 + a_1t + a_2t^2 + a_3t^3) = a_1 + 2a_2t + 3a_3t^2$. Is the linear operator T diagonalizable?

8. Which of the following matrices are diagonalizable over $\mathbb{C}$ and which are diagonalizable over $\mathbb{R}$?
$$\begin{bmatrix} 0 & 0 & 1 \\ 1 & 0 & 0 \\ 0 & 1 & 0 \end{bmatrix}, \quad \begin{bmatrix} 0 & 0 & 1 \\ 0 & 0 & 0 \\ 0 & 0 & 0 \end{bmatrix}, \quad \begin{bmatrix} 0 & 0 & -1 \\ 0 & 0 & 0 \\ 1 & 0 & 0 \end{bmatrix}, \quad \begin{bmatrix} 0 & 1 & 0 & 0 \\ 0 & 0 & 1 & 0 \\ 0 & 0 & 0 & 1 \\ 1 & 0 & 0 & 0 \end{bmatrix}.$$
9. Show that the linear operator $T : \mathbb{R}^{3\times 1} \to \mathbb{R}^{3\times 1}$ corresponding to each of the following matrices is diagonalizable. Determine a basis of eigenvectors of T for $\mathbb{R}^{3\times 1}$ which diagonalizes the given matrix.
$$\begin{bmatrix} 1 & 1 & 1 \\ 1 & -1 & 1 \\ 1 & 1 & -1 \end{bmatrix}, \quad \begin{bmatrix} 1 & 1 & 1 \\ 0 & 1 & 1 \\ 0 & 0 & 1 \end{bmatrix}, \quad \begin{bmatrix} 1 & 0 & 1 \\ 1 & 1 & 0 \\ 0 & 1 & 1 \end{bmatrix}, \quad \begin{bmatrix} 3/2 & -1/2 & 0 \\ -1/2 & 3/2 & 0 \\ 1/2 & -1/2 & 1 \end{bmatrix}.$$
10. Let $T : \mathbb{R}^4 \to \mathbb{R}^4$ be given by $T(a, b, c, d) = (0, \alpha b, \beta c, \gamma d)$. Under what conditions on $\alpha, \beta, \gamma \in \mathbb{R}$, T is diagonalizable?
11. Let π be a permutation on $\{1, \ldots, n\}$, that is a bijection on this set. Let a linear operator T on $\mathbb{C}^n$ be given by $T(a_1, \ldots, a_n) = (a_{\pi(1)}, \ldots, a_{\pi(n)})$. Find a basis B of $\mathbb{C}^n$ so that the matrix $[T]_{B,B}$ is a diagonal matrix.
12. Prove or disprove: If an upper triangular matrix is similar to a diagonal matrix, then it is necessarily diagonal.
13. Prove or disprove: If $A \in \mathbb{C}^{n\times n}$ satisfies $A^2 = A$, then A is diagonalizable.
14. Let $A \in \mathbb{F}^{n\times n}$. Show that if $A^n = I$ for some $n \in \mathbb{N}$, then A is diagonalizable.
15. Let A be a real symmetric matrix with $A^3 = 0$. Show that $A = 0$.

6.2 Triangularizability and Block-Diagonalization

Analogous to the case of a diagonal matrix, eigenvalues of a triangular matrix can be read off its diagonal. Due to the constraints on diagonalizability, our next interest is in representing a linear operator by a triangular matrix.

Definition 6.7 Let T be a linear operator on a finite dimensional vector space V.

(a) T is said to be **triangularizable** if there exists a basis B for V such that $[T]_{B,B}$ is an upper triangular matrix.
(b) When V is an inner product space, T is said to be **unitarily triangularizable** if there exists an orthonormal basis B for V such that $[T]_{B,B}$ is an upper triangular matrix.

Observe that we have used upper triangular matrices in the definition of triangularizability. Alternatively, one may define triangularizability using lower triangular matrices. To derive one form from the other, one has to reorder the basis vectors from last to first. We continue with the upper triangular form.

Interpreting the definitions for matrices, it turns out that a square matrix is called *triangularizable* if and only if it is similar to an upper triangular matrix. Similarly, a square matrix A is called *unitarily triangularizable* if and only if there exists a unitary matrix P such that $P^{-1}AP$ is upper triangular.

The following theorem connects the notions of invariant subspaces and triangularizability.

Theorem 6.8 *Let $B = \{v_1, \ldots, v_n\}$ be a basis for a vector space V. Let T be a linear operator on V. Then the following statements are equivalent:*

(1) $[T]_{B,B}$ *is upper triangular.*
(2) $Tv_k \in \text{span}\{v_1, \ldots, v_k\}$ *for each $k \in \{1, \ldots, n\}$.*
(3) $\text{span}\{v_1, \ldots, v_k\}$ *is T-invariant for each $k \in \{1, \ldots, n\}$.*

Proof (1) $\Rightarrow$ (2) : Let $[T]_{B,B}$ be upper triangular. There are scalars a_{ij} such that

$$[T]_{B,B} = \begin{bmatrix} a_{11} & a_{12} & \cdots & a_{1n} \\ 0 & a_{22} & \cdots & a_{2n} \\ & & \vdots & \\ 0 & 0 & \cdots & a_{nn} \end{bmatrix}$$

As $B = \{v_1, \ldots, v_n\}$, we have $Tv_k = a_{1k}v_1 + a_{2k}v_2 + \cdots + a_{kk}v_k$. Therefore, for each $k \in \{1, \ldots, n\}$, $Tv_k \in \text{span}\{v_1, \ldots, v_k\}$

(2) $\Rightarrow$ (3) : Assume that $Tv_k \in \text{span}\{v_1, \ldots, v_k\}$ for each $k \in \{1, \ldots, n\}$. Let $v \in \text{span}\{v_1, \ldots, v_k\}$. Then $Tv \in \text{span}\{Tv_1, \ldots, Tv_k\}$. Since

$$Tv_1 \in \text{span}\{v_1\}, \ldots, Tv_k \in \text{span}\{v_1, \ldots, v_k\},$$

we see that $Tv \in \text{span}\{v_1, \ldots, v_k\}$. Therefore, $\text{span}\{v_1, \ldots, v_k\}$ is T-invariant.

(3) $\Rightarrow$ (1) : Suppose $\text{span}\{v_1, \ldots, v_k\}$ is T-invariant for each $k \in \{1, \ldots, n\}$. Notice that $\{v_1, \ldots, v_k\}$ is a basis for $\text{span}\{v_1, \ldots, v_k\}$. The invariance condition implies that there are scalars b_{ij} such that

$$\begin{aligned} Tv_1 &= b_{11}v_1 \\ Tv_2 &= b_{12}v_1 + b_{22}v_2 \\ &\vdots \\ Tv_n &= b_{1n}v_1 + b_{2n}v_2 + \cdots + b_{nn}v_n \end{aligned}$$

Therefore, $[T]_{B,B}$ is upper triangular. ■

Recall that if the characteristic polynomial of T splits, then T can be written as a direct sum of its restrictions to the generalized eigenspaces of its distinct eigenvalues. Also, the generalized eigenspaces are T-invariant subspaces of V; see Theorem 5.21(4). This means that if we take a basis which is a union of the bases of the generalized eigenspaces (keeping those basis vectors together in the union), then with respect to this basis, the matrix of T will be a block-diagonal matrix. Then we need to check whether each such block can be triangularized. Let us look at the ensuing particular case first.

Lemma 6.9 *Let T be a linear operator on a vector space V of dimension n over $\mathbb{F}$. Suppose that T has a single eigenvalue λ with algebraic multiplicity n. Then there exists a basis B for V such that $[T]_{B,B} \in \mathbb{F}^{n\times n}$ is an upper triangular matrix with each diagonal entry as λ.*

Proof We wish to construct a basis B for V so that $[T]_{B,B}$ is upper triangular. For this, let ℓ be the ascent of the eigenvalue λ. Consider the sequence

$$N(T-\lambda I) \subsetneqq \cdots \subsetneqq N(T-\lambda I)^j \subsetneqq N(T-\lambda I)^{j+1} \subsetneqq \cdots \subsetneqq N(T-\lambda I)^\ell = G(\lambda).$$

Since λ is the only eigenvalue of T, by Theorem 5.29, $G(\lambda) = V$. We wish to extend a basis of $N(T-\lambda I)$ to a basis for $N(T-\lambda I)^\ell$ in a specific manner.

For this, start with a basis B_1 for $N(T-\lambda I)$. Suppose $B_1, \ldots, B_j$ have already been constructed so that $B_1 \cup \cdots \cup B_j$ is a basis for $N(T-\lambda I)^j$, then select a set B_{j+1} of vectors from $N(T-\lambda I)^{j+1} \setminus N(T-\lambda I)^j$ so that $B_1 \cup \cdots \cup B_{j+1}$ is a basis for $N(T-\lambda I)^{j+1}$. The construction results in a basis $B := B_1 \cup \cdots \cup B_\ell$ for the generalized eigenspace $G(\lambda)$. In the ordered set B, the vectors from B_{j+1} succeed the vectors from B_j.

Now, if $u \in B_1 \subseteq N(T-\lambda I)$, then $Tu = \lambda u$. And, if $v \in B_{j+1} \subseteq N(T-\lambda I)^{j+1}$, then $(T-\lambda I)v \in N(T-\lambda I)^j = \text{span}(B_1 \cup \cdots \cup B_j)$. Thus, $Tv = \lambda v + w$, where $w \in \text{span}(B_1 \cup \cdots \cup B_j)$.

Therefore, the T-image of the first vector in the ordered basis B is λ times that vector, and each other vector in B is equal to λ times that vector plus one from the span of the previous vectors appearing in B. By Theorem 6.8, $[T]_{B,B}$ is upper triangular with each diagonal entry equal to λ. ■

Theorem 6.10 (Block-diagonalization) *Let T be a linear operator on a finite dimensional vector space V over $\mathbb{F}$. Suppose that the characteristic polynomial of T splits over $\mathbb{F}$. Then there exist a basis B for V and distinct scalars $\lambda_1, \ldots, \lambda_k$ such that $[T]_{B,B}$ is a block-diagonal matrix of the form $[T]_{B,B} = \text{diag}(A_1, \ldots, A_k)$, where each block A_i is an upper triangular matrix with diagonal entries as λ_i.*

Proof Let $\lambda_1, \ldots, \lambda_k$ be the distinct eigenvalues of T. By Theorems 5.27 and 5.29, $V = G(\lambda_1) \oplus \cdots \oplus G(\lambda_k)$, and $G(\lambda_1), \ldots, G(\lambda_k)$ are T-invariant subspaces of V. Then for each $i \in \{1, \ldots, k\}$, the restriction operator $T_i := T|_{G(\lambda_i)}$ maps $G(\lambda_i)$ to $G(\lambda_i)$.

The operator T_i has the single eigenvalue λ_i with its algebraic multiplicity as $\mu_i := \dim(G(\lambda_i))$. By Lemma 6.9, there exists an ordered basis B_i for $G(\lambda_i)$ such that $[T_i]_{B_i,B_i}$ is an upper triangular matrix of size $\mu_i \times \mu_i$ with each diagonal entry as λ_i. We denote $[T_i]_{B_i,B_i}$ as A_i.

Next, we construct the ordered set $B = B_1 \cup \cdots \cup B_k$, where we keep the vectors of B_1 with their own ordering, then the vectors of B_2 with their own ordering, and so on. Now, the matrix $[T]_{B,B}$ is a block-diagonal matrix with the diagonal blocks as $A_1, \ldots, A_k$. ■

Example 6.11 Let $T : \mathbb{R}^3 \to \mathbb{R}^3$ be given by $T(a, b, c) = (6a + 3b, 2a + b, -a + b + 7c)$. The matrix of T with respect to the standard basis is

$$[T] = \begin{bmatrix} 6 & 3 & 0 \\ 2 & 1 & 0 \\ -1 & 1 & 7 \end{bmatrix}.$$

The characteristic polynomial of A is $p(t) = t(t-7)^2$. The eigenvalues are 7 and 0, with respective multiplicities as 2 and 1. For the eigenvalue 7,

$$(T - 7I)(a, b, c) = (-a + 3b, 2a - 6b, -a + b).$$

Now, $(T - 7I)(a, b, c) = (0, 0, 0)$ if and only if $a = b = 0$. So, $N(T - 7I) = \{(0, 0, c) : c \in \mathbb{R}\}$. A basis for $N(T - 7I)$ is $\{(0, 0, 1)\}$. Further,

$$(T - 7I)^2(a, b, c) = (7a - 21b, -14a + 42b, 3a - 9b) = (a - 3b)(7, -14, 3).$$

Here, $(T - 7I)^2(a, b, c) = 0$ if and only if $a = 3b$. That is, $N(T - 7I)^2 = G(7) = \{(3b, b, c) : b, c \in \mathbb{R}\}$. A basis for $G(7)$ is $\{(0, 0, 1),\ (3, 1, 0)\}$, which is an extension of $\{(0, 0, 1)\}$.

For the eigenvalue 0, $(T - 0I)(a, b, c) = 0$ gives $7c = 3a,\ b = -2a$. A basis for $N(T - 0I)$ is $\{(7, -14, 3)\}$. Notice that since the eigenvalue 0 has algebraic multiplicity 1, $G(0) = N(T - 0I)$.

Putting the bases together we obtain the following basis B for $\mathbb{R}^3$:

$$B = \{(0, 0, 1),\ (3, 1, 0),\ (7, -14, 3)\}.$$

Computing the T-images of the basis vectors, we find that

$$\begin{aligned} T(0, 0, 1) &= \quad (0, 0, 7) \quad = 7(0, 0, 1) \\ T(3, 1, 0) &= (21, 7, -2) = -2(0, 0, 1) + 7(3, 1, 0) \\ T(7, -14, 3) &= \quad (0, 0, 0) \quad = 0(7, -14, 3). \end{aligned}$$

Hence the block upper triangular matrix representation of T is given by

$$[T]_{B,B} = \begin{bmatrix} 7 & -2 & \\ & 7 & \\ & & 0 \end{bmatrix}.$$

Notice that the resulting block-diagonal matrix each of whose blocks is upper triangular is itself upper triangular. □

Exercises for Sect. 6.2

1. If $(1, 2)^T$ and $(2, 1)^T$ are eigenvectors corresponding to eigenvalues 1 and 2 of a 2×2 matrix A, then what is A?
2. Is the matrix $\begin{bmatrix} 0 & 1 & 0 \\ 2 & -2 & 2 \\ 2 & -3 & 2 \end{bmatrix}$ similar to a triangular matrix with real entries?
3. Let T be a linear operator on a finite dimensional vector space. Prove that there exists a nonnegative integer k such that $N(T^k) \cap R(T^k) = \{0\}$.
4. Let A be a block-diagonal matrix with two blocks. Show that A is diagonalizable if and only if the two blocks are diagonalizable.

6.3 Schur Triangularization

We may interpret Theorem 6.10 for matrices. It says that each square matrix of complex numbers is similar to a block-diagonal matrix with upper triangular blocks. The change of basis involved in this similarity transformation is an isomorphism. In fact, we can choose a unitary similarity transformation. The trick is to orthonormalize the basis obtained for block-diagonalization.

Theorem 6.12 (Schur triangularization) *Let T be a linear operator on a finite dimensional inner product space V over $\mathbb{F}$. If the characteristic polynomial of T splits over $\mathbb{F}$, then T is unitarily triangularizable.*

Proof Due to Theorem 6.10, we have a basis B with respect to which $[T]_{B,B}$ is block-diagonal, each block corresponds to a particular eigenvalue of T; and each block is upper triangular. Hence $[T]_{B,B}$ is upper triangular. By Theorem 6.8, if $B = \{u_1, \ldots, u_n\}$, then for each $j \in \{1, \ldots, n\}$,

$$Tu_j \in \text{span}\{u_1, \ldots, u_j\}.$$

Now, apply Gram–Schmidt procedure to orthonormalize B. If the resulting basis is $E = \{v_1, \ldots, v_n\}$, then for each $j \in \{1, \ldots, n\}$,

$$\text{span}\{v_1, \ldots, v_j\} = \text{span}\{u_1, \ldots, u_j\}.$$

Then $v_j \in \text{span}\{u_1, \ldots, u_j\}$ implies that

$$Tv_j \in \text{span}\{Tu_1, \ldots, Tu_j\} \subseteq \text{span}\{u_1, \ldots, u_j\} = \text{span}\{v_1, \ldots, v_j\}.$$

Therefore, $[T]_{E,E}$ is upper triangular with respect to the orthonormal basis E. That is, T is unitarily triangularizable. ■

Recall that two square matrices A and B are called *unitarily similar* if there exists a unitary matrix U such that $B = U^{-1}AU = U^*AU$. Theorem 6.12 says the following:

Each square matrix is unitarily similar to an upper triangular matrix.

The proof of Theorem 6.12 does not say that each block in the block-diagonal form is triangularized. In fact, orthonormalization of the basis vectors may destroy the block-diagonal form.

Example 6.13 Consider the matrix of Example 6.11 for triangularization:

$$A = \begin{bmatrix} 6 & 3 & 0 \\ 2 & 1 & 0 \\ -1 & 1 & 7 \end{bmatrix}.$$

It has eigenvalues 0 and 7. The eigenvalue 7 has algebraic multiplicity 2. In that example, we had already obtained a basis with respect to which the linear map has the block-diagonal form. Since the space here is $\mathbb{R}^{3\times 1}$, we use the transpose instead of adjoint. Our computation in Example 6.11 shows that $[A]_{B,B}$ is block-diagonal, where

$$B = \{(0,0,1)^T,\ (3,1,0)^T,\ (7,-14,3)^T\}.$$

Using Gram–Schmidt process on B, we obtain the orthonormal basis

$$E = \left\{(0,0,1)^T,\ \left(\tfrac{3}{\sqrt{10}}, \tfrac{1}{\sqrt{10}}, 0\right)^T,\ \left(\tfrac{1}{\sqrt{10}}, -\tfrac{3}{\sqrt{10}}, 0\right)^T\right\}.$$

Expressing A with respect to this basis we get the Schur triangular form:

$$[A]_{E,E} = \begin{bmatrix} 7 & -2/\sqrt{10} & -4/\sqrt{10} \\ 0 & 7 & -1 \\ 0 & 0 & 0 \end{bmatrix}.$$

Notice that the block-diagonal form is destroyed in triangularization. □

We give another proof of Theorem 6.12 without using block-diagonalization.

An Alternative Proof of Theorem 6.12:

Let $T : V \to V$, where V is a finite dimensional inner product space. We use strong induction on $\dim(V)$. If $\dim(V) = 1$, then $[T]_{B,B}$ is of order one, which is upper triangular, whichever basis B we choose. Lay out the induction hypothesis that the result is true for all inner product spaces of dimension less than n. Let $\dim(V) = n$. Since the characteristic polynomial of T splits, let λ be an eigenvalue of T.

We choose an invariant subspace of smaller dimension and use the induction hypothesis. For this purpose, consider the subspace $U := R(T - \lambda I)$ of V. Since $T - \lambda I$ is not one-to-one,

$$k := \dim(U) = \dim(R(T - \lambda I)) < \dim(V) = n.$$

If $u \in U$, then there exists $v \in V$ such that $u = (T - \lambda I)v$. Now,

$$Tu = T(T - \lambda I)v = (T - \lambda I)(Tv) \in U.$$

That is, U is T-invariant.

By Theorem 5.20, the characteristic polynomial of $T|_U$ splits. By the induction hypothesis and Theorem 6.8, there exists a basis $B := \{u_1, \ldots, u_k\}$ of U such that $[T|_U]_{B,B}$ is upper triangular. Extend this basis to a basis $E := B \cup \{v_1, \ldots, v_m\}$ for V. Notice that $k + m = n$.

Since $[T|_U]_{B,B}$ is upper triangular,

$$Tu_i = (T|_U)u_i \in \text{span}\{u_1, \ldots, u_i\} \quad \text{for each } i \in \{1, \ldots, k\}.$$

Next, for any $j \in \{1, \ldots, m\}$,

$$Tv_j = (T - \lambda I)v_j + \lambda v_j.$$

Since $(T - \lambda I)v_j \in U$ and $\lambda v_j \in \text{span}\{v_j\}$, we see that

$$Tv_j \in \text{span}\{u_1, \ldots, u_k, v_j\} \subseteq \text{span}\{u_1, \ldots, u_k, v_1, \ldots, v_j\}.$$

Therefore, by Theorem 6.8, $[T]_{E,E}$ is upper triangular.

To see that T is unitarily triangularizable, use Gram–Schmidt orthonormalization on the basis E to obtain an orthonormal basis $F := \{w_1, \ldots, w_k, x_1, \ldots, x_m\}$. Now, for any $i \in \{1, \ldots, k\}$ and for any $j \in \{1, \ldots, m\}$,

$$\text{span}\{w_1, \ldots, w_i\} = \text{span}\{u_1, \ldots, u_i\},$$
$$\text{span}\{x_1, \ldots, x_j\} = \text{span}\{u_1, \ldots, u_k, v_1, \ldots, v_j\}.$$
$$w_i \in \text{span}\{w_1, \ldots, w_i, \}, \quad x_j \in \text{span}\{w_1, \ldots, w_k, x_1, \ldots, x_j\}.$$

Therefore, $[T]_{F,F}$ is also upper triangular. ■

The matrix form of Schur Triangularization can be proved without using the range space of $T - \lambda I$ explicitly. Let v_1 be an eigenvector corresponding to an eigenvalue λ of T. Now,

$$Tv_1 = \lambda v_1.$$

Construct an orthonormal basis $\{v_1, \ldots, v_n\}$ for V by extending $\{v_1\}$ and then using Gram–Schmidt orthonormalization. In this basis, the matrix of T looks like:

$$\begin{bmatrix} \lambda & x \\ 0 & C \end{bmatrix}$$

where $x \in \mathbb{F}^{1\times(n-1)}$ and $C \in \mathbb{F}^{(n-1)\times(n-1)}$. Use the induction hypothesis that C can be triangularized. Then join the pieces together.

The inductive proof reveals that in the resulting upper triangular matrix representation, we may have the eigenvalues in any preferred order on the diagonal.

Example 6.14 We illustrate the inductive proof given for Schur triangularization for the matrix of Example 6.13:

$$A = \begin{bmatrix} 6 & 3 & 0 \\ 2 & 1 & 0 \\ -1 & 1 & 7 \end{bmatrix}.$$

This matrix has eigenvalues 0 and 7. For the eigenvalue 0,

$$R(A - 0I) = R(A) = \text{span}\{(6, 2, -1)^T,\ (3, 1, 1)^T,\ (0, 0, 7)^T\}.$$

A basis for $R(A - 0\,I)$ is $B_1 := \{(0, 0, 1)^T,\ (3, 1, 0)^T\}$ since $(3, 1, 0)^T$ is not a scalar multiple of $(0, 0, 1)^T$. And,

$$\begin{aligned} (6, 2, -1)^T &= -1(0, 0, 1)^T + 2(3, 1, 0)^T \\ (3, 1, 1)^T &= (1, 0, 0)^T + (3, 1, 0)^T \\ (0, 0, 7)^T &= 7(0, 0, 1)^T. \end{aligned}$$

A basis for $\mathbb{R}^3$ containing B_1 is $\{(0, 0, 1)^T,\ (3, 1, 0)^T,\ (1, -3, 0)^T\}$. Orthonormalizing this basis yields the following orthonormal basis for $\mathbb{R}^{3\times 1}$:

$$\left\{(0, 0, 1)^T,\ \left(\tfrac{3}{\sqrt{10}}, \tfrac{1}{\sqrt{10}}, 0\right)^T,\ \left(\tfrac{1}{\sqrt{10}}, -\tfrac{3}{\sqrt{10}}, 0\right)^T\right\}.$$

Incidentally, this basis is same as E that we constructed earlier. Next, we take the matrix Q_1 formed by the basis vectors:

$$Q_1 = \begin{bmatrix} 0 & 3/\sqrt{10} & 1/\sqrt{10} \\ 0 & 1/\sqrt{10} & -3/\sqrt{10} \\ 1 & 0 & 0 \end{bmatrix}.$$

Then we form the product

$$Q_1^T A Q_1 = \begin{bmatrix} 7 & -2/\sqrt{10} & -4/\sqrt{10} \\ 0 & 7 & -1 \\ 0 & 0 & 0 \end{bmatrix} = U.$$

Since it is already upper triangular, there is no need to execute the inductive step. We have obtained the orthogonal matrix $P = Q_1$ and the upper triangular matrix $U = P^T A P$.

In the following, we illustrate the alternate method of taking an eigenvalue and starting the inductive construction with a single eigenvector, as discussed earlier.

For variety, let us consider this time the eigenvalue 7 of A. we see that $(0, 0, 1)^T$ is a corresponding eigenvector. Choosing additional vectors arbitrarily to form an orthonormal basis and assembling them into a matrix, we obtain

$$Q_1 = \begin{bmatrix} 0 & 0 & 1 \\ 0 & 1 & 0 \\ 1 & 0 & 0 \end{bmatrix}.$$

We form the product

$$Q_1^T A Q_1 = \begin{bmatrix} -1 & 1 & 7 \\ 2 & 1 & 0 \\ 6 & 3 & 0 \end{bmatrix} Q_1 = \begin{bmatrix} 7 & 1 & -1 \\ 0 & 1 & 2 \\ 0 & 3 & 6 \end{bmatrix}.$$

Notice that the entries in the first column of $Q_1^T A Q_1$ below the diagonal are all 0. Next, we take the 2×2 submatrix

$$A_2 = \begin{bmatrix} 1 & 2 \\ 3 & 6 \end{bmatrix}.$$

The eigenvalues of A_2 are 7 and 0 as expected. An eigenvector corresponding to the eigenvalue 7 is $(1, 3)^T$. A matrix with columns consisting of vectors of an orthonormal basis for $\mathbb{R}^{2\times 1}$ obtained from extending this vector is:

$$Q_2 = \begin{bmatrix} 1/\sqrt{10} & -3/\sqrt{10} \\ 3/\sqrt{10} & 1/\sqrt{10} \end{bmatrix}.$$

Next, we form the product

$$Q_2^T A_2 Q_2 = Q_2^T \begin{bmatrix} 7/\sqrt{10} & -1/\sqrt{10} \\ 21/\sqrt{10} & -3/\sqrt{10} \end{bmatrix} = \begin{bmatrix} 7 & 0 \\ 0 & 0 \end{bmatrix}.$$

Then we assemble Q_1 and Q_2 to obtain the orthogonal matrix P as follows:

$$P = Q_1 \begin{bmatrix} 1 & 0 \\ 0 & Q_2 \end{bmatrix} = Q_1 \begin{bmatrix} 1 & 0 & 0 \\ 0 & 1/\sqrt{10} & -3/\sqrt{10} \\ 0 & 3/\sqrt{10} & 1/\sqrt{10} \end{bmatrix} = \begin{bmatrix} 0 & 3/\sqrt{10} & 1/\sqrt{10} \\ 0 & 1/\sqrt{10} & -3/\sqrt{10} \\ 1 & 0 & 0 \end{bmatrix}.$$

As earlier, we see that $P^T A P$ is the upper triangular matrix U. □

In general, Schur triangularization neither yields a unique P nor a unique U. Nonuniqueness stems from choosing an orthonormal basis and the order of the vectors in such a basis. However, the set of entries on the diagonal of U is the same since the diagonal elements are precisely the eigenvalues of A.

Exercises for Sect. 6.3

1. Let $A = \begin{bmatrix} 1 & 2 \\ -1 & -2 \end{bmatrix}$, $B = \begin{bmatrix} -1 & 0 \\ 0 & 0 \end{bmatrix}$ and let $C = \begin{bmatrix} 0 & -2 \\ 0 & -1 \end{bmatrix}$. Find orthogonal matrices P and Q such that $P^T AP = B$ and $Q^T AQ = C$. This shows the nonuniqueness in Schur triangularization.
2. Show that Schur triangularization of $\begin{bmatrix} 1 & 2 \\ 3 & 6 \end{bmatrix}$ yields $\begin{bmatrix} 0 & 1 \\ 0 & 7 \end{bmatrix}$ but not $\begin{bmatrix} 0 & 2 \\ 0 & 7 \end{bmatrix}$.
3. Are $\begin{bmatrix} 1 & 1 \\ 0 & 1 \end{bmatrix}$ and $\begin{bmatrix} 0 & 0 \\ 1 & 0 \end{bmatrix}$ unitarily similar?
4. Determine Schur triangularization of the matrix $\begin{bmatrix} 2 & 3 & 3 \\ 1 & 3 & 3 \\ -1 & -2 & -3 \end{bmatrix}$.
5. Let $A \in \mathbb{R}^{3\times 3}$ be such that for every invertible $P \in \mathbb{R}^{3\times 3}$, $P^{-1}AP$ fails to be upper triangular. Prove that there exists an invertible matrix $Q \in \mathbb{C}^{3\times 3}$ such that $Q^{-1}AQ$ is a diagonal matrix.
6. Let $\lambda_1, \ldots, \lambda_n$ be eigenvalues of a linear operator T on a complex inner product space of dimension n. Let $A = [a_{ij}]$ be the matrix of T with respect to some orthonormal basis. Show that $\sum_{i=1}^n |\lambda_i|^2 \le \operatorname{tr}(T^*T) = \sum_{i=1}^n \sum_{j=1}^n |a_{ij}|^2$.
7. Prove that any square matrix with complex entries is unitarily similar to a lower triangular matrix. Deduce that any square matrix with real entries having all its eigenvalues real is orthogonally similar to a lower triangular matrix.

6.4 Jordan Block

The triangular blocks in a block-diagonal form can still be simplified by choosing a basis for the generalized eigenspaces in a specified way. First we consider the particular case of an operator having a single eigenvalue. We will show that such operators can be represented by block-diagonal matrices consisting of the so-called Jordan blocks.

Definition 6.15 A **Jordan block** of order m, denoted by $J(\lambda, m)$, is an $m \times m$ matrix that satisfies the following:

(1) Each diagonal entry is equal to λ.
(2) Each entry on the super-diagonal is equal to 1.
(3) All other entries are equal to 0.

By not showing the zero entries a typical Jordan block is written as

$$J(\lambda, m) = \begin{bmatrix} \lambda & 1 & & & \\ & \lambda & 1 & & \\ & & \ddots & \ddots & \\ & & & & 1 \\ & & & & \lambda \end{bmatrix}.$$

The Jordan block $J(\lambda, 1)$ is the 1×1 matrix $[\lambda]$. Notice that $J(\lambda, m) - \lambda I = J(0, m)$. When the order m of the Jordan block $J(\lambda, m)$ is not of particular interest, we write the Jordan block as $J(\lambda)$.

Example 6.16 Consider the matrix $J = \mathrm{diag}(J(0,6),\ J(0,5),\ J(0,2))$. What is rank (J^4)?

We see that

$$J(0,6) = \begin{bmatrix} 0&1&0&0&0&0 \\ 0&0&1&0&0&0 \\ 0&0&0&1&0&0 \\ 0&0&0&0&1&0 \\ 0&0&0&0&0&1 \\ 0&0&0&0&0&0 \end{bmatrix}, \ (J(0,6))^2 = \begin{bmatrix} 0&0&1&0&0&0 \\ 0&0&0&1&0&0 \\ 0&0&0&0&1&0 \\ 0&0&0&0&0&1 \\ 0&0&0&0&0&0 \\ 0&0&0&0&0&0 \end{bmatrix}, \ (J(0,6))^4 = \begin{bmatrix} 0&0&0&0&1&0 \\ 0&0&0&0&0&1 \\ 0&0&0&0&0&0 \\ 0&0&0&0&0&0 \\ 0&0&0&0&0&0 \\ 0&0&0&0&0&0 \end{bmatrix}.$$

Thus, rank $(J(0,6))^4 = 2$. Similarly, rank $(J(0,5))^4 = 1$ and rank $(J(0,2))^4 = 0$. Then

$$J^4 = \mathrm{diag}((J(0,6))^4, (J(0,5))^4, (J(0,2))^4); \text{ and rank } (J^4) = 2 + 1 + 0 = 3. \square$$

The pattern in the powers of $J(0, m)$ is easy to see. In $J(0, m)$, the number of 1s on the super-diagonal shows that it has rank $m - 1$. In $(J(0, m))^2$, the number of 1s on the second-super-diagonal shows that it has rank $m - 2$ and so on. In general,

$$\text{rank } (J(0,m))^i = \begin{cases} m - i & \text{for } 1 \le i \le m \\ 0 & \text{for } i > m. \end{cases}$$

Example 6.17 Consider the matrix

$$A = \mathrm{diag}(J(2,3),\ J(2,2)) = \begin{bmatrix} 2&1&0&0&0 \\ 0&2&1&0&0 \\ 0&0&2&0&0 \\ 0&0&0&2&1 \\ 0&0&0&0&2 \end{bmatrix}.$$

The only eigenvalue of A is 2; it has algebraic multiplicity 5. We have

$$A - 2I = \begin{bmatrix} 0&1&0&0&0 \\ 0&0&1&0&0 \\ 0&0&0&0&0 \\ 0&0&0&0&1 \\ 0&0&0&0&0 \end{bmatrix}, \quad (A - 2I)^2 = \begin{bmatrix} 0&0&1&0&0 \\ 0&0&0&0&0 \\ 0&0&0&0&0 \\ 0&0&0&0&0 \\ 0&0&0&0&0 \end{bmatrix}, \quad (A - 2I)^3 = 0.$$

We find that

$$N(A-2I) = \{(a,0,0,d,0)^T : a,d \in \mathbb{R}\}; \quad \text{null}\,(A-2I) = 2.$$
$$N(A-2I)^2 = \{(a,b,0,d,e)^T : a,b,d,e \in \mathbb{R}\}; \quad \text{null}\,(A-2I)^2 = 4.$$
$$N(A-2I)^3 = \mathbb{R}^{3\times 1}; \quad \text{null}\,(A-2I)^3 = 5.$$

There is $5-4=1$ linearly independent vector in $N(A-2I)^3 \setminus N(A-2I)^2$. Similarly, there are 2 linearly independent vectors in $N(A-2I)^2 \setminus N(A-2I)$, and 2 linearly independent vectors in $N(A-2I)$. These vectors together form a basis for $\mathbb{R}^{5\times 1}$. Let us choose these vectors.

Now, $e_3 \in N(A-2I)^3 \setminus N(A-2I)^2$. $(A-2I)e_3 = e_2 \in N(A-2I)^2 \setminus N(A-2I)$. We choose one more vector as $e_5 \in N(A-2I)^2 \setminus N(A-2I)$. Next, we obtain $(A-2I)e_2 = e_1$ and $(A-2I)e_5 = e_4$ as the vectors in $N(A-2I)$. The basis vectors e_1, e_2, e_3, e_4, e_5 of $\mathbb{R}^{5\times 1}$ have been obtained the following way (read from top to bottom, and right to left):

$N(A-2I)$	$N(A-2I)^2 \setminus N(A-2I)$	$N(A-2I)^3 \setminus N(A-2I)^2$
$e_1 = (A-2I)e_2$	$e_2 = (A-2I)e_3$	e_3
$e_4 = (A-2I)e_5$	e_5	

Starting from $e_3 \in N(A-2I)^3 \setminus N(A-2I)^2$, we generate $(A-2I)e_3$, $(A-2I)^2e_3$, and so on. In the first line, it stops at e_1, which is equal to $(A-2I)^2e_3$ since $(A-2I)^3e_3 = 0$. In the last line, starting with $e_5 \in N(A-2I)^2 \setminus N(A-2I)$, which is linearly independent with the earlier obtained vector $e_2 \in N(A-2I)^2 \setminus N(A-2I)$, we generate $(A-2I)e_5$, $(A-2I)^2e_5$, and so on. Again, it stops at $(A-2I)e_5$, since $(A-2I)^2e_5 = 0$. The process stops once sufficient linearly independent vectors from $N(A-2I)$ have been generated. This way, we obtain an ordered basis for $\mathbb{R}^{5\times 1}$, which happens to be the standard basis $\{e_1, e_2, e_3, e_4, e_5\}$.

Notice the ordering of the basis vectors that has been compiled from the above construction. We put together the vectors in first row from left to right; next the vectors from the second row, left to right, and so on. Clearly, in this basis the matrix of A is A itself.

To check how our choices of vectors affect later computation, suppose we choose $e_2 + e_5$ as the vector in $N(A-2I)^2 \setminus N(A-2I)$, instead of e_5. We already had obtained the vectors e_3 and e_2. The chosen vector $e_2 + e_5$ is linearly independent with e_2. Taking their $(A-2I)$-images we get the vectors e_1 and $e_1 + e_4$ in $N(A-2I)$. Then we have the basis vectors as

$N(A-2I)$	$N(A-2I)^2 \setminus N(A-2I)$	$N(A-2I)^3 \setminus N(A-2I)^2$
$e_1 = (A-2I)e_2$	$e_2 = (A-2I)e_3$	e_3
$e_1 + e_4 = (A-2I)(e_2 + e_5)$	$e_2 + e_5$	

Putting together the basis vectors row-wise from left to right, we obtain the ordered basis $E = \{e_1, e_2, e_3, e_1 + e_4, e_2 + e_5\}$. We then see that

$$\begin{aligned} Ae_1 &= 2\,e_1 \\ Ae_2 &= 1\,e_1 + 2\,e_2 \\ Ae_3 &= 1\,e_2 + 2\,e_3 \\ A(e_1 + e_4) &= 2(e_1 + e_4) \\ A(e_2 + e_5) &= 1\,(e_1 + e_4) + 2\,(e_2 + e_5). \end{aligned}$$

Therefore, $[A]_{E,E} = A$. Observe that our choice does not produce any change in the final matrix representation. □

In what follows, we use the convention that for any $B \in \mathbb{C}^{n\times n}$, $B^0 = I_n$.

Theorem 6.18 (Jordan block representation) *Let T be a linear operator on a vector space V of dimension n. Let λ be the only eigenvalue of T with $\mu(\lambda) = n$. Then there exists a basis B of V such that the following are true:*

(1) $[T]_{B,B} = \operatorname{diag}(J(\lambda, m_1), \ldots, J(\lambda, m_k))$,
where $k = \gamma(\lambda)$, $\ell(\lambda) = m_1 \geq \cdots \geq m_k \geq 1$ and $m_1 + \cdots + m_k = \mu(\lambda) = n$.

(2) *For $i \in \{1, \ldots, \ell(\lambda)\}$, if n_i denotes the number of Jordan blocks $J(\lambda, i)$ of order i that appear in $[T]_{B,B}$, then*

$$n_i = 2\,\text{null}\,(T - \lambda I)^i - \text{null}\,(T - \lambda I)^{i-1} - \text{null}\,(T - \lambda I)^{i+1}.$$

Further, such a representation of T as a block-diagonal matrix with Jordan blocks is unique up to a permutation of the blocks.

Proof Let ℓ be the ascent of the eigenvalue λ. If $\ell = 1$, then $\mu(\lambda) = \gamma(\lambda)$; thus T is diagonalizable. The diagonal matrix that represents T is equal to $\operatorname{diag}(\lambda, \ldots, \lambda)$. It is a block-diagonal matrix with each block as $J(\lambda, 1)$.

Next, assume that $\ell \geq 2$. Write $G_j := N(T - \lambda I)^j$. Then

$$G_0 := \{0\} \subsetneqq E(\lambda) = G_1 \subsetneqq \cdots \subsetneqq G_j \subsetneqq \cdots \subsetneqq G_\ell = G(\lambda) = V.$$

Here, $G(\lambda) = V$ since λ is the only eigenvalue of T. For $1 \leq j \leq \ell$, write

$$\gamma_j := \dim(G_j); \quad k_j := \gamma_j - \gamma_{j-1}.$$

As $\gamma_0 = \text{null}\,(T - \lambda I)^0 = \text{null}\,(I) = 0$, we have $k_1 = \gamma_1$. For $j \geq 2$, Theorem 5.27(7) implies that

$$k_1 \geq \cdots \geq k_j \geq \cdots \geq k_\ell \geq 1.$$

For $j = \ell$, we see that there exists a linearly independent subset

$$B_\ell := \{u^\ell_{k_1}, \ldots, u^\ell_{k_\ell}\}$$

of G_ℓ such that $G_\ell = \text{span}(B_\ell) \oplus G_{\ell-1}$. Define vectors

$$u_i^{\ell-1} := (T - \lambda I)u_i^{\ell} \quad \text{for } i \in \{1, \ldots, k_\ell\}.$$

Write $C_{\ell-1} := \{u_i^{\ell-1} : 1 \le i \le k_\ell\}$. By Theorem 5.27(6), $u_1^{\ell-1}, \ldots, u_{k_\ell}^{\ell-1}$ are linearly independent vectors in $G_{\ell-1}$ and $\text{span}(C_{\ell-1}) \cap G_{\ell-2} = \{0\}$. Then there exists a subspace $U_{\ell-1}$ of $G_{\ell-1}$ such that $G_{\ell-1} = G_{\ell-2} \oplus U_{\ell-1}$ and $C_{\ell-1} \subseteq U_{\ell-1}$. Notice that $\dim(U_{\ell-1}) = \gamma_{\ell-1} - \gamma_{\ell-2} = k_{\ell-1}$. We extend the linearly independent set $C_{\ell-1}$ to a basis

$$B_{\ell-1} := u_1^{\ell-1}, \ \ldots, \ u_{k_\ell}^{\ell-1}, \ u_{k_\ell+1}^{\ell-1}, \ \ldots, \ u_{k_{\ell-1}}^{\ell-1}.$$

for $U_{\ell-1}$. We obtain $G_{\ell-1} = G_{\ell-2} \oplus \text{span}(B_{\ell-1})$ with $C_{\ell-1} \subseteq B_{\ell-1}$. Next, we define

$$u_i^{\ell-2} := (T - \lambda I)u_i^{\ell-1} \quad \text{for } i \in \{1, \ldots, k_{\ell-1}\}.$$

Write $C_{\ell-1}$ as the set of these $u_i^{\ell-2}$, and extend it to a basis $B_{\ell-2}$ for a subspace $U_{\ell-2}$ so that $G_{\ell-2} = G_{\ell-3} \oplus \text{span}(B_{\ell-2})$. We continue this process to construct the bases $B_\ell, B_{\ell-1}, \ldots, B_2, B_1$. The set $B := B_1 \cup \cdots \cup B_\ell$ is a basis for V. Next, we order the vectors in B by first taking the first elements of $B_1, \ldots, B_\ell$, in that order; next, taking the second elements of each, and so on. It is

$$B = \{u_1^1, u_1^2, \ldots, u_1^\ell; \ u_2^1, u_2^2, \ldots, u_2^\ell; \ \ldots, u_{k_1}^1\}.$$

The ordering may be given schematically as follows. We write the elements in $B_1, \ldots, B_\ell$ vertically as listed below and then visit them horizontally in the two-dimensional array:

$$\begin{array}{lllll}
B_1 & B_2 & \cdots & B_{\ell-1} & B_\ell \\
u_1^1 & u_1^2 & \cdots & u_1^{\ell-1} & u_1^\ell \\
\vdots & \vdots & \cdots & \vdots & \vdots \\
u_{k_\ell}^1 & u_{k_\ell}^2 & \cdots & u_{k_\ell}^{\ell-1} & u_{k_\ell}^\ell \\
 & & & & \\
u_{k_\ell+1}^1 & u_{k_\ell+1}^2 & \cdots & u_{k_\ell+1}^{\ell-1} & \\
\cdots & \vdots & \cdots & \vdots & \\
u_{k_{\ell-1}}^1 & u_{k_{\ell-1}}^2 & \cdots & u_{k_{\ell-1}}^{\ell-1} & \\
\vdots & \vdots & & & \\
u_{k_2-1}^1 & u_{k_2-1}^2 & & & \\
 & & & & \\
u_{k_2}^1 & u_{k_2}^2 & & & \\
 & & & & \\
u_{k_2+1}^1 & & & & \\
\vdots & & & & \\
u_{k_1}^1 & & & &
\end{array}$$

This is the required basis for V.

(1) To determine the matrix representation of T with respect to this basis B, we compute the T-images of the basis vectors as in the following:

$$
\begin{array}{llll}
T(u_1^1) & & = \lambda u_1^1 & \text{since } u_1^1 \in N(T-\lambda I). \\
T(u_1^2) & = (T-\lambda I)u_1^2 + \lambda u_1^2 & = u_1^1 + \lambda u_1^2 & \\
 & \vdots & & \\
T(u_1^\ell) & = (T-\lambda I)u_1^\ell + \lambda u_1^\ell & = u_1^{\ell-1} + \lambda u_1^\ell & \text{First line in } B \\
T(u_2^1) & & = \lambda u_2^1 & \text{since } u_2^1 \in N(T-\lambda I). \\
T(u_2^2) & = (T-\lambda I)u_2^2 + \lambda u_2^2 & = u_2^1 + \lambda u_2^2 & \\
 & \vdots & & \\
T(u_2^\ell) & = (T-\lambda I)u_2^\ell + \lambda u_2^\ell & = u_2^{\ell-1} + \lambda u_2^\ell & \text{Second line in } B \\
 & \vdots & & \\
T(u_{k_1}^1) & & = \lambda u_{k_1}^1 & \text{Last line in } B
\end{array}
$$

Observe that the matrix $[T]_{B,B}$ is block-diagonal with the diagonal blocks as Jordan blocks. Further, it is clear from the construction that the blocks satisfy the required properties.

(2) To compute the formula for n_i, look at the way we wrote the basis B as a two-dimensional array. Each Jordan block in $[T]_{B,B}$ corresponds to a row of vectors in that array. For example, the first row corresponds to the first Jordan block of order k_ℓ. The number of vectors in B_ℓ, which are linearly independent vectors from $G_\ell \setminus G_{\ell-1}$, is the number of Jordan blocks of order k_ℓ. Thus,

$$n_{k_\ell} = k_\ell = \gamma_\ell - \gamma_{\ell-1} = 2\gamma_\ell - \gamma_{\ell-1} - \gamma_{\ell+1}.$$

The last equality follows since $\gamma_\ell = \gamma_{\ell+1}$.

In general, scanning the columns in that array from right to left, we see that the number of rows containing exactly i number of vectors is the number n_i of Jordan blocks in $[T]_{B,B}$. However, the number of such rows is the number of vectors which we had chosen in extending the vectors in B_{i+1} to those in B_i. Such vectors are the vectors in B_i which are not in the form $(T - \lambda I)v$ for vectors v in B_{i+1}. That is, the number of such vectors is $|B_i| - |B_{i+1}|$. Therefore,

$$n_i = k_i - k_{i+1} = (\gamma_i - \gamma_{i-1}) - (\gamma_{i+1} - \gamma_i) = 2\gamma_i - \gamma_{i-1} - \gamma_{i+1}.$$

This proves (2).

Notice that the numbers n_i do not depend on the particular basis B. The formula in (2) shows that n_is are uniquely determined by the dimensions of the generalized eigenspaces of T. Therefore, the Jordan block representation of T is unique up to a permutation of the blocks. ■

We will call the matrix $[T]_{B,B}$ in Theorem 6.18 as *the Jordan block representation* of the linear operator T. The theorem is stated informally as follows:

> If T is a linear operator on a complex vector space of dimension n with a single eigenvalue λ, then there exists an $n \times n$ block-diagonal matrix representation of T with Jordan blocks on the diagonal, in which the maximum block size is the ascent of λ, the total number of blocks is the geometric multiplicity of λ; and this representation is unique up to a permutation of the blocks.

Since λ is the only eigenvalue of T, by Theorem 5.29, $G(\lambda) = V$ and hence the linear operator T in Theorem 6.18 satisfies the equation $(T - \lambda I)^n = 0$. Recall that a linear operator S on V with $S^k = 0$ for some $k \in \mathbb{N}$, is a *nilpotent* operator. If S is a nilpotent operator on V and $\dim(V) = n$, then necessarily $S^n = 0$ since every vector in V is a generalized eigenvector of S. The *index of nilpotency* of a nilpotent operator S is a natural number $m \le n$ such that $S^m = 0$ but $S^{m-1} \ne 0$. Since 0 is the only eigenvalue of a nilpotent operator, Theorem 6.18 speaks of a canonical form of a nilpotent operator, which may be stated as follows:

> If S is a nilpotent operator on an n-dimensional vector space V, with index of nilpotency m, then a basis B for V can be chosen in such a way that $[S]_{B,B} = \text{diag}\big(J(0, m_1), \ldots, J(0, m_k)\big)$, where $m_1 \ge \cdots \ge m_k \ge 1$, k is the geometric multiplicity and m_1 is the ascent of the eigenvalue 0 of S.

Any row of the two-dimensional array of the basis elements in B is a set of vectors generated by the rightmost vector v in that row by taking successive images of v under $(T - \lambda I)$. The set of all vectors in any such row is called a **Jordan chain.** The subspace spanned by a Jordan chain is T-invariant. The matrix of the restriction of T to this invariant subspace is the corresponding Jordan Block.

Exercises for Sect. 6.4

1. Let $A \in \mathbb{C}^{2\times 2}$ with $A \ne 0$ but $A^2 = 0$. Show that A is similar to $\begin{bmatrix} 0 & 1 \\ 0 & 0 \end{bmatrix}$.
2. Prove that any $A \in \mathbb{C}^{2\times 2}$ is similar to a matrix in the form $\begin{bmatrix} a & 0 \\ 0 & b \end{bmatrix}$ or $\begin{bmatrix} a & 1 \\ 0 & a \end{bmatrix}$.
3. Show that the matrix $\begin{bmatrix} 0 & a & b \\ 0 & 0 & c \\ 0 & 0 & 0 \end{bmatrix}$ is nilpotent of index 3 for any nonzero numbers a, b, c.
4. Describe the actions of all 2×2 matrices A on $\mathbb{R}^2$ if $A^2 = I$.
5. Show that $A = [a_{ij}] \in \mathbb{R}^{3\times 3}$ with $a_{ij} = (-1)^{i+1}$ is nilpotent, and determine its Jordan block representation.
6. Show that the Jordan block $J(\lambda, m)$ has the only eigenvalue λ and the associated eigenspace $E(\lambda)$ has dimension 1. Also, show that $(J(\lambda, m) - \lambda I)^m = 0$ and $(J(\lambda, m) - \lambda I)^k \ne 0$ for $k < m$.
7. Show that rank $(J(\lambda, k))^i$ is k if $\lambda \ne 0$, and it is $\max\{0, k - i\}$ if $\lambda = 0$.
8. Show that if a linear operator T on a finite dimensional vector space has a single eigenvalue λ, then the number of 1s in the super-diagonal of its Jordan block representation is equal to $\mu(\lambda) - \gamma(\lambda)$.
9. Show that the differentiation operator on $\mathcal{P}_n(\mathbb{C})$ is nilpotent. What is its index of nilpotency? What is its Jordan block representation?

10. If an $n \times n$ matrix A has trace 0 and rank 1, then show that A is nilpotent.
11. Show that each nilpotent $n \times n$ matrix is similar to a matrix with possibly nonzero entries only on the super-diagonal.
12. Let A be an $n \times n$ matrix with nonzero super-diagonal entries, all other entries being 0. Show that A is nilpotent of index n.
13. Let T be a nilpotent operator on a vector space of dimension n. Show that $T^n = 0$ and that 0 is the only eigenvalue of T.
14. Let T be a linear operator on an n-dimensional vector space V. Prove that T is nilpotent if and only if the characteristic polynomial of T is t^n.
15. Show that if a matrix $B \in \mathbb{C}^{m \times m}$ commutes with a Jordan block $J(0, m)$, then there exists $p(t) \in \mathcal{P}_m(\mathbb{C})$ such that $B = p(J(0, m))$.

6.5 Jordan Normal Form

The general case of a linear operator having many eigenvalues is addressed in the following theorem.

Theorem 6.19 (Jordan normal form) *Let T be a linear operator on a vector space V of dimension n over $\mathbb{F}$. Suppose the characteristic polynomial of T splits over $\mathbb{F}$. Then there exists a basis B for V such that the following are true:*

(1) $[T]_{B,B}$ is a block-diagonal matrix having Jordan blocks on its diagonal.
(2) Let λ be an eigenvalue of T with ascent $\ell(\lambda)$ and let $\gamma_i = \text{null}\,(T - \lambda I)^i$. If n_i is the number of Jordan blocks $J(\lambda, i)$ that appear in $[T]_{B,B}$, then

$$n_i = 2\gamma_i - \gamma_{i-1} - \gamma_{i+1} \quad \textit{for } i \in \{1, \ldots, \ell(\lambda)\}.$$

Further, such a representation of T as a block-diagonal matrix with Jordan blocks is unique up to a permutation of the Jordan blocks.

Proof Suppose $\lambda_1, \ldots, \lambda_r$ are the distinct eigenvalues of T. By Theorems 5.27 and 5.29, $V = G(\lambda_1) \oplus \cdots G(\lambda_r)$ where each generalized eigenspace $G(\lambda_i)$ is a T-invariant subspace.

Consider the restriction linear operators $T_i := T|_{G(\lambda_i)}$. For each $i \in \{1, \ldots, r\}$, T_i is a linear operator on $G(\lambda_i)$, and it has only one eigenvalue λ_i with algebraic multiplicity $\mu(\lambda_i) = \dim(G(\lambda_i))$. Due to Theorem 6.18, there exists a basis B_i such that $[T_i]_{B_i, B_i}$ is a block-diagonal matrix J_i whose blocks are the Jordan blocks of the form $J(\lambda_i)$. Take the basis $B = B_1 \cup \cdots \cup B_r$ for V.

(1) Clearly, $[T]_{B,B}$ is the required block-diagonal matrix.

(2) Let $\lambda = \lambda_j$ be an eigenvalue for some $j \in \{1, \ldots, r\}$. We compute n_i and γ_i for this eigenvalue. The number n_i is the number of Jordan blocks $J(\lambda_j, i)$ that appear in $[T]_{B,B}$. Notice that n_i is equal to the number of Jordan blocks $J(\lambda_j, i)$ that appear on the Jordan block representation of $[T_j]_{B_j, B_j}$. As $\gamma_i = \text{null}\,(T - \lambda_j I)^i = \text{null}\,(T_j - \lambda_j I)^i$, by Theorem 6.18, it follows that $n_i = 2\gamma_i - \gamma_{i-1} - \gamma_{i+1}$.

The uniqueness of the block-diagonal matrix up to a permutation of the Jordan blocks follows from the formula for n_i. ■

If T is a linear operator on a finite dimensional vector space V, then any basis B for V, with respect to which $[T]_{B,B}$ is in Jordan normal form, is called a **Jordan basis**.

Due to Theorem 6.18, our proof says that $[T]_{B,B} = \operatorname{diag}(J_1, \ldots, J_r)$, where J_i is a block-diagonal matrix having blocks as Jordan blocks of the form $J(\lambda_i)$ for $i \in \{1, \ldots, r\}$. In addition, the following are true:

1. The distinct numbers on the diagonal of $[T]_{B,B}$ are $\lambda_1, \ldots, \lambda_r$, where each λ_i occurs $\mu(\lambda_i)$ times.
2. The number of Jordan blocks whose diagonal entries are λ_i is $\gamma(\lambda_i)$.
3. The largest Jordan block whose diagonal entries are λ_i has order $\ell(\lambda_i)$.
4. The characteristic polynomial of T is $\Pi_{i=1}^{r}(t - \lambda_i)^{\mu(\lambda_i)}$.
5. The minimal polynomial of T is $\Pi_{i=1}^{r}(t - \lambda_i)^{\ell(\lambda_i)}$; see Theorem 5.34.
6. The number of 1s in the super-diagonal of the Jordan normal form is equal to $\sum_{i=1}^{r}(\mu(\lambda_i) - \gamma(\lambda_i))$.

As our construction shows, the Jordan blocks in each J_i read from left top to right bottom are nonincreasing in order. We can, of course, choose any ordering of these J_is. That is, a Jordan normal form is unique provided we fix some ordering of the eigenvalues $\lambda_1, \ldots, \lambda_r$. In this sense, Jordan normal form is canonical; we thus speak of *the* Jordan normal form of a linear operator.

Example 6.20 Let $T : \mathbb{R}^6 \to \mathbb{R}^6$ be defined by $T(a_1, a_2, a_3, a_4, a_5, a_6) = (b_1, b_2, b_3, b_4, b_5, b_6)$, where

$$b_1 = 2a_1, \quad b_2 = a_1 + 2a_2, \quad b_3 = -a_1 + 2a_3,$$
$$b_4 = a_2 + 2a_4, \quad b_5 = a_1 + a_2 + a_3 + a_4 + 2a_5, \quad b_6 = a_5 - a_6.$$

The matrix of T in the standard basis is lower triangular; its diagonal entries are the eigenvalues. The characteristic polynomial of T is $(t-2)^5(t+1)$.

For the eigenvalue -1, we see that

$$(T + I)(a_1, a_2, a_3, a_4, a_5, a_6) = (3a_1, \ a_1 + 3a_2, \ -a_1 + 3a_3, \ a_2 + 3a_4, \ a_1 + a_2 + a_3 + a_4 + 3a_5, \ a_5).$$

Then $(T + I)(a_1, a_2, a_3, a_4, a_5, a_6) = 0$ when $a_1 = a_2 = a_3 = a_4 = a_5 = 0$ and a_6 arbitrary. A corresponding eigenvector is clearly e_6. Since $N(T + I)^2 = N(T + I)$, $G(-1) = N(T + I) = \operatorname{span}\{e_6\}$. Also, $\gamma(-1) = \mu(-1) = \ell(-1) = 1$. The restriction operator $T|_{G(-1)}$ is given by

$$T|_{G(-1)}(0, 0, 0, 0, 0, a_6) = (0, 0, 0, 0, 0, -a_6).$$

Notice that $\dim(G(-1)) = 1$. Thus the matrix of $T|_{G(-1)}$ in the basis $\{e_6\}$ is a 1×1 matrix having the single entry as -1. It is the Jordan block $J(-1, 1)$.

We see that the algebraic multiplicity of the eigenvalue 2 is $\mu(2) = 5$. We compute its geometric multiplicity and ascent as in the following.
$(T - 2I)(a_1, a_2, a_3, a_4, a_5, a_6)$
$= (0,\ a_1,\ -a_1,\ a_2,\ a_1 + a_2 + a_3 + a_4,\ a_5 - 3a_6)$
$T(T - 2I)^2(a_1, a_2, a_3, a_4, a_5, a_6)$
$= (T - 2I)(0, a_1, -a_1, a_2, a_1 + a_2 + a_3 + a_4, a_5 - 3a_6)$
$= (0,\ 0,\ 0,\ a_1,\ a_2,\ a_1 + a_2 + a_3 + a_4 - 3a_5 + 9a_6)$
$T(T - 2I)^3(a_1, a_2, a_3, a_4, a_5, a_6)$
$= (0, 0, 0, 0, a_1, -3a_1 - 2a_2 - 3a_3 - 3a_4 + 9a_5 - 27a_6)$
$T(T - 2I)^4(a_1, a_2, a_3, a_4, a_5, a_6)$
$= (0, 0, 0, 0, 0, -3(10a_1 + 6a_2 + 9a_3 + 9a_4 - 27a_5 + 81a_6))$
$(T - 2I)^5(a_1, a_2, a_3, a_4, a_5, a_6)$
$= (0, 0, 0, 0, 0, -3(10a_1 + 6a_2 + 9a_3 + 9a_4 - 27a_5 + 81a_6))$.

$G_1 = N(T - 2I) = \{(0, 0, a_3, a_4, a_5, a_6) : a_3 + a_4 = a_5 - 3a_6 = 0\}$,
$G_2 = N(T - 2I)^2 = \{(0, 0, a_3, a_4, a_5, a_6) : a_3 + a_4 - 3a_5 + 9a_6 = 0\}$,
$G_3 = N(T - 2I)^3$
$\quad = \{(0, a_2, a_3, a_4, a_5, a_6) : 2a_2 + 3a_3 + 3a_4 - 9a_5 + 27a_6 = 0\}$,
$G_4 = N(T - 2I)^4$
$\quad = \{(a_1, a_2, a_3, a_4, a_5, a_6) : 10a_1 + 6a_2 + 9a_3 + 9a_4 - 27a_5 + 81a_6 = 0\}$
$G_5 = N(T - 2I)^5 = N(T - 2I)^4 = G_4$.

Hence $\ell(2) = 4$ and $G(2) = G_4 = N(T - 2I)^4$. Write γ_i for $\gamma_i(2)$. Thus, we obtain

$$\gamma_0 = N(T - 2I)^0 = N(I) = 0, \quad \gamma_1 = \gamma = \dim(G_1) = 2, \quad \gamma_2 = \dim(G_2) = 3,$$
$$\gamma_3 = \dim(G_3) = 4, \quad \gamma_4 = \dim(G_4) = 5, \quad \gamma_{4+j} = 5 \text{ for } j \geq 1.$$

The number of Jordan blocks $J(2, i)$ appearing in the Jordan form is given by $n_i = 2\gamma_i - \gamma_{i-1} - \gamma_{i+1}$. It gives

$$n_1 = 1, \quad n_2 = 0, \quad n_3 = 0, \quad n_4 = 1, \quad n_{4+j} = 0 \text{ for } j \geq 1.$$

Therefore, there are two Jordan blocks for the eigenvalue 2, namely $J(2, 4)$ and $J(2, 1)$. Along with the one for $\lambda = -1$, the Jordan normal form of T is the block-diagonal matrix diag($J(2, 4), J(2, 1), J(-1, 1)$).

To illustrate the construction in the proof of Theorem 6.18, we consider the restriction operator $T|_{G(2)}$, which is same as T with domain and co-domain as $G(2)$. Thus, for ease in notation, we continue using T instead of $T|_{G(2)}$. Recall that $k_i = \gamma_i - \gamma_{i-1}$ implies that $k_1 = 2,\ k_2 = 1,\ k_3 = 1$ and $k_4 = 1$. The next step in the construction is to start with $k_4 = 1$ linearly independent vectors from $G_4 \setminus G_3$. Such a vector $u_1^4 = (a_1, a_2, a_3, a_4, a_5, a_6)$ satisfies

$$10a_1 + 6a_2 + 9a_3 + 9a_4 - 27a_5 + 81a_6 = 0$$

but does not satisfy

$$a_1 = 0 \quad \text{and} \quad 2a_2 + 3a_3 + 3a_4 - 9a_5 + 27a_6 = 0.$$

We choose one by taking a_1 nonzero and satisfying the first equation in the simplest manner, say, by choosing a_2 nonzero and all others zero:

$$u_1^4 = (3, \ -5, \ 0, \ 0, \ 0, \ 0).$$

Since $k_4 = 1$, there is no need to extend $\{u_1^4\}$ to a basis for U_4 in order that $G_4 = G_3 \oplus U_4$ happens. We see that such a U_4 is given by span$\{u_1^4\}$.

Next, we take

$$u_1^3 := (T - 2I)u_1^4 = (T - 2I)(3, \ -5, \ 0, \ 0, \ 0, \ 0) = (0, 3, -3, -5, -2, 0).$$

Since all vectors in G_2 have second component equal to 0, $u_1^3 \notin G_2$ as stated in Theorem 6.18. Further, $u_1^3 \in G_3$; its first component is 0 and the rest of the components satisfy:

$$2a_2 + 3a_3 + 3a_4 - 9a_5 + 27a_6 = 0.$$

Thus $u_1^3 \in G_3$. Since $k_3 = 1$, we take $U_3 = \text{span}\{u_1^3\}$ and we have $G_3 = G_2 \oplus U_3$.

In the next step, we take

$$u_1^2 = (T - 2I)u_1^3 = (T - 2I)(0, 3, -3, -5, -2, 0) = (0, 0, 0, 3, -5, -2).$$

Check that $u_1^2 \in G_2 \setminus G_1$. as $k_2 = 1$, $G_2 = G_1 \oplus U_2$, where $U_2 = \text{span}\{u_1^2\}$.

Finally, we take

$$u_1^1 = (T - 2I)u_1^2 = (T - 2I)(0, 0, 0, 3, -5, -2) = (0, 0, 0, 0, 3, 1).$$

It is easy to check that $u_1^1 \in G_1 \setminus G_0$, recalling that $G_0 = \{0\}$. Since $k_1 = 2$, we need to extend $\{u_1^1\}$ to a linearly independent set by adding one more vector from G_1. In u_1^1, we have the fourth component 0, so we choose a vector with its fourth component nonzero for keeping linear independence. Our choice is

$$u_2^1 = (0, 0, -1, 1, 0, 0).$$

Now, we must order the basis vectors $u_1^1, u_2^1, u_1^2, u_1^3, u_1^4$. For this purpose, we write them vertically for the respective spaces

$$\begin{array}{cccc} G_1 & U_2 & U_3 & U_4 \\ u_1^1 & u_1^2 & u_1^3 & u_1^4 \\ u_2^1 & & & \end{array}$$

We visit them horizontally to get the basis for $G(2)$ as

$$B_1 = \{u_1^1,\ u_1^2,\ u_1^3,\ u_1^4,\ u_2^1\}.$$

As discussed earlier, $G(-1)$ has the basis $\{e_6\}$. Since $\mathbb{R}^6 = G(2) \oplus G(-1)$, we have the basis for $\mathbb{R}^6$ as

$$B = \{u_1^1,\ u_1^2,\ u_1^3,\ u_1^4,\ u_2^1, e_6\}.$$

This is the Jordan basis in which $[T]_{B,B}$ should be in the Jordan normal form. We verify it as follows:

$$\begin{array}{lcllll}
Tu_1^1 &=& T(0,0,0,0,3,1) &= (0,0,0,0,6,2) &= 2u_1^1\\
Tu_1^2 &=& T(0,0,0,3,-5,-2) &= (0,0,0,6,-7,-3) &= u_1^1 + 2u_1^2\\
Tu_1^3 &=& T(0,3,-3,-5,-2,0) &= (0,6,-6,-7,-9,-2) &= u_1^2 + 2u_1^3\\
Tu_1^4 &=& T(3,-5,0,0,0,0) &= (6,-7,-3,-5,-2,0) &= u_1^3 + 2u_1^4\\
Tu_2^1 &=& T(0,0,-1,1,0,0) &= (0,0,-2,2,0,0) &= 2u_2^1\\
Te_6 &=& T(0,0,0,0,0,1) &= (0,0,0,0,0,-1) &= -e_6
\end{array}$$

Then $[T]_{E,E} = J$ as given below. Further, construct the matrix P by taking the transposes of the basis vectors as its columns, and take $[T]$ as the matrix of T with respect to the standard basis of $\mathbb{R}^6$. Then the following verification shows that $P^{-1}[T]P = J$.

$$T = \begin{bmatrix} 2&0&0&0&0&0\\ 1&2&0&0&0&0\\ -1&0&2&0&0&0\\ 0&1&0&2&0&0\\ 1&1&1&1&2&0\\ 0&0&0&0&1&-1 \end{bmatrix},\quad P = \begin{bmatrix} 0&0&0&6&0&0\\ 0&0&3&-5&0&0\\ 0&0&-3&0&-1&0\\ 0&3&-5&0&1&0\\ 3&-5&-2&0&0&0\\ 1&-2&0&0&0&1 \end{bmatrix},$$

$$J = \begin{bmatrix} 2&1&&&&\\ &2&1&&&\\ &&2&1&&\\ &&&2&&\\ &&&&2&\\ &&&&&-1 \end{bmatrix},\quad TP = PJ = \begin{bmatrix} 0&0&0&6&0&0\\ 0&0&6&-7&0&0\\ 0&0&-6&-3&-2&0\\ 0&6&-7&-5&2&0\\ 6&-7&-9&-2&0&0\\ 2&-3&-2&0&0&-1 \end{bmatrix}.$$

Notice that the columns of the matrix P are the transposes of the vectors of the Jordan basis.

In general, if T is a given matrix of order n, then it is considered as a linear operator on $\mathbb{F}^{n\times 1}$. In that case, vectors in the Jordan basis are from $\mathbb{F}^{n\times 1}$; consequently, they become the columns of P instead of their transposes. □

From the uniqueness of Jordan normal form, it follows that two $n \times n$ matrices are similar if and only if they have the same Jordan normal form, up to permutations of Jordan blocks. Moreover, the Jordan normal form of a linear operator T on a finite dimensional complex vector space is determined by the dimensions of the generalized eigenspaces, equivalently, from the ranks of the operators $(T - \lambda I)^j$ for $j = 0, 1, 2, \ldots,$ for eigenvalues λ of T, due to the rank-nullity theorem.

For any matrix $A \in \mathbb{C}^{n\times n}$, equality of the row rank with the rank implies that $\operatorname{rank}(A^T - \lambda I)^j = \operatorname{rank}(A - \lambda I)^j$ for $j = 0, 1, 2, \ldots$. Therefore, A^T is similar to A.

To see how A^T is similar to A explicitly, go back to the basis B as written down in the proof of Theorem 6.18. Instead of reading the two-dimensional list from left to right, read them from right to left, row after row from top to bottom. For example, the first two rows in the new ordering would now look like:

$$u_1^\ell \ldots, u_1^2,\ u_1^1;$$

$$u_2^\ell \ldots, u_2^2,\ u_2^1;$$

We call this new ordering of B as the *reverse order basis.* In the reverse order basis, each Jordan block will be a lower triangular matrix. Consequently, the Jordan matrix in Theorem 6.19 will be a lower triangular matrix with Jordan blocks having 1s on the subdiagonal instead of the super-diagonal. Notice that in the reverse order basis, the Jordan matrix is exactly J^T, where J is the Jordan matrix of Theorem 6.19. Therefore, A is similar to J as well as to J^T. However, if A has the Jordan representation $A = P^{-1}JP$, then $A^T = P^T J^T (P^T)^{-1}$. That is, A^T is similar to J^T. Consequently, A is similar to A^T.

Exercises for Sect. 6.5

1. Determine a Jordan basis and a Jordan normal form of each of the following matrices:

(a) $\begin{bmatrix} 0&1&1&0&1\\ 0&0&1&1&1\\ 0&0&0&0&0\\ 0&0&0&0&0\\ 0&0&0&0&0 \end{bmatrix}$ (b) $\begin{bmatrix} 1&-1&0\\ -1&4&-1\\ -4&13&-3 \end{bmatrix}$ (c) $\begin{bmatrix} 1&0&3&0\\ 1&3&0&3\\ 0&0&1&0\\ 0&0&3&1 \end{bmatrix}$

(d) $\begin{bmatrix} 1&-1&-2&3\\ 0&0&-2&3\\ 0&1&1&-1\\ 0&0&-1&2 \end{bmatrix}$ (e) $\begin{bmatrix} 1&0&0&0\\ 0&0&1&0\\ 0&0&0&1\\ 1&6&-1&-4 \end{bmatrix}$ (f) $\begin{bmatrix} 5&-1&0&0\\ 9&-1&0&0\\ 0&0&7&-2\\ 0&0&12&-3 \end{bmatrix}$.

2. Show that the following matrices have the same characteristic polynomial, the same minimal polynomial, but they have different Jordan normal forms:

(a) $\begin{bmatrix} 0&1&0&0\\ 0&0&0&0\\ 0&0&0&0\\ 0&0&0&0 \end{bmatrix}$ (b) $\begin{bmatrix} 0&1&0&1\\ 0&0&0&0\\ 0&0&0&1\\ 0&0&0&0 \end{bmatrix}$.

3. Show that $\begin{bmatrix} 1&1&1\\ -1&-1&-1\\ 1&1&0 \end{bmatrix}$ and $\begin{bmatrix} 1&1&1\\ -1&-1&-1\\ 1&0&0 \end{bmatrix}$ are not similar.

4. In each of the following cases, determine if the two matrices similar.

(a) $\begin{bmatrix} 1 & 1 & 1 \\ 1 & 1 & 1 \\ 1 & 1 & 1 \end{bmatrix}$ and $\begin{bmatrix} 3 & 0 & 0 \\ 0 & 0 & 0 \\ 0 & 0 & 0 \end{bmatrix}$.

(b) $\begin{bmatrix} 0 & 1 & \alpha \\ 0 & 0 & 1 \\ 0 & 0 & 0 \end{bmatrix}$ and $\begin{bmatrix} 1 & 1 & 0 \\ 0 & 0 & 1 \\ 0 & 0 & 0 \end{bmatrix}$, where α is any scalar.

(c) $\begin{bmatrix} 0 & 1 & 0 \\ 0 & 0 & 1 \\ 1 & 0 & 0 \end{bmatrix}$ and $\begin{bmatrix} 1 & 0 & 0 \\ 0 & \omega & 0 \\ 0 & 0 & \omega^2 \end{bmatrix}$, where $\omega^3 = 1$ and $\omega \neq 1$.

5. Let $A = [a_{ij}]$, $B = [b_{ij}] \in \mathbb{C}^{n\times n}$ with $a_{ij} = 1$ for all i, j; $b_{11} = n$, and $b_{ij} = 0$ for all other i, j. Show that A and B are similar.
6. Let $J \in \mathbb{C}^{n\times n}$ be a block-diagonal matrix with blocks as Jordan blocks. Suppose that the distinct diagonal entries in J are $\lambda_1, \ldots, \lambda_k$, where λ_i appears m_i times. Also, suppose that the largest Jordan block whose diagonal entries are λ_i is of order ℓ_i, and that the number of Jordan blocks whose diagonal entries are λ_i is γ_i. Then show the following:

   (a) The characteristic polynomial of J is $(t-\lambda_1)^{m_1}\cdots(t-\lambda_k)^{m_k}$.
   (b) The minimal polynomial of J is $(t-\lambda_1)^{\ell_1}\cdots(t-\lambda_k)^{\ell_k}$.
   (c) $\operatorname{null}(J-\lambda_i I) = \gamma_i$.

7. How many nonsimilar 3×3 matrices with real entries having characteristic polynomial $(t-1)^3$ exist?
8. Let T be a linear operator on a complex vector space of dimension 5. Suppose $\chi_T(t) = (t-2)^2(t-3)^3$, $\operatorname{null}(T-2I) = 1$, and $\operatorname{null}(T-3I) = 2$. What is the Jordan normal form of T?
9. Let T be a linear operator on a complex vector space of dimension 5. Suppose $\chi_T(t) = (t-\lambda)^5$ and $\operatorname{rank}(T-\lambda I) = 2$. What are the possible Jordan normal forms of T?
10. Let $A \in \mathbb{C}^{6\times 6}$. Suppose we are given the characteristic polynomial, the minimal polynomial, and also the geometric multiplicities of each eigenvalue of A. Show that the Jordan normal form of A is uniquely determined.
11. Let $A \in \mathbb{R}^{7\times 7}$ have the characteristic polynomial $(t-2)^4(t-3)^3$ and minimal polynomial $(t-2)^2(t-3)^2$. Show that A is similar to one of two matrices in Jordan normal form depending on $\gamma(2)$.
12. Give two matrices in $\mathbb{C}^{7\times 7}$ having the same characteristic polynomial, the same minimal polynomial, and the same geometric multiplicity of each eigenvalue, but which have distinct Jordan normal forms.
13. Find all possible Jordan normal forms of an 8×8 matrix whose characteristic polynomial is $(t-6)^4(t-7)^4$, minimal polynomial is $(t-6)^2(t-7)^2$, and $\gamma(6) = 3$.
14. Determine all possible Jordan normal forms of an 8×8 complex matrix with minimal polynomial $t^2(t-1)^3$.

15. Find all possible Jordan normal forms for 10×10 complex matrices having minimal polynomial as $t^2(t-1)^2(t+1)^3$.
16. What are the invariant subspaces of a linear operator whose Jordan normal form consists of a single Jordan block?
17. Show that $(J(0,m))^m = 0$. Show also that $p(J(\lambda, m)) = 0$, where $p(t)$ is the characteristic polynomial of the Jordan block $J(\lambda, m)$. Use this result to give another proof of Cayley–Hamilton theorem.
18. Let J_m be a Jordan block of order m. Let P be an $m \times m$ matrix whose anti-diagonal entries are all 1 and all other entries are 0. The anti-diagonal entries are the entries in the positions $(1, m), (2, m-1), \ldots, (m, 1)$. Show that $P^{-1} = P$ and that $P^{-1} J_m P = J_m^T$. Use this to prove that any square matrix is similar to its transpose.
19. Let $A \in \mathbb{C}^{n \times n}$ have only real eigenvalues. Show that there exists an invertible matrix $P \in \mathbb{R}^{n \times n}$ such that $P^{-1} A P \in \mathbb{R}^{n \times n}$.

6.6 Problems

1. How many orthogonal matrices P exist such that $P^{-1} \begin{bmatrix} 0 & 1 \\ 1 & 0 \end{bmatrix} P$ is diagonal?
2. Find a matrix P such that $P^T \begin{bmatrix} 1 & 5 \\ 5 & 26 \end{bmatrix} P = I$ and $P^T \begin{bmatrix} 1 & 8 \\ 8 & 56 \end{bmatrix}$ is diagonal.
3. Determine a matrix P such that $P^T \begin{bmatrix} -2 & 1 \\ 1 & -1 \end{bmatrix} P$ and $P^T \begin{bmatrix} 5 & 2 \\ 2 & -2 \end{bmatrix}$ are diagonal
4. Under what conditions on the complex numbers $\alpha_1, \ldots, \alpha_n$ the $n \times n$ matrix $\begin{bmatrix} & & \alpha_1 \\ & \alpha_2 & \\ \cdot^{\cdot^{\cdot}} & & \\ \alpha_n & & \end{bmatrix}$ is diagonalizable?
5. Are the following statements true? Give reasons.

   (a) Any 2×2 matrix with a negative determinant is diagonalizable.
   (b) If T is a linear operator on $\mathbb{C}^n$ with $T^k = 0$ for some $k > 0$, then T is diagonalizable.
   (c) If the minimal polynomial of a linear operator T on $\mathbb{C}^n$ has degree n, then T is diagonalizable.

6. Let T be a linear operator on a finite dimensional complex vector space V. Prove that T is diagonalizable if and only if T is annihilated by a polynomial having no multiple zeros.
7. Let T be an invertible linear operator on a finite dimensional complex vector space. If T^m is diagonalizable for some $m \in \mathbb{N}$, then prove that T is diagonalizable.

8. Let T be a diagonalizable operator on a finite dimensional vector space V. Prove that if U is a T-invariant subspace of V, then $T|_U$ is also diagonalizable.
9. Let T be a linear operator on a real vector space V. Prove that if V has a basis of eigenvectors of T, then $\text{tr}(T^2) \geq 0$.
10. Let S and T be linear operators on a finite dimensional vector space V. If $ST = TS$, prove that there exists a basis of V with respect to which both $[S]$ and $[T]$ are upper triangular.
11. Let T be a linear operator on a vector space. Prove that if for each nonzero vector v, there exists a scalar λ such that $Tv = \lambda v$, then T is a scalar operator.
12. Show that any matrix of trace zero is similar to a matrix with each diagonal entry zero.
13. Let T be a linear operator on a vector space V of dimension n. If $\text{tr}(T) = n\lambda$, prove that there exists a basis of V with respect to which all diagonal entries of the matrix $[T]$ are λ.
14. Let T be a linear operator on a finite dimensional inner product space. Prove that T is nilpotent if and only if $[T]_{B,B}$ is upper triangular with all diagonal entries 0 with respect to some basis B of V.
15. Let T be a linear operator on a finite dimensional vector space V over $\mathbb{F}$. Let U be a T-invariant proper subspace of V. Prove that T is triangularizable if and only if the minimal polynomial of T splits over $\mathbb{F}$.
16. Let T be a linear operator on a finite dimensional complex vector space. Prove that if 0 is the only eigenvalue of T, then T is nilpotent. This is not necessarily true for a real vector space.
17. Prove that if the minimal polynomial of a matrix $A \in \mathbb{C}^{3\times 3}$ having a single eigenvalue λ is known, then the Jordan block representation of A is uniquely determined. Also show that this is no longer true if $A \in \mathbb{C}^{4\times 4}$.
18. Let T be a nilpotent operator on a finite dimensional vector space. Suppose that in the Jordan block representation of T, the longest string of consecutive 1s on the super-diagonal has length m. Show that the minimal polynomial of T is t^{m+1}.
19. What is the minimal polynomial of the zero (square) matrix?
20. Show that the minimal polynomial of a nonzero square matrix is a nonzero polynomial.
21. Show that the relation "unitarily similar" is an equivalence relation.
22. Are A^*A and AA^* unitarily similar?
23. Are A^* and A unitarily similar?
24. Let $A \in \mathbb{C}^{n\times n}$ be such that $\text{tr}(A^k) = 0$ for each $k \in \mathbb{N}$. Prove that A is nilpotent.
25. Let S and T be nilpotent linear operators on a finite dimensional vector space V. Prove that if ST is nilpotent, then so is TS.
26. Let $A \in \mathbb{F}^{n\times n}$ be a nilpotent matrix of index n. Determine a matrix B such that $B^2 = I + A$.
27. Let T be a linear operator on a finite dimensional vector space. Prove that there are subspaces U and W of V such that $V = U \oplus W$, $T|_U$ is nilpotent, and $T|_W$ is invertible.
28. Let T be a nilpotent operator on a vector space of dimension n. Prove that $\det(I + T) = 1$.

29. Let T be a nilpotent operator on a vector space of dimension n. Suppose $\text{null}\,(T^{n-1}) \neq \text{null}\,(T^n)$. Prove that for each $j \in \{0, 1, \ldots, n\}$, $\text{null}\,(T^j) = j$.
30. Let T be a nilpotent operator on a finite dimensional vector space V. Let m be the index of nilpotency of T. Prove that for each $k \in \{1, \ldots, m-1\}$, $\text{null}\,(T^{k+1}) + \text{null}\,(T^{k-1}) \le 2\,\text{null}\,(T^k)$.
31. Theorem 6.18 can also be proved by induction on $n = \dim(V)$. Towards this end, assume that T is a nilpotent linear operator on a vector space V of finite dimension whose characteristic polynomial splits. First, show that if $\dim(V) = 1$, then there exists a basis $B = \{v\}$ of V such that $[T]_{B,B} = [\lambda]$. Next, assume that each nilpotent linear operator on any vector space of dimension at most $n-1$, has a basis in which the matrix of the linear operator is a block-diagonal matrix with Jordan blocks on its diagonal. Let $\dim(V) = n$. Prove the following:

    (a) Let $U := R(T)$ and let $S = T|_U$. Then U is an S-invariant subspace of V, and $S(u) = T(u)$ for each $u \in U$. Further, there exists a basis B for U with $B = B_1 \cup \cdots \cup B_k$, where

    $$B_i = \{u_1^i, \ldots, u_{n_i}^i\} \text{ with } S(u_1^i) = 0,\ T(u_j^i) = u_{j-1}^i \text{ for } 2 \le j \le n_i.$$

    (b) Write $N(T) = (R(T) \cap U) \oplus W$, $n_i = |B_i|$, $r := n_1 + \cdots + n_k$, and $s := \dim(R(T) \cap U)$. The last vector $u_{n_i}^i$ in each B_i is in $N(T)$. Write $u_{n_i+1}^i := Tu_{n_i}^i$. Construct $E_i = B_i \cup \{u_{n_i+1}^i\}$.
    (c) Choose a basis $\{z_1, \ldots, z_{n-r-s}\}$ for W. Then the set $E := E_1 \cup \cdots \cup E_k \cup \{z_1, \ldots, z_{n-r-s}\}$ is linearly independent.
    (d) E is a basis for V and $[T]_{E,E}$ is a block-diagonal matrix with Jordan blocks on its diagonal.

32. Let a square matrix A have the Jordan normal form J so that there exists an invertible matrix P where $AP = PJ$. Show that the columns of P are either eigenvectors of A or solutions x of the linear equation of the form $Ax = u + \lambda x$ for a known vector u and an eigenvalue λ of A.
33. Assume that the matrix A has the Jordan normal form J, where

    $$A = \begin{bmatrix} -2 & -1 & -3 \\ 4 & 3 & 3 \\ -2 & 1 & -1 \end{bmatrix}, \quad J = P^{-1}AP = \begin{bmatrix} -4 & 0 & 0 \\ 0 & 2 & 1 \\ 0 & 0 & 2 \end{bmatrix}.$$

    Use the ideas in Problem 32 to determine the matrix P.
34. Let J be a Jordan block representation of a linear operator T on an n-dimensional complex vector space V having a single eigenvalue λ. Show that the uniqueness of Jordan block representation in Theorem 6.18 follows from each one of the following.

    (a) $\text{null}\,(J - \lambda I)^i - \text{null}\,(J - \lambda I)^{i-1} =$ the number of Jordan blocks of order at least i.

(b) $\dim(N(A - \lambda I) \cap R(A - \lambda I)^{j-1})$ = the number of Jordan blocks of order at least j.
(c) rank $(J - \lambda I)^k = \sum_{i=1}^{s}(m_i - k)$, if $J = \mathrm{diag}(J(\lambda, m_1), \ldots, J(\lambda, m_r))$ with $m_1 \geq \cdots \geq m_r$ and s is the largest index in $\{1, \ldots, r\}$ such that $m_s > k$.

35. Another way to show uniqueness of a Jordan normal form is given below; prove it.
Let T be a linear operator on a vector space of dimension n over $\mathbb{F}$. Suppose that the characteristic polynomial of T splits over $\mathbb{F}$. For an eigenvalue λ of T, let $k = \ell(\lambda)$; and $m = \mu(\lambda)$. For $j \in \mathbb{N}$, write $r_j = \text{rank }(T - \lambda I)^j$, and let n_j denote the number of Jordan blocks of the form $J(\lambda, i)$ that appear in the Jordan normal form of T. Then

(a) $n_1 + 2n_2 + 3n_3 + \cdots + kn_k = m.$
(b) $r_j = n - m$ for $j \geq k$, and $r_j > n - m$ for $1 \leq j < k$.
(c) $r_{k-1} = n_k + n - m,\ r_{k-2} = 2n_k + n_{k-1} + n - m,$
$r_{k-3} = 3n_k + 2n_{k-1} + n_{k-2} + n - m,\ \ldots$
$r_1 = (k-1)n_k + (k-2)n_{k-1} + \cdots + 2n_3 + n_2 + n - m.$
(d) $n_j = r_{j+1} - 2r_j + r_{j-1} = 2\gamma_j - \gamma_{j-1} - \gamma_{j+1}.$

36. Prove or disprove: Let $A, B \in \mathbb{F}^{n\times n}$. If the matrices $\begin{bmatrix} A & 0 \\ 0 & A \end{bmatrix}$ and $\begin{bmatrix} B & 0 \\ 0 & B \end{bmatrix}$ are similar, then A and B are similar.
37. Let $A, B \in \mathbb{C}^{n\times n}$. Prove that if $AB - BA$ commutes with A, then $AB - BA$ is nilpotent.
38. Let $A \in \mathbb{F}^{n\times n}$. Define $T : \mathbb{F}^{n\times n} \to \mathbb{F}^{n\times n}$ by $T(X) = AX - XA$. Prove that if A is a nilpotent matrix, then T is a nilpotent operator.
39. Let T be a linear operator on a complex vector space V of dimension n. Prove that V cannot be written as $U \oplus W$ for T-invariant proper subspaces U, W if and only if the minimal polynomial of T is of the form $(t - \lambda)^n$.
40. Let T be a linear operator on a finite dimensional vector space V over $\mathbb{F}$. Let U be a T-invariant proper subspace of V. Suppose that the minimal polynomial of T splits over $\mathbb{F}$. Prove that there exists a vector $x \in V \setminus U$ and there exists an eigenvalue λ of T such that $(T - \lambda I)v \in U$.
41. Prove Cayley–Hamilton theorem and the spectral mapping theorem using Schur triangularization.
42. (*Real Shur Form*) Let A be an $n \times n$ matrix with real entries. Prove that there exists an orthogonal matrix P such that $P^T AP$ is real block upper triangular; i.e., it is in the form

$$\begin{bmatrix} D_1 & \star & & \star \\ 0 & D_2 & & \star \\ & & \ddots & \\ 0 & 0 & & D_r \end{bmatrix},$$

where each block D_j on the diagonal is either a 1×1 block with a real eigenvalue of A or a 2×2 block whose eigenvalues consist of a pair of complex conjugate characteristic roots of A.

43. (*Real Jordan Form*) Let T be a linear operator on a real vector space V of dimension n, with characteristic polynomial

$$\chi_T(t) = (t - \lambda_1)^{i_1} \cdots (t - \lambda_k)^{i_k} ((t - \alpha_1)^2 + \beta_1)^{j_1} \cdots ((t - \alpha_m)^2 + \beta_m)^{j_m},$$

where λs, αs, and βs are real numbers. Let $U_1, \ldots, U_k$ be the generalized eigenspaces corresponding to the distinct eigenvalues $\lambda_1, \ldots, \lambda_k$, respectively. Let $V_1 = N(((t - \alpha_1)^2 + \beta_1)^n), \ldots, V_m = N(((t - \alpha_m)^2 + \beta_m)^n)$. Prove the following:

(a) $V = U_1 \oplus \cdots \oplus U_k \oplus V_1 \oplus \cdots \oplus V_m$.
(b) Each of $U_1, \ldots, U_k, V_1, \ldots, V_m$ is a T-invariant subspace of V.
(c) Each $(T - \lambda_j I)|_{U_j}$ and each $((T - \alpha_j)^2 + \beta_j))|_{V_j}$ is nilpotent.

44. Let S and T be linear operators on $\mathbb{C}^n$. Let $p(t)$ be a polynomial such that $p(ST) = 0$. Then prove that $TSp(TS) = 0$. What can you conclude about the minimal polynomials of ST and of TS?
45. Let S and T be linear operators on a finite dimensional inner product space V. Prove that there exists an orthonormal basis for V with respect to which both S and T have diagonal matrix representations if and only if $ST = TS$. Interpret the result for matrices.
46. Let S and T be linear operators on a finite dimensional vector space V. Show that $I - ST + TS$ is not nilpotent.
47. An operator T is called *unipotent* if $T - I$ is nilpotent. Determine the characteristic polynomial of a unipotent operator. What are the possible eigenvalues of a unipotent operator?
48. (*Jordan Decomposition*) Prove that each complex $n \times n$ matrix is similar to a matrix $D + N$, where D is diagonal, N is nilpotent, $DN = ND$, each of D and N can be expressed as polynomials in A; and that such D and N are unique.
49. Prove that each complex $n \times n$ invertible matrix is similar to a matrix DN, where D is diagonal, N is unipotent (i.e. $N - I$ is nilpotent), $DN = ND$, each of D and N can be expressed as polynomials in A; and that such D and N are unique.
50. Let $p(t)$ be the characteristic polynomial of an $n \times n$ matrix A. Let C_p be the companion matrix of the polynomial $p(t)$. Prove that A is similar to C_p if and only if $p(t)$ is the minimal polynomial of A.

Chapter 7
Spectral Representation

7.1 Playing with the Adjoint

Recall that if T is a linear operator on a finite dimensional inner product space V, then there exists a unique linear operator $T^* : V \to V$, called the adjoint of T, which satisfies

$$\langle Tx, y \rangle = \langle x, T^* y \rangle \quad \text{for all vectors } x, y \in V.$$

In applications we come across many types of linear operators which satisfy some properties involving the adjoint. Some of them are listed in the following definition.

Definition 7.1 A linear operator T on a finite dimensional inner product space V over $\mathbb{F}$ is called

(a) **unitary** if $T^*T = TT^* = I$;
(b) **orthogonal** if $\mathbb{F} = \mathbb{R}$ and T is unitary;
(c) **isometric** if $\|Tx\| = \|x\|$ for each $x \in V$;
(d) **normal** if $T^*T = TT^*$;
(e) **self-adjoint** if $T^* = T$;
(f) **positive semi-definite** if T is self-adjoint and $\langle Tx, x \rangle \geq 0$ for all $x \in V$;
(g) **positive definite** if T is self-adjoint and $\langle Tx, x \rangle > 0$ for all nonzero $x \in V$.

An isometric operator is also called an *isometry*, and a positive semi-definite operator is also called a *positive operator*. A self-adjoint operator $T : V \to V$ is called *negative definite* if $\langle Tx, x \rangle < 0$ for each nonzero $x \in V$.

For any linear operator T on a finite dimensional complex inner product space,

$$T = T_1 + iT_2 \quad \text{with} \quad T_1 := \frac{1}{2}(T + T^*), \quad T_2 = \frac{1}{2i}(T - T^*).$$

Here, both T_1 and T_2 are self-adjoint. The linear operator T_1 is called the *real part*, and the linear operator T_2 is called the *imaginary part* of the linear operator T.

M. T. Nair and A. Singh, *Linear Algebra*,
https://doi.org/10.1007/978-981-13-0926-7_7

As matrices in $\mathbb{F}^{n\times n}$ are linear operators on $\mathbb{F}^{n\times 1}$, the same terminology applies to matrices also. Further, when $\mathbb{F} = \mathbb{R}$, the adjoint of a matrix in $\mathbb{F}^{n\times n}$ coincides with its transpose. In addition to the above, the following types of matrices come up frequently in applications:

A matrix $A \in \mathbb{F}^{n\times n}$ is called

hermitian if $A^* = A$;
real symmetric if $\mathbb{F} = \mathbb{R}$ and $A^T = A$;
skew-hermitian if $A^* = -A$;
real skew-symmetric if $\mathbb{F} = \mathbb{R}$ and $A^T = -A$.

For any square matrix A, the matrix $(A + A^*)/2$ is hermitian, and the matrix $(A - A^*)/2$ is skew-hermitian. Thus any matrix can be written as the sum of a hermitian matrix and a skew-hermitian matrix. Similarly, a real square matrix can be written as the sum of a real symmetric matrix and a real skew-symmetric matrix by using transpose instead of the adjoint.

Example 7.2 Let V be an inner product space, and let $\{u_1, \ldots, u_n\}$ be an orthonormal subset of V. Corresponding to scalars $\lambda_1, \ldots, \lambda_n$ in $\mathbb{F}$, define $T : V \to V$ by

$$Tx = \sum_{j=1}^{n} \lambda_j \langle x, u_j\rangle u_j, \quad x \in V.$$

Then for every $x, y \in V$, we have

$$\langle Tx, y\rangle = \sum_{j=1}^{n} \lambda_j \langle x, u_j\rangle\langle u_j, y\rangle = \sum_{j=1}^{n} \langle x, \overline{\lambda}_j \langle y, u_j\rangle u_j\rangle = \Big\langle x, \sum_{j=1}^{n} \overline{\lambda}_j \langle y, u_j\rangle u_j\Big\rangle.$$

Thus,

$$T^* y = \sum_{j=1}^{n} \overline{\lambda}_j \langle y, u_j\rangle u_j, \quad y \in V.$$

Therefore, T is self-adjoint if and only if $\lambda_j \in \mathbb{R}$ for each $j \in \{1, \ldots, n\}$.

Observe that $Tu_i = \lambda_i u_i$ and $T^* u_i = \overline{\lambda}_i u_i$ for $i \in \{1, \ldots, n\}$. Hence, for every $x, y \in V$, we have

$$T^*Tx = \sum_{j=1}^{n} \lambda_j \langle Tx, u_j\rangle T^* u_j = \sum_{j=1}^{n} |\lambda_j|^2 \langle x, u_j\rangle u_j,$$

$$TT^*y = \sum_{j=1}^{n} \overline{\lambda}_j \langle T^*y, u_j\rangle Tu_j = \sum_{j=1}^{n} |\lambda_j|^2 \langle y, u_j\rangle u_j.$$

Therefore, T is a normal operator. We will see that every normal operator on a finite dimensional inner product space is of the above form.

If $\{u_1, \ldots, u_n\}$ is an orthonormal basis of V, then by Fourier expansion, we have $x = \sum_{j=1}^{n} \langle x, u_j \rangle u_j$. In this case, the above equalities imply that T is unitary if and only if $|\lambda_j| = 1$ for each $j \in \{1, \ldots, n\}$. □

The following result characterizes unitary operators.

Theorem 7.3 *Let T be a linear operator on a finite dimensional inner product space V. Then the following are equivalent:*

(1) *T is unitary.*
(2) *$T^*T = I$.*
(3) *$\langle Tx, Ty \rangle = \langle x, y \rangle$ for all $x, y \in V$.*
(4) *T is an isometry.*
(5) *If a list of vectors $v_1, \ldots, v_n$ is orthonormal in V, then the list $Tv_1, \ldots, Tv_n$ is also orthonormal in V.*
(6) *If $\{v_1, \ldots, v_n\}$ is an orthonormal basis of V, then $\{Tv_1, \ldots, Tv_n\}$ is also an orthonormal basis of V.*

Proof (1) ⇒ (2) : Trivial.

(2) ⇒ (3) : $T^*T = I$ implies that $\langle Tx, Ty \rangle = \langle x, T^*Ty \rangle = \langle x, y \rangle$ for all $x, y \in V$.

(3) ⇒ (4) : By (3), $\|Tx\|^2 = \langle Tx, Tx \rangle = \langle x, x \rangle = \|x\|^2$ for all $x \in V$.

(4) ⇒ (5) : Suppose that $\|Tx\| = \|x\|$ for each $x \in V$. Let $B := \{v_1, \ldots, v_n\}$ be an orthonormal list of vectors in V. Let $i, j \in \{1, \ldots, n\}$. Due to Theorem 4.8, we see that $\langle Tv_i, Tv_j \rangle = \langle v_i, v_j \rangle$. Hence $E := \{Tv_1, \ldots, Tv_n\}$ is an orthonormal set. Now, for any $x \in V$, if $Tx = 0$, then $\|x\| = \|Tx\| = 0$ implies that T is injective. In particular, $v_i \neq v_j$ implies that $Tv_i \neq Tv_j$. That is, E is an orthonormal list.

(5) ⇒ (6) : Assume (5). If $\{v_1, \ldots, v_n\}$ is an orthonormal basis of V, then the orthonormal set $\{Tv_1, \ldots, Tv_n\}$ with n distinct vectors is also an orthonormal basis of V.

(6) ⇒ (1) : Let $E := \{v_1, \ldots, v_n\}$ be an orthonormal basis of V. By (6), the set $\{Tv_1, \ldots, Tv_n\}$ is an orthonormal basis for V. Let $v_j \in E$. Then for each $i \in \{1, \ldots, n\}$,

$$\langle v_i, T^*Tv_j \rangle = \langle Tv_i, Tv_j \rangle = \delta_{ij} = \langle v_i, v_j \rangle.$$

That is, $\langle v_i, (T^*T - I)v_j \rangle = 0$ for each basis vector v_i. So, $(T^*T - I)v_j = 0$. Since this holds for each basis vector v_j, we conclude that $T^*T = I$.

As T maps a basis of V onto a basis of V, it is invertible. Therefore, $T^* = T^{-1}$; and $TT^* = I$. ■

The following implications for the statements in Theorem 7.3 can be verified independently:

$$(2) \Rightarrow (1), \quad (3) \Rightarrow (2), \quad (4) \Rightarrow (3), \quad (3) \Leftrightarrow (5), \quad (3) \Leftrightarrow (6).$$

Further, it is easy to see that T is unitary if and only if T^* is unitary. The equivalent statements using T^* instead of T in Theorem 7.3 may then be formulated and proved. Since orthogonal operators are unitary operators with added constraint that the underlying field is $\mathbb{R}$, all the statements in Theorem 7.3(2)–(6) with suitable changes are equivalent to the statement that T is orthogonal. In particular, the sixth statement there says that all orthogonal and unitary operators are isomorphisms. The fifth statement asserts that

> the matrix representation of an orthogonal (unitary) operator with respect to an orthonormal basis is an orthogonal (unitary) matrix.

The matrix version of Theorem 7.3 thus looks as follows.

Theorem 7.4 *For a matrix $A \in \mathbb{F}^{n\times n}$ the following statements are equivalent:*

(1) *A is unitary.*
(2) $A^*A = I$.
(3) $AA^* = I$.
(4) $(Ay)^*Ax = y^*x$ *for all* $x, y \in \mathbb{F}^{n\times 1}$.
(5) $(Ax)^*Ax = x^*x$ *for each* $x \in \mathbb{F}^{n\times 1}$.
(6) *The columns of A form an orthonormal basis for $\mathbb{F}^{n\times 1}$.*
(7) *The rows of A form an orthonormal basis for $\mathbb{F}^{1\times n}$.*

For any matrix $A \in \mathbb{R}^{n\times n}$, the above theorem gives equivalent statements for the orthogonality of A; we just replace the *adjoint* by *transpose.*

In proving that two operators act the same way on each vector, a condition on the inner products of the form $\langle Tv, v\rangle$ is helpful.

Theorem 7.5 *Let T be a linear operator on a finite dimensional inner product space V over $\mathbb{F}$. Let $\langle Tv, v\rangle = 0$ for all $v \in V$. If $\mathbb{F} = \mathbb{C}$ or T is self-adjoint, then $T = 0$.*

Proof We consider two cases.
(1) Suppose $\mathbb{F} = \mathbb{C}$. Let $u, w \in V$. A straightforward calculation shows that

$$4\langle Tu, w\rangle = \langle T(u+w), u+w\rangle - \langle T(u-w), u-w\rangle \\ + i\,\langle T(u+iw), u+iw\rangle - i\,\langle T(u-iw), u-iw\rangle = 0.$$

In particular, with $w = Tu$, we have $\langle Tu, Tu\rangle = 0$ for each $u \in V$. Hence $T = 0$.

(2) Suppose T is self-adjoint and $\mathbb{F} = \mathbb{R}$. Then $\langle Tu, w\rangle = \langle u, T^*w\rangle = \langle u, Tw\rangle = \langle Tw, u\rangle$ for all $u, w \in V$. Thus

$$4\langle Tu, w\rangle = \langle T(u+w), u+w\rangle - \langle T(u-w), u-w\rangle = 0.$$

Once again, taking $w = Tu$, we have $Tu = 0$ for all $u \in V$. ■

There are non-self-adjoint operators on real inner product spaces such that $\langle Tv, v\rangle = 0$ for all $v \in V$. For instance, the linear operator $T : \mathbb{R}^2 \to \mathbb{R}^2$ defined by $T(a, b) = (-b, a)$ satisfies

$$\langle T(a, b), (a, b)\rangle = \langle(-b, a), (a, b)\rangle = 0 \quad \text{for all} \quad (a, b) \in \mathbb{R}^2$$

but $T \neq 0$. Here, of course, $T^* = -T$.

Theorem 7.6 *Let T be a linear operator on a finite dimensional real inner product space V. If $T^* = -T$, then $\langle Tx, x\rangle = 0$ for all $x \in V$. In particular, if $A \in \mathbb{R}^{n\times n}$ is real skew-symmetric, then $x^T Ax = 0$ for each $x \in \mathbb{R}^n$.*

Proof Suppose $T^* = -T$. Then for each $x \in V$, we have

$$\langle Tx, x\rangle = \langle x, T^*x\rangle = -\langle x, Tx\rangle = -\langle Tx, x\rangle$$

so that $\langle Tx, x\rangle = 0$. The particular case is obvious. ■

A useful characterization of self-adjoint operators on complex inner product spaces is as follows.

Theorem 7.7 *Let T be a linear operator on a finite dimensional complex inner product space V. Then T is self-adjoint if and only if $\langle Tv, v\rangle \in \mathbb{R}$ for each $v \in V$.*

Proof If T is self-adjoint, then for each $v \in V$,

$$\langle Tv, v\rangle = \langle T^*v, v\rangle = \langle v, Tv\rangle = \overline{\langle Tv, v\rangle}.$$

Therefore, $\langle Tv, v\rangle \in \mathbb{R}$ for each $v \in V$.

Conversely, suppose that $\langle Tv, v\rangle \in \mathbb{R}$ for each $v \in V$. Then

$$\langle Tv, v\rangle = \overline{\langle Tv, v\rangle} = \langle v, Tv\rangle = \langle T^*v, v\rangle.$$

That is, $\langle (T - T^*)v, v\rangle = 0$ for all $v \in V$. By Theorem 7.5, $T - T^* = 0$. Therefore, T is self-adjoint. ■

Certain properties of eigenvalues follow from the nature of the linear operator.

Theorem 7.8 *Let λ be an eigenvalue of a linear operator T on a finite dimensional inner product space V.*

(1) *If T is self-adjoint, then $\lambda \in \mathbb{R}$.*
(2) *If T is positive semi-definite, then $\lambda \geq 0$.*
(3) *If T is positive definite, then $\lambda > 0$.*
(4) *If T is unitary, then $|\lambda| = 1$.*

Proof Let v be an eigenvector of T corresponding to the eigenvalue λ.

(1) Suppose T is self-adjoint. Then

$$\lambda\langle v, v\rangle = \langle \lambda v, v\rangle = \langle Tv, v\rangle = \langle v, Tv\rangle = \langle v, \lambda v\rangle = \overline{\lambda}\langle v, v\rangle.$$

As $v \neq 0$, $\lambda = \overline{\lambda}$. that is, $\lambda \in \mathbb{R}$.

(2) Let T be positive semi-definite. As $v \neq 0$, $0 \leq \frac{\langle Tv, v\rangle}{\|v\|^2} = \frac{\langle \lambda v, v\rangle}{\|v\|^2} = \lambda$.

(3) Similar to (2).

(4) For unitary T, $Tv = \lambda v$ implies $\|v\| = \|Tv\| = |\lambda|\,\|v\|$. As $v \neq 0$, $|\lambda| = 1$. ■

Theorem 7.9 *Let T be a self-adjoint operator on a finite dimensional inner product space V. Then each characteristic value of T is real, and it is an eigenvalue of T.*

Proof Let λ be a characteristic value of T. In general, $\lambda \in \mathbb{C}$. Let B be an orthonormal basis of V, and let $A := [T]_{B,B}$. Then A is hermitian, and λ is a characteristic value of A. Since A is a linear operator on $\mathbb{C}^{n\times 1}$, the scalar λ is an eigenvalue of A. Theorem 7.8(1) implies that $\lambda \in \mathbb{R}$. Then λ is a real characteristic value of T. Due to Theorem 5.11, λ is an eigenvalue of T. ■

Then the following existence result is immediate.

Theorem 7.10 *Each self-adjoint operator on a finite dimensional inner product space has an eigenvalue.*

Theorem 7.10 can be proved without using matrices. For this purpose, we use the following result.

Lemma 7.11 *Let T be a linear operator on a finite dimensional inner product space V. Then for each nonzero real number β, the operator $T^*T + \beta^2 I$ is injective.*

Proof Let $u \in V$. Then

$$\langle (T^*T + \beta^2 I)u,\ u\rangle = \langle T^*Tu,\ u\rangle + \langle \beta^2 u,\ u\rangle = \langle Tu,\ Tu\rangle + \beta^2\langle u,\ u\rangle.$$

Now, if $(T^*T + \beta^2)u = 0$ then $u = 0$. Therefore, $T^*T + \beta^2 I$ is injective. ■

An alternative proof of Theorem 7.10: Let T be a self-adjoint operator on an inner product space V over $\mathbb{F}$. If $\mathbb{F} = \mathbb{C}$, then Theorem 5.4 implies that T has an eigenvalue. If $T = 0$, then 0 is an eigenvalue of T. So, suppose that $\mathbb{F} = \mathbb{R}$, $T \neq 0$, and $\dim(V) = n \geq 1$. Then there exists a vector $v \in V$ such that $v \neq 0$ and $Tv \neq 0$. The list $v,\ Tv,\ \ldots,\ T^n v$, having $n + 1$ vectors, is linearly dependent. Then there exist scalars $a_0, \ldots, a_n$ not all zero such that

$$a_0 v + a_1 Tv + \cdots a_n T^n v = 0.$$

Let k be the maximum index such that $a_k \neq 0$. We obtain a polynomial $p(t) = a_0 + a_1 t + \cdots + a_k t^k$ with $a_0, \ldots, a_k \in \mathbb{R}$, such that

$$p(T) = a_0 I + a_1 T + \cdots + a_k T^k, \quad a_k \neq 0, \quad p(T)(v) = 0$$

for a nonzero vector v. Thus $p(T)$ is not injective.

By Proposition 5.1,

$$p(t) = a_k(t - \lambda_1)\cdots(t - \lambda_j)((t - \alpha_1)^2 + \beta_1^2)\cdots((t - \alpha_m)^2 + \beta_m^2),$$

where λs, αs, and βs are real numbers. Then

$$p(T) = a_k(T - \lambda_1 I)\cdots(T - \lambda_j I)((T - \alpha_1 I)^2 + \beta_1^2 I)\cdots((T - \alpha_m I)^2 + \beta_m^2 I).$$

If no quadratic factor is present in this product, then we have at least one linear factor. Since $p(T)$ is not injective, one such linear factor is not injective.

Otherwise, there is at least one quadratic factor present in the product, so that the corresponding β is nonzero. Since T is self-adjoint, so is $T - \alpha I$ for any $\alpha \in \mathbb{R}$. Due to Lemma 7.11, each operator of the form $(T - \alpha I)^2 + \beta^2 I$ in the above composition is injective. But $p(T)$ is not injective. Hence in the above product, there exists a linear factor which is not injective.

In any case, at least one of the maps of the form $T - \lambda_j I$ is not injective. Then λ_j is an eigenvalue of T. ■

In contrast, an isometry on a real inner product space need not have an eigenvalue. For example, the linear operator $T : \mathbb{R}^2 \to \mathbb{R}^2$ given by $T(a, b) = (b, -a)$ is an isometry, since

$$\|T(a, b)\| = \|(b, -a)\| = (b^2 + a^2)^{1/2} = \|(a, b)\|.$$

But T does not have an eigenvalue.

Exercises for Sect. 7.1

1. For which complex numbers α the following matrices are unitary?
   (a) $\begin{bmatrix} \alpha & 1/2 \\ -1/2 & \alpha \end{bmatrix}$ (b) $\begin{bmatrix} \alpha & 0 \\ 1 & 1 \end{bmatrix}$.
2. Find a 3×3 unitary matrix, where all entries on its first row are equal.
3. Let $A \in \mathbb{R}^{n \times n}$. Show that if A is skew-symmetric, then $\langle Ax, x\rangle = 0$ for each $x \in \mathbb{R}^{n \times 1}$. Is the converse true?
4. Let $A \in \mathbb{R}^{n \times n}$. Are the following statements true?
   (a) If A is symmetric or skew-symmetric, then for each $x \in \mathbb{R}^{n \times 1}$, $A^2 x = 0$ implies $Ax = 0$.
   (b) If A is skew-symmetric and n is odd, then $\det(A) = 0$.
   (c) If A is skew-symmetric, then rank (A) is even.
5. Let the linear operator $T : \mathbb{R}^3 \to \mathbb{R}^3$ be given by

$$T(a, b, c) = (b + c,\ \ -a + 2b + c,\ a - 3b - 2c)\ \text{ for } a, b, c, \in \mathbb{R}.$$

Check whether $T : \mathbb{R}^3 \to \mathbb{R}^3$ is self-adjoint if the inner product is given by

(a) $\langle (a, b, c), (\alpha, \beta, \gamma)\rangle = a\alpha + b\beta + c\gamma$.
(b) $\langle (a, b, c), (\alpha, \beta, \gamma)\rangle = 2a\alpha + 3b\beta + 2c\gamma + a\gamma + c\alpha + 2b\gamma + 2c\beta$.

6. Consider the inner product space $\mathcal{P}(\mathbb{C})$ with $\langle p, q\rangle = \int_0^1 p(t)\overline{q(t)}dt$. Define the linear operator T on $\mathcal{P}(\mathbb{C})$ by $(Tp)(t) = p(-t)$. Is T an isometry?
7. Show that each self-adjoint operator is normal, and also that each unitary operator is normal.
8. Give an example of a normal operator which is neither unitary nor self-adjoint.
9. Show that for each $\alpha \in \mathbb{R}$, the 2×2 matrix $\begin{bmatrix} \alpha & -\alpha \\ \alpha & \alpha \end{bmatrix}$ is normal.
10. Let $A \in \mathbb{C}^{n\times n}$. Write $H = (A + A^*)/2$ and $K = (A - A^*)/2$. Show that A is normal if and only if $HK = KH$.
11. Consider $\mathcal{P}_2(\mathbb{R})$ as an inner product space with $\langle f, g\rangle = \int_0^1 f(t)g(t)dt$. Let the linear operator T on $\mathcal{P}_2(\mathbb{R})$ be defined by $T(a_0 + a_1t + a_2t^2) = a_1t$. Prove that T is not self-adjoint but $[T]_{B,B}$ is real symmetric, where $B = \{1, t, t^2\}$.
12. Let T be a linear operator on a real inner product space V of dimension n. Show that T is self-adjoint if and only if $[T]_{B,B}$ is real symmetric for any orthonormal basis B of V.
13. Let T be a linear operator on a finite dimensional inner product space V. Prove that $\det(T^*) = \overline{\det(T)}$.
14. Show that if A is a hermitian matrix, then $\det(A) \in \mathbb{R}$.
15. Let $T, D : \mathcal{P}_n(\mathbb{F}) \to \mathcal{P}_n(\mathbb{F})$ be given by $(Tp)(t) = tp(t)$ and $Dp = dp/dt$. Check whether T and D are self-adjoint if the inner product is given by
(a) $\langle p, q\rangle = \int_0^1 p(t)\overline{q(t)}dt$. (b) $\langle p, q\rangle = \sum_{j=0}^{n} p(j/n)\overline{q(j/n)}$.
16. Consider the inner product space $\mathcal{P}(\mathbb{C})$ with $\langle p, q\rangle = \int_0^1 p(t)\overline{q(t)}dt$. Is the linear operator T on $\mathcal{P}(\mathbb{C})$ given by $(Tp)(t) = p(-t)$ self-adjoint?
17. Which of the following matrices are positive semi-definite?

(a) $\begin{bmatrix} 1 & 1 \\ 1 & 0 \end{bmatrix}$ (b) $\begin{bmatrix} 0 & 1 \\ -1 & 0 \end{bmatrix}$ (c) $\begin{bmatrix} 0 & i \\ -i & 0 \end{bmatrix}$ (d) $\begin{bmatrix} 1 & 1 & 1 \\ 1 & 1 & 1 \\ 1 & 1 & 1 \end{bmatrix}$

(e) $\begin{bmatrix} 0 & 0 & 1 \\ 0 & 1 & 1 \\ 1 & 1 & 1 \end{bmatrix}$ (f) $\begin{bmatrix} 1 & 2 & 0 \\ 2 & 6 & 1 \\ 0 & 1 & 2 \end{bmatrix}$ (g) $\begin{bmatrix} 1 & i & 0 & i \\ -i & 1 & 2 & 0 \\ 0 & 2 & 1 & 0 \\ 1 & 0 & 0 & 1 \end{bmatrix}$.

18. For which value(s) of α is the matrix $\begin{bmatrix} 1 & \alpha & 1 \\ 0 & 1 & 0 \\ 0 & 1 & 0 \end{bmatrix}$ positive semi-definite?
19. Show that if T is a hermitian operator on a complex inner product space of finite dimension, then $I + iT$ is invertible.
20. Show that if S is positive semi-definite and T is hermitian, then $S + iT$ is invertible.
21. Let A and B be positive definite matrices. Which among A^2, A^{-1}, AB, and $A + B$ are hermitian; and which are positive definite?

22. Give an example of a linear operator which is not positive definite, but all its eigenvalues are positive.

7.2 Projections

The word *projection* is used for projection of vectors on a subspace as in the notation $\text{proj}_U(y)$, and also it is used for operators as in the following definition.

Definition 7.12 A linear operator P on a vector space V is called a **projection operator** or a **projection** on V if

$$Px = x \quad \text{for each } x \in R(P).$$

If P is a projection on V with $U = R(P)$ and $W = N(P)$, then we say that P is a **projection onto** U **along** W, and we denote it by $P_{U,W}$.

Example 7.13 Let $P : \mathbb{R}^2 \to \mathbb{R}^2$ be given by

$$P(a,b) = (a,a) \quad \text{for } (a,b) \in \mathbb{R}^2.$$

It is easy to see that P is a linear operator on $\mathbb{R}^2$ and $R(P) = \{(a,a) : a \in \mathbb{R}\}$. Note that $P(u) = u$ for each $u \in R(P)$ so that P is a projection. □

Here is a characterization of projection operators.

Theorem 7.14 *Let P be a linear operator on a vector space V. Then, P is a projection if and only if $P^2 = P$; in that case,*

$$V = R(P) \oplus N(P), \quad R(P) = N(I-P), \quad \textit{and} \quad N(P) = R(I-P).$$

Proof Suppose P is a projection, that is, $Px = x$ for every $x \in R(P)$. Then for every $x \in V$, we have $P(I-P)x = Px - P(Px) = Px - Px = 0$ so that $P^2 = P$. Conversely, suppose $P^2 = P$. Let $x \in R(P)$. Then $x = Pu$ for some $u \in V$. Hence, $Px = P^2u = Pu = x$. Thus, $Px = x$ for every $x \in R(P)$ so that P is a projection.

Next assume that $P^2 = P$. We show the three equalities.

For every $x \in V$, we have $x = Px + (I-P)x$. Here, $Px \in R(P)$ and

$$P(I-P)x = Px - P^2x = 0.$$

Hence, $V = R(P) + N(P)$. Moreover, if $x \in R(P) \cap N(P)$, then $x = Px = 0$. That is, $R(P) \cap N(P) = \{0\}$. Hence $V = R(P) \oplus N(P)$.

Since $P^2 = P$, we have $(I-P)P = 0$ and $P(I-P) = 0$. Hence, $R(P) \subseteq N(I-P)$ and $R(I-P) \subseteq N(P)$. Further, $x \in N(I-P)$ implies $x = Px \in R(P)$, and $x \in N(P)$ implies $x = x - Px \in R(I-P)$. Thus, $N(I-P) \subseteq R(P)$ and $N(P) \subseteq R(I-P)$. ■

We observe that if $T : V \to V$ is a linear operator such that $V = R(T) \oplus N(T)$, then it is not necessary that T is a projection. For example, consider the linear operator $T : \mathbb{R}^3 \to \mathbb{R}^3$ defined by

$$T(a, b, c) = (a, 2b, 0) \quad \text{for } (a, b) \in \mathbb{R}^2.$$

We have $R(T) = \{(a, b, 0) : a, b \in \mathbb{R}\}$ and $N(T) = \{(0, 0, c) : c \in \mathbb{R}\}$ so that $V = R(T) \oplus N(T)$. But T is not a projection.

However, we have the following result.

Theorem 7.15 *Let U and W be subspaces of a vector space V such that $V = U \oplus W$. Then there exists a unique projection $P : V \to V$ such that $R(P) = U$ and $N(P) = W$.*

Proof Let $v \in V$. Then there are unique vectors $u \in U$ and $w \in W$ such that $v = u + w$. Define the map $P : V \to V$ by

$$Pv = u \quad \text{for } v \in V.$$

To see that P is a linear operator, let $x, y \in V$ and $\alpha \in \mathbb{F}$. Then there exist unique vectors $x_1, y_1 \in U$ and $x_2, y_2 \in W$ such that $x = x_1 + x_2$ and $y = y_1 + y_2$. Then

$$x + \alpha y = (x_1 + \alpha y_1) + (x_2 + \alpha y_2)$$

with $x_1 + \alpha y_1 \in U$ and $x_2 + \alpha y_2 \in W$. Since $V = U \oplus W$, the vectors $(x_1 + \alpha y_1) \in U$ and $(x_2 + \alpha y_2) \in W$ are uniquely determined. Therefore,

$$P(x + \alpha y) = x_1 + \alpha y_1 = Px + \alpha Py.$$

This shows that P is a linear operator on V. Further, for every $x \in R(P) = U$, since $x = x + 0$, we have $Px = x$ so that P is a projection.

Next, we show that $R(P) = U$ and $N(P) = W$. By the definition of P, we have $R(P) \subseteq U$. Also, for $u \in U$, $Pu = P(u + 0) = u$ so that $U \subseteq R(P)$. Thus, $R(P) = U$. For the other equality, let $x \in N(P)$. If $x = u + w$ for $u \in U$ and $w \in W$, then $0 = Px = u$. Thus, $x = w \in W$. Therefore, $N(P) \subseteq W$. On the other hand, if $w \in W$, then $w = 0 + w$, where $0 \in U$ and $w \in w$. We have $Pw = 0$. Thus, $W \subseteq N(P)$.

For uniqueness, suppose that P and Q are projections with the said properties. Let $x \in V$. Now, there exist unique $u \in U$ and $w \in W$ such that $x = u + w$. Since $R(P) = U = R(Q)$ and $N(P) = W = N(Q)$, and since P and Q are projections, we have $Pu = u = Qu$ and $Pw = 0 = Qw$. Hence, $Px = Pu + Pw = u = Qu + Qw = Qx$. Therefore, $P = Q$. ■

As a corollary to Theorem 7.15, we have the following result.

Theorem 7.16 *For any subspace U of a vector space V, there exists a projection P on V such that $R(P) = U$.*

Proof Let U be a subspace of a vector space V. If $U = V$, then the identity operator on V is the projection onto U, and if $U = \{0\}$, then the zero operator $P = 0$ is the projection onto U. Next, suppose that $U \neq V$ and $U \neq \{0\}$. By Theorem 1.46, there exists a subspace W of V such that $V = U \oplus W$. Due to Theorem 7.15, there exists a unique projection $P : V \to V$ with $R(P) = U$ and $N(P) = W$. ■

Example 7.17 Consider the subspaces $U = \{(a, b, a+b) : a, b \in \mathbb{R}\}$ and $W = \{(c, c, c) : c \in \mathbb{R}\}$ of $\mathbb{R}^3$. First we write any vector in $\mathbb{R}^3$ as a sum of a vector from U and one from W. Towards this, we write

$$(a, b, c) = (\alpha, \beta, \alpha + \beta) + (\gamma, \gamma, \gamma).$$

This leads to the linear system of equations

$$\alpha + \gamma = a, \ \ \beta + \gamma = b, \ \ \alpha + \beta + \gamma = c.$$

The linear system has a unique solution

$$\alpha = c - b, \ \ \beta = c - a, \ \ \gamma = a + b - c.$$

Hence each vector in $\mathbb{R}^3$ can be written uniquely as a sum of a vector from U and one from W as follows:

$$(a, b, c) = (c - b, \ c - a, \ 2c - a - b) + (a + b - c, \ a + b - c, \ a + b - c).$$

In fact, if $(a, b, c) \in U \cap W$, then there exist α, β, γ such that

$$(a, b, c) = (\alpha, \beta, \alpha + \beta) = (\gamma, \gamma, \gamma),$$

so that $\gamma = \alpha = \beta = \alpha + \beta$; consequently, $(a, b, c) = (0, 0, 0)$. Thus, the projection $P_{U,W} : \mathbb{R}^3 \to \mathbb{R}^3$ is given by

$$P_{U,W}(a, b, c) = (c - b, \ c - a, \ 2c - a - b) \quad \text{for } (a, b, c) \in \mathbb{R}^3.$$

Observe that

$$P_{U,W}(a, b, a + b) = (a + b - b, \ a + b - a, \ 2a + 2b - a - b) = (a, b, a + b).$$

That is, $P_{U,W}(u) = u$ for $u \in U$. Similarly, it may be verified that $P_{U,W}^2(a, b, c) = P_{U,W}(a, b, c)$ for any $(a, b, c) \in \mathbb{R}^3$. □

A subspace U of a vector space V can have many complementary subspaces, and hence, by Theorem 7.15, there can be more than one projections whose ranges coincide with U. The following example illustrates this.

Example 7.18 Let $V = \mathbb{R}^2$ and let $U = \{(a, b) \in \mathbb{R}^2 : b = 0\}$. Then we see that the subspaces $W_1 = \{(a, b) \in \mathbb{R}^2 : a = 0\}$ and $W_2 = \{(a, b) \in \mathbb{R}^2 : a = b\}$ satisfy

$$V = U \oplus W_1, \quad V = U \oplus W_2.$$

Indeed, for any $(a, b) \in \mathbb{R}^2$, we have

$$(a, b) = (a, 0) + (0, b), \quad (a, b) = (a - b, 0) + (b, b).$$

So, $V = U + W_1$ and $V = U + W_2$. Clearly, $U \cap W_1 = \{0\} = U + W_2$. Corresponding to the subspaces W_1 and W_2, we have the projections P_1 and P_2 defined by

$$P_1(a, b) = (a, 0), \quad P_2(a, b) = (a - b, 0)$$

onto U along W_1 and W_2, respectively. Clearly, $P_1 \neq P_2$.

In fact, there are infinitely many projections onto U. To see this, for each $\alpha \in \mathbb{R}$, consider the subspace

$$W_\alpha = \{(a, b) \in \mathbb{R}^2 : a = \alpha b\}.$$

Then we see that $U \cap W_\alpha = \{0\}$ for every $\alpha \in \mathbb{R}$. Further, for any $(a, b) \in \mathbb{R}^2$,

$$(a, b) = (a - \alpha b, 0) + (\alpha b, b)$$

with $(a - \alpha b, 0) \in U$ and $(\alpha b, b) \in W_\alpha$. The linear operator $P_\alpha : V \to V$ defined by

$$P_\alpha(a, b) = (a - \alpha b, 0) \quad (a, b) \in V,$$

is a projection onto U along W_α. □

For a nonzero vector v and a projection P, we have $(I - P)Pv = P(I - P)v = 0$. This means that the eigenvalues of a projection can be 0 or 1. For example, the zero map has the only eigenvalue 0; the identity map has the only eigenvalue 1; and the map P on $\mathbb{R}^2$ with $P(a, b) = (a, 0)$ has both 0 and 1 as eigenvalues.

In fact, a projection cannot have any eigenvalue other than 0 or 1, as the following theorem shows.

Theorem 7.19 *Suppose $P : V \to V$ is a projection operator such that $P \neq 0$ and $P \neq I$. Then the eigenspectrum of P is $\{0, 1\}$.*

Proof Since $P \neq 0$, there exists $x \neq 0$ such that $Px \neq 0$. Since $P(Px) = Px$, 1 is an eigenvalue of P with corresponding eigenvector Px. Next, since $P \neq I$, there exists $v \neq 0$ such that $(I - P)v \neq 0$. Since $P(I - P)v = 0$, 0 is an eigenvalue of P with corresponding eigenvector $(I - P)v$.

On the other hand, suppose λ is an eigenvalue of P with a corresponding eigenvector u. Since $Pu = \lambda u$, we have $P^2 u = \lambda P u$ so that

$$\lambda u = Pu = P^2 u = \lambda P u = \lambda^2 u.$$

Thus, $(1 - \lambda)\lambda u = 0$. Since $u \neq 0$, it follows that $\lambda \in \{0, 1\}$. ∎

Projections become more useful when the subspaces in the decomposition are orthogonal to each other.

Definition 7.20 Let V be an inner product space. A projection P on V is called an **orthogonal projection** if $R(P) \perp N(P)$. With $U := R(P)$, such an orthogonal projection is denoted by P_U.

Example 7.21 Consider the subspace $U = \{(a, b, a + b) : a, b \in \mathbb{R}\}$ of $\mathbb{R}^3$ as in Example 7.17. A basis for U is given by $\{(1, 0, 1),\ (0, 1, 1)\}$. If $(\alpha, \beta, \gamma) \in U^\perp$, then

$$\langle(\alpha, \beta, \gamma),\ (1, 0, 1)\rangle = 0, \quad \langle(\alpha, \beta, \gamma),\ (0, 1, 1)\rangle = 0.$$

These equations lead to $\gamma = -\alpha$ and $\beta = \alpha$. Thus

$$U^\perp = \{(\alpha, \alpha, -\alpha) : \alpha \in \mathbb{R}\}.$$

To write any vector in $\mathbb{R}^3$ as a sum of a vector from U and one from $U^\perp$, let

$$(a, b, c) = (\alpha, \beta, \alpha + \beta) + (\gamma, \gamma, -\gamma).$$

The ensuing linear system has the unique solution

$$\alpha = \frac{1}{3}\big(2a - b + c\big),\ \beta = \frac{1}{3}\big(2b - a + c\big),\ \gamma = \frac{1}{3}\big(a + b - c\big).$$

Then $\mathbb{R}^3 = U \oplus U^\perp$ and for every $(a, b, c) \in \mathbb{R}^3$,

$$\begin{aligned}(a, b, c) &= \frac{1}{3}\big(2a - b + c,\ 2b - a + c,\ a + b + 2c\big)\\ &\quad + \frac{1}{3}\big(a + b - c,\ a + b - c,\ -a - b + c\big).\end{aligned}$$

Therefore, the orthogonal projection $P_U : \mathbb{R}^3 \to \mathbb{R}^3$ is given by

$$P_U(a, b, c) = \frac{1}{3}\big(2a - b + c,\ 2b - a + c,\ a + b + 2c\big).$$

Clearly, $P_U(a, b, c) \in U$. Conversely, to see that any vector $(a, b, a + b) \in U$ is equal to $P_U(\alpha, \beta, \gamma)$ for some $(\alpha, \beta, \gamma) \in R^3$, we set up the linear equations

$$(a, b, a+b) = \frac{1}{3}(2\alpha - \beta + \gamma,\ 2\beta - \alpha + \gamma,\ \alpha + \beta + 2\gamma).$$

Solving the linear system for α, β, γ, we obtain

$$\alpha = 2a + b - c,\ \beta = a + 2b - c,\ \gamma = c,$$

where c is arbitrary. We may fix c to some real number for a particular solution. For instance, take $c = 0$ to obtain

$$\alpha = 2a + b,\ \beta = a + 2b,\ \gamma = 0.$$

It may be verified that $P_U(\alpha, \beta, \gamma) = (a, b, a+b)$. This proves that $R(P_U) = U$.

To determine $N(P_U)$, we set up

$$P_U(a, b, c) = \frac{1}{3}(2a - b + c,\ 2b - a + c,\ a + b + 2c) = (0, 0, 0).$$

Solving the linear equations, we obtain $a = b = -c$. That is,

$$N(P_U) = \{(c, c, -c) : c \in \mathbb{R}\} = U^{\perp}. \qquad \Box$$

Our terminology says that an orthogonal projection P_U is the projection onto U along $U^{\perp}$, where $U = R(P)$ and $U^{\perp} = N(P)$. That is,

$$P_U = P_{R(P)} = P_{U,U^{\perp}} = P_{R(P),R(P)^{\perp}}.$$

An orthogonal projection P_U satisfies the following properties:

1. $P_U^2 = P_U$.
2. $R(P_U) = U,\ N(P_U) = U^{\perp},\ U \oplus U^{\perp} = V$.
3. For each $v \in V,\ v = P_U(v) + \big(v - P_U(v)\big),\ P_U(v) \in U,\ P_U(v) \perp \big(v - P_U(v)\big)$.
4. For each $v \in V,\ \|P_U(v)\| \le \|v\|$.

Since $P_U(v) \perp v - P_U(v)$, the last property is obtained by applying Pythagoras theorem on the equality $v = P_U(v) + \big(v - P_U(v)\big)$.

The projection theorem implies that if U is a finite dimensional subspace of an inner product space, then the orthogonal projection P_U exists. Moreover, an explicit formula for P_U can be given in terms of an orthonormal basis of U.

Theorem 7.22 *Let U be a finite dimensional subspace of an inner product space V. If $\{u_1, \ldots, u_k\}$ is an orthonormal basis of U, then P_U is given by*

$$P_U(v) = \mathrm{proj}_U(v) = \sum_{j=1}^{k} \langle v, u_j \rangle u_j \quad \textit{for each } v \in V.$$

Proof Let $\{u_1, \ldots, u_k\}$ be an orthonormal basis of U. Define $P : V \to V$ by

$$P(v) = \text{proj}_U(v) = \sum_{j=1}^{k} \langle v, u_j \rangle u_j \quad \text{for } v \in V.$$

As we have already verified, P is a linear operator. Also, $Pu_i = u_i$ for each $i \in \{1, \ldots, k\}$. Hence $P^2 = P$ and $R(P) = U$. That is, P is a projection onto U. It remains to show that $R(P) \perp N(P)$. For this, let $x \in R(P)$ and $y \in N(P)$. Then

$$x = Px = \sum_{j=1}^{k} \langle x, u_j \rangle u_j \quad \text{and} \quad Py = \sum_{j=1}^{k} \langle y, u_j \rangle u_j = 0.$$

Being orthonormal, $\{u_1, \ldots, u_k\}$ is linearly independent. Hence, $\langle y, u_j \rangle = 0$ for all $j \in \{1, \ldots, k\}$. Then $\langle x, y \rangle = \sum_{j=1}^{k} \langle x, u_j \rangle \langle u_j, y \rangle = 0$. Therefore, $R(P) \perp N(P)$. ■

However, if U is an infinite dimensional subspace of an inner product space V, then there need not exist an orthogonal projection P on V such that $R(P) = U$. Look at the following example.

Example 7.23 Let $V = \ell^2$, the inner product space of all square-summable sequences of scalars. Recall that for $x = (\alpha_1, \alpha_2, \ldots)$ and $y = (\beta_1, \beta_2, \ldots)$ in ℓ^2, the inner product $\langle x, y \rangle$ is given by

$$\langle x, y \rangle = \sum_{n=1}^{\infty} \alpha_i \overline{\beta}_i.$$

Let $U = c_{00}$, the subspace of ℓ^2 of all square-summable sequences of scalars having only a finite number of nonzero entries. We observe that if $x \in \ell^2$ satisfies $\langle x, u \rangle = 0$ for all $u \in U$, then $x = 0$. Indeed, if $x = (\alpha_1, \alpha_2, \ldots)$ and $\langle x, u \rangle = 0$ for all $u \in U$, then $\alpha_k = \langle x, e_k \rangle = 0$ for all $k \in \mathbb{N}$ so that $x = 0$. Hence, $U^\perp$ is the zero subspace, and hence $V \neq U \oplus U^\perp$. □

The following results show some connection between projections and self-adjoint operators, in the case of finite dimensional inner product spaces.

Theorem 7.24 *Let P be a projection on a finite dimensional inner product space V. Then the following are equivalent:*

(1) *P is an orthogonal projection.*
(2) *P is self-adjoint.*
(3) *P is positive semi-definite.*
(4) *For each $x \in V$, $\|Px\| \leq \|x\|$.*

Moreover, if $x \in R(P)$ then $\langle Px, x \rangle = \|x\|^2$.

Proof (1) ⇒ (2) Let P be an orthogonal projection. If $v \in V$, then it can be written uniquely as

$$v = Pv + (v - Pv) \quad \text{where } Pv \perp v - Pv.$$

Then

$$\langle Pv, v\rangle = \langle Pv, Pv\rangle + \langle Pv, v - Pv\rangle = \langle Pv, Pv\rangle \in \mathbb{R}.$$

By Theorem 7.7, P is self-adjoint.

(2) ⇒ (3) Let the projection P be self-adjoint. Then, for each $x \in V$,

$$\langle Px, x\rangle = \langle P^2x, x\rangle = \langle Px, P^*x\rangle = \langle Px, Px\rangle \geq 0. \tag{7.1}$$

That is, P is positive semi-definite.

(3) ⇒ (4) Suppose that the projection P is positive semi-definite. Then it is self-adjoint. Let $x \in V$. As in (7.1), $\langle Px, x - Px\rangle = 0$. By Pythagoras theorem,

$$\|x\|^2 = \|Px + x - Px\|^2 = \|Px\|^2 + \|x - Px\|^2 \geq \|Px\|^2.$$

(4) ⇒ (1) Suppose that for each $x \in V$, $\|Px\| \leq \|x\|$. We show that $R(P) = N(P)^{\perp}$. For this purpose, let $v \in N(P)^{\perp}$. Write $w := Pv - v$. Then

$$Pw = P(Pv - v) = P^2v - Pv = Pv - Pv = 0.$$

That is, $w \in N(P)$. So, $\langle v, w\rangle = 0$ and $Pv = v + w$. Using Pythagoras theorem, we obtain

$$\|v\|^2 \leq \|v\|^2 + \|w\|^2 = \|v + w\|^2 = \|Pv\|^2 \leq \|v\|^2.$$

Therefore, $\|w\|^2 = 0$, which implies that $w = 0$. Consequently $v = Pv$. That is, $v \in R(P)$. This shows that $N(P)^{\perp} \subseteq R(P)$.

For the other containment, let $v \in R(P)$. Then $Pv = v$. Since $V = R(P) + N(P)^{\perp}$, we write $v = u + w$ for some $u \in N(P)$ and $w \in N(P)^{\perp}$. We already know that $N(P)^{\perp} \subseteq R(P)$. So, $w \in R(P)$; which implies that $Pw = w$. Then

$$v = Pv = Pu + Pw = Pw = w \in N(P)^{\perp}.$$

Therefore, $R(P) \subseteq N(P)^{\perp}$.

Moreover, if $x \in R(P)$, then $Px = x$. Hence $\langle Px, x\rangle = \langle x, x\rangle = \|x\|^2$. ■

The following theorem lists some relations between two orthogonal projections.

Theorem 7.25 *Let U and W be subspaces of a finite dimensional inner product space V. Let P_U and P_W be the respective orthogonal projections. Then the following are equivalent:*

(1) *$P_W - P_U$ is positive semi-definite.*
(2) *$\|P_U(x)\| \leq \|P_W(x)\|$ for each $x \in V$.*

(3) *U is a subspace of W.*
(4) $P_W P_U = P_U$.
(5) $P_U P_W = P_U$.

Proof (1) $\Rightarrow$ (2) : Let $x \in V$. Since P_U is an orthogonal projection, it is self-adjoint. As in (7.1), we have $\langle P_U x, x\rangle = \|P_U x\|^2$. Similarly, $\langle P_W x, x\rangle = \|P_W x\|^2$. Assume that $P_W - P_U$ is positive semi-definite. Then

$$0 \le \langle (P_W - P_U)(x), x\rangle = \langle P_W x, x\rangle - \langle P_U x, x\rangle = \|P_W x\|^2 - \|P_U x\|^2.$$

(2) $\Rightarrow$ (3) : Assume that $\|P_U x\| \le \|P_W x\|$ for each $x \in V$. Let $y \in U$. Then

$$\langle y, y\rangle = \|y\|^2 = \|P_U y\|^2 \le \|P_W y\|^2 = \langle P_W y, y\rangle \le \|y\|^2.$$

Hence $\langle y, y\rangle - \langle P_W y, y\rangle = 0$. Or that $\langle (I - P_W)y, y\rangle = 0$.

Notice that $R(I - P_W) = N(P_W)$ and $N(I - P_W) = R(P_W)$. Thus $I - P_W$ is also an orthogonal projection. As in (7.1), we have $\langle (I - P_W)y, y\rangle = \|(I - P_W)y\|^2$.

We thus conclude that $(I - P_W)y = 0$. That is, $P_W y = y$. Therefore, $y \in W$.

(3) $\Rightarrow$ (4) : Let $x \in V$. Then $P_U x \in U$. If $U \subseteq W$, then $P_U x \in W$. It follows that $P_W(P_U x) = P_U x$.

(4) $\Rightarrow$ (5) : If $P_W P_U = P_U$, then $P_U^* P_W^* = P_U^*$. As all orthogonal projections are self-adjoint, we have $P_U P_W = P_U$.

(5) $\Rightarrow$ (1) : Suppose $P_U P_W = P_U$. Then $P_W^* P_U^* = P_U^*$ implies that $P_W P_U = P_U$. Write $Q := P_W - P_U$. Then

$$Q^2 = P_W^2 + P_U^2 - P_W P_U - P_U P_W = P_W + P_U - P_U - P_U = Q.$$

So, Q is a projection. Since P_W and P_u are self-adjoint, so is Q. Further, $\langle Qx, x\rangle = \|Qx\|^2 \ge 0$ for each $x \in V$. Therefore, $Q = P_W - P_U$ is positive semi-definite. ■

We will again use these facts in proving that sum of two orthogonal projections is an orthogonal projection provided their compositions are identically zero.

Theorem 7.26 *Let P and Q be two orthogonal projections on a finite dimensional inner product space V. Then $P + Q$ is an orthogonal projection on V if and only if $PQ = QP = 0$ if and only if $R(P) \perp R(Q)$.*

Proof Let P, Q, and $P + Q$ be orthogonal projections on V. Let $v \in V$. Write $x := Pv$. Then

$$\begin{aligned}\|x\|^2 &= \|Px\|^2 \le \|Px\|^2 + \|Qx\|^2 = \langle Px, x\rangle + \langle Qx, x\rangle \\ &= \langle (P + Q)x, x\rangle \le \|(P + Q)x\|^2 = \|x\|^2.\end{aligned}$$

Hence, $\|Px\|^2 = \|Px\|^2 + \|Qx\|^2$; which implies that $Qx = 0$. So, $QPv = 0$ for all $v \in V$. That is, $QP = 0$. Interchanging P and Q, we also have $PQ = 0$.

Conversely, suppose P and Q are orthogonal projections. If $PQ = QP = 0$, then $(P+Q)^2 = P^2 + Q^2 = P + Q$ shows that $P + Q$ is a projection. Moreover, $(P+Q)^* = P^* + Q^* = P + Q$. Using Theorem 7.24, we conclude that $P + Q$ is an orthogonal projection.

For the second equivalence, assume that $PQ = QP = 0$. Let $x \in R(P)$ and let $y \in R(Q)$. Then

$$\langle x, y\rangle = \langle Px, Qy\rangle = \langle x, P^*Qy\rangle = \langle x, PQy\rangle = \langle x, 0\rangle = 0.$$

That is, $R(P) \perp R(Q)$.

Conversely, suppose $R(P) \perp R(Q)$. Let $v \in V$. Then

$$Qv \in R(Q) \subseteq R(P)^{\perp} = N(P).$$

That is, $PQv = 0$. Therefore, $PQ = 0$; $QP = Q^*P^* = (PQ)^* = 0$. ■

Theorem 7.26 can be generalized to more than two orthogonal projections in an obvious way; see Exercises 13–14.

Exercises for Sect. 7.2

1. Let $U := \text{span}(1, -1, 1)$. Find a vector $u \in U$ so that $(1, 1, 1) - u \perp U$.
2. Give an example of a projection that is not an orthogonal projection.
3. Construct two projections P and Q such that $PQ = 0$ but $QP \neq 0$.
4. Determine the matrices that represent all orthogonal projections on $\mathbb{C}^{3\times 1}$.
5. Let P be an orthogonal projection on an inner product space V, and let $x \in V$. Prove the following:
   (a) $\sup\{\|Py\| : \|y\| = 1,\ y \in V\} = 1$.
   (b) If $u \in R(P)$, then $\|x - Px\| \leq \|x - u\|$.
   (c) $\|x - Px\| = \inf\{\|x - u\| : u \in R(P)\}$.
6. Let U and W be subspaces of a finite dimensional inner product space V. Let P be a projection onto U along W. Prove the following:
   (a) $I - P$ is a projection onto W along U.
   (b) $U = \{x \in V : Px = x\}$ and $W = \{x \in V : Px = 0\}$.
7. Let $P : \mathbb{R}^{3\times 1} \to \mathbb{R}^{2\times 1}$ be a projection, where $P(e_1) = \alpha e_1$ for some $\alpha \in \mathbb{R}$, and $P(e_1), P(e_2), P(e_3)$ form an equilateral triangle. Determine $[P]$.
8. Let P be a projection on an inner product space V. Prove that there exists a basis B of V such that $[P]_{B,B}$ has diagonal entries as 0 or 1, and all off-diagonal entries 0.
9. Suppose that T is a diagonalizable operator having eigenvalues in $\{0, 1\}$. Does it mean that T is a projection?
10. Let P and Q be projections on an inner product space V such that $P + Q = I$. Prove that $PQ = 0$.

11. Let U_1, U_2, W_1, W_2 be subspaces of a finite dimensional inner product space V. Suppose that P_1 is a projection on U_1 along W_1 and P_2 is a projection on U_2 along W_2. Prove the following:

    (a) $P_1 + P_2$ is a projection if and only if $P_1P_2 = P_2P_1$. In such a case, $P_1 + P_2$ is a projection on $U_1 + U_2$ along $W_1 \cap W_2$; and $U_1 \cap U_2 = \{0\}$.
    (b) $P_1 - P_2$ is a projection if and only if $P_1P_2 = P_2P_1 = P_2$. In such a case, $P_1 - P_2$ is a projection on $U_1 \cap W_2$ along $U_2 + W_1$, and $U_2 \cap W_1 = \{0\}$.
    (c) If $P_1P_2 = P_2P_1 = P$, then P is a projection on $U_1 \cap U_2$ along $W_1 + W_2$.

12. Prove that if P is a projection, then $I + P$ is invertible. Find $(I + P)^{-1}$.
13. Let $T_1, \ldots, T_k$ be projections on an inner product space V. Prove that if $T_1 + \cdots + T_k = I$, then $T_iT_j = 0$ for all $i \neq j$.
14. Let $T_1, \ldots, T_k$ be linear operators on an inner product space V. Suppose that $T_iT_j = 0$ for all $i \neq j$, and that $T_1 + \cdots + T_k = I$. Prove that each T_i is a projection.

7.3 Normal Operators

Recall that a linear operator T on a finite dimensional inner product space is normal if and only if $T^*T = TT^*$. Though self-adjoint operators are normal, a normal operator need not be self-adjoint. A look-alike result is as follows.

Theorem 7.27 *Let V be a finite dimensional inner product space. A linear operator T on V is normal if and only if for each $v \in V$, $\|Tv\| = \|T^*v\|$.*

Proof T is a normal operator

$$\begin{aligned}
&\text{if and only if } T^*T - TT^* = 0\\
&\text{if and only if for each } v \in V,\ \langle (T^*T - TT^*)v, v\rangle = 0\\
&\text{if and only if for each } v \in V,\ \langle T^*Tv, v\rangle = \langle TT^*v, v\rangle\\
&\text{if and only if for each } v \in V,\ \langle Tv, Tv\rangle = \langle T^*v, T^*v\rangle\\
&\text{if and only if for each } v \in V,\ \|Tv\| = \|T^*v\|.
\end{aligned}$$

■

If T is a linear operator on a finite dimensional inner product space, then λ is an eigenvalue of T if and only if $\overline{\lambda}$ is an eigenvalue of T^*. However, the corresponding eigenvectors need not be equal. That is, if $Tv = \lambda v$, then $T^*v = \overline{\lambda}v$ need not hold. For example, let $T : \mathbb{R}^2 \to \mathbb{R}^2$ be defined by

$$T(a, b) = (a + b, b), \quad (a, b) \in \mathbb{R}^2.$$

We see that $T(1, 0) = 1\,(1, 0)$. That is, 1 is an eigenvalue of T with a corresponding eigenvector $(1, 0)$. Now, $T^* : \mathbb{R}^2 \to \mathbb{R}^2$ is given by

$$T^*(a, b) = (a, a + b), \quad (a, b) \in \mathbb{R}^2.$$

And, $T^*(0, 1) = 1\,(0, 1)$. That is, $1 = \overline{1}$ is an eigenvalue of T^* with a corresponding eigenvector $(0, 1)$. But $(1, 0)$, which was found to be an eigenvector of T is not an eigenvector T^* as $T^*(1, 0) = (1, 1)$, which cannot be a scalar multiple of $(1, 0)$. However, for normal operators, the following stronger result holds.

Theorem 7.28 *Let T be a normal operator on an inner product space V.*

(1) *For any $\lambda \in \mathbb{F}$ and any $v \in V$, $Tv = \lambda v$ if and only if $T^*v = \overline{\lambda}v$.*
(2) *Eigenvectors corresponding to distinct eigenvalues are orthogonal.*

Proof (1) Let $\lambda \in \mathbb{F}$ and let $v \in V$. Since $T^*T = TT^*$,

$$(T - \lambda I)^*(T - \lambda I) = T^*T - (\lambda + \overline{\lambda})I + |\lambda|^2 I = (T - \lambda I)(T - \lambda I)^*.$$

That is, the linear operator $T - \lambda I$ is normal. By Theorem 7.27,

$$\|(T^* - \overline{\lambda} I)v\| = \|(T - \lambda I)^* v\| = \|(T - \lambda I)v\|.$$

It then follows that $Tv = \lambda v$ if and only if $T^*v = \overline{\lambda}v$.

(2) Let λ and μ be distinct eigenvalues of T with corresponding eigenvectors u and v, respectively. Then $Tu = \lambda u$ and $Tv = \mu v$. From (1), we obtain $T^*v = \overline{\mu}\,v$. Now,

$$\langle Tu, v\rangle = \langle \lambda u, v\rangle = \lambda\langle u, v\rangle, \quad \langle u, T^*v\rangle = \langle u, \overline{\mu}\,v\rangle = \mu\langle u, v\rangle.$$

Since $\langle Tu, v\rangle = \langle u, T^*v\rangle$, we obtain $(\lambda - \mu)\langle u, v\rangle = 0$. But $\lambda \neq \mu$. Therefore, $u \perp v$. ■

We shall make use of the following general results in due course.

Theorem 7.29 *Let T be a linear operator on a finite dimensional inner product space V and U be a subspace of V. Then, U is T-invariant if and only if $U^\perp$ is T^*-invariant. Further, if U is invariant under T and T^*, then the operators $T_U := T|_U : U \to U$ and $(T^*)_U := T^*|_U : U \to U$ satisfy $(T_U)^* = (T^*)_U$.*

Proof By the definition of adjoint, we have

$$\langle Tu, v\rangle = \langle u, T^*v\rangle \quad \text{for all } u, v \in V.$$

Suppose $T(U) \subseteq U$. Then for all $v \in U^\perp$,

$$\langle u, T^*v\rangle = \langle Tu, v\rangle = 0 \quad \text{for all } u \in U.$$

Consequently, $T^*v \in U^\perp$. Conversely, let $T^*(U^\perp) \subseteq U^\perp$. Then for all $u \in U$,

$$\langle Tu, v\rangle = \langle u, T^*v\rangle = 0 \quad \text{for all } v \in U^\perp.$$

Hence $Tu \in U^{\perp\perp} = U$.

Next assume that U is invariant under T and T^*. Then, for all $u, v \in U$,

$$\langle u, (T_U)^* v\rangle = \langle T_U u, v\rangle = \langle Tu, v\rangle = \langle u, T^* v\rangle = \langle u, (T^*)_U v\rangle.$$

Therefore, $(T_U)^* = (T^*)_U$. ■

Theorem 7.30 *Let T be a normal operator on a finite dimensional inner product space V.*

(1) *If $p(t)$ is a polynomial, then $p(T)$ is normal.*
(2) *$N(T^*) = N(T)$ and $R(T^*) = R(T)$.*
(3) *For all $x, y \in V$, $\langle Tx, Ty\rangle = \langle T^*x, T^*y\rangle$.*
(4) *For each $k \in \mathbb{N}$, $N(T^k) = N(T)$ and $R(T^k) = R(T)$.*

Proof (1) For any two operators T and S on V if $TS = ST$, then

$$T^2S^2 = T(TS)S = T(ST)S = (TS)(TS) = (ST)(ST) = S(TS)T = S(ST)T = S^2T^2.$$

In particular, taking $S = T^*$, we obtain $(T^*)^2T^2 = T^2(T^*)^2$. It follows by induction that $(p(T))^*p(T) = p(T)(p(T))^*$.

(2) By Theorem 4.54 (2)–(3), we obtain $N(T^*) = N(TT^*) = N(T^*T) = N(T)$ and $R(T^*) = N(T)^\perp = N(T^*)^\perp = R(T)$.

(3) Let $x, y \in V$. Then $\langle Tx, Ty\rangle = \langle x, T^*Ty\rangle = \langle x, TT^*y\rangle = \langle T^*x, T^*y\rangle$.

(4) We know that $N(T) \subseteq N(T^k)$ for all $k \in \mathbb{N}$. Now, let $i \geq 1$ and let $v \in V$. By Theorem 7.27,

$$\|T^{i+1}v\| = \|T(T^iv)\| = \|T^*(T^iv)\| = \|(T^*T)(T^{i-1}v)\|.$$

If $v \in N(T^{i+1})$, then $T^{i-1}v \in N(T^*T) = N(T)$; consequently, $v \in N(T^i)$. (Here, we use the convention that $T^0 = I$.) That is, $N(T^{i+1}) \subseteq N(T^i)$. Hence, $N(T^k) \subseteq N(T^{k-1}) \subseteq \cdots \subseteq N(T)$. It then follows that $N(T^k) = N(T)$.

Next, $R(T^k) = N(T^k)^\perp = N(T)^\perp = R(T^*) = R(T)$. ■

Theorem 7.31 *Let T be a normal operator on a finite dimensional inner product space V. Then the following are true:*

(1) *The ascent of every eigenvalue of T is 1. In particular, the algebraic multiplicity and geometric multiplicity of each eigenvalue are equal.*
(2) *For each $\lambda \in \mathbb{F}$, $N(T - \lambda I)$ and $N(T - \lambda I)^\perp$ are both T-invariant and T^*-invariant.*

Proof (1) Let $\lambda \in \mathbb{F}$ be an eigenvalue of T. Since $T - \lambda I$ is also a normal operator, by Theorem 7.30 (4), $N(T - \lambda I) = N(T - \lambda I)^2$. Hence, the ascent of λ is 1. Consequently, the algebraic multiplicity and geometric multiplicity of each eigenvalue are equal.

(2) Let $\lambda \in \mathbb{F}$. Clearly, $N(T - \lambda I)$ is T-invariant, and $N(T^* - \overline{\lambda} I)$ is T^*-invariant. By Theorem 7.29, $N(T - \lambda I)^\perp$ is T^*-invariant, and $N(T^* - \overline{\lambda} I)^\perp$ is T-invariant. ($T^{**} = T$.) Since $T - \lambda I$ is a normal operator, by Theorem 7.28 (1), we have $N(T^* - \overline{\lambda} I) = N(T - \lambda I)$. We conclude that $N(T - \lambda I)$ and $N(T - \lambda I)^\perp$ are invariant under both T and T^*. ■

If T is not a normal operator, then $N(T - \lambda I)$ can fail to be T^*-invariant and $N(T - \lambda I)^\perp$ can fail to be T-invariant.

Example 7.32 Consider $T : \mathbb{R}^2 \to \mathbb{R}^2$ given by $T(a, b) = (a + b, b)$ for all $a, b \in \mathbb{R}$. The matrix of T with respect to the standard basis of $\mathbb{R}^2$ is $[T] = \begin{bmatrix} 1 & 1 \\ 0 & 1 \end{bmatrix}$. Thus T^* is given by $T^*(a, b) = (a, a + b)$. Now, 1 is an eigenvalue of T since $T(1, 0) = 1\,(1, 0)$. Then

$$N(T - I) = \{(a, 0) : a \in \mathbb{R}\}, \quad N(T - I)^\perp = \{(0, b) : b \in \mathbb{R}\}.$$

Since $T(a, 0) = (a, 0) \in N(T - I)$ and $T^*(0, b) = (0, b)$, we conclude that $N(T - I)$ is T-invariant and $N(T - I)^\perp$ is T^*-invariant, verifying Theorem 5.21 (2)–(3). But

$$T^*(1, 0) = (1, 1) \notin N(T - I), \quad T(0, 1) = (1, 1) \notin N(T - I)^\perp.$$

Therefore, neither $N(T - I)$ is T^*-invariant nor $N(T - I)^\perp$ is T-invariant. Note that T is not a normal operator. □

As a consequence of Theorem 7.31, we obtain the spectral theorem for normal operators on a finite dimensional vector space. Recall that when λ is an eigenvalue of T, we denote the eigenspace $N(T - \lambda I)$ by $E(\lambda)$.

Theorem 7.33 (Spectral theorem for normal operators) *Let T be a normal operator on a finite dimensional inner product space V. Suppose the characteristic polynomial of T splits over $\mathbb{F}$. Let $\lambda_1, \ldots, \lambda_k$ be the distinct eigenvalues of T. Then the following are true:*

(1) *$V = E(\lambda_1) \oplus \cdots \oplus E(\lambda_k)$, and there exists an orthonormal basis B for V such that $[T]_{B,B}$ is a diagonal matrix.*
(2) *Let $P_1, \ldots, P_k$ be the orthogonal projections onto $E(\lambda_1), \ldots, E(\lambda_k)$, respectively. Then $P_i P_j = 0$ for all $i, j \in \{1, \ldots, k\}$, $i \neq j$, and*

$$T = \lambda_1 P_1 + \cdots + \lambda_k P_k, \quad P_1 + \cdots + P_k = I.$$

(3) *If $\{u_{ij} : j = 1, \ldots, n_i\}$ is an orthonormal basis for $E(\lambda_i)$ for $i = 1, \ldots, k$, then*

$$Tx = \sum_{i=1}^{k} \lambda_i \Big(\sum_{j=1}^{n_i} \langle x, u_{ij} \rangle\, u_{ij} \Big) \quad \text{for each } x \in V.$$

Proof (1) By Theorems 7.31 and 5.29, we have $V = E(\lambda_1) \oplus \cdots \oplus E(\lambda_k)$. Let $B_1, \ldots, B_k$ be orthonormal bases of $E(\lambda_1), \ldots, E(\lambda_k)$, respectively. Then their union $B := \bigcup_{i=1}^{k} B_i$ is an orthonormal basis of V and $[T]_{B,B}$ is a diagonal matrix.

(2) Let $x \in V$. Since $V = E(\lambda_1) \oplus \cdots \oplus E(\lambda_k)$, the vector x can be written as

$$x = x_1 + \cdots + x_k,$$

where each $x_i \in E(\lambda_i)$ is uniquely determined. Since

$$P_i(x) = x_i \quad \text{and} \quad Tx_i = \lambda_i x_i \quad \text{for } i \in \{1, \ldots, k\},$$

we obtain

$$\begin{aligned} x &= P_1 x + \cdots + P_k x = (P_1 + \cdots P_k)x; \\ Tx &= Tx_1 + \cdots + Tx_k = \lambda_1 x_1 + \cdots + \lambda_k x_k \\ &= \lambda_1 P_1 x + \cdots + \lambda_k P_k x = (\lambda_1 P_1 + \cdots + \lambda_k P_k)x. \end{aligned}$$

Thus, $T = \lambda_1 P_1 + \cdots + \lambda_k P_k$ and $P_1 + \cdots + P_k = I$. Also, since $E_{\lambda_i} \perp E_{\lambda_j}$ for $i \neq j$, it follows that $P_i P_j = 0$ for all $i, j \in \{1, \ldots, k\}$, $i \neq j$.

(3) For $i \in \{1, \ldots, k\}$, let $B_i := \{u_{ij} : j = 1, \ldots, n_i\}$ be an orthonormal basis of $E(\lambda_i)$. Then P_i is given by

$$P_i x = \sum_{j=1} \langle x, u_{ij} \rangle \, u_{ij} \quad \text{for each } x \in V.$$

Hence, from (2) we have

$$Tx = \lambda_1 P_1 x + \cdots + \lambda_k P_k x = \sum_{i=1}^{k} \lambda_i \Big(\sum_{j=1}^{n_i} \langle x, u_{ij} \rangle \, u_{ij} \rangle \Big)$$

for each $x \in V$. ■

The proof of Theorem 7.33 uses the decomposition $V = E(\lambda_1) \oplus \cdots \oplus E(\lambda_k)$ of the space V, which again uses Theorem 5.29. In fact, the decomposition of V can be proved without relying on Theorem 5.29.

Lemma 7.34 *Let T be a normal operator on a finite dimensional inner product space V. Let U be invariant under T and T^*. Then $U^\perp$ is invariant under T, and the restriction operators $T\big|_U : U \to U$ and $T|_{U^\perp} : U^\perp \to U^\perp$ are normal operators.*

Proof Since U is invariant under T^*, by Theorem 7.29, $U^\perp$ is invariant under $T^{**} = T$ and $(T_U)^* = T^*_U$, where $T_U := T\big|_U : U \to U$ and $T^*_U := T^*\big|_U : U \to U$. Next, using the fact that T is a normal operator, for every $u, v \in U$, we have

$$\langle T_U^* T_U u, v\rangle = \langle T_U u, T_U v\rangle = \langle Tu, Tv\rangle = \langle T^*Tu, v\rangle = \langle TT^*u, v\rangle = \langle T_U T_U^* u, v\rangle.$$

Hence, $T_U^* T_U = T_U T_U^*$. Thus, $T\big|_U : U \to U$ is a normal operator on U. Similarly, it can be shown that $T|_{U^\perp} : U^\perp \to U^\perp$ is a normal operator on $U^\perp$. ■

Proof of Theorem 7.33 *without using Theorem* 5.29:
Let $U := E(\lambda_1) \oplus \cdots \oplus E(\lambda_k)$. We prove $U^\perp = \{0\}$ so that by projection theorem, $V = U$. By Theorem 7.31 (1) $E(\lambda_i) := N(T - \lambda_i I)$ is invariant under both T and T^*. Hence, U is also invariant under T and T^*. Therefore, by Lemma 7.34, $U^\perp$ is invariant under T, and both $T\big|_U : U \to U$ and $T|_{U^\perp} : U^\perp \to U^\perp$ are normal operators. Now, if $U^\perp \neq \{0\}$, then $T\big|_{U^\perp}$ has an eigenvalue, say λ with a corresponding eigenvector $x \in U^\perp$. Since λ is also an eigenvalue of T, $x \in E(\lambda_i) \subseteq W$ for some $i \in \{1, \ldots, k\}$. This is not possible. Therefore, $U^\perp = \{0\}$, and hence $V = E(\lambda_1) \oplus \cdots \oplus E(\lambda_k)$. ■

The following theorem is an obvious matrix interpretation of Theorem 7.33 by constructing the unitary matrix P whose columns are the basis vectors from B.

Theorem 7.35 *If $A \in \mathbb{C}^{n\times n}$ is a normal matrix, then there exists a unitary matrix $P \in \mathbb{C}^{n\times n}$ such that $P^{-1}AP$ is a diagonal matrix.*

Proof of the spectral theorem provides a way to diagonalize a normal matrix A. Assume further that, if $\mathbb{F} = \mathbb{R}$, then the characteristic polynomial of A splits. First, get the distinct eigenvalues $\lambda_1, \ldots, \lambda_k$ of A. Corresponding to each eigenvalue λ_i of multiplicity m_i, determine m_i orthonormal eigenvectors. Existence of such eigenvectors is guaranteed by the theorem. Then put the eigenvectors together as columns keeping the eigenvectors together corresponding to same eigenvalue. Call the new matrix as P. Then $P^{-1} = P^*$, that is, P is a unitary matrix, and $P^{-1}AP$ is the diagonal matrix whose diagonal entries are the eigenvalues $\lambda_1, \ldots, \lambda_k$ repeated $m_1, \ldots, m_k$ time, respectively.

Moreover, if all eigenvalues of A are real, then we can choose real eigenvectors corresponding to them. The construction shows that the unitary matrix U can be chosen as a real orthogonal matrix. Therefore, if a normal matrix is similar to a real matrix over $\mathbb{C}$, then it is also similar to the same real matrix over $\mathbb{R}$. Self-adjoint matrices fall under this category.

The representation

$$T = \lambda_1 P_1 + \cdots + \lambda_k P_k$$

of a normal operator with orthogonal projections $P_1, \ldots, P_k$ onto the eigenspaces $E(\lambda_1), \ldots, E(\lambda_k)$, respectively, is called the *spectral form* of T. Notice that the projections P_i in the spectral form of T behave as identity on the eigenspaces and take the value zero elsewhere.

If T is a linear operator on a real inner product space, and the characteristic polynomial $\chi_T(t)$ of T does not split over $\mathbb{R}$, then it is of the form

$$\chi_T(t) = (t - \lambda_1)\cdots(t - \lambda_j)((t - \alpha_1)^2 + \beta_1^2)\cdots((t - \alpha_m)^2 + \beta_m^2),$$

where at least one quadratic factor is present. We guess that there exists an orthonormal basis for the real inner product space with respect to which the normal operator has a matrix representation as a block-diagonal matrix in the form

$$\text{diag}\left(\lambda_1, \ \cdots, \ \lambda_j, \ \begin{bmatrix}\alpha_1 & -\beta_1\\ \beta_1 & \alpha_1\end{bmatrix}, \ \cdots, \ \begin{bmatrix}\alpha_m & -\beta_m\\ \beta_m & \alpha_m\end{bmatrix}\right).$$

This result is true; and it can be proved by using Lemma 7.34 or using complexification of the vector space.

Exercises for Sect. 7.3

1. For each (normal) matrix A as given below, find a unitary matrix P so that P^*AP is diagonal:

   (a) $\begin{bmatrix}1 & i\\ -i & 1\end{bmatrix}$ (b) $\begin{bmatrix}1 & 2\\ 2 & 1\end{bmatrix}$ (c) $\begin{bmatrix}1 & 0 & 1\\ 0 & 1 & 0\\ 1 & 0 & 0\end{bmatrix}$.
2. Are the following statements true for a linear operator T on a finite dimensional complex inner product space V?

   (a) Let $\alpha, \beta \in \mathbb{C}$. Then $\alpha T + \beta T^*$ is normal.
   (b) T is normal if and only if $T(U) \subseteq U$ implies that $T(U^\perp) \subseteq U^\perp$ for all subspaces U of V.
   (c) If T is normal and idempotent ($T^2 = T$), then T is self-adjoint.
   (d) If T is normal and nilpotent (for some $n \in \mathbb{N}$, $T^n = 0$), then $T = 0$.
   (e) If T is normal and $T^3 = T^2$, then T is idempotent.
   (f) If T is self-adjoint and $T^n = I$ for some $n \geq 1$ then $T^2 = I$.

3. Is it true that the sum of two normal operators on a complex inner product space is also a normal operator?
4. Show that every real skew-symmetric matrix is normal.
5. Let T be a normal operator on a finite dimensional inner product space. Prove the following:

   (a) T is self-adjoint if and only if each eigenvalue of T is real.
   (b) T is positive semi-definite if and only if each eigenvalue of T is nonnegative.
   (c) T is positive definite if and only if each eigenvalue of T is positive.
   (d) T is unitary if and only if each eigenvalue of T is of absolute value one.
   (e) T is invertible if and only if each eigenvalue of T is nonzero.
   (f) T is idempotent if and only if each eigenvalue of T is zero or one.

6. Let T be a normal operator on a finite dimensional inner product space. Let λ be an eigenvalue of T. Prove that both $E(\lambda)$ and $E(\lambda)^\perp$ are T-invariant.
7. Show that the matrix $A = [a_{ij}] \in \mathbb{C}^{n\times n}$ is normal if and only if $\sum_{i,j=1}^n a_{ij}^2 = \sum_{i=1}^n |\lambda_i|^2$, where $\lambda_1, \ldots, \lambda_n$ are the eigenvalues of A repeated according to their algebraic multiplicities.
8. Let $A = B^{-1}B^*$ for some invertible $B \in \mathbb{C}^{n\times n}$. Show that A is unitary if and only if B is normal.

9. Let $A \in \mathbb{C}^{n \times n}$ be invertible. Suppose that A can be written as $A = HNH$, where H is hermitian and N is normal. Show that $B := A^{-1}A^*$ is similar to a unitary matrix.
10. Prove that the cyclic shift operator T on $\mathbb{C}^{n \times 1}$ defined by

$$Te_1 = e_n, Te_2 = e_1, Te_3 = e_2, \ldots, Te_n = e_{n-1}$$

is a normal operator. Diagonalize it.
11. A *circulant matrix* is a matrix in the following form:

$$\begin{bmatrix} c_0 & c_1 & c_2 & \cdots & c_n \\ c_n & c_0 & c_1 & \cdots & c_{n-1} \\ \vdots & & & & \vdots \\ c_1 & c_2 & c_3 & \cdots & c_0 \end{bmatrix}.$$

Show that every circulant matrix is a normal matrix.
12. Let T be a linear operator on a finite dimensional complex inner product space. Prove that T can be written as $T = T_R + iT_I$, where $T_R^* = T_R$ and $T_I^* = T_I$. Then show that T is normal if and only if $T_R T_I = T_I T_R$.
13. Let T be a linear operator on a finite dimensional complex inner product space V. Prove that S is an isometry if and only if there exists an orthonormal basis for V consisting of eigenvectors of T with corresponding eigenvalues having absolute value 1.

7.4 Self-adjoint Operators

Most properties of a self-adjoint operator follow from the fact that it is a normal operator. However, self-adjoint operators enjoy some nice properties that are not shared by a general normal operator.

For a self-adjoint operator on a finite dimensional inner product space, the spectral mapping theorem (Theorem 5.17) can be strengthened as follows.

Theorem 7.36 (Spectral mapping theorem) *Let V be a finite dimensional inner product space over $\mathbb{F}$. Let T be a self-adjoint operator on V. Let $p(t)$ be a polynomial with coefficients from $\mathbb{F}$. Then*

$$\text{eig}(p(T)) = \{p(\lambda) : \lambda \in \text{eig}(T)\}.$$

Proof If $\mathbb{F} = \mathbb{C}$, the statement follows from Theorem 5.17(2). So, let $\mathbb{F} = \mathbb{R}$. By Theorem 5.17(1), $\{p(\lambda) : \lambda \in \text{eig}(T)\} \subseteq \text{eig}(p(T))$. We need to show that $\text{eig}(p(T)) \subseteq \{p(\lambda) : \lambda \in \text{eig}(T)\}$. Notice that since T is self-adjoint, so is $p(T)$, and it has an eigenvalue.

Let a real number $\gamma \in \text{eig}(p(T))$. By Proposition 5.1, the polynomial $p(t) - \gamma$ can be written as

$$p(t) - \gamma = a_k(t - \lambda_1) \cdots (t - \lambda_j)((t - \alpha_1)^2 + \beta_1^2) \cdots ((t - \alpha_m)^2 + \beta_m^2),$$

where λs, αs, and βs are real numbers. Moreover, if any quadratic factor is present in this product, then the corresponding βs are nonzero. Consequently,

$$p(T) - \gamma I = a_k(T - \lambda_1 I) \cdots (T - \lambda_j I)((T - \alpha_1 I)^2 + \beta_1^2 I) \cdots ((T - \alpha_m I)^2 + \beta_m^2 I).$$

Since $p(T) - \gamma I$ is not injective, at least one of the factors on the right hand side is not injective. However, each map of the form $(T - \alpha I)^2 + \beta^2 I$ is injective due to Lemma 7.11. Hence $T - \lambda I$ is not injective for some $\lambda \in \{\lambda_1, \ldots, \lambda_j\}$. Such a λ is in $\text{eig}(T)$. Therefore, $p(\lambda) - \gamma = 0$. That is, $\gamma \in \{p(\lambda) : \lambda \in \text{eig}(T)\}$. Therefore, $\text{eig}(p(T)) \subseteq \{p(\lambda) : \lambda \in \text{eig}(T)\}$. ■

Self-adjoint operators are normal operators, and also they have eigenvalues. Thus, the assumption that "the characteristic polynomial splits" holds true for a self-adjoint operator. We thus obtain the following result.

Theorem 7.37 (Spectral theorem for self-adjoint operators) *Let T be a self-adjoint operator on a finite dimensional inner product space V. Then there exists an orthonormal basis B for V such that $[T]_{B,B}$ is a diagonal matrix with real entries. Further, the conclusions in* (1)–(3) *of Theorem 7.33 hold.*

The spectral theorem is also called as the *Diagonalization Theorem* due to obvious reasons. The matrix interpretation of the spectral theorem for self-adjoint operators takes the following form:

Theorem 7.38 *Every hermitian matrix is unitarily similar to a real diagonal matrix. Every real symmetric matrix is orthogonally similar to a real diagonal matrix.*

Proof Any hermitian matrix $A \in \mathbb{C}^{n \times n}$ is a self-adjoint operator on $\mathbb{C}^{n \times 1}$. Thus we have an orthonormal basis B for $\mathbb{C}^{n \times 1}$ such that $[A]_{B,B} = D$ is a diagonal matrix. Construct the matrix $U \in \mathbb{C}^{n \times n}$ by taking its columns as the vectors from B in the same order. Then U is unitary and $U^*AU = U^{-1}AU = D$. Since the diagonal entries are the eigenvalues of the hermitian matrix A, they are real, so $D \in \mathbb{R}^{n \times n}$.

For the second statement, notice that a real symmetric matrix in $\mathbb{R}^{n \times n}$ is a self-adjoint operator on $\mathbb{R}^{n \times 1}$ and a unitary matrix with real entries is an orthogonal matrix. ■

Alternatively, the spectral theorem for self-adjoint operators can be deduced from Schur's theorem. It is as follows: if A is a hermitian matrix, i.e. if $A^* = A$, then by Theorem 6.12, there exists a unitary matrix Q such that Q^*AQ is upper triangular. Thus, $Q^*A^*Q = (Q^*AQ)^*$ is lower triangular. On the other hand, $Q^*A^*Q = Q^*AQ$ is upper triangular. Therefore, Q^*AQ is diagonal.

Example 7.39 Consider the real symmetric matrix

$$A = \begin{bmatrix} 2 & -1 & 0 & 0 \\ -1 & 2 & 0 & 0 \\ 0 & 0 & 2 & -1 \\ 0 & 0 & -1 & 2 \end{bmatrix}.$$

This is the representation of the linear operator $T : \mathbb{R}^4 \to \mathbb{R}^4$ defined by

$$T(a,\ b,\ c,\ d) = \big(2a - b,\ -a + 2b,\ 2c - d,\ -c + 2d\big)$$

with respect to the standard basis. Its characteristic equation is

$$\det(A - tI) = (t-1)^2(t-3)^2 = 0.$$

We determine the eigenvectors corresponding to eigenvalues 1 and 3. For the eigenvalue 1, let $(a, b, c, d) \in \mathbb{R}^4$ satisfy $T(a, b, c, d) = (a, b, c, d)$. Then we have

$$2a - b = a, \quad -a + 2b = b, \quad 2c - d = 2c, \quad -c + 2d = 2d.$$

From these equations we obtain $a = b$ and $c = d$. Thus, $(1, 1, 0, 0)$ and $(0, 0, 1, 1)$ are eigenvectors of T corresponding to the eigenvalue 1.

For the eigenvalue 3, we have $T(a, b, c, d) = 3(a, b, c, d)$, i.e. the equations

$$2a - b = 3a, \quad -a + 2b = 3b, \quad 2c - d = 3c, \quad -c + 2d = 3d.$$

It leads to $a = b$ and $c = d$ so that $(1, -1, 0, 0)$ and $(0, 0, -1, 1)$ are eigenvectors of T corresponding to the eigenvalue 3. Notice that the eigenvectors

$$(1, 1, 0, 0), \quad (0, 0, 1, 1), \quad (1, -1, 0, 0), \quad (0, 0, 1, -1)$$

are orthogonal. The normalized forms of these eigenvectors are

$$\Big(\frac{1}{\sqrt{2}}, \frac{1}{\sqrt{2}}, 0, 0\Big), \quad \Big(0, 0, \frac{1}{\sqrt{2}}, \frac{1}{\sqrt{2}}\Big), \quad \Big(\frac{1}{\sqrt{2}}, -\frac{1}{\sqrt{2}}, 0, 0\Big), \quad \Big(0, 0, \frac{1}{\sqrt{2}}, -\frac{1}{\sqrt{2}}\Big).$$

The matrix of these eigenvectors in that order is:

$$Q := \begin{bmatrix} 1/\sqrt{2} & 0 & 1/\sqrt{2} & 0 \\ 1/\sqrt{2} & 0 & -1/\sqrt{2} & 0 \\ 0 & 1/\sqrt{2} & 0 & 1/\sqrt{2} \\ 0 & 1/\sqrt{2} & 0 & -1/\sqrt{2} \end{bmatrix}.$$

It can be verified that $Q^{-1} = Q^T$ and

$$Q^{-1} A Q = \begin{bmatrix} 1 & 0 & 0 & 0 \\ 0 & 1 & 0 & 0 \\ 0 & 0 & 3 & 0 \\ 0 & 0 & 0 & 3 \end{bmatrix}.$$

The eigenvalues of T are on the diagonal of $Q^{-1} A Q$ as they should. □

As in the case of normal operators, the representation of a self-adjoint operator T as

$$T = \lambda_1 P_1 + \cdots + \lambda_k P_k$$

through the orthogonal projections $P_1, \ldots, P_k$ onto the eigenspaces $E(\lambda_1), \ldots, E(\lambda_k)$, respectively, is also called the *spectral form* of T.

We show here a connection between self-adjoint operators and positive definite operators.

Theorem 7.40 (Sylvester inertia) *Let T be a self-adjoint operator on a finite dimensional inner product space V. Then there exist T-invariant subspaces V_+, V_0 and V_- of V of unique dimensions such that $V = V_+ \oplus V_0 \oplus V_-$. Moreover, $T|_{V_+}$ is positive definite, $T|_{V_0}$ is the zero operator, $T|_{V_-}$ is negative definite, and T can be written as $T = T|_{V_+} \oplus T|_{V_0} \oplus T|_{V_-}$, in the sense that for every $x \in V$, if $x = u + v + w$ with $u \in V_+$, $v \in V_0$, $w \in V_-$, then*

$$Tx = T|_{V_+}u + T|_{V_0}v + T|_{V_-}w.$$

Proof Since T is self-adjoint, all its eigenvalues are real numbers. Write the distinct eigenvalues of T as in the following:

$$\lambda_1, \ldots, \lambda_k > 0, \quad \lambda_{k+1} = 0, \quad \lambda_{k+2}, \ldots, \lambda_{k+1+r} < 0,$$

where k is the number of distinct positive eigenvalues and r is the number of distinct negative eigenvalues. (If 0 is not an eigenvalue of T, then ignore λ_{k+1}; similarly, any other group is ignored if the case so demands.) Construct the subspaces V_+, V_0 and V_- as follows:

$$V_+ := E(\lambda_1) \oplus \cdots \oplus E(\lambda_k), \quad V_0 := E(\lambda_{k+1}), \quad V_- := E(\lambda_{k+2}) \oplus \cdots \oplus E(\lambda_{k+1+r}).$$

Since $V = E(\lambda_1) \oplus \cdots \oplus E(\lambda_{k+1+r})$, it follows that $V = V_+ \oplus V_0 \oplus V_-$, and the dimensions of V_+, V_0 and V_- are, respectively, the number of positive, zero, and negative eigenvalues of T, counting multiplicities, which are uniquely determined from T.

Due to Theorems 7.33–7.37, the corresponding restriction maps are given by

$$T_{V_+} := P_1 + \cdots + P_k, \quad T_{V_0} := P_{k+1}, \quad T_{V_-} := P_{k+2} + \cdots + P_{k+1+r}.$$

so that $T = T|_{V_+} + T|_{V_0} + T|_{V_-}$ in the sense spelled out in the theorem. Notice that the only eigenvalue of T_0 is 0, all eigenvalues of $T|_{V_+}$ are $\lambda_1, \ldots, \lambda_k > 0$, and those of T_{V_-} are $\lambda_{k+2}, \ldots, \lambda_{k+1+r} < 0$. Due to Theorem 7.24, each orthogonal projection is self-adjoint. Hence all of T_+, T_0, T_- are self-adjoint.

By Theorem 7.37, T_{V_+} is diagonalizable. So, let $\{v_1, \ldots, v_m\}$ be an orthonormal basis of V_+ with $T_{V_+} v_i = \alpha_i v_i$ for $i = 1, \ldots, m$. Here, each α_i is in $\{\lambda_1, \ldots, \lambda_k\}$ and, hence, positive. Let $x \in V_+$. By the Fourier expansion, and by Theorem 7.33(3), we have

$$x = \sum_{i=1}^{m} \langle x, v_i \rangle v_i, \quad T_{V_+} x = \sum_{i=1}^{m} \alpha_i \langle x, v_i \rangle v_i.$$

Now, if $x \neq 0$, then for at least one $i \in \{1, \ldots, m\}$, $\langle x, v_i \rangle \neq 0$. Then

$$\langle T_{V_+} x, x \rangle = \Big\langle \sum_{i=1}^{m} \alpha_i \langle x, v_i \rangle v_i, \ \sum_{i=1}^{m} \langle x, v_i \rangle v_i \Big\rangle = \sum_{i=1}^{m} \alpha_i |\langle x, v_i \rangle|^2 > 0.$$

Hence, T_{V_+} is positive definite. Similarly, it follows that $T|_{V_0}$ is the zero operator, and $T|_{V_-}$ is negative definite. ■

Let $A \in \mathbb{C}^{m \times m}$ be a hermitian matrix. Its diagonalization yields an invertible matrix Q such that $Q^* A Q = \text{diag}(\lambda_1, \ldots, \lambda_p, 0, \ldots, 0, -\lambda_{p+k+1}, \ldots, -\lambda_{p+k+n})$, where $m = p + k + n$, $Q^{-1} = Q^*$, and each λ_j is positive. Also,

$$Q = [w_1 \ \cdots \ w_m]$$

is the matrix formed by taking the orthonormal eigenvectors w_j corresponding to the eigenvalues. Define a diagonal matrix $D \in \mathbb{R}^{m \times m}$ as follows:

$$D = \text{diag}\Big(\frac{1}{\sqrt{\lambda_1}}, \ldots, \frac{1}{\sqrt{\lambda_p}}, 1, \ldots, 1, -\frac{1}{\sqrt{\lambda_{p+k+1}}}, \ldots, -\frac{1}{\sqrt{\lambda_{p+k+n}}} \Big).$$

We see that

$$DQ^* AQD = D \,\text{diag}(\lambda_1, \ldots, \lambda_p, 0, \ldots, 0, -\lambda_{n+k+1}, \ldots, -\lambda_{n+k+p})\, D = \text{diag}(I_p, 0, -I_n).$$

Since D is a real diagonal matrix, with $P = QD$, we have

$$P^* AP = \text{diag}(I_p, 0, -I_n).$$

Thus the matrix version of Theorem 7.40 may be stated as follows.

Theorem 7.41 (Sylvester form) *Let A be a hermitian matrix. Then there exists an invertible matrix P such that $P^* AP = \text{diag}(I_p, 0, -I_n)$, where p is the number of positive eigenvalues, and n is the number of negative eigenvalues of A, counting multiplicities.*

Observe that the P^*AP is not a similarity transform, in general, since P^* need not be equal to P^{-1}. However, each invertible matrix is a product of elementary matrices; therefore, it says that the diagonal matrix in the Sylvester form may be obtained by a sequence of *symmetric* elementary operations. That is, whenever we apply an elementary row operation, it must be immediately followed by a corresponding column operation so that the sequence of operations becomes symmetric. If in the sequence of operations, we use $E[i,j]$, and then following it, we must immediately apply $E'[i,j]$; if we use $E_\alpha[i]$, then soon after we must use $E'_{\overline{\alpha}}[i]$; and if we use the row operation $E_\alpha[i,j]$, the next column operation must be $E'_{\overline{\alpha}}[i,j]$. The goal of applying such elementary operations is to reduce the off-diagonal entries in A to 0. The symmetric operations would imply that once a row operation reduces an entry a_{ij} to 0, the succeeding column operation would reduce a_{ji} to 0.

Example 7.42 Let $A = \begin{bmatrix} 3 & -2 & -2i \\ -2 & 2 & 2i \\ 2i & -2i & -2 \end{bmatrix}$. We use symmetric elementary operations to reduce it to its Sylvester form as follows:

$$\begin{bmatrix} 3 & -2 & -2i \\ -2 & 2 & 2i \\ 2i & -2i & -2 \end{bmatrix} \xrightarrow{E_1[1,2]} \begin{bmatrix} 1 & 0 & 0 \\ -2 & 2 & 2i \\ 2i & -2i & -2 \end{bmatrix} \xrightarrow{E'_1[1,2]} \begin{bmatrix} 1 & 0 & 0 \\ 0 & 2 & 2i \\ 0 & -2i & -2 \end{bmatrix}$$

$$\xrightarrow{E_i[3,2]} \begin{bmatrix} 1 & 0 & 0 \\ 0 & 2 & 2i \\ 0 & 0 & -2 \end{bmatrix} \xrightarrow{E'_{-i}[3,2]} \begin{bmatrix} 1 & 0 & 0 \\ 0 & 2 & 0 \\ 0 & 0 & -2 \end{bmatrix} \xrightarrow{E_{1/\sqrt{2}}[2]} \begin{bmatrix} 1 & 0 & 0 \\ 0 & \sqrt{2} & 0 \\ 0 & 0 & -2 \end{bmatrix}$$

$$\xrightarrow{E'_{1/\sqrt{2}}[2]} \begin{bmatrix} 1 & 0 & 0 \\ 0 & 1 & 0 \\ 0 & 0 & -2 \end{bmatrix} \xrightarrow{E_{1/\sqrt{2}}[3]} \begin{bmatrix} 1 & 0 & 0 \\ 0 & 1 & 0 \\ 0 & 0 & -\sqrt{2} \end{bmatrix} \xrightarrow{E'_{1/\sqrt{2}}[3]} \begin{bmatrix} 1 & 0 & 0 \\ 0 & 1 & 0 \\ 0 & 0 & -1 \end{bmatrix}.$$

□

In view of Theorems 7.40 and 7.41, let p be the number of positive eigenvalues and n be the number of negative eigenvalues, counting multiplicities, of the self-adjoint operator T. Then the number $n + p$ is the rank of T. The number $p - n$, which is the difference between the number of positive and negative eigenvalues, counting multiplicities, is called the *signature* of T. The diagonal matrix $\mathrm{diag}(I_p, 0, -I_n)$ in Theorem 7.41 is called the *Sylvester matrix*. The Sylvester matrix is completely determined from the rank and the signature of a linear operator.

We say that two matrices $A, B \in \mathbb{C}^{m \times m}$ are *congruent* if there exists an invertible matrix P such that $B = P^*AP$. Notice that under a congruence transform, eigenvalues of a matrix may change unless the matrix P is unitary. *Sylvester's Law of Inertia* states that under a congruence transform, the signs of eigenvalues of a hermitian matrix do not change.

Exercises for Sect. 7.4

1. Diagonalize the matrices $A = \begin{bmatrix} 4 & 1 \\ 1 & 4 \end{bmatrix}$ and $B = \begin{bmatrix} -2 & 6 \\ 6 & -3 \end{bmatrix}$. Use the results to diagonalize the 4×4 matrix $C = \begin{bmatrix} A & 0 \\ 0 & B \end{bmatrix}$.
2. Find the diagonal matrix similar to the 6×6 matrix with all entries as 1.
3. Find an orthogonal matrix P so that $P^T AP$ is diagonal, where

   (a) $A = \begin{bmatrix} 2 & 1 & 1 \\ 1 & 2 & -1 \\ 1 & -1 & 2 \end{bmatrix}$ (b) $A = \begin{bmatrix} \cos\theta & \sin\theta \\ \sin\theta & -\cos\theta \end{bmatrix}$
4. Let v be an eigenvector of a self-adjoint operator T on an inner product space V. Show that $v^\perp$ is invariant under T.
5. Show that there does not exist a self-adjoint operator T on $\mathbb{R}^3$ with $T(1, 2, 3) = (1, 2, 3)$ and $T(4, 5, 6) = (0, 0, 0)$.
6. Prove that the zero operator is the only nilpotent self-adjoint operator on a finite dimensional inner product space.
7. Let T be a linear operator on an inner product space V over $\mathbb{F}$. Prove the following:

   (a) $T = 0$ if and only if $\langle Tx, y\rangle = 0$ for all $x, y \in V$.
   (b) If T is self-adjoint, then $T = 0$ if and only if $\langle Tx, x\rangle = 0$ for each $x \in V$.
   (c) If $\mathbb{F} = \mathbb{C}$, then $T = 0$ if and only if $\langle Tx, x\rangle = 0$ for each $x \in V$.
   (d) If $\mathbb{F} = \mathbb{C}$, then T self-adjoint if and only if $\langle Tx, x\rangle \in \mathbb{R}$ for each $x \in V$.
   (e) In case $\mathbb{F} = \mathbb{R}$, neither (c) nor (d) is true, in general.

8. Find an invertible matrix P such that $P^T AP$ is in Sylvester form, and then determine the rank and signature of A, where

$$A = \begin{bmatrix} -1 & -1 & -1 \\ -1 & 1 & 0 \\ -1 & 0 & 1 \end{bmatrix}.$$

7.5 Singular Value Decomposition

We had seen many factorizations of matrices, such as the QR factorization, the LU decomposition, the rank factorization, Schur triangularization, Jordan normal form, and diagonalization, whenever possible. We add to our cart another such factorization, often used in many applications of matrices and linear transformations.

Recall that if $T : U \to V$ is a linear transformation on a finite dimensional inner product space U, then its adjoint T^* is a map from V to U. Therefore, we cannot talk about eigenvalues, eigenvectors, etc., of T when U and V are different. However, the composition operator T^*T is an operator on U, so that we can make use of those concepts for the operator T^*T and make some inferences on T.

Notice that T^*T is a self-adjoint operator on U. It is positive semi-definite, since

$$\langle T^*Tx, x\rangle = \langle Tx, Tx\rangle = \|Tx\|^2 \geq 0$$

for all $x \in U$. Hence all its eigenvalues are real and nonnegative, and their algebraic and geometric multiplicities are the same. Thus the following definition makes sense.

Definition 7.43 Let U and V be inner product spaces of dimensions n and m, respectively. Let $T : U \to V$ be a linear transformation. The nonnegative square root of any eigenvalue of the linear operator $T^*T : U \to U$ is called a **singular value** of T.

Since $T^*T : U \to U$ is self-adjoint, it is diagonalizable. Hence, there exists an orthonormal basis $B = \{u_1, \ldots, u_n\}$ of eigenvectors of T^*T, where each basis vector corresponds to a nonnegative eigenvalue. Let $\lambda_1, \ldots, \lambda_n$ be the nonnegative eigenvalues corresponding to the eigenvectors $u_1, \ldots, u_n$, respectively, that is,

$$T^*Tu_i = \lambda_i u_i \quad \text{for } i = 1, \ldots, n.$$

Assume that $T \neq 0$. Then at least one λ_i is nonzero. Let s_i be the nonnegative square root of λ_i for $i = 1, \ldots, n$. Then $s_1, \ldots, s_n$ are the singular values of T. Without loss of generality, we assume that

$$s_1 \geq s_2 \geq \cdots \geq s_r > s_{r+1} = \cdots = s_n = 0.$$

Here, $s_1, \ldots, s_r$, with possible repetitions, are the positive singular values of T. In case, $T = 0$, each λ_i is equal to 0, so that all singular values are also equal to 0. We include this case in the above arrangement by allowing r to take the value 0. In this notation,

$$T^*Tu_i = s_i^2 u_i \quad \text{for } i = 1, \ldots, n.$$

Theorem 7.44 (Singular value representation) *Let U and V be inner product spaces of dimensions n and m, respectively. Let $T : U \to V$ be a nonzero linear transformation. Let B be an orthonormal basis of U consisting of eigenvectors $u_1, \ldots, u_n$ of T^*T with corresponding eigenvalues $s_1^2, \ldots, s_n^2$ such that $s_1^2 \geq \cdots \geq s_r^2 > s_{r+1}^2 = \cdots = s_n^2 = 0$. For $i = 1, \ldots, r$, let $v_i := \frac{1}{s_i} Tu_i$. Then the following are true:*

(1) $Tx = \sum_{i=1}^{r} s_i \langle x, u_i\rangle v_i$ *for each $x \in U$.*
(2) $T^*y = \sum_{j=1}^{r} s_j \langle y, v_j\rangle u_j$ *for each $y \in V$.*
(3) $\{u_1, \ldots, u_r\}$ *is an orthonormal basis of $N(T)^\perp$.*
(4) $\{v_1, \ldots, v_r\}$ *is an orthonormal basis of $R(T)$.*
(5) $\{u_{r+1}, \ldots, u_n\}$ *is an orthonormal basis of $N(T)$.*
(6) $r = \text{rank}\,(T) \leq \min\{m, n\}$.

Proof (1) Let $x \in U$. Since $x = \sum_{i=1}^{n} \langle x, u_i \rangle u_i$, and $s_i = 0$ for $i > r$, we have

$$Tx = \sum_{i=1}^{n} \langle x, u_i \rangle T u_i = \sum_{i=1}^{n} \langle x, u_i \rangle s_i v_i = \sum_{i=1}^{r} \langle x, u_i \rangle s_i v_i.$$

(2) Let $y \in V$ and let $x \in U$. Then using (1), we have

$$\langle x, T^* y \rangle = \langle Tx, y \rangle = \Big\langle \sum_{i=1}^{r} s_i \langle x, u_i \rangle v_i, y \Big\rangle = \sum_{i=1}^{r} s_i \langle x, u_i \rangle \langle v_i, y \rangle = \Big\langle x, \sum_{i=1}^{r} s_i \langle y, v_i \rangle u_i \Big\rangle.$$

Hence, $T^* y = \sum_{i=1}^{r} s_i \langle y, v_i \rangle u_i$ for each $y \in V$.

(3)–(4) By (1) and (2), and using the facts that $T u_i = s_i v_i$, $T^* v_i = s_i u_i$ and $s_i \neq 0$ for $i = 1, \ldots, r$, we have

$$R(T) = \text{span}(\{v_1, \ldots, v_r\}) \quad \text{and} \quad R(T^*) = \text{span}(\{u_1, \ldots, u_r\}).$$

Since $\{u_1, \ldots, u_r\}$ is an orthonormal set and $R(T^*) = N(T)^{\perp}$, $\{u_1, \ldots, u_r\}$ is an orthonormal basis of $N(T)^{\perp}$. Notice that for $i, j = 1, \ldots, r$,

$$\langle v_i, v_j \rangle = \frac{1}{s_i s_j} \langle T u_i, T u_j \rangle = \frac{1}{s_i s_j} \langle u_i, T^* T u_j \rangle = \frac{s_j^2}{s_i s_j} \langle u_i, u_j \rangle = \frac{s_j}{s_i} \langle u_i, u_j \rangle.$$

Since $\{u_1, \ldots, u_n\}$ is an orthonormal set, so is $\{v_1, \ldots, v_r\}$. It follows that $\{v_1, \ldots, v_n\}$ is an orthonormal basis of $R(T)$.

(5) Since $T^* T u_i = s_i^2 u_i = 0$ for $i > r$, we have $u_i \in N(T^* T) = N(T)$ for $i > r$. Since $U = N(T)^{\perp} + N(T)$ and $\{u_1, \ldots, u_r\}$ is an orthonormal basis of $N(T)^{\perp}$, it follows that $\{u_{r+1}, \ldots, u_n\}$ is an orthonormal basis of $N(T)$.

(6) It follows from (3). ■

In Schur triangularization, we had a linear operator and we chose a basis for the space in which the matrix is upper triangular. If we are free to choose different bases for the domain space and the co-domain space, then we would end up in a still nicer form of the matrix. The following theorem does exactly that.

Theorem 7.45 (Singular value decomposition) *Let U and V be inner product spaces of dimensions n and m, respectively. Let $T : U \to V$ be a linear transformation. Let $s_1, \ldots, s_n$ be the singular values of T, where $s_1 \geq \cdots \geq s_r > s_{r+1} = \cdots = s_n = 0$. Then there exist orthonormal bases B for U and E for V so that the following statements hold:*

(1) $[T]_{E,B} = \begin{bmatrix} S & 0 \\ 0 & 0 \end{bmatrix} \in \mathbb{R}^{m \times n}$ *and* $[T^*]_{B,E} = \begin{bmatrix} S & 0 \\ 0 & 0 \end{bmatrix} \in \mathbb{R}^{n \times m}$, *where* $S = \text{diag}(s_1, \ldots, s_r)$ *and* $r \leq \min\{m, n\}$.

(2) $[T^* T]_{B,B} = \text{diag}(s_1^2, \ldots, s_r^2, 0, \ldots, 0) \in \mathbb{R}^{n \times n}$.

(3) $[T T^*]_{E,E} = \text{diag}(s_1^2, \ldots, s_r^2, 0, \ldots, 0) \in \mathbb{R}^{m \times m}$.

Proof Let $u_1, \ldots, u_n$ and $v_1, \ldots, v_r$ be as in Theorem 7.44. Let $B = \{u_1, \ldots, u_n\}$ and let $E = \{v_1, \ldots, v_m\}$ be an orthonormal basis of V obtained by extending the orthonormal set $\{v_1, \ldots, v_r\}$.

(1) We know that $Tu_j = s_j v_j$ for $j \in \{1, \ldots, r\}$ and $Tu_j = 0$, for $j > r$. Hence $[T]_{E,B}$ is in the required form. Also, we have $T^* v_j = s_j u_j$ for $j \in \{1, \ldots, r\}$. By Theorem 7.44(2), $T^* v_j = 0$ for $j > r$. Hence, we obtain the representation for T^*.

(2) $[T^*T]_{B,B} = [T^*]_{B,E}[T]_{E,B} = \begin{bmatrix} S & 0 \\ 0 & 0 \end{bmatrix} \begin{bmatrix} S & 0 \\ 0 & 0 \end{bmatrix} = \text{diag}(s_1^2, \ldots, s_r^2, 0, \ldots, 0)$.

(3) Similar to (2). ■

Notice that TT^* and T^*T have the same nonzero eigenvalues (Theorem 7.45 (2)) and they are all real numbers. Depending on the dimensions of the spaces U and V, there are more or less zero eigenvalues of TT^* compared with those of T^*T.

The matrix interpretation of Theorem 7.45 is as follows. We use the phrase **a full orthonormal set of eigenvectors** of a linear operator $T : U \to U$ to mean the following:

If an eigenvalue λ of T has algebraic multiplicity μ, then the orthonormal set contains exactly μ number of eigenvectors corresponding to λ.

Theorem 7.46 (Singular value decomposition of matrices) *Let $A \in \mathbb{F}^{m \times n}$. Let the positive singular values of A be $s_1 \geq \cdots \geq s_r > 0$. Then there exist unitary matrices $P \in \mathbb{F}^{n \times n}$, $Q \in \mathbb{F}^{m \times m}$ such that A can be written as $A = Q\,\Sigma\,P^*$, where*

$$\Sigma = \begin{bmatrix} S & 0 \\ 0 & 0 \end{bmatrix} \in \mathbb{R}^{m \times n}, \quad S = \text{diag}(s_1, \ldots, s_r), \quad r \leq \min\{m, n\}.$$

Moreover, the columns of P form a full orthonormal set of eigenvectors of A^*A, and the columns of Q form a full orthonormal set of eigenvectors of AA^*.

Proof The matrix A is a linear transformation from $\mathbb{F}^{n \times 1}$ to $\mathbb{F}^{m \times 1}$. Due to Theorem 7.45, we have orthonormal bases $B = \{u_1, \ldots, u_n\}$ for $\mathbb{F}^{n \times 1}$ and $E = \{v_1, \ldots, v_m\}$ for $\mathbb{F}^{m \times 1}$ such that

$$[A]_{E,B} = \begin{bmatrix} S & 0 \\ 0 & 0 \end{bmatrix}.$$

Construct the matrices $P := \begin{bmatrix} u_1 & \cdots & u_n \end{bmatrix}$ and $Q := \begin{bmatrix} v_1 & \cdots & v_m \end{bmatrix}$. Since B and E are orthonormal bases, $Q^{-1} = Q^*$ and $P^{-1} = P^*$. Theorem 2.52 implies that

$$\Sigma := [A]_{E,B} = Q^{-1}AP = Q^*AP = \begin{bmatrix} S & 0 \\ 0 & 0 \end{bmatrix}.$$

Here, $S = \text{diag}(s_1, \ldots, s_r)$ is an $r \times r$ matrix, and $r = \text{rank}\,(A) \leq \min\{m, n\}$. Notice that $\Sigma^* = \Sigma$. Further, A^*A has positive eigenvalues $s_1^2, \ldots, s_r^2$ (with possible repetitions) and the eigenvalue 0 has multiplicity $n - r$. Next, $Q^*AP = \Sigma$ implies that

$A = Q\Sigma P^*$. Set $s_i = 0$ for $i \in \{r+1, \ldots, n\}$. Now,

$$A^*AP = P\Sigma Q^*Q\Sigma P^*P = P\Sigma^2 I_n = P\,\mathrm{diag}(s_1^2, \ldots, s_r^2, s_{r+1}^2, \ldots, s_n^2).$$

That is, for each $i \in \{1, \ldots, n\}$, $A^*Au_i = s_i^2 u_i$. Since B is an orthonormal basis of $\mathbb{F}^{n\times 1}$, it follows that the columns of P form a full orthonormal set of eigenvectors of A^*A.

For the eigenvectors of AA^*, notice that the nonzero eigenvalues of AA^* are $s_1^2, \ldots, s_r^2$ (with possible repetitions), and the eigenvalue 0 has multiplicity $m - r$. Set $s_j = 0$ for $j \in \{r+1, \ldots, m\}$. Then

$$AA^*Q = Q\Sigma P^*P\Sigma Q^*Q = Q\Sigma^2 I_m = Q\,\mathrm{diag}(s_1^2, \ldots, s_r^2, s_{r+1}^2, \ldots, s_m^2).$$

That is, for each $j \in \{1, \ldots, m\}$, $AA^*v_j = s_j^2 v_j$. Since E is an orthonormal basis of $\mathbb{F}^{m\times 1}$, it follows that the columns of Q form a full orthonormal set of eigenvectors of AA^*. ■

Example 7.47 Let $A = \begin{bmatrix} 2 & -2 & 4 \\ -1 & 1 & -2 \end{bmatrix}$. Then $A^*A = \begin{bmatrix} 5 & -5 & 10 \\ -5 & 5 & -10 \\ 10 & -10 & 20 \end{bmatrix}$, $AA^* = \begin{bmatrix} 24 & -12 \\ -12 & 6 \end{bmatrix}$.

The eigenvalues of A^*A are $30,\ 0,\ 0$ and of AA^* are $30,\ 0$. Thus the matrix S in Theorem 7.46 is $S = \mathrm{diag}(\sqrt{30}, 0)$ and $\Sigma \in \mathbb{F}^{2\times 3}$ with first 2×2 block as S and other entries being 0.

The vector $u_1 = (1/\sqrt{6},\ -1/\sqrt{6},\ 2/\sqrt{6})^T$ is an eigenvector of norm 1 corresponding to the eigenvalue 30 of A^*A. Eigenvectors of A^*A corresponding to 0 satisfy $x_1 - x_2 + 2x_3 = 0$. Thus we choose two linearly independent eigenvectors as $u_2 = (1/\sqrt{2},\ 1/\sqrt{2},\ 0)^T$ and $u_3 = (1/\sqrt{3},\ -1/\sqrt{3},\ -1/\sqrt{3})^T$ so that $B = \{u_1, u_2, u_3\}$ is an orthonormal set. Then $P = \begin{bmatrix} u_1 & u_2 & u_3 \end{bmatrix}$.

Similarly, the vector $v_1 = (2/\sqrt{5},\ -1/\sqrt{5})^T$ is an eigenvector of AA^* corresponding to the eigenvalue 30. For the eigenvalue 0, we then take the eigenvector as $v_2 = (1/\sqrt{5},\ 2/\sqrt{5})^T$ so that $E = \{v_1, v_2\}$ is an orthonormal set. Then $Q = \begin{bmatrix} v_1 & v_2 \end{bmatrix}$. We see that

$$Q\Sigma P^* = \begin{bmatrix} 2/\sqrt{5} & 1/\sqrt{5} \\ -1/\sqrt{5} & 2/\sqrt{5} \end{bmatrix} \begin{bmatrix} \sqrt{30} & 0 & 0 \\ 0 & 0 & 0 \end{bmatrix} \begin{bmatrix} 1/\sqrt{6} & -1/\sqrt{6} & 2/\sqrt{6} \\ 1/\sqrt{2} & 1/\sqrt{2} & 0 \\ 1/\sqrt{3} & -1/\sqrt{3} & -1/\sqrt{3} \end{bmatrix} = \begin{bmatrix} 2 & -2 & 4 \\ -1 & 1 & -2 \end{bmatrix} = A.$$

$$R(A) = \mathrm{span}\{(2,-1)^T,\ (-2,1)^T,\ (4,-2)^T\} = \mathrm{span}\{(2,-1)^T\} = \mathrm{span}\{v_1\}.$$

$$Au_2 = \begin{bmatrix} 2 & -2 & 4 \\ -1 & 1 & -2 \end{bmatrix} \begin{bmatrix} 1/\sqrt{2} \\ 1/\sqrt{2} \\ 0 \end{bmatrix} = \begin{bmatrix} 0 \\ 0 \end{bmatrix}.$$

$$Au_3 = \begin{bmatrix} 2 & -2 & 4 \\ -1 & 1 & -2 \end{bmatrix} \begin{bmatrix} 1/\sqrt{3} \\ -1/\sqrt{3} \\ -1/\sqrt{3} \end{bmatrix} = \begin{bmatrix} 0 \\ 0 \end{bmatrix}.$$

Since $\dim(N(A)) = 3 - \dim(R(A) = 2,\ \{u_2, u_3\}$ is a basis for $N(A)$.

Let $x = (a, b, c)^T \in \mathbb{R}^{3\times 1}$. Then $\langle x, u_1\rangle = u_1^T x = \frac{1}{\sqrt{6}}(a - b + 2c)$. We have

$$Ax = \begin{bmatrix} 2a - 2b + 4c \\ -a + b - 2c \end{bmatrix} = \sqrt{30}\,\frac{1}{\sqrt{6}}(a - b + 2c)\begin{bmatrix} 2/\sqrt{5} \\ -1/\sqrt{5} \end{bmatrix} = \langle x, u_1\rangle v_1 = \sum_{i=1}^{r} s_i\langle x, u_i\rangle v_i.$$

Let $y = (a, b)^T \in \mathbb{R}^{2\times 1}$. Then $\langle y, v_1\rangle = v_1^T y = \frac{1}{\sqrt{5}}(2a - b)$. We have

$$A^* y = \begin{bmatrix} 2a - b \\ -2a + b \\ 4a - 2b \end{bmatrix} = \sqrt{30}\,\frac{1}{\sqrt{5}}(2a - b)\begin{bmatrix} 1/\sqrt{6} \\ -1/\sqrt{6} \\ 2/\sqrt{6} \end{bmatrix} = \langle y, v_1\rangle u_1 = \sum_{i=1}^{r} s_i\langle y, v_i\rangle u_i.$$

With this, we have verified all the statements in Theorems 7.44–7.46. □

For a matrix $A \in \mathbb{F}^{m\times n}$ of rank r, let $S = \operatorname{diag}(s_1, \ldots, s_r)$. Then the singular value decomposition of A is $A = Q\,\Sigma\,P^*$, where $\Sigma \in \mathbb{R}^{m\times n}$ has one of the following forms:

If $r < \min(m, n)$, then $\Sigma = \begin{bmatrix} S & 0 \\ 0 & 0 \end{bmatrix}$.

If $r = m < n$, then $\Sigma = \begin{bmatrix} S & 0 \end{bmatrix}$.

If $r = n < m$, then $\Sigma = \begin{bmatrix} S \\ 0 \end{bmatrix}$.

If $r = m = n$, then $\Sigma = S$.

Moreover, if $A \in \mathbb{R}^{m\times n}$, then the matrices P and Q can be chosen as real orthogonal matrices.

A linear operator maps subspaces to subspaces. In $\mathbb{R}^2$, we see that a circle may become an ellipse under a linear operator since scaling and rotation can happen to the axes. Singular value decomposition shows that this is true in higher dimensions. We use the notation of Theorem 7.45 implicitly.

An n-dimensional unit sphere is defined as the set of all unit vectors in $\mathbb{R}^{n\times 1}$. Let x be a point on the n-dimensional unit sphere, i.e. $x \in \mathbb{R}^{n\times 1}$ and $\|x\| = 1$. Since $\{u_1, \ldots, u_n\}$ is an orthonormal basis for $\mathbb{R}^{n\times 1}$, $x = \sum_{i=1}^{n}\langle x, u_i\rangle u_i$. Theorem 7.44 (3) implies that $Ax = \sum_{i=1}^{r} s_i\langle x, u_i\rangle v_i$. Since v_is form an orthonormal set, Parseval's identity gives

$$\|Ax\|^2 = \sum_{i=1}^{r} s_i^2 |\langle x, u_i\rangle|^2 \quad\text{and}\quad \sum_{i=1}^{n} |\langle x, u_i\rangle|^2 = \|x\|^2 = 1.$$

Thus, writing $y_i = s_i|\langle x, u_i\rangle|$, we see that

$$\frac{y_1^2}{s_1^2} + \cdots + \frac{y_r^2}{s_r^2} \le 1.$$

That is, the image of x lies in the interior of the hyper-ellipsoid with semi-axes as $s_i v_i$. Notice that when $r = n$, the image is the surface of the hyper-ellipsoid.

Singular value decomposition can be used to solve least squares problem arising out of linear systems. In the least squares problem for a linear system

$$Ax = b,$$

we try to determine an x that minimizes the square of the residual: $\|b - Ax\|^2$. Recall that such a minimizer x is a *least squares solution* of the linear system $Ax = b$. In the following, we use the Euclidean norm of a vector, i.e.

$$\|(a_1, \ldots, a_k)^T\|^2 = |a_1|^2 + \cdots |a_k|^2.$$

Suppose $A = Q\,\Sigma\,P^*$ is the singular value decomposition of $A \in \mathbb{F}^{m\times n}$. Then

$$\|b - Ax\|^2 = \|b - Q\,\Sigma\,P^*x\|^2 = \|Q(Q^*b - \Sigma\,P^*x)\|^2 = \|Q^*b - \Sigma\,P^*x\|^2.$$

The last equality follows since for any $u \in \mathbb{F}^{n\times 1}$,

$$\|Qu\|^2 = \langle Qu, Qu\rangle = \langle u, Q^*Qu\rangle = \langle u, u\rangle = \|u\|^2.$$

With S as the nonzero block of order r in Σ, write

$$y = P^*x = \begin{bmatrix} y_1 \\ y_2 \end{bmatrix}, \quad \hat{b} = Q^*b = \begin{bmatrix} b_1 \\ b_2 \end{bmatrix}, \quad \Sigma = \begin{bmatrix} S & 0 \\ 0 & 0 \end{bmatrix},$$

where $S = \mathrm{diag}(s_1, \ldots, s_r) \in \mathbb{R}^{r\times r}$, $y_1, b_1 \in \mathbb{F}^{r\times 1}$, $y_2 \in \mathbb{F}^{(n-r)\times 1}$ and $b_2 \in \mathbb{F}^{(m-r)\times 1}$. Then

$$\|b - Ax\|^2 = \|Q^*b - \Sigma P^*x\|^2 = \|\hat{b} - \Sigma y\|^2 = \left\| \begin{bmatrix} b_1 - Sy_1 \\ b_2 \end{bmatrix} \right\|^2 = \|b_1 - Sy_1\|^2 + \|b_2\|^2.$$

So, $\|b - Ax\|^2$ is minimized for $b_1 = Sy_1$, i.e. when

$$x = Py = P\begin{bmatrix} y_1 \\ y_2 \end{bmatrix} = P\begin{bmatrix} S^{-1}b_1 \\ y_2 \end{bmatrix} \quad \text{for arbitrary } y_2.$$

Again, since $P^*P = I$,

$$\|x\|^2 = \left(P\begin{bmatrix} S^{-1}b_1 \\ y_2 \end{bmatrix}\right)^* \left(P\begin{bmatrix} S^{-1}b_1 \\ y_2 \end{bmatrix}\right) = \|S^{-1}b_1\|^2 + \|y_2\|^2.$$

Therefore, a least squares solution of $Ax = b$ which also has minimum norm is $x = Py$ with $y_2 = 0$. Explicitly, such an x is given by

$$x = P\begin{bmatrix} S^{-1}b_1 \\ 0 \end{bmatrix} = P\begin{bmatrix} S^{-1} & 0 \\ 0 & 0 \end{bmatrix}\begin{bmatrix} b_1 \\ b_2 \end{bmatrix} = P\begin{bmatrix} S^{-1} & 0 \\ 0 & 0 \end{bmatrix} Q^*b.$$

Definition 7.48 Let $A \in \mathbb{F}^{m\times n}$ and let $A = Q\begin{bmatrix} S & 0 \\ 0 & 0 \end{bmatrix} P^*$ be the singular value decomposition of A. Then the matrix

$$A^\dagger := P\begin{bmatrix} S^{-1} & 0 \\ 0 & 0 \end{bmatrix} Q^*$$

in $\mathbb{F}^{m\times n}$ is called the **generalized inverse** or the **Moore–Penrose inverse** of A.

The discussion preceding the above definition shows that a minimum norm least squares solution of the system $Ax = b$ is given by $x = A^\dagger b$. Of course, when A is invertible, $A^\dagger$ coincides with A^{-1}.

Example 7.49 Consider the matrix $A = \begin{bmatrix} 2 & -2 & 4 \\ -1 & 1 & -2 \end{bmatrix}$ of Example 7.47. Its singular value decomposition was computed as $A = Q\,\Sigma\,P^*$, where

$$Q = \begin{bmatrix} 2/\sqrt{5} & 1/\sqrt{5} \\ -1/\sqrt{5} & 2/\sqrt{5} \end{bmatrix}, \quad \Sigma = \begin{bmatrix} \sqrt{30} & 0 & 0 \\ 0 & 0 & 0 \end{bmatrix}, \quad P = \begin{bmatrix} 1/\sqrt{6} & 1/\sqrt{2} & 1/\sqrt{3} \\ -1/\sqrt{6} & 1/\sqrt{2} & -1/\sqrt{3} \\ 2/\sqrt{6} & 0 & -1/\sqrt{3} \end{bmatrix}.$$

Then

$$A^\dagger = \begin{bmatrix} 1/\sqrt{6} & 1/\sqrt{2} & 1/\sqrt{3} \\ -1/\sqrt{6} & 1/\sqrt{2} & -1/\sqrt{3} \\ 2/\sqrt{6} & 0 & -1/\sqrt{3} \end{bmatrix} \begin{bmatrix} 1/\sqrt{30} & 0 \\ 0 & 0 \\ 0 & 0 \end{bmatrix} \begin{bmatrix} 2/\sqrt{5} & -1/\sqrt{5} \\ 1/\sqrt{5} & 2/\sqrt{5} \end{bmatrix} = \frac{1}{30}\begin{bmatrix} 2 & -1 \\ -2 & 1 \\ 4 & -2 \end{bmatrix}.$$

We find that $(AA^\dagger)^* = AA^\dagger$, $(A^\dagger A)^* = A^\dagger A$, $AA^\dagger A = A$ and $A^\dagger AA^\dagger = A^\dagger$. □

Theorem 7.50 *Let $A \in \mathbb{F}^{m\times n}$. Then $A^\dagger$ is the unique matrix in $\mathbb{F}^{n\times m}$ that satisfies*

$$(AA^\dagger)^* = AA^\dagger,\ (A^\dagger A)^* = A^\dagger A,\ AA^\dagger A = A, \text{ and } A^\dagger AA^\dagger = A^\dagger.$$

Proof Let $A = Q\,\Sigma\,P^*$ be the singular value decomposition of A, where the $n \times n$ matrix P and the $m \times m$ matrix Q are unitary, and $\Sigma = \begin{bmatrix} S & 0 \\ 0 & 0 \end{bmatrix} \in \mathbb{F}^{m\times n}$. Here, $S = \operatorname{diag}(s_1, \ldots, s_r)$, with $s_1 \geq \cdots \geq s_r > 0$ and $r = \operatorname{rank}(A)$. Construct $\hat{\Sigma} \in \mathbb{F}^{n\times m}$ as follows:

$$\hat{\Sigma} = \begin{bmatrix} S^{-1} & 0 \\ 0 & 0 \end{bmatrix}.$$

Observe that

$$\hat{\Sigma}^*\,\Sigma^* = \Sigma\,\hat{\Sigma}, \quad \Sigma^*\,\hat{\Sigma}^* = \hat{\Sigma}\,\Sigma, \quad \hat{\Sigma}\,\Sigma\,\hat{\Sigma} = \hat{\Sigma}, \quad \Sigma\,\hat{\Sigma}\,\Sigma = \Sigma, \quad A^\dagger = P\,\hat{\Sigma}\,Q^*.$$

We verify the four required identities as follows:

$$\begin{aligned}
(AA^\dagger)^* &= (Q\,\Sigma\,P^*\,P\,\hat{\Sigma}\,Q^*)^* = Q\,\hat{\Sigma}^*\,\Sigma^*\,Q^* = Q\,\Sigma\,\hat{\Sigma}\,Q^* = Q\,\Sigma\,P^*\,P\,\hat{\Sigma}\,Q^* = AA^\dagger.\\
(A^\dagger A)^* &= (P\,\hat{\Sigma}\,Q^* Q\,\Sigma\,P^*)^* = P\,\Sigma^*\,\hat{\Sigma}^*\,P^* = P\,\hat{\Sigma}\,\Sigma\,P^* = P\,\hat{\Sigma}\,Q^*\,Q\,\Sigma\,P^* = A^\dagger A.\\
AA^\dagger A &= Q\,\Sigma\,P^*\,P\,\hat{\Sigma}\,Q^*\,Q\,\Sigma\,P^* = Q\,\Sigma\,\hat{\Sigma}\,\Sigma\,P^* = Q\,\Sigma\,P^* = A.\\
A^\dagger AA^\dagger &= P\,\hat{\Sigma}\,Q^*\,Q\,\Sigma\,P^*\,P\,\hat{\Sigma}\,Q^* = P\,\hat{\Sigma}\,\Sigma\,\hat{\Sigma}\,Q^* = P\,\hat{\Sigma}\,Q^* = A^\dagger.
\end{aligned}$$

For uniqueness, suppose $B, C \in \mathbb{F}^{m\times n}$ such that

$$\begin{aligned}
&(AB)^* = AB,\ (BA)^* = BA,\ ABA = A,\ BAB = B,\\
&(AC)^* = AC,\ (CA)^* = CA,\ ACA = A,\ CAC = C.
\end{aligned}$$

Then

$$\begin{aligned}
B &= BAB = B(AB)^* = BB^*A^* = BB^*(ACA)^* = BB^*A^*C^*A^* = BB^*A^*(AC)^*\\
&= BB^*A^*AC = B(AB)^*AC = BABAC = B(ABA)C = BAC = B(ACA)C\\
&= BACAC = BA(CA)C = BA(CA)^*C = BAA^*C^*C = (BA)^*A^*C^*C\\
&= A^*B^*A^*C^*C = (ABA)^*C^*C = A^*C^*C = (CA)^*C = CAC = C.
\end{aligned}$$

■

There are other types of generalized inverses for rectangular matrices. For a matrix $A \in \mathbb{F}^{m\times n}$, any matrix $X \in \mathbb{F}^{n\times m}$ that satisfies $XA = I_n$ is called a *left inverse* of A; similarly, any matrix $Y \in \mathbb{F}^{n\times m}$ that satisfies $AY = I_m$, is called a *right inverse* of A. In general, left and right inverses of a matrix are not unique.

If A has m linearly independent columns, then A^*A is invertible, and then $B = (A^*A)^{-1}A^*$ is a left inverse of A. Similarly, if A has m linearly independent rows, then AA^* is invertible, and $C = A^*(AA^*)^{-1}$ is a right inverse of A.

Exercises for Sect. 7.5

1. Let A be a square matrix. Prove or disprove:

   (a) If A is self-adjoint, then $\operatorname{tr}(A)$ is real.
   (b) If A is positive semi-definite, then $\operatorname{tr}(A) \geq 0$.

2. Let $A \in \mathbb{C}^{m \times n}$. Show that A^*A and AA^* have the same nonzero eigenvalues, counting multiplicities.
3. Determine the singular value decomposition of the matrices

$$\text{(a)} \begin{bmatrix} 2 & -2 \\ 1 & 1 \\ 2 & 2 \end{bmatrix} \quad \text{(b)} \begin{bmatrix} 2 & -1 & 1 \\ -2 & 1 & -1 \end{bmatrix} \quad \text{(c)} \begin{bmatrix} 2 & 0 & -1 & 0 \\ 1 & 2 & 2 & 3 \\ 2 & -5 & 4 & 0 \end{bmatrix}.$$

4. Let T be a self-adjoint operator on a finite dimensional inner product space. Show that the singular values of T are precisely the absolute values of the eigenvalues of T.
5. Show that the singular values of T^2 need not be the squares of the singular values of T.
6. Show that the matrices $\begin{bmatrix} 2 & -1 \\ 1 & 0 \end{bmatrix}$ and $\begin{bmatrix} 1 & 1 \\ 0 & 1 \end{bmatrix}$ are similar to each other, but they do not have the same singular values.
7. Let T be a linear operator on a finite dimensional inner product space. Show that T is an isometry if and only if all singular values of T are 1.
8. Let $A \in \mathbb{R}^{m \times n}$ be a matrix of rank r. Show that there exist real orthogonal matrices $P \in \mathbb{R}^{n \times n}$ and $Q \in \mathbb{R}^{m \times m}$ such that A can be written as $Q\,\Sigma\,P^T$, where the first r diagonal entries of Σ are positive real numbers $s_1 \geq \cdots \geq s_r$ and all other entries of Σ are 0.
9. Determine the generalized inverse of the matrices in Exercise 3.
10. Let the columns of $A \in \mathbb{F}^{m \times n}$ be linearly independent. Show the following:

    (a) A^*A is invertible. (b) $A^\dagger = (A^*A)^{-1}A^*$.

11. Let the rows of $A \in \mathbb{F}^{m \times n}$ be linearly independent. Show the following:

    (a) AA^* is invertible. (b) $A^\dagger = A^*(AA^*)^{-1}$.

12. Let $A \in \mathbb{C}^{m \times n}$ have n linearly independent columns. Is it true that $A^T A$ is invertible?
13. Construct a matrix having at least two left inverses.
14. Construct a matrix which admits of at least two right inverses.

7.6 Polar Decomposition

Recall that a complex number z can be written as $z = |z|e^{i\theta}$, where $|z|$ is a nonnegative real number and the absolute value of $e^{i\theta}$ is 1. It presupposes that each nonnegative real number has a nonnegative square root. Similar representations are possible for linear transformations as well. Of course, we should have the notion of a square root of a positive semi-definite operator.

Definition 7.51 Let T be a linear operator on a finite dimensional inner product space V. We say that a linear operator $S : V \to V$ is a **square root of** T if $S^2 = T$.

Example 7.52 (1) Define linear operators $S,\ T : \mathbb{F}^2 \to \mathbb{F}^2$ by

$$S(a, b) = (a + b, 2a + b), \quad T(a, b) = (3a + 2b, 4a + 3b) \quad \text{for } (a, b) \in \mathbb{F}^2.$$

Then $S^2(a, b)=S(a + b, 2a + b)=(a + b + 2a + b, 2(a + b) + 2a + b)=T(a, b)$. Hence S is a square root of T. Also, the linear operator $-S$ is a square root of T.

(2) The linear operator $T : \mathbb{F}^2 \to \mathbb{F}^2$ defined by

$$T(a, b) = (b, 0), \quad \text{for } (a, b) \in \mathbb{F}^2$$

does not have a square root. On the contrary, if $S : \mathbb{R}^2 \to \mathbb{R}^2$ is a square root of T, then $S^2 = T$. Notice that $T^2 = 0$. Therefore, $S^4 = 0$. Now, S is a nilpotent operator on a 2-dimensional vector space; hence $S^2 = 0$. (See Exercise 13 of Sect. 6.4.) This is a contradiction, since $T \neq 0$. □

Notice that if $T^* = T$ and $\langle Tv, v\rangle \geq 0$ for each basis vector v in some (any) basis of V, then T is positive semi-definite. We show that each positive semi-definite operator has a square root.

Theorem 7.53 *Let T be a positive semi-definite linear operator on an inner product space V of dimension n. Then T has a unique positive semi-definite square root S. Further, T and S have the following representations:*

$$Tx = \sum_{i=1}^{n} \lambda_i \langle x, u_i\rangle u_i, \quad Sx = \sum_{i=1}^{n} \sqrt{\lambda_i}\, \langle x, u_i\rangle u_i, \quad \text{for } x \in V,$$

where $\{u_1, \ldots, u_n\}$ is an orthonormal basis of V, and u_i is an eigenvector corresponding to the eigenvalue $\lambda_i \geq 0$ of T for $i = 1, \ldots, n$.

Proof Since T is self-adjoint, spectral theorem implies that

$$Tx = \sum_{i=1}^{n} \lambda_i \langle x, u_i\rangle u_i \quad \text{for } x \in V,$$

where $\{u_1, \ldots, u_n\}$ is an orthonormal basis of V and $Tu_i = \lambda_i u_i$ for $i = 1, \ldots, n$. Since T is positive semi-definite each $\lambda_i \geq 0$. Define $S : V \to V$ by

$$Sx = \sum_{i=1}^{n} \sqrt{\lambda_i}\, \langle x, u_i\rangle u_i \quad \text{for } x \in V.$$

It is easily verified that S is a positive semi-definite linear operator and that $Su_i = \sqrt{\lambda_i}\, u_i$ for $i = 1, \ldots, n$. Now, for any $x \in X$,

$$S^2 x = S(Sx) = \sum_{i=1}^{n} \sqrt{\lambda_i}\, \langle x, u_i\rangle S u_i = \sum_{i=1}^{n} \lambda_i \langle x, u_i\rangle u_i = Tx.$$

Thus, S is a positive semi-definite square root of T.

For uniqueness of S, suppose Q is also a positive semi-definite square root of T. We claim that

(λ, v) is an eigenpair of T if and only if $(\sqrt{\lambda}, v)$ is an eigenpair of Q.

To prove the claim, assume that $Tv = \lambda v$ and $v \neq 0$. As T is self-adjoint, $\lambda \geq 0$. We consider the case $\lambda = 0$ and $\lambda > 0$ separately.

If $\lambda = 0$, then $Q^2 v = Tv = 0$. Consequently, $\langle Qv, Qv\rangle = \langle v, Q^2 v\rangle = \langle v, 0\rangle = 0$. That is, $Qv = 0 = \sqrt{\lambda}\, v$.

Otherwise, let $\lambda > 0$. Now, $Q^2 v = \lambda v$ implies that $(Q + \sqrt{\lambda} I)(Q - \sqrt{\lambda} I)v = 0$. As Q is positive semi-definite, $-\sqrt{\lambda}$ is not an eigenvalue of Q. Thus $Q + \sqrt{\lambda} I$ is invertible. Then $(Q - \sqrt{\lambda} I)v = 0$. That is, $Qv = \sqrt{\lambda}\, v$.

Conversely, if $Qv = \sqrt{\lambda}\, v$ then $Tv = Q^2 v = Q\big(\sqrt{\lambda}\, v\big) = \sqrt{\lambda}\, Qv = \lambda v$. This proves our claim.

Notice that our claim is also true for S in place of Q. Now, since $\lambda_1, \ldots, \lambda_n$ are the eigenvalues of T with corresponding eigenvectors $u_1, \ldots, u_n$, our claim for Q and S imply that for each $j \in \{1, \ldots, n\}$,

$$Qu_j = \sqrt{\lambda_j}\, u_j \quad \text{if and only if} \quad Tu_j = \lambda_j v_j \quad \text{if and only if} \quad Su_j = \sqrt{\lambda_j}\, u_j.$$

That is, $Qu_j = Su_j$ for each $j \in \{1, \ldots, n\}$. As $\{u_1, \ldots, u_n\}$ is a basis of V, we conclude that $Q = S$. ■

Notation: The positive semi-definite square root of a positive semi-definite operator T is written as $\sqrt{T}$.

In the general case, let $T : U \to V$ be a linear transformation, where U and V are inner product spaces of dimensions n and m, respectively. The linear operators $T^*T : U \to U$ and $TT^* : V \to V$ are positive semi-definite. We see that for each $u \in U$,

$$\langle Tu, Tu\rangle = \langle u, T^*Tu\rangle = \big\langle u, \sqrt{T^*T}\,\sqrt{T^*T}u\big\rangle = \big\langle \sqrt{T^*T}\, u, \sqrt{T^*T}\, u\big\rangle.$$

That is, $\|Tu\| = \|\sqrt{T^*T}\, u\|$ for each $u \in U$. It says that the linear operators T and $\sqrt{T^*T}$ have the same effect on the length of a vector. Now, $Tu = 0$ if and only if $\sqrt{T^*T}\, u = 0$. Using the rank-nullity theorem, we have

$$\text{rank}\,(\sqrt{T^*T}) = \text{rank}\,(T).$$

Further, by Theorem 7.53, T^*T and $\sqrt{T^*T}$ have the representations

$$T^*T(x) = \sum_{j=1}^{n} \lambda_j \langle x, u_j \rangle u_j, \quad \sqrt{T^*T}(x) = \sum_{j=1}^{n} s_j \langle x, u_j \rangle u_j \quad \text{for } x \in U,$$

where $\lambda_1, \ldots, \lambda_n$ are the eigenvalues of T^*T and $s_1, \ldots, s_n$ are their nonnegative square roots. Similar results hold true for the linear operator $\sqrt{TT^*}$. However, due to singular value decomposition, $\sqrt{TT^*}$ has the representation

$$\sqrt{TT^*}(y) = \sum_{i=1}^{m} s_i \langle y, v_i \rangle v_j, \quad \text{for } y \in U,$$

where $v_i = (1/\sqrt{\lambda_i})Tu_i$ for $i = 1, \ldots, r$ with $r := \text{rank}\,(T)$, and $\{v_1, \ldots, v_m\}$ is an orthonormal basis of V obtained by extending the orthonormal set $\{v_1, \ldots, v_r\}$.

In the following theorem, we use the linear operators $\sqrt{T^*T}$ and $\sqrt{TT^*}$ for representing T and T^*.

Theorem 7.54 (Polar decomposition) *Let U and V be inner product spaces of dimensions n and m, respectively. Let $T : U \to V$ be a linear transformation. Then there exists a linear transformation $P : U \to V$ with $\|Pu\| \le \|u\|$ for each $u \in U$ such that*

$$T = P\sqrt{T^*T} = \sqrt{TT^*}\,P \quad \text{and} \quad T^* = \sqrt{T^*T}\,P^* = P^*\sqrt{TT^*}.$$

Further, if $m = n$, then such a linear operator P can be chosen as an isometry.

Proof Let rank $(T) = r$. The singular value representation of T is given by

$$Tx = \sum_{i=1}^{r} s_i \langle x, u_i \rangle v_i \quad \text{for } x \in U$$

with suitable orthonormal bases $\{u_1, \ldots, u_n\}$ for U and $\{v_1, \ldots, v_n\}$ for V, and positive singular values $s_1 \ge \cdots \ge s_r$ of T. Then (See (4.2) in Sect. 4.8.)

$$T^*y = \sum_{i=1}^{r} s_i \langle y, v_i \rangle u_i \quad \text{for } y \in V.$$

Consequently,

$$T^*Tx = \sum_{i=1}^{r} s_i^2 \langle x, u_i \rangle u_i, \quad \sqrt{T^*T}(x) = \sum_{i=1}^{r} s_i \langle x, u_i \rangle u_i \quad \text{for } x \in U;$$

$$TT^*y = \sum_{i=1}^{r} s_i^2 \langle y, v_i \rangle u_i, \quad \sqrt{TT^*}(y) = \sum_{i=1}^{r} s_i \langle y, v_i \rangle v_i \quad \text{for } y \in V.$$

Define the linear operator $P : U \to V$ by

$$Px := \sum_{i=1}^{r} \langle x, u_i \rangle v_i \quad \text{for } x \in U. \tag{7.2}$$

Then $Pu_j = v_j$ and $\langle Px, v_j \rangle = \langle x, u_j \rangle$ for $j = 1, \ldots, r$ and $x \in U$. Hence, for every $x \in U$,

$$P\sqrt{T^*T}x = \sum_{i=1}^{r} s_i \langle x, u_i \rangle Pu_i = \sum_{i=1}^{r} s_i \langle x, u_i \rangle v_i = Tx;$$

$$\sqrt{TT^*}Px = \sum_{i=1}^{r} s_i \langle Px, v_i \rangle v_i = \sum_{i=1}^{r} s_i \langle x, u_i \rangle v_i = Tx.$$

Therefore, $T = P\sqrt{T^*T} = \sqrt{TT^*}P$ and $T^* = \sqrt{T^*T}\,P^* = P^*\sqrt{TT^*}$.

Using (7.2), we see that for each $x \in U$,

$$\|Px\|^2 = \sum_{i=1}^{r} |\langle x, u_i \rangle|^2 \le \|x\|^2.$$

If $k = m = n$, then $\|Px\| = \|x\|$ for all $x \in U$. That is, P is an isometry. ■

According to the position of the square root operators, the polar decompositions are called left or right. That is, $T = P\sqrt{T^*T}$ is called a *right polar decomposition*, and $T = \sqrt{TT^*}\,P$ is called a *left polar decomposition* of T.

The positive semi-definite operator $\sqrt{T^*T}$ is usually denoted by $|T|$. When $U = V$ and $k = n$, then the unitary matrix P is written as U, and then we have polar decomposition as

$$T = U|T|.$$

This sloppy notation is used for the obvious reason of writing a complex number z as $e^{i\theta}|z|$ and also as $|z|\,e^{i\theta}$.

We interpret Theorem 7.54 for matrices. Suppose a matrix $A \in \mathbb{F}^{m\times n}$ has rank $r \le \min\{m, n\}$. Let $\{u_1, \ldots, u_n\}$ be an orthonormal basis of $\mathbb{F}^{n\times 1}$ of eigenvectors of A^*A corresponding to the eigenvalues $s_1^2, \ldots, s_n^2$, and let $\{v_1, \ldots, v_m\}$ be an orthonormal basis of eigenvectors of AA^* corresponding to the eigenvalues $s_1^2, \ldots, s_m^2$. Then the operator P in Theorem 7.54 can be represented as

$$Px = \sum_{i=1}^{r} (u_i^* x) v_i = \sum_{i=1}^{r} (v_i u_i^*) x, \quad x \in \mathbb{F}^n.$$

Recall that for a column vector u, u^* denotes the conjugate transpose of u. Thus, $v_i u_i^*$ is a matrix of rank 1, and hence $P \in \mathbb{F}^{m\times n}$ is the matrix

$$P = \sum_{i=1}^{r} (v_i u_i^*) = [v_1 \ \cdots \ v_r] \begin{bmatrix} u_1^* \\ \vdots \\ u_r^* \end{bmatrix}.$$

If $m = n$, then P is a unitary matrix. It results in the following theorem.

Theorem 7.55 (Polar decomposition of matrices) *Let $A \in \mathbb{F}^{m\times n}$. There exists $P \in \mathbb{F}^{m\times n}$ such that $A = P\sqrt{A^*A} = \sqrt{AA^*}\,P$, where*

(1) *if $m < n$, then P has orthonormal rows;*
(2) *if $m > n$, then P has orthonormal columns; and*
(3) *if $m = n$, then P is unitary.*

Example 7.56 Consider the matrix $A = \begin{bmatrix} 2 & -2 & 4 \\ -1 & 1 & -2 \end{bmatrix}$ of Example 7.47. We have seen that

$$A^*A = \begin{bmatrix} 5 & -5 & 10 \\ -5 & 5 & -10 \\ 10 & -10 & 20 \end{bmatrix}, \quad AA^* = \begin{bmatrix} 24 & -12 \\ -12 & 6 \end{bmatrix}.$$

For A^*A, the eigenvectors were denoted by $u_1,\ u_2,\ u_3$ corresponding to the eigenvalue $30,\ 0,\ 0$; for AA^*, the orthonormal eigenvectors were $v_1,\ v_2$ corresponding to the eigenvalues $30,\ 0$. These vectors were found to be

$$u_1 = \frac{1}{\sqrt{6}}\begin{bmatrix} 1 \\ -1 \\ 2 \end{bmatrix},\ u_2 = \frac{1}{\sqrt{2}}\begin{bmatrix} 1 \\ 1 \\ 0 \end{bmatrix},\ u_3 = \frac{1}{\sqrt{3}}\begin{bmatrix} 1 \\ -1 \\ -1 \end{bmatrix};\ v_1 = \frac{1}{\sqrt{5}}\begin{bmatrix} 2 \\ -1 \end{bmatrix},\ v_2 = \frac{1}{\sqrt{5}}\begin{bmatrix} 1 \\ 2 \end{bmatrix}.$$

Observe that $A^*A = P_1 \Sigma P_1^*$, where $P_1 = \begin{bmatrix} u_1 \ u_2 \ u_3 \end{bmatrix}$ and $\Sigma = \mathrm{diag}(30,\ 0,\ 0)$. Hence $\sqrt{A^*A} = P_1\,\mathrm{diag}(\sqrt{30}, 0\,, 0)\,P_1^*$. Similarly, $\sqrt{AA^*} = Q_1\,\mathrm{diag}(\sqrt{30},\ 0)\,Q_1^*$, where $Q_1 = \begin{bmatrix} v_1 \ v_2 \end{bmatrix}$. Thus, we obtain

$$\begin{aligned}\sqrt{A^*A} &= \begin{bmatrix} 1/\sqrt{6} & 1/\sqrt{2} & 1/\sqrt{3} \\ -1/\sqrt{6} & 1/\sqrt{2} & -1/\sqrt{3} \\ 2/\sqrt{6} & 0 & -1/\sqrt{3} \end{bmatrix} \begin{bmatrix} \sqrt{30} & 0 & 0 \\ 0 & 0 & 0 \\ 0 & 0 & 0 \end{bmatrix} \begin{bmatrix} 1/\sqrt{6} & -1/\sqrt{6} & 2/\sqrt{6} \\ 1/\sqrt{2} & 1/\sqrt{2} & 0 \\ 1/\sqrt{3} & -1/\sqrt{3} & -1/\sqrt{3} \end{bmatrix} \\ &= \frac{\sqrt{5}}{\sqrt{6}} \begin{bmatrix} 1 & -1 & 2 \\ -1 & 1 & -2 \\ 2 & -2 & 4 \end{bmatrix}.\end{aligned}$$

$$\sqrt{AA^*} = \begin{bmatrix} 2/\sqrt{5} & 1/\sqrt{5} \\ -1/\sqrt{5} & 2/\sqrt{5} \end{bmatrix} \begin{bmatrix} \sqrt{30} & 0 \\ 0 & 0 \end{bmatrix} \begin{bmatrix} 2/\sqrt{5} & -1/\sqrt{5} \\ 1/\sqrt{5} & 2/\sqrt{5} \end{bmatrix} = \frac{\sqrt{6}}{\sqrt{5}} \begin{bmatrix} 4 & -2 \\ -2 & 1 \end{bmatrix}.$$

Since $A \in \mathbb{F}^{2\times 3}$, by Theorem 7.55 (1), the matrix P has orthonormal rows, and it is given by (See the discussion previous to that theorem.)

$$P = \begin{bmatrix} v_1 & v_2 \end{bmatrix} \begin{bmatrix} u_1^* \\ u_2^* \end{bmatrix} = \begin{bmatrix} 2/\sqrt{5} & 1/\sqrt{5} \\ -1/\sqrt{5} & 2/\sqrt{5} \end{bmatrix} \begin{bmatrix} 1/\sqrt{6} & -1/\sqrt{6} & 2/\sqrt{6} \\ 1/\sqrt{2} & 1/\sqrt{2} & 0 \end{bmatrix}$$
$$= \frac{1}{\sqrt{30}} \begin{bmatrix} 2+\sqrt{3} & -2+\sqrt{3} & 4 \\ -1+2\sqrt{3} & 1+2\sqrt{3} & -2 \end{bmatrix}.$$

Then, we see that $\sqrt{AA^*}\, P = A = P\, \sqrt{A^*A}$ as required. □

Exercises for Sect. 7.6

1. Construct a positive semi-definite matrix with some negative entries.
2. Give an example of a matrix with all positive entries, which is not positive semi-definite.
3. Prove or disprove: If a 2×2 matrix $A = [a_{ij}]$ is positive semi-definite, then $a_{11} \geq 0,\ a_{22} \geq 0$ and $a_{11}a_{22} - a_{12}a_{21} \geq 0$. Is the converse true?
4. How many square roots the identity operator can have?
5. Let $T : \mathbb{C}^3 \to \mathbb{C}^3$ be defined by $T(a, b, c) = (b, c, 0)$. Show that T does not have a square root.
6. Let $A, B \in \mathbb{C}^{n\times n}$ be hermitian matrices. Show that if A is positive definite, then all zeros of the polynomial $\det(B - t\,A)$ are real.
7. Determine the polar decompositions of the matrix $\begin{bmatrix} 1 & -1 \\ 1 & 1 \end{bmatrix}$.
8. Taking cue from the computations in Example 7.56, determine the polar decompositions of the matrix $\begin{bmatrix} 2 & -1 \\ -2 & 1 \\ 4 & -2 \end{bmatrix}$.
9. Let $T : \mathbb{F}^3 \to \mathbb{F}^3$ be defined by $T(a, b, c) = (c, 2a, 3b)$. Determine an isometry S on $\mathbb{F}^3$ such that $T = S\sqrt{T^*T}$.
10. Let T be a linear operator on a finite dimensional inner product space. Prove the following:

    (a) If T is a normal operator and $T = U|T|$ is a left polar decomposition of T, then $|T|$ commutes with every such U.
    (b) If $T = U|T|$ is a left polar decomposition of T and $|T|$ commutes with one such U, then T is a normal operator.

11. Let $T : V \to V$ be a linear operator, where V is a finite dimensional inner product space. Suppose T can be written as $T = PU$, where P is positive semi-definite and U is unitary. Prove that in the decomposition $T = PU$, the positive semi-definite operator P and the unitary operator U are unique if and only if T is invertible.

12. Do the statements in Exercise 10(a)–(b) hold for a right polar decomposition?
13. Let $A \in \mathbb{F}^{m\times n}$. Derive the singular value decomposition of A from the polar decomposition of A.

7.7 Problems

1. Let U be a two-dimensional subspace of $\mathbb{R}^{3\times 1}$. Let P be the orthogonal projection of $\mathbb{R}^{3\times 1}$ onto U. Let $\{v_1, v_2, v_3\}$ be an orthonormal basis of $\mathbb{R}^{3\times 1}$. Let $P(v_i) = (a_i, b_i)$ for $i = 1, 2, 3$. Prove that $\{(a_1, a_2, a_3), (b_1, b_2, b_3)\}$ is an orthonormal set.
2. Let $n \in \mathbb{N}$ and let $\alpha = e^{2\pi i/n}$. Let $B = [b_{jk}] \in \mathbb{C}^{n\times n}$, where $b_{jk} = \alpha^{jk}/\sqrt{n}$. Prove that B is a unitary matrix.
3. Let A be a real orthogonal matrix. Let $x \in \mathbb{C}^{n\times 1}$ be an eigenvector of A corresponding to an eigenvalue λ. Then

$$x^T A^T x = (Ax)^T x = \lambda x^T x; \quad x^T A^T x = x^T (A^{-1}x) = \lambda^{-1} x^T x.$$

Therefore, $\lambda = \pm 1$. But the reflection operator on the plane has eigenvalues as $e^{\pm i\theta}$. Is there anything wrong with the above argument?
4. Prove that if columns of a square matrix are orthogonal to each other, then the rows of the matrix are also orthogonal to each other.
5. Is the composition of two self-adjoint operators on a finite dimensional inner product space self-adjoint?
6. Let T be a linear operator on a finite dimensional vector space V. Prove that T is diagonalizable if and only if there exists an inner product on V such that T is a self-adjoint operator on the inner product space V.
7. Let U and V be inner product spaces of dimensions n and m, with orthonormal ordered bases B and E, respectively. Let $T : U \to V$ be a linear transformation. Write $A = [T]_{E,B}$. Show that

$$\langle Tx, y\rangle = \langle A[x]_B, [y]_E\rangle \quad \text{for all } (x, y) \in U \times V.$$

Deduce that $\langle Tx, y\rangle = \langle x, Ty\rangle$ for all $(x, y) \in U \times V$ if and only $[T]_{E,B}$ is hermitian.
8. Prove that the linear factors of the minimal polynomial of a hermitian matrix are all distinct. Does the converse hold?
9. Let S and T be two normal operators on a finite dimensional inner product space. If $ST = 0$, does it follow that $TS = 0$?
10. Let $A, B \in \mathbb{F}^{n\times n}$. Prove that if both A, B are hermitian, or if both are skew-hermitian, then $AB + BA$ is hermitian and $AB - BA$ is skew-hermitian. What happens if one of A, B is hermitian and the other is skew-hermitian?

11. Let T be a linear operator on a finite dimensional inner product space V over $\mathbb{F}$. Prove the following versions of the spectral theorem:

    (a) Let $\mathbb{F} = \mathbb{C}$. Then, V has an orthonormal basis consisting of eigenvectors of T if and only if T is normal.
    (b) Let $\mathbb{F} = \mathbb{R}$. Then, V has an orthonormal basis consisting of eigenvectors of T if and only if T is self-adjoint.

12. Let V be an inner product space of dimension $n \geq 2$. Show that the set of all normal operators on V does not form a subspace of $\mathcal{L}(V, V)$.
13. Let S be the set of all self-adjoint operators on a finite dimensional nontrivial vector space V. Show the following:

    (a) If $\mathbb{F} = \mathbb{R}$, then S is a subspace of $\mathcal{L}(V, V)$.
    (b) If $\mathbb{F} = \mathbb{C}$, then S is not a subspace of $\mathcal{L}(V, V)$.

14. Let A be a hermitian matrix. Prove that the following are equivalent:

    (a) A is positive definite.
    (b) Each principal minor of A is positive.
    (c) $A = PP^*$ for some invertible matrix P.
    (d) Each eigenvalue of A is positive.

15. If $A = [a_{ij}]$ is a positive definite matrix, then show that $a_{ii} > 0$ for each i and that $a_{ii}a_{jj} > |a_{ii}|^2$ for all $i \neq j$.
16. Let A, B, C be linear operators on an inner product space, where $B - A$ and C are positive semi-definite. Prove that if C commutes with both A and B, then $BC - AC$ is positive semi-definite.
17. If T is a positive definite operator on a finite dimensional inner product space and B is a linear operator such that $B - A$ is positive semi-definite, then prove that $A^{-1} - B^{-1}$ is positive semi-definite. (Hint: Try $A = I$ first.)
18. Let A and B be linear operators on a finite dimensional inner product space. If A and $B - A$ are positive semi-definite, then show that $\det(A) \leq \det(B)$. (Hint: if $\det(B) \neq 0$, then $\sqrt{B}$ is invertible.)
19. Does every invertible linear operator on a finite dimensional vector space have a square root?
20. Let A be a real symmetric positive definite matrix. Show that there exist a positive definite matrix B and an orthogonal matrix Q such that $A = BQ$.
21. Let $A \in \mathbb{R}^{n \times n}$ be positive definite. Let $b \in \mathbb{R}^n$. Prove that $\mathrm{Sol}(A, b)$ is the set of values of x in $\mathbb{R}^n$ that minimizes the functional $x \mapsto x^T Ax - 2x^T b$.
22. Let P be a projection. Let $f(t)$ be a polynomial. Suppose $f(P) = aI + bP$ for some scalars a and b. Can you express a and b in terms of the coefficients of $f(t)$?
23. Let T be a linear operator on a finite dimensional inner product space V. Let U be a subspace of V. Prove the following:

    (a) If U is T-invariant, then $PTP = TP$ for each projection P on U.
    (b) If $PTP = TP$ for some projection P on U, then U is T-invariant.

24. Let T be a linear operator on a finite dimensional vector space V. Prove that there exist invertible operators P and Q on V such that both PT and TQ are projections.
25. Let U be a finite dimensional subspace of an inner product space V. Show that the orthogonal projection $P_U : V \to V$ is self-adjoint. Determine the eigenvalues and the eigenspaces of P_U.
26. Prove that the only real square matrix which is symmetric, orthogonal, and positive definite is the identity matrix.
27. Prove that if a real square matrix A is unitarily similar to a real square matrix B, then A is orthogonally similar to B.
28. Let $A \in \mathbb{R}^{2\times 2}$. Prove that A has characteristic polynomial $(t-\alpha)^2 + \beta^2$ for nonzero real numbers α and β if and only if there exists an orthogonal matrix $P \in \mathbb{R}^{2\times 2}$ such that $P^T AP = \begin{bmatrix} \alpha & -\beta \\ \beta & \alpha \end{bmatrix}$.
29. Let $x, y \in \mathbb{C}^{n\times 1}$; $x \neq 0$. Prove that there exists a matrix $A \in \mathbb{C}^{n\times n}$ such that $A^T = A$ and $Ax = y$.
30. Let $A \in \mathbb{C}^{n\times n}$ be symmetric and invertible. Prove that there exists a matrix P such that $A = P^T P$.
31. Let $A \in \mathbb{C}^{n\times n}$. Prove that $I + A^*A$ and $\begin{bmatrix} I & A^* \\ -A & I \end{bmatrix}$ are invertible.
32. Prove or disprove the following:

    (a) $A \in \mathbb{C}^{n\times n}$ is skew-symmetric if and only if $x^T Ax = 0$ for all $x \in \mathbb{C}^{n\times 1}$.
    (b) A skew-symmetric $n \times n$ matrix is not invertible if and only if n is odd.
    (c) The eigenvalues of a skew-symmetric matrix are purely imaginary.

33. Let A be a real skew-symmetric matrix. Prove the following:

    (a) $I + A$ is invertible.
    (b) $(I - A)(I + A)^{-1}$ is an orthogonal matrix.
    (c) $\det(A) \geq 0$.
    (d) If all entries of A are integers, then $\det(A)$ is the square of an integer.

34. Let $A, B \in \mathbb{R}^{n\times n}$. Determine which of the following cases ensure that at least one of $AB + BA$ or $AB - BA$ is symmetric:

    (a) A and B are symmetric.
    (b) A and B are skew-symmetric.
    (c) One of A, B is symmetric, and the other is skew-symmetric.

35. Let x and y be vectors in a finite dimensional inner product space. Is it true that there exists a positive semi-definite operator T on V with $Tx = y$ if and only if $\langle x, y\rangle > 0$?
36. Let T be a linear operator on a finite dimensional inner product space V. Prove that the following are equivalent:

    (a) T is positive semi-definite.
    (b) T is self-adjoint, and all eigenvalues of T are nonnegative.

(c) T has a positive semi-definite square root.
(d) T has a self-adjoint square root.
(e) $T = S^*S$ for some operator S on V.

37. Let T be a positive semi-definite operator on a finite dimensional inner product space. Prove that if $\mathrm{tr}(T) = 0$, then $T = 0$.
38. Let A and B be hermitian matrices such that $AB = BA$. Prove that there exists a matrix P such that both P^*AP and P^*BP are diagonal.
39. Let S and T be positive semi-definite operators on a finite dimensional inner product space. Prove that ST is positive semi-definite if and only if $ST = TS$.
40. Let V be a vector space. For $v_1, \ldots, v_m \in V$, define $G := G(v_1, \ldots, v_m)$ as the $m \times m$ matrix $[g_{ij}]$, where $g_{ij} = \langle v_i, v_j \rangle$. Prove that G is positive semi-definite. Such a matrix is called a *Gramian*.
41. Let $u := (x_1, \ldots, x_n)$ and $v := (y_1, \ldots, y_n)$ be n-tuples of vectors from a finite dimensional inner product space V. Prove or disprove: there exists an isometry T on V such that $Tx_i = y_i$ for each $i \in \{1, \ldots, n\}$ if and only if u and v have the same Gramian.
42. In Problem 41, if $x_1, \ldots, x_n$ are linearly independent, then show that the Gramian is nonsingular.
43. Prove that every positive semi-definite matrix is a Gramian.
44. Let S and T be linear operators on a finite dimensional inner product space V. Suppose that S and ST are self-adjoint and $N(S) \subseteq N(T)$. Does it follow that there exists a self-adjoint operator A on V with $AS = T$?
45. Let V be a finite dimensional complex vector space. Prove the following:

(a) If $T \in \mathcal{L}(V, V)$, then $T = 0$ if and only if $\mathrm{tr}(T^*T) = 0$.
(b) If $T_1, \ldots, T_m \in \mathcal{L}(V, V)$ satisfy $\sum_{i=1}^{m} T_i^* T_i = 0$, then $T_1 = \cdots = T_m = 0$.
(c) If $T \in \mathcal{L}(V, V)$ satisfy $T^*T = S^*S - SS^*$ for an $S \in \mathcal{L}(V, V)$, then $T = 0$.
(d) If $S, T \in \mathcal{L}(V, V)$ satisfy $S^*S = SS^*$ and $ST = TS$, then $S^*T = TS^*$.

46. Let V be a finite dimensional inner product space. Define $\|T\| = \sqrt{\mathrm{tr}(T^*T)}$ for $T \in \mathcal{L}(V, V)$. Show that this defines a norm on $\mathcal{L}(V, V)$. Does this norm come from an inner product on $\mathcal{L}(V, V)$?
47. Prove that each Jordan block $J(\lambda, m)$ with $\lambda \neq 0$ has a square root. Then show that each invertible matrix has a square root.
48. Let T be a nilpotent operator on a finite dimensional vector space. Prove that there exist scalars $a_1, \ldots, a_{k-1}$ such that $(I + a_1 T + \cdots a_{k-1} T^{k-1})^2 = I + T$.
49. Using Problem 48, prove that every invertible operator on a finite dimensional complex vector space has a square root.
50. Determine the square root of the linear operator T on $\mathbb{R}^5$ given by

$$T(a, b, c, d, e) = (a + 2b, b + 3c, c - d, d + 4e, e).$$

51. Let P_U be an orthogonal projection on a subspace U of a finite dimensional inner product space V. Let $T : V \to V$ be a linear operator. Prove:

(a) U is T-invariant if and only if $P_U T P_U = T P_U$.

(b) Both U and $U^{\perp}$ are T-invariant if and only if $P_U T = T P_U$.

52. Let T be a linear operator on a two-dimensional real inner product space V. Prove that the following are equivalent:

    (a) T is normal but not self-adjoint.

    (b) $[T]_{B,B}$ has the form $\begin{bmatrix} a & -b \\ b & a \end{bmatrix}$, $b > 0$ for an orthonormal basis B of V.

    (c) $[T]_{E,E}$ has the form $\begin{bmatrix} a & -b \\ b & a \end{bmatrix}$, $b \neq 0$ for all orthonormal bases E of V.

53. Let T be a linear operator on a finite dimensional real inner product space V. Prove that T is normal if and only if there exists an orthonormal basis B of V such that $[T]_{B,B}$ is a block-diagonal matrix where each block is either a 1×1 matrix or a 2×2 matrix of the form $\begin{bmatrix} a & -b \\ b & a \end{bmatrix}$ with $b > 0$.
54. Is it true that a triangular normal matrix is necessarily diagonal?
55. Let T be a linear operator on a finite dimensional inner product space. Prove that $\det(\sqrt{T^*T}) = |\det(T)|$.
56. Prove that the following are equivalent:

    (a) T is an isometry.
    (b) $\langle Tu, Tv \rangle = \langle u, v \rangle$ for all $u, v \in V$.
    (c) $T^*T = I$.
    (d) If $u_1, \ldots, u_m$ are orthonormal, then so are $Tu_1, \ldots, Tu_m$.
    (e) There exists an orthonormal basis $\{v_1, \ldots, v_n\}$ of V such that the vectors $Tv_1, \ldots, Tv_n$ are orthonormal.
    (f) T^* is an isometry.
    (g) $\langle T^*u, T^*v \rangle = \langle u, v \rangle$ for all $u, v \in V$.
    (h) $TT^* = I$.
    (i) If $u_1, \ldots, u_m$ are orthonormal, then so are $T^*u_1, \ldots, T^*u_m$.
    (j) There exists an orthonormal basis $\{v_1, \ldots, v_n\}$ of V such that the vectors $T^*v_1, \ldots, T^*v_n$ are orthonormal.

57. Let V be a finite dimensional inner product space. An *involution* is a linear operator T on V such that $T^2 = I$. Show that $S = 2T - I$ gives a bijection between the set of all projections and the set of all involutions.
58. How does a matrix of an involution look like with respect to a suitable basis?
59. Let P_1, P_2, and P_3 be projections on some subspaces of a finite dimensional inner product space. Show that if $P_1 + P_2 + P_3 = I$, then $P_i P_j = 0$ for all $i, j \in \{1, 2, 3\}$, $i \neq j$. Does it work for four instead of three projections?
60. It is claimed that if a linear operator satisfies two of the following, then it satisfies the third. Is it correct?
    (a) T is self-adjoint. (b) T is an isometry. (c) T is an involution.
61. Let V be an n-dimensional inner product space over $\mathbb{F}$. Consider the inner product space $\mathbb{F}^n$ with the standard inner product. Prove that there exists an isometry from V onto $\mathbb{F}^n$. [Both the norms are denoted by $\|\cdot\|$.]

62. Let $\{u_1, \ldots, u_n\}$ be an orthonormal basis of an inner product space V. Show that for any $x, y \in V$, $\langle x, y\rangle = \sum_{k=1}^{n} \langle x, u_k\rangle\langle u_k, y\rangle$. Then deduce that there exists an isometry from V to $\mathbb{F}^n$.
63. Let T be a linear operator on a finite dimensional real vector space V. Prove that T is an isometry if and only if there exists an orthonormal basis B of V such that $[T]_{B,B}$ is a block-diagonal matrix where each block is a 1×1 matrix containing ± 1 or a 2×2 matrix of the form $\begin{bmatrix} \cos\theta & -\sin\theta \\ \sin\theta & \cos\theta \end{bmatrix}$ for some $\theta \in (0, \pi)$.
64. Let T be an isometry on a finite dimensional inner product space. Prove that $|\det(T)| = 1$.
65. Let T be a linear operator on a vector space V of dimension n. Prove that if each subspace of dimension $n - 1$ is T-invariant, then $T = \alpha I$ for some $\alpha \in \mathbb{F}$.
66. Let T be a linear operator on a vector space V. Prove that if $P^2 = P$, then $V = R(P) \oplus N(P)$.
67. Suppose $V = U \oplus W$, where U and W are nontrivial subspaces of a vector space V. Then each $v \in V$ is uniquely written as $v = u + w$, where $u \in U$ and $w \in W$. Define the linear operator P on V by $P(v) = u$. This is a projection operator whose null space is W. Find all eigenvalues and eigenvectors of P.
68. Prove that every normal operator on a finite dimensional complex inner product space has a square root.
69. Give an example of a linear operator on a finite dimensional complex inner product space which has no square root.
70. Prove that every positive semi-definite operator on a finite dimensional inner product space has the same rank as that of its square root.
71. Prove that each self-adjoint operator on a finite dimensional inner product space has a unique cube root. Find the cube root of the 3×3 matrix $\text{diag}(1, -1, 8)$.
72. Let T be a normal operator on a finite dimensional complex inner product space. Prove that if $T^{k+1} = T^k$ for some $k \in \mathbb{N}$, then T is an orthogonal projection.
73. Let T be a self-adjoint operator on a finite dimensional inner product space V. Let $\delta \in \mathbb{F}$ and let $\varepsilon > 0$. Suppose that there exists a unit vector $v \in V$ such that $\|Tv - \delta v\| < \varepsilon$. Prove that there exists an eigenvalue λ of T with $|\delta - \lambda| < \varepsilon$.
74. Let T be a linear operator on a finite dimensional real vector space V. Prove that there exists a basis B for V consisting of eigenvectors of T if and only if there exists an inner product on V so that T is self-adjoint.
75. Give an example to show that orthogonal complement of a T-invariant subspace need not be T-invariant.
76. Prove that the sum of two positive semi-definite operators on an inner product space is positive semi-definite.
77. Prove that positive powers of positive semi-definite operators are positive semi-definite.
78. Prove that a positive semi-definite operator on a finite dimensional inner product space is positive definite if and only if it is invertible.
79. Let $\{v_1, \ldots, v_n\}$ be an orthonormal basis of an inner product space V. Let T be a linear operator on V with $\|Tv_j\| = 1$ for each $j \in \{1, \ldots, n\}$. Show by an example that T need not be an isometry.

80. Let $T : \mathbb{R}^3 \to \mathbb{R}^3$ be an isometry. Prove that there exists a nonzero vector $v \in \mathbb{R}^3$ such that $T^2 v = v$.
81. Let R, S, and T be linear operators on a finite dimensional inner product space. Suppose S is an isometry, R is positive semi-definite, and $T = SR$. Prove that $R = \sqrt{T^*T}$.
82. Let T be a linear operator on a finite dimensional inner product space V. Define $f : V \times V \to \mathbb{F}$ by $f(x, y) = \langle Tx, y\rangle$. Under what conditions on T, the function f is an inner product?
83. Let T b a linear operator on a finite dimensional inner product space V. Prove that the following are equivalent:

    (a) T is invertible.
    (b) All singular values of T are nonzero.
    (c) There exists a unique isometry S on V such that $T = S\sqrt{T^*T}$.

84. Prove that two linear operators S and T on a finite dimensional inner product space V have the same singular values (with respectively equal multiplicities) if and only if there exist isometries P and Q on V such that $T = PSQ$.
85. Let T be a linear operator on an inner product space V of dimension n. Let s_1 be the smallest singular value of T, and let s_n be the largest singular value of T. Prove that for each $v \in V$, $\|s_1 v\| \le \|Tv\| \le s_n\|v\|$.
86. Let T be an invertible linear operator on a finite dimensional inner product space V. Suppose $\{u_1, \ldots, u_n\}$ and $\{v_1, \ldots, v_n\}$ are orthonormal bases for V and $s_1, \ldots, s_n$, all positive, are the singular values of T so that $Tv = \sum_{j=1}^{n} s_j \langle v, u_j\rangle v_j$ for each $v \in V$. Then prove that $T^{-1}v = \sum_{j=1}^{n} \frac{\langle v, v_j\rangle}{s_j} u_j$.
87. Let T be a linear operator on a finite dimensional real vector space V. Prove that if T has no eigenvalues, then every T-invariant subspace of V is of even dimension.
88. Let T be a linear operator on a finite dimensional real vector space V of dimension n. Suppose that with respect to some basis of V, the matrix of T is of the form $\begin{bmatrix} A_1 & & \star \\ & \ddots & \\ 0 & & A_k \end{bmatrix}$, where the blocks $A_1, \ldots, A_k$ on the diagonal are either 1×1 matrices or 2×2 matrices with no eigenvalues. Prove the following:

    (a) The 1×1 matrices among $A_1, \ldots, A_k$ are the eigenvalues of T. Moreover, the number of times such an eigenvalue $\lambda \in \mathbb{R}$ occurs among $A_1, \ldots, A_k$ is equal to the algebraic multiplicity of λ.
    (b) The characteristic polynomial of any 2×2 matrix among $A_1, \ldots, A_k$ is of the form $(t - \alpha)^2 + \beta^2$ for real α, β and nonzero β. Moreover for any such given α, β, twice the number of 2×2 matrices among $A_1, \ldots, A_k$ having characteristic polynomial $(t - \alpha)^2 + \beta^2$ is given by $\text{null}\,((T - \alpha I)^2 + \beta^2 I)$.

89. Let T be a linear operator on a finite dimensional real vector space. Let α and β be nonzero real numbers. Prove that null $((T - \alpha I)^2 + \beta^2 I)^k$ is even for any $k \in \mathbb{N}$.
90. Let S and T be linear operators on a finite dimensional inner product space. If σ, τ are the largest singular values of S, T, respectively, then show that the largest singular value of $S + T$ is at most $\sigma + \tau$.
91. Give a linear operator T on $\mathbb{C}^5$ such that $T^2 + T + I$ is nilpotent.
92. Prove that if T is a linear operator on $\mathbb{R}^5$, then $T^2 + T + I$ is never nilpotent.
93. Let T be a linear operator on a finite dimensional inner product space V. Prove that if $\|T^*v\| \le \|Tv\|$ for each $v \in V$, then T is a normal operator.
94. Let T be a surjective linear operator on an inner product space satisfying $\langle Tx, Ty \rangle = \langle x, y \rangle$ for all vectors x and y. Prove the following:

    (a) T is an isomorphism.
    (b) $\langle T^{-1}x, T^{-1}y \rangle = \langle x, y \rangle$ and $\langle Tx, y \rangle = \langle x, T^{-1}y \rangle$ for all x, y.
    (c) For any fixed y, the map f defined by $f(x) = \langle x, T^{-1}y \rangle$ is linear in x.

95. Let V be a finite dimensional inner product space. Show that $\langle S, T \rangle = \operatorname{tr}(ST^*)$ defines an inner product on $\mathcal{L}(V, V)$. In case $V = \mathbb{F}^{n\times 1}$, see that this inner product on $\mathbb{F}^{n\times n}$ is simply the dot product of two vectors in $\mathbb{F}^{n^2}$.
96. Let $A \in \mathbb{C}^{n\times n}$ with $I + A$ invertible. The *Cayley transform* $K : \mathbb{C}^{n\times n} \to \mathbb{C}^{n\times n}$ is defined by $K(A) = (I + A)^{-1}(I - A)$. Show the following:

    (a) If A is skew-hermitian, then $K(A)$ is unitary.
    (b) If A is unitary, then $K(A)$ is skew-hermitian.

    What would happen for $A \in \mathbb{R}^{n\times n}$?

Note: A and $K(A)$ are called *Cayley transforms* of each other. This has its origin in the map that takes z to $(z + 1)/(z - 1)$ in the complex plane, which transforms the imaginary axis once around the unit circle leaving 1.

97. Let $A \in \mathbb{F}^{n\times n}$ have trace 0. Prove that there exists an isometry T on $\mathbb{F}^{n\times 1}$ such that $[T]^{-1}A[T]$ has all diagonal entries 0. (See Problem 12 of Chap. 6.)
98. Let T be a linear operator on an inner product space of dimension n with (not necessarily distinct) eigenvalues $\lambda_1, \ldots, \lambda_n$. Prove that $\sum_{i=1}^{n} |\lambda_i|^2 \le \operatorname{tr}(T^*T)$. Prove also that equality holds if and only if T is normal.
99. Let T be a linear operator on a finite dimensional inner product space V over $\mathbb{C}$. The *numerical range* of T is defined to be the set $NR(T) := \{\langle Tx, x \rangle \in \mathbb{C} : x \in V,\ \|x\| = 1\}$. Prove the following:

    (a) If T is normal, then $NR(T)$ is convex. That is, if $x, y \in NR(T)$ and $0 \le \alpha \le 1$, then $\alpha x + (1 - \alpha)y \in NR(T)$.
    (b) If T is normal, then each extreme point of $NR(T)$ is an eigenvalue of T. (An extreme point of $NR(T)$ is a point which is not equal to $\alpha x + (1 - \alpha)y$ for $x, y \in NR(T)$ and for $0 < \alpha < 1$.) This conclusion can be false when T is not normal.

100. Let S and T be self-adjoint operators on a finite dimensional inner product space V. Show that the set of points of the form $(\langle Sx, x\rangle, \langle Tx, x\rangle) \in \mathbb{R}^2$ with $\|x\| = 1$ is convex. (A subset U of V is called *convex* if for all $x, y \in U$, and for all α with $0 \le \alpha \le 1$, $\alpha x + (1-\alpha)y \in U$.)
101. Let T be an orthogonal operator on a finite dimensional inner product space V. Prove that there exists a basis B of V such that $[T]_{B,B}$ is a block-diagonal matrix in which each block is either a single entry 1, or a single entry -1, or a 2×2 matrix of the following form, where $\theta \in \mathbb{R}$. (Different 2×2 blocks may have different θ.)
$$\begin{bmatrix} \cos\theta & -\sin\theta \\ \sin\theta & \cos\theta \end{bmatrix}.$$
102. Let T be a positive semi-definite operator. Are the following true?
   (a) If $\langle Ax, x\rangle = 0$ for some vector x, then $Ax = 0$.
   (b) $|\langle Ax, y\rangle|^2 \le \langle Ax, y\rangle\langle Ay, y\rangle$ for all vectors x and y.
103. Show that if the linear operators S and $T - S$ are positive semi-definite, then $\sqrt{T} - \sqrt{S}$ is positive semi-definite.
104. Let A and B be positive semi-definite matrices with complex entries. Show that if A^2 and B^2 are unitarily similar, then so are A and B.
105. Let $A \in \mathbb{C}^{n\times n}$ be a positive definite matrix. Show the following:
   (a) $\det(A) \le \big(\mathrm{tr}(A)/n\big)^n$.
   (b) $\det(A) \le a_{11} \times \cdots \times a_{nn}$.
106. The *Rayleigh quotient* of a hermitian (real symmetric) matrix A at any nonzero vector v is defined as $\rho(v) = (v^*Av)/(v^*v)$.
   (a) Show that if v is an eigenvector of A, then $\rho(v)$ is its corresponding eigenvalue.
   (b) Show that $\rho(v)$ lies between the minimum and the maximum of eigenvalues of A for each nonzero v.
   (c) Taking $v = (1, t, 1)^T$ for some $t \in \mathbb{R}$, estimate the largest and the smallest eigenvalues of the matrix $A = \begin{bmatrix} 5 & 4 & -4 \\ 4 & 5 & 4 \\ -4 & 4 & 5 \end{bmatrix}$.
107. *Householder's reflections*: Let $w \in \mathbb{R}^{n\times 1}$ with $\|w\| = 1$. Let $P = I - 2ww^T$.
   (a) Prove that P is an orthogonal matrix.
   (b) Prove that P is a reflection through $w^\perp$; that is, if we write any $v \in \mathbb{R}^{n\times 1}$ as $v = \alpha w + w'$, where $w' \in w^\perp$, then $Pv = -\alpha w + w'$.
   (c) Notice that in (a)–(b), the matrix P depends on the given vector w. Let $x, y \in \mathbb{R}^{n\times 1}$ such that $\|x\| = \|y\|$. Determine a vector w such that $Px = y$.
108. Use Householder's reflections to prove that every real orthogonal $n \times n$ matrix is a product of at most n reflections.

109. *Rayleigh Principle*: Let $A \in \mathbb{C}^{n\times n}$ be hermitian. Let $\lambda_1 \le \cdots \le \lambda_n$ be the eigenvalues of A with corresponding eigenvectors $v_1, \ldots, v_n$. Prove the following:

   (a) For $k > 1$, $\lambda_k = \min\{\rho(v) : v \neq 0 \text{ and } v^*v_i = 0 \text{ for } i = 1, \ldots, k-1\}$.
   (b) For $k < n$, $\lambda_k = \max\{\rho(v) : v \neq 0 \text{ and } v^*v_j = 0 \text{ for } j = k+1, \ldots, n\}$.

110. Let $A = \begin{bmatrix} 1 & i & i \\ -i & 2 & -1 \\ -i & -1 & 2 \end{bmatrix}$. Given that $\lambda_3 = 3$ is the smallest eigenvalue of A with corresponding eigenvector $v_3 = (0, 1, -1)^T$, use Rayleigh principle to estimate the next largest eigenvalue λ_2.
111. *Courant–Fisher Theorem*: Let $A \in \mathbb{C}^{n\times n}$ be hermitian. Let $\lambda_1 \le \cdots \le \lambda_n$ be its eigenvalues. Prove that for each $k \in \{1, \ldots, n\}$, $\lambda_k = \min\max \rho(v)$, where the maximum and the minimum are taken as follows: Choose any set $S = \{u_1, \ldots, u_{n-k}\}$ of vectors in $\mathbb{F}^n$, take the maximum of $\rho(v)$ over $S^\perp$, and then take the minimum as $u_1, \ldots, u_{n-k}$ vary over $\mathbb{F}^n$.
112. Let A and B be $n \times n$ hermitian matrices. Let $\lambda_k(C)$ be the kth smallest eigenvalue of the hermitian matrix C, where C is either A, B, or $A + B$, and $k \in \{1, \ldots, n\}$. Prove the following for each such k:

   (a) $\lambda_1(A) + \lambda_k(B) \le \lambda_k(A+B) \le \lambda_n(A) + \lambda_k(B)$.
   (b) If $v^*Av \le v^*Bv$ for all $v \in \mathbb{C}^n$, then $\lambda_k(A) \le \lambda_k(B)$.

113. *Hardamard Inequality*: Prove that if $A = [a_{ij}] \in \mathbb{F}^{n\times n}$, then

$$\det(A) \le \Pi_{i=1}^n \Big(\sum_{j=1}^n |a_{ij}|^2\Big).$$

114. *Matrix exponential*: Let $A \in \mathbb{C}^{n\times n}$. Its exponential is defined by

$$\exp(A) = I + A + \frac{1}{2!}A^2 + \frac{1}{3!}A^3 + \cdots.$$

   Suppose this infinite sum is meaningful. Show that if $D = \operatorname{diag}(d_1, \ldots, d_n)$, then $\exp(D) = \operatorname{diag}(e^{d_1}, \ldots, e^{d_n})$.
115. Let $A \in \mathbb{R}^{2\times 2}$ have two distinct eigenvalues λ and μ. Show that

$$\exp(A) = \frac{\lambda e^{\mu} - \mu e^{\lambda}}{\lambda - \mu} I + \frac{e^{\lambda} - e^{\mu}}{\lambda - \mu} A.$$

   Obtain a correct formula when A has only one eigenvalue repeated twice.

116. Let $J(\lambda, m)$ be a Jordan block of order m. Show that

$$\exp(J(\lambda, m)) = e^{\lambda} \begin{bmatrix} 1 & 1 & 1/2! & 1/3! & \cdots & 1/(m-2)! & 1/(m-1)! \\ 0 & 1 & 1 & 1/2! & \cdots & 1/(m-3)! & 1/(m-2)! \\ 0 & 0 & 1 & 1 & \cdots & 1/(m-4)! & 1/(m-3)! \\ & & & & \cdots & & \\ 0 & 0 & 0 & 0 & \cdots & 0 & 1 \end{bmatrix}.$$

117. Using the previous problem show how to compute $\exp(A)$ for $A \in \mathbb{C}^{n\times n}$.
118. Prove that if A is a skew-symmetric matrix, then $\exp(A)$ is unitary.
119. Prove that corresponding to each unitary matrix U, there exists a hermitian matrix H such that $U = \exp(iH)$.
120. An $n \times n$ matrix is called a *stochastic matrix* if all its entries are nonnegative and sum of entries in each column is 1. Show that if λ is an eigenvalue of a stochastic matrix, then $|\lambda| \leq 1$.

References

1. Artin, M.: Algebra, 2nd edn. Pearson Education, Essex (2011)
2. Axler, S.: Linear Algebra Done Right, 2nd edn. Springer International Student Edition (1997)
3. Blass, A.: Existence of bases implies the axiom of choice. Axiomatic Set Theory (Boulder, Colo., 1983). Contemporary Mathematics, vol. 31, pp. 31–33. American Mathematical Society, Providence (1984)
4. Choi, Man-Duen: Tricks or treats with the Hilbert matrix. Am. Math. Mon. **90**, 301–312 (1983)
5. Friedberg, S.H., Insel, A., Spence, L.: Linear Algebra, 4th edn. Pearson Education, Essex (2014)
6. Gelfand, I.M.: Lectures on Linear Algebra. Dover Publications Inc, New York (1990)
7. Golub, G., Van Loan, C.F.: Matrix Computations, 4th edn. SIAM, Johns Hopkins University Press, Baltimore (2013)
8. Halmos, P.R.: Finite Dimensional Vector Spaces. Springer, New York (1987)
9. Hoffman, K., Kunze, R.: Linear Algebra, 2nd edn. Prentice Hall Int. Inc., Englewood Cliffs (1971)
10. Horn, R.A., Johnson, C.R.: Matrix Analysis, 2nd edn. Cambridge University Press, Cambridge (2012)
11. Jänich, K.: Linear Algebra. Undergraduate Texts in Mathematics. Springer, New York (1994)
12. Lewis, D.W.: Matrix Theory. World Scientific, Singapore (1991)
13. Meyer, C.: Matrix Analysis and Applied Linear Algebra. SIAM, Philadelphia (2000)
14. Nair, M.T.: Multiplicities of an eigenvalue: some observations. Resonance **7**, 31–41 (2002)
15. Nair, M.T.: Functional Analysis - A First Course. Prentice Hall of India, New Delhi (2002)
16. Poole, D.: Linear Algebra: A Modern Introduction, 3rd edn. Cengage Learning, Boston (2010)
17. Roman, S.: Advanced Linear Algebra, 3rd edn. Springer, London (2010)
18. Shilov, G.E.: In: Silverman, R.A. (ed.) Linear Algebra. Dover Publications Inc, New York (1977)
19. Singh, A.: Introduction to Matrix Theory. Ane Books, New Delhi (2018)
20. Strang, G.: Introduction to Linear Algebra, 5th edn. Wellesley-Cambridge Press, Wellesley (2016)

M. T. Nair and A. Singh, *Linear Algebra*,
https://doi.org/10.1007/978-981-13-0926-7

Index

Symbols
$\ell^1(\mathbb{N}, \mathbb{F})$, 48
$\ell^\infty(S, \mathbb{F})$, 48
$\mathcal{C}(I, \mathbb{R})$, 7
$\mathcal{C}^k([a, b], \mathbb{F})$, 7
$\mathcal{F}(S, V)$, 6
$\mathcal{P}([a, b], \mathbb{R})$, 12
$\mathcal{P}(\mathbb{F})$, 5
$\mathcal{P}_n(\mathbb{F})$, 5
$\mathcal{R}([a, b], \mathbb{R})$, 7
$c_{00}(\mathbb{N}, \mathbb{F})$, 16
$\mathbb{F}^\infty$, 6
$\mathbb{F}^n$, 4
$\mathbb{F}^{m \times n}$, 4

A
Additive inverse, 3
Algebra, 93
Angle between vectors, 170
Annihilates, 234
Annihilator, 104
Appolonius identity, 172
Ascent, 226

B
Basic variables, 150
Basis, 27
  dual, 95
  orthogonal, 182
  orthonormal, 182
Bessel inequality, 174
Best approximate solution, 194
Best approximation, 190
Box in $\mathbb{R}^n$, 159

C
Canonical basis isomorphism, 76
Cauchy–Schwarz inequality, 168
Cayley transform, 331
Characteristic
  equation, 214
  polynomial, 214
  value, 214
Complementary subspace, 36
Complexification, 213
Congruence, 307
Congruence modulo, 44
Convergence, 208
Coordinate
  functionals, 53
  vector, 76

D
Determinant, 132, 140
Diagonalizable, 243
Diagonalization
  of self-adjoint operators, 303
Dimension, 29
Direct sum
  of maps, 101
  of subspaces, 36, 37
Distance, 170

E
Eigenpair, 209
Eigenspace, 219
Eigenspectrum, 219
Eigenvalue, 209
Eigenvector, 209
Elementary

M. T. Nair and A. Singh, *Linear Algebra*,
https://doi.org/10.1007/978-981-13-0926-7

column operation, 128
operation, symmetric, 307
row operation, 108
Endomorphism, 52
Euclidean space, 164
Evaluation functional, 53
Exchange lemma, 19

F
Finite dimensional space, 29
Finite rank, 104
Fixed point, 156
Fourier
coefficient, 175
expansion, 174
minimum property, 205
Fredholm alternative, 207
Free variables, 150
Full orthonormal set of eigenvectors, 311
Full rank factorization, 130
Functional
conjugate linear, 196
linear, 51
Function space, 6
Fundamental theorem
of algebra, 210

G
Gaussian elimination, 151
Gauss–Jordan elimination, 153
Generalized eigenspace, 226
Generalized eigenvector, 226
Gramian, 327
Gram–Schmidt
orthogonalization, 177
othonormalization, 180

H
Hardamard inequality, 333
Homogeneous system, 142
Homomorphism, 52
Householder's reflection, 332
Hyperspace, 103

I
Identity map, 52
Index of nilpotency, 61, 263
Infinite dimensional space, 29
Inner product, 164
standard, 164
Inner product space, 164
complex, 164
real, 164
Invariant
subspace, 222
Isometry, 277
Isomorphic, 72
Isomorphism, 72

J
Jordan
basis, 265
block, 257
block representation, 262
chain, 263
decomposition, 275

K
Kronecker's delta, 15

L
Lagrange polynomials, 104
Least squares solution, 193, 314
Legendre polynomials, 179
Linear
combination, 15
dependence, 21, 22
extension, 56
independence, 21, 22
maps, 52
span, 15
Linear system, 141
consistent, 142
has a solution, 142
solution set, 141
solvable, 141
Linear transformation, 51
adjoint, 200
induced by a matrix, 58
inverse, 72
invertible, 72
product of, 59

M
Matrix
adjoint, 140
adjugate, 136
augmented, 120
Cauchy, 161
change of basis, 86

circulant, 160, 302
classical Vandermonde, 160
co-factor of, 136
column equivalence, 128
combinatorial, 160
companion, 217
elementary, 108
equivalent, 89
exponential, 333
generalized inverse, 315
Gram, 193
Hadamard, 161
hermitian, 140, 164, 278
Hilbert, 160
idempotent, 239
inverse of, 72
invertible, 72
left eigenvalue, 241
left eigenvector, 241
left inverse, 316
minor of, 136
Moore–Penrose inverse, 315
orthogonal, 140
permutation, 114, 239
positive definite, 164
rank echelon, 91
real symmetric, 278
representation, 77
right inverse, 316
row equivalence, 111
similar, 89
skew hermitian, 278
skew-symmetric, 14, 278
stochastic, 334
Sylvester, 307
symmetric, 14
trace, 132
transpose of, 7
triangularizable, 248
unitarily similar, 253
unitarily triangularizable, 249
unitary, 140
Vandermonde, 160
Maximal linearly independent set, 28
Minimal polynomial, 235
Minimal spanning set, 28
Monic polynomial, 214
Multi-linear map, 136
Multiplicity
algebraic, 215
counting, 215
geometric, 219
of a zero, 211

N
Norm of a vector, 168
Nullity, 65
Null space, 65

O
Operation
column, 128
row, 111
Operator
backward shift, 53
commute, 59
differentiation, 61
exponential, 208
forward shift, 53
has an eigenvalue, 210
idempotent, 101
identity, 52
imaginary part, 277
involution, 328
isometric, 277
left shift, 53, 99
linear, 51
negative definite, 277
nilpotent, 61, 263
normal, 277
numerical range, 331
orthogonal, 277
positive definite, 277
positive semi-definite, 277
real part, 277
right shift, 53, 99
scalar, 52
self-adjoint, 277
square root of, 318
triangularizable, 248
unipotent, 275
unitarily triangularizable, 248
unitary, 277
zero, 52
Ordered basis, 27
Orthogonal
complement, 186
projection, 289
set, 172
to a set, 186
vectors, 171
Orthonormal set, 173

P
Parallelogram law, 169
Parseval identity, 174

Pivot, 111
Polar decomposition, 320
  left, 321
  of matrices, 322
  right, 321
Polynomial in T, 61
Positive, 277
Powers of T, 60
Product space, 8
Projection, 285
  vector, 175
$\text{Proj}_U(y)$, 175
Proper orthogonal set, 173

Q
Quadrature formula, 54
Quotient space, 45

R
Range space, 65
Rank, 65
Rank theorem, 90
Rayleigh principle, 333
Rayleigh quotient, 332
Real Jordan Form, 275
Real Schur form, 274
Reduction
  row echelon form, 113
  row reduced echelon form, 119
Reflection, 246
Residual, 193
Reverse order basis, 269
Riesz representer, 197
Rigid motion, 207
Rotation, 246
Row echelon form, 111
Row rank, 91, 107, 124
Row reduced echelon form, 118
Row space, 124

S
Same cardinality, 34
Scalars, 3
Sequence space, 6
Shear, 246
Signature, 307
Singular value, 309
  decomposition, 310
  decomposition of matrices, 311
  representation, 309
Space
  dual, 94
  $\mathcal{L}(V, W)$, 93
Span, 15
Spanned by, 18
Spanning set, 18
Spans, 18
Spectral form, 305
Spectral theorem
  for normal operators, 298
Splits, polynomial, 215
Standard basis, 27, 94
Sum of subsets, 12
Sylvester form, 306
Sylvester law of inertia, 307

T
Theorem
  Block-diagonalization, 250
  Cayley–Hamilton, 234
  Courant–Fisher, 333
  Cramer's rule, 142
  Diagonalizability, 244
  Jordan block representation, 260
  Jordan normal form, 264
  LU-factorization, 115
  Projection theorem, 188
  Pythagoras, 171
  QR-factorization, 180
  Rank decomposition, 130
  Rank factorization, 130
  Rank-nullity, 69
  Riesz representation, 196
  Row echelon form, 112
  Row reduced echelon form, 118
  Schur triangularization, 252
  Spectral mapping, 220, 302
  Sylvester inertia, 305
  well-ordering, 35
Triangle inequality, 169

U
Unitary space, 164
Unit vector, 168

V
Vectors, 3
  column, 6
  row, 6
Vector space, 2
  complex, 3
  nonzero, 4

real, 3
subspace of, 10
trivial, 4
zero, 4
Volume of a box, 159

Z

Zero map, 52
Zero vector, 2, 3
Zorn's Lemma, 33

Printed in the United States
By Bookmasters